Gregory J. Miller
232-84-8883

**TIME-HARMONIC
ELECTROMAGNETIC FIELDS**

McGRAW-HILL ELECTRICAL AND ELECTRONIC ENGINEERING SERIES

FREDERICK EMMONS TERMAN, *Consulting Editor*
W. W. HARMAN AND J. G. TRUXAL,
Associate Consulting Editors

AHRENDT AND SAVANT · Servomechanism Practice
ANGELO · Electronic Circuits
ASELTINE · Transform Method in Linear System Analysis
ATWATER · Introduction to Microwave Theory
BAILEY AND GAULT · Alternating-current Machinery
BERANEK · Acoustics
BRACEWELL · The Fourier Transform and Its Applications
BRACEWELL · Transform Methods in Linear Systems
BRENNER AND JAVID · Analysis of Electric Circuits
BROWN · Analysis of Linear Time-invariant Systems
BRUNS AND SAUNDERS · Analysis of Feedback Control Systems
CAGE · Theory and Application of Industrial Electronics
CAUER · Synthesis of Linear Communication Networks
CHEN · The Analysis of Linear Systems
CHEN · Linear Network Design and Synthesis
CHIRLIAN · Analysis and Design of Electronic Circuits
CHIRLIAN AND ZEMANIAN · Electronics
CLEMENT AND JOHNSON · Electrical Engineering Science
COTE AND OAKES · Linear Vacuum-tube and Transistor Circuits
CUCCIA · Harmonics, Sidebands, and Transients in Communication Engineering
CUNNINGHAM · Introduction to Nonlinear Analysis
D'AZZO AND HOUPIS · Feedback Control System Analysis and Synthesis
EASTMAN · Fundamentals of Vacuum Tubes
ELGERD · Control Systems Theory
FEINSTEIN · Foundations of Information Theory
FITZGERALD AND HIGGINBOTHAM · Basic Electrical Engineering
FITZGERALD AND KINGSLEY · Electric Machinery
FRANK · Electrical Measurement Analysis
FRIEDLAND, WING, AND ASH · Principles of Linear Networks
GHAUSI · Principles and Design of Linear Active Circuits
GHOSE · Microwave Circuit Theory and Analysis
GREINER · Semiconductor Devices and Applications
HAMMOND · Electrical Engineering
HANCOCK · An Introduction to the Principles of Communication Theory
HAPPELL AND HESSELBERTH · Engineering Electronics
HARMAN · Fundamentals of Electronic Motion
HARMAN · Principles of the Statistical Theory of Communication
HARMAN AND LYTLE · Electrical and Mechanical Networks
HARRINGTON · Introduction to Electromagnetic Engineering
HARRINGTON · Time-harmonic Electromagnetic Fields
HAYASHI · Nonlinear Oscillations in Physical Systems
HAYT · Engineering Electromagnetics
HAYT AND KEMMERLY · Engineering Circuit Analysis
HILL · Electronics in Engineering
JAVID AND BRENNER · Analysis, Transmission, and Filtering of Signals
JAVID AND BROWN · Field Analysis and Electromagnetics
JOHNSON · Transmission Lines and Networks
KOENIG AND BLACKWELL · Electromechanical System Theory
KOENIG, TOKAD, AND KESAVAN · Analysis of Discrete Physical Systems
KRAUS · Antennas
KRAUS · Electromagnetics
KUH AND PEDERSON · Principles of Circuit Synthesis
KUO · Linear Networks and Systems
LEDLEY · Digital Computer and Control Engineering
LEPAGE · Analysis of Alternating-current Circuits

LePage · Complex Variables and the Laplace Transform for Engineering
LePage and Seely · General Network Analysis
Levi and Panzer · Electromechanical Power Conversion
Ley, Lutz, and Rehberg · Linear Circuit Analysis
Linvill and Gibbons · Transistors and Active Circuits
Littauer · Pulse Electronics
Lynch and Truxal · Introductory System Analysis
Lynch and Truxal · Principles of Electronic Instrumentation
Lynch and Truxal · Signals and Systems in Electrical Engineering
Manning · Electrical Circuits
McCluskey · Introduction to the Theory of Switching Circuits
Meisel · Principles of Electromechanical-energy Conversion
Millman · Vacuum-tube and Semiconductor Electronics
Millman and Seely · Electronics
Millman and Taub · Pulse and Digital Circuits
Millman and Taub · Pulse, Digital, and Switching Waveforms
Mishkin and Braun · Adaptive Control Systems
Moore · Traveling-wave Engineering
Nanavati · An Introduction to Semiconductor Electronics
Pettit · Electronic Switching, Timing, and Pulse Circuits
Pettit and McWhorter · Electronic Amplifier Circuits
Pfeiffer · Concepts of Probability Theory
Pfeiffer · Linear Systems Analysis
Reza · An Introduction to Information Theory
Reza and Seely · Modern Network Analysis
Rogers · Introduction to Electric Fields
Ruston and Bordogna · Electric Networks: Functions, Filters, Analysis
Ryder · Engineering Electronics
Schwartz · Information Transmission, Modulation, and Noise
Schwarz and Friedland · Linear Systems
Seely · Electromechanical Energy Conversion
Seely · Electron-tube Circuits
Seely · Electronic Engineering
Seely · Introduction to Electromagnetic Fields
Seely · Radio Electronics
Seifert and Steeg · Control Systems Engineering
Siskind · Direct-current Machinery
Skilling · Electric Transmission Lines
Skilling · Transient Electric Currents
Spangenberg · Fundamentals of Electron Devices
Spangenberg · Vacuum Tubes
Stevenson · Elements of Power System Analysis
Stewart · Fundamentals of Signal Theory
Storer · Passive Network Synthesis
Strauss · Wave Generation and Shaping
Su · Active Network Synthesis
Terman · Electronic and Radio Engineering
Terman and Pettit · Electronic Measurements
Thaler · Elements of Servomechanism Theory
Thaler and Brown · Analysis and Design of Feedback Control Systems
Thaler and Pastel · Analysis and Design of Nonlinear Feedback Control Systems
Thompson · Alternating-current and Transient Circuit Analysis
Tou · Digital and Sampled-data Control Systems
Tou · Modern Control Theory
Truxal · Automatic Feedback Control System Synthesis
Tuttle · Electric Networks: Analysis and Synthesis
Valdes · The Physical Theory of Transistors
Van Bladel · Electromagnetic Fields
Weinberg · Network Analysis and Synthesis
Williams and Young · Electrical Engineering Problems

McGRAW-HILL TEXTS IN ELECTRICAL ENGINEERING
By Authors at Syracuse University

GLASFORD · Fundamentals of Television Engineering
GOLDMAN · Frequency Analysis, Modulation, and Noise
HARRINGTON · Introduction to Electromagnetic Engineering
HARRINGTON · Time-harmonic Electromagnetic Fields
LEPAGE · Analysis of Alternating-current Circuits
LEPAGE · Complex Variables and the LaPlace Transform for Engineers
LEPAGE AND SEELY · General Network Analysis
NANAVATI · An Introduction to Semiconductor Electronics
REZA · An Introduction to Information Theory
REZA AND SEELY · Modern Network Analysis

TIME-HARMONIC ELECTROMAGNETIC FIELDS

Roger F. Harrington

Professor of Electrical Engineering
Syracuse University

McGRAW-HILL BOOK COMPANY

NEW YORK   TORONTO   LONDON

1961

TIME-HARMONIC ELECTROMAGNETIC FIELDS

Copyright © 1961 by McGraw-Hill, Inc. All Rights Reserved. Printed in the United States of America. This book, or parts thereof, may not be reproduced in any form without permission of the publishers. *Library of Congress Catalog Card Number* 60-14221

7 8 9 10 11 12 – MPMB – 1

ISBN 07-026745-6

PREFACE

This book was written primarily as a graduate-level text, but it should also be useful as a reference book. The organization is somewhat different from that normally found in engineering books. The material is arranged according to similarity of mathematical techniques instead of according to devices (antennas, waveguides, cavities, etc.). This organization reflects the main purpose of the book—to present mathematical techniques for handling electromagnetic engineering problems. In the sense that theorems are proved and formulas derived, the book is theoretical. However, numerous practical examples illustrate the theory, and in this sense the book is practical. The experimental aspect of the subject is not considered explicitly.

The term *time-harmonic* has been used in the title to indicate that only sinusoidally time-varying fields are considered. To describe such fields, the adjective *a-c* (alternating-current) has been borrowed from the corresponding specialization of circuit theory. Actually, much of the theory can easily be extended to arbitrarily time-varying fields by means of the Fourier or Laplace transformations.

The nomenclature and symbolism used is essentially the same as that of the author's earlier text, "Introduction to Electromagnetic Engineering," except for the following change. Boldface script letters denote instantaneous vector quantities and boldface block letters denote complex vectors. This is a departure from the confusing convention of using the same symbol for the two different quantities, instantaneous and complex. Also, the complex quantities are chosen to have rms (root-mean-square) amplitudes, which corresponds to the usual a-c circuit theory convention.

The many examples treated in the text are intended to be simple treatments of practical problems. Most of the complicated formulas are illustrated by numerical calculations or graphs. To augment the examples, there is an extensive set of problems at the end of each chapter. Many of these problems are of theoretical or practical significance, and are therefore listed in the index. Answers are given for most of the problems.

Some of the material of the text appears in book form for the first time. References are given to the original sources when they are known.

However, it has not been possible to trace each concept back to its original inventor; hence many references have probably been omitted. For this the author offers his apologies. Credit has also been given to persons responsible for the original calculations of curves whenever possible. A bibliography of books for supplemental reading is given at the end of the text.

The book has been used for a course directly following an introductory course and also for a course following an intermediate one. On the former level, the progress was slower than on the latter, but the organization of the book seemed satisfactory in both cases. There is more than enough material for a year's work, and the teacher will probably want to make his own choice of topics.

The author expresses his sincere appreciation to everyone who in any way contributed to the creation of this book. Thanks to W. R. LePage, whose love for learning and teaching inspired the author; to V. H. Rumsey, from whom the author learned many of his viewpoints; to H. Gruenberg, who read the galleys; to colleagues and students, for their many valuable comments and criticisms; and, finally, to the several secretaries who so expertly typed the manuscript.

Roger F. Harrington

CONTENTS

Preface vii

Chapter 1. Fundamental Concepts

1-1. Introduction. 1
1-2. Basic Equations. 1
1-3. Constitutive Relationships. 5
1-4. The Generalized Current Concept 7
1-5. Energy and Power 9
1-6. Circuit Concepts 12
1-7. Complex Quantities. 13
1-8. Complex Equations. 16
1-9. Complex Constitutive Parameters 18
1-10. Complex Power 19
1-11. A-C Characteristics of Matter 23
1-12. A Discussion of Current 26
1-13. A-C Behavior of Circuit Elements 29
1-14. Singularities of the Field 32

Chapter 2. Introduction to Waves

2-1. The Wave Equation 37
2-2. Waves in Perfect Dielectrics 41
2-3. Intrinsic Wave Constants 48
2-4. Waves in Lossy Matter 51
2-5. Reflection of Waves 54
2-6. Transmission-line Concepts 61
2-7. Waveguide Concepts 66
2-8. Resonator Concepts 74
2-9. Radiation 77
2-10. Antenna Concepts 81
2-11. On Waves in General 85

Chapter 3. Some Theorems and Concepts

3-1. The Source Concept 95
3-2. Duality 98
3-3. Uniqueness 100
3-4. Image Theory 103
3-5. The Equivalence Principle. 106
3-6. Fields in Half-space 110

CONTENTS

3-7. The Induction Theorem 113
3-8. Reciprocity 116
3-9. Green's Functions 120
3-10. Tensor Green's Functions 123
3-11. Integral Equations 125
3-12. Construction of Solutions 129
3-13. The Radiation Field 132

Chapter 4. Plane Wave Functions

4-1. The Wave Functions 143
4-2. Plane Waves 145
4-3. The Rectangular Waveguide 148
4-4. Alternative Mode Sets 152
4-5. The Rectangular Cavity 155
4-6. Partially Filled Waveguide 158
4-7. The Dielectric-slab Guide 163
4-8. Surface-guided Waves 168
4-9. Modal Expansions of Fields 171
4-10. Currents in Waveguides 177
4-11. Apertures in Ground Planes 180
4-12. Plane Current Sheets 186

Chapter 5. Cylindrical Wave Functions

5-1. The Wave Functions 198
5-2. The Circular Waveguide 204
5-3. Radial Waveguides 208
5-4. The Circular Cavity 213
5-5. Other Guided Waves 216
5-6. Sources of Cylindrical Waves 223
5-7. Two-dimensional Radiation 228
5-8. Wave Transformations 230
5-9. Scattering by Cylinders 232
5-10. Scattering by Wedges 238
5-11. Three-dimensional Radiation 242
5-12. Apertures in Cylinders 245
5-13. Apertures in Wedges 250

Chapter 6. Spherical Wave Functions

6-1. The Wave Functions 264
6-2. The Spherical Cavity 269
6-3. Orthogonality Relationships 273
6-4. Space as a Waveguide 276
6-5. Other Radial Waveguides 279
6-6. Other Resonators 283
6-7. Sources of Spherical Waves 286
6-8. Wave Transformations 289
6-9. Scattering by Spheres 292
6-10. Dipole and Conducting Sphere 298

6-11.	Apertures in Spheres	301
6-12.	Fields External to Cones	303
6-13.	Maximum Antenna Gain	307

Chapter 7. Perturbational and Variational Techniques

7-1.	Introduction	317
7-2.	Perturbations of Cavity Walls	317
7-3.	Cavity-material Perturbations	321
7-4.	Waveguide Perturbations	326
7-5.	Stationary Formulas for Cavities	331
7-6.	The Ritz Procedure	338
7-7.	The Reaction Concept	340
7-8.	Stationary Formulas for Waveguides	345
7-9.	Stationary Formulas for Impedance	348
7-10.	Stationary Formulas for Scattering	355
7-11.	Scattering by Dielectric Obstacles	362
7-12.	Transmission through Apertures	365

Chapter 8. Microwave Networks

8-1.	Cylindrical Waveguides	381
8-2.	Modal Expansions in Waveguides	389
8-3.	The Network Concept	391
8-4.	One-port Networks	393
8-5.	Two-port Networks	398
8-6.	Obstacles in Waveguides	402
8-7.	Posts in Waveguides	406
8-8.	Small Obstacles in Waveguides	411
8-9.	Diaphragms in Waveguides	414
8-10.	Waveguide Junctions	420
8-11.	Waveguide Feeds	425
8-12.	Excitation of Apertures	428
8-13.	Modal Expansions in Cavities	431
8-14.	Probes in Cavities	434
8-15.	Aperture Coupling to Cavities	436

Appendix A.	Vector Analysis	447
Appendix B.	Complex Permittivities	451
Appendix C.	Fourier Series and Integrals	456
Appendix D.	Bessel Functions	460
Appendix E.	Legendre Functions	465

Bibliography 471

Index 473

CHAPTER 1

FUNDAMENTAL CONCEPTS

1-1. Introduction. The topic of this book is the theory and analysis of electromagnetic phenomena that vary sinusoidally in time, henceforth called a-c (alternating-current) phenomena. The fundamental concepts which form the basis of our study are presented in this chapter. It is assumed that the reader already has some acquaintance with electromagnetic field theory and with electric circuit theory. The vector analysis concepts that we shall need are summarized in Appendix A.

We shall view electromagnetic phenomena from the "macroscopic" standpoint, that is, linear dimensions are large compared to atomic dimensions and charge magnitudes are large compared to atomic charges. This allows us to neglect the granular structure of matter and charge. We assume all matter to be stationary with respect to the observer. No treatment of the mechanical forces associated with the electromagnetic field is given.

The rationalized mksc system of units is used throughout. In this system the unit of length is the meter, the unit of mass is the kilogram, the unit of time is the second, and the unit of charge is the coulomb. We consider these units to be *fundamental units*. The units of all other quantities depend upon this choice of fundamental units, and are called *secondary units*. The mksc system of units is particularly convenient because the electrical units are identical to those used in practice.

The concepts necessary for our study are but a few of the many electromagnetic field concepts. We shall start with the familiar Maxwell equations and specialize them to our needs. New notation and nomenclature, more convenient for our purposes, will be introduced. For the most part, these innovations are extensions of a-c circuit concepts.

1-2. Basic Equations. The usual electromagnetic field equations are expressed in terms of six quantities. These are

$\mathcal{E}$, called the *electric intensity* (volts per meter)
$\mathcal{H}$, called the *magnetic intensity* (amperes per meter)
$\mathcal{D}$, called the *electric flux density* (coulombs per square meter)
$\mathcal{B}$, called the *magnetic flux density* (webers per square meter)
$\mathcal{J}$, called the *electric current density* (amperes per square meter)
q_v, called the *electric charge density* (coulombs per cubic meter)

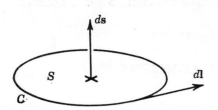
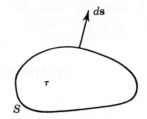

Fig. 1-1. $d\mathbf{l}$ and $d\mathbf{s}$ on an open surface.   Fig. 1-2. $d\mathbf{s}$ on a closed surface.

We shall call a quantity *well-behaved* wherever it is a continuous function and has continuous derivatives. Wherever the above quantities are well-behaved, they obey the *Maxwell equations*

$$\nabla \times \mathcal{E} = -\frac{\partial \mathcal{B}}{\partial t} \qquad \nabla \cdot \mathcal{B} = 0$$
$$\nabla \times \mathcal{H} = \frac{\partial \mathcal{D}}{\partial t} + \mathcal{J} \qquad \nabla \cdot \mathcal{D} = q_v \tag{1-1}$$

These equations include the information contained in the *equation of continuity*

$$\nabla \cdot \mathcal{J} = -\frac{\partial q_v}{\partial t} \tag{1-2}$$

which expresses the conservation of charge. Note that we have used boldface script letters for the various vector quantities, since we wish to reserve the usual boldface roman letters for complex quantities, introduced in Sec. 1-7.

Corresponding to each of Eqs. (1-1) are the integral forms of Maxwell's equations

$$\oint \mathcal{E} \cdot d\mathbf{l} = -\frac{d}{dt}\iint \mathcal{B} \cdot d\mathbf{s} \qquad \oiint \mathcal{B} \cdot d\mathbf{s} = 0$$
$$\oint \mathcal{H} \cdot d\mathbf{l} = \frac{d}{dt}\iint \mathcal{D} \cdot d\mathbf{s} + \iint \mathcal{J} \cdot d\mathbf{s} \qquad \oiint \mathcal{D} \cdot d\mathbf{s} = \iiint q_v \, d\tau \tag{1-3}$$

These are actually more general than Eqs. (1-1) because it is no longer required that the various quantities be well-behaved. In the equations of the first column, we employ the usual convention that $d\mathbf{l}$ encircles $d\mathbf{s}$ according to the right-hand rule of Fig. 1-1. In the equations of the last column, we use the convention that $d\mathbf{s}$ points outward from a closed surface, as shown in Fig. 1-2. The circle on a line integral denotes a closed contour; the circle on a surface integral denotes a closed surface. The integral form of Eq. (1-2) is

$$\oiint \mathcal{J} \cdot d\mathbf{s} = -\frac{d}{dt}\iiint q_v \, d\tau \tag{1-4}$$

where the same convention applies. This is the statement of conservation of charge as it applies to a region.

We shall use the name *field quantity* to describe the quantities discussed above. Associated with each field quantity there is a *circuit quantity*, or integral quantity. These circuit quantities are

  v, called the *voltage* (volts)
  i, called the *electric current* (amperes)
  q, called the *electric charge* (coulombs)
  ψ, called the *magnetic flux* (webers)
  ψ^e, called the *electric flux* (coulombs)
  u, called the *magnetomotive force* (amperes)

The explicit relationships of the field quantities to the circuit quantities can be summarized as follows:

$$v = \int \mathcal{E} \cdot dl \qquad \psi = \iint \mathcal{B} \cdot ds$$
$$i = \iint \mathcal{J} \cdot ds \qquad \psi^e = \iint \mathcal{D} \cdot ds \qquad (1\text{-}5)$$
$$q = \iiint q_v \, d\tau \qquad u = \int \mathcal{H} \cdot dl$$

All the circuit quantities are algebraic quantities and require reference conditions when designating them. Our convention for a "line-integral" quantity, such as voltage, is positive reference at the start of the path of integration. This is illustrated by Fig. 1-3. Our convention for a "surface-integral" quantity, such as current, is positive reference in the direction of ds. This is shown in Fig. 1-4. Charge is a "net-amount" quantity, being the amount of positive charge minus the amount of negative charge.

We shall call Eqs. (1-1) to (1-4) *field equations*, since all quantities appearing in them are field quantities. Corresponding equations written in terms of circuit quantities we shall describe as *circuit equations*. Equa-

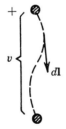

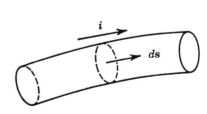

Fig. 1-3. Reference convention for voltage.

Fig. 1-4. Reference convention for current.

tions (1-3) are commonly written in mixed field and circuit form as

$$\oint \mathcal{E} \cdot dl = -\frac{d\psi}{dt} \qquad \oint \mathcal{B} \cdot d\mathbf{s} = 0$$
$$\oint \mathcal{H} \cdot dl = \frac{d\psi^e}{dt} + i \qquad \oint \mathcal{D} \cdot d\mathbf{s} = q \qquad (1\text{-}6)$$

Similarly, the equation of continuity in mixed field and circuit form is

$$\oint \mathcal{J} \cdot d\mathbf{s} = -\frac{dq}{dt} \qquad (1\text{-}7)$$

Finally, the various equations can be written entirely in terms of circuit quantities. For this, we shall use the notation that Σ denotes summation over a closed contour for a line-integral quantity, and summation over a closed surface for a surface-integral quantity. In this notation, the circuit forms of Eqs. (1-6) are

$$\sum v = -\frac{d\psi}{dt} \qquad \sum \psi = 0$$
$$\sum u = \frac{d\psi^e}{dt} + i \qquad \sum \psi^e = q \qquad (1\text{-}8)$$

and the circuit form of Eq. (1-7) is

$$\sum i = -\frac{dq}{dt} \qquad (1\text{-}9)$$

Note that the first of Eqs. (1-8) is a generalized form of *Kirchhoff's voltage law*, and Eq. (1-9) is a generalized form of *Kirchhoff's current law*.

It is apparent from the preceding summary that many mathematical forms can be used to present a single physical concept. An understanding of the concepts is an invaluable aid to remembering the equations. While an extensive exposition of these concepts properly belongs in an introductory textbook, let us here summarize them. Consider the sets of Eqs. (1-1), (1-3), (1-6), and (1-8). The first equation in each set is essentially *Faraday's law of induction*. It states that a changing magnetic flux induces a voltage in a path surrounding it. The second equation in each set is essentially *Ampère's circuital law*, extended to the time-varying case. It is a partial definition of magnetic intensity and magnetomotive force. The third equation of each set states that magnetic flux has no "flux source," that is, lines of $\mathcal{B}$ can have no beginning or end. The fourth equation in each set is *Gauss' law* and states that lines of $\mathcal{D}$ begin and end on electric charge. It is essentially a partial definition of electric flux. Finally, Eqs. (1-2), (1-4), (1-7), and (1-9) are all forms of the *law of conservation of charge*. They state that charge can be neither created nor destroyed, merely transported. Lines of current must begin and end at points of increasing or decreasing charge density.

1-3. Constitutive Relationships. In addition to the equations of Sec. 1-2 we need equations specifying the characteristics of the medium in which the field exists. We shall consider the domain of $\mathcal{E}$ and $\mathcal{H}$ as the electromagnetic field and express $\mathfrak{D}$, $\mathfrak{B}$, and $\mathfrak{J}$ in terms of $\mathcal{E}$ and $\mathcal{H}$. Equations of the general form

$$\begin{aligned} \mathfrak{D} &= \mathfrak{D}(\mathcal{E},\mathcal{H}) \\ \mathfrak{B} &= \mathfrak{B}(\mathcal{E},\mathcal{H}) \\ \mathfrak{J} &= \mathfrak{J}(\mathcal{E},\mathcal{H}) \end{aligned} \tag{1-10}$$

are called *constitutive relationships*. Explicit forms for these can be found by experimentation or deduced from atomic considerations.

The term *free space* will be used to denote vacuum or any other medium having essentially the same characteristics as vacuum (such as air). The constitutive relationships assume the particularly simple forms

$$\left. \begin{aligned} \mathfrak{D} &= \epsilon_0 \mathcal{E} \\ \mathfrak{B} &= \mu_0 \mathcal{H} \\ \mathfrak{J} &= 0 \end{aligned} \right\} \quad \text{in free space} \tag{1-11}$$

where ϵ_0 is the *capacitivity* or *permittivity* of vacuum, and μ_0 is the *inductivity* or *permeability* of vacuum. It is a mathematical consequence of the field equations that $(\epsilon_0 \mu_0)^{-\frac{1}{2}}$ is the velocity of propagation of an electromagnetic disturbance in free space. Light is electromagnetic in nature, and this velocity is called the *velocity of light c*. Measurements have established that

$$c = \frac{1}{\sqrt{\epsilon_0 \mu_0}} = 2.99790 \times 10^8 \approx 3 \times 10^8 \text{ meters per second} \tag{1-12}$$

The choice of either ϵ_0 or μ_0 determines a system of electromagnetic units according to our equations. By international agreement, the value of μ_0 has been chosen as

$$\mu_0 = 4\pi \times 10^{-7} \text{ henry per meter} \tag{1-13}$$

for the mksc system of units. It then follows from Eq. (1-12) that

$$\epsilon_0 = 8.854 \times 10^{-12} \approx \frac{1}{36\pi} \times 10^{-9} \text{ farad per meter} \tag{1-14}$$

for the mksc system of units.

Under certain conditions, the constitutive relationships become simple proportionalities for many materials. We say that such matter is linear in the simple sense, and call it *simple matter* for short. Thus

$$\left. \begin{array}{l} \mathcal{D} = \epsilon \mathcal{E} \\ \mathcal{B} = \mu \mathcal{H} \\ \mathcal{J} = \sigma \mathcal{E} \end{array} \right\} \quad \text{in simple matter} \qquad (1\text{-}15)$$

where, as in the free-space case, ϵ is called the capacitivity of the medium and μ is called the inductivity of the medium. The parameter σ is called the *conductivity* of the medium. We originally made the qualifying statement that Eqs. (1-15) hold "under certain conditions." They may not hold if $\mathcal{E}$ or $\mathcal{H}$ are very large, or if time derivatives of $\mathcal{E}$ or $\mathcal{H}$ are very large.

Matter is often classified according to its values of σ, ϵ, and μ. Materials having large values of σ are called *conductors* and those having small values of σ are called *insulators* or *dielectrics*. For analyses, it is often convenient to approximate good conductors by *perfect conductors*, characterized by $\sigma = \infty$, and to approximate good dielectrics by *perfect dielectrics*, characterized by $\sigma = 0$. The capacitivity ϵ of any material is never less than that of vacuum ϵ_0. The ratio $\epsilon_r = \epsilon/\epsilon_0$ is called the *dielectric constant* or *relative capacitivity*. The dielectric constant of a good conductor is hard to measure but appears to be unity. For most linear matter, the inductivity μ is approximately that of free space μ_0. There is a class of materials, called *diamagnetic*, for which μ is slightly less than μ_0 (of the order of 0.01 per cent). There is a class of materials, called *paramagnetic*, for which μ is slightly greater than μ_0 (again of the order of 0.01 per cent). A third class of materials, called *ferromagnetic*, has values of μ much larger than μ_0, but these materials are often nonlinear. For our purposes, we shall call all materials except the ferromagnetic ones *nonmagnetic* and take $\mu = \mu_0$ for them. The ratio $\mu_r = \mu/\mu_0$ is called the *relative inductivity* or *relative permeability* and is, of course, essentially unity for nonmagnetic matter.

Quite often the restriction on the time rate of change of the field, made on the validity of Eqs. (1-15), can be overcome by extending the definition of linearity. We say that matter is linear in the general sense, and call it *linear matter*, when the constitutive relationships are the following linear differential equations:

$$\left. \begin{array}{l} \mathcal{D} = \epsilon \mathcal{E} + \epsilon_1 \dfrac{\partial \mathcal{E}}{\partial t} + \epsilon_2 \dfrac{\partial^2 \mathcal{E}}{\partial t^2} + \cdots \\[6pt] \mathcal{B} = \mu \mathcal{H} + \mu_1 \dfrac{\partial \mathcal{H}}{\partial t} + \mu_2 \dfrac{\partial^2 \mathcal{H}}{\partial t^2} + \cdots \\[6pt] \mathcal{J} = \sigma \mathcal{E} + \sigma_1 \dfrac{\partial \mathcal{E}}{\partial t} + \sigma_2 \dfrac{\partial^2 \mathcal{E}}{\partial t^2} + \cdots \end{array} \right\} \quad \text{in linear matter} \qquad (1\text{-}16)$$

Even more complicated formulas for the constitutive relationships may be necessary in some cases, but Eqs. (1-16) are the most general that we shall consider. Note that Eqs. (1-16) reduce to Eqs. (1-15) when the time derivatives of $\mathcal{E}$ and $\mathcal{H}$ become sufficiently small.

The physical significance of the extended definition of linearity is as follows. The atomic particles of matter have mass as well as charge, so when the field changes rapidly the particles cannot "follow" the field. For example, suppose an electron has been accelerated by the field, and then the direction of $\mathcal{E}$ changes. There will be a time lag before the electron can change direction, because of its momentum. Such a picture holds for $\mathcal{J}$ if the electron is a free electron. It holds for $\mathcal{D}$ if the electron is a bound electron. A similar picture holds for $\mathcal{B}$ except that the magnetic moment of the electron is the contributing quantity. We shall not attempt to give significance to each term of Eqs. (1-16). It will be shown in Sec. 1-9 that all terms of Eqs. (1-16) contribute to an "admittivity" and an "impedivity" of a material in the time-harmonic case.

1-4. The Generalized Current Concept. It was Maxwell who first noted that Ampère's law for statics, $\nabla \times \mathcal{H} = \mathcal{J}$, was incomplete for time-varying fields. He amended the law to include an *electric displacement current* $\partial \mathcal{D}/\partial t$ in addition to the conduction current. He visualized this displacement current in free space as a motion of bound charge in an "ether," an ideal weightless fluid permeating all space. We have since discarded the concept of an ether, for it has proved undetectable and even somewhat illogical in view of the theory of relativity. In dielectrics, part of the term $\partial \mathcal{D}/\partial t$ is a motion of the bound particles and is thus a current in the true sense of the word. However, it is convenient to consider the entire $\partial \mathcal{D}/\partial t$ term as a current. In view of the symmetry of Maxwell's equations, it also is convenient to consider the term $\partial \mathcal{B}/\partial t$ as a *magnetic displacement current*. Finally, to represent sources, we amend the field equations to include *impressed currents*, electric and magnetic. These are the currents we view as the cause of the field. We shall see in the next section that the impressed currents represent energy sources.

The symbols $\mathcal{J}$ and $\mathcal{M}$ will be used to denote electric and magnetic currents in general, with superscripts indicating the type of current. As discussed above, we define total currents

$$\begin{aligned} \mathcal{J}^t &= \frac{\partial \mathcal{D}}{\partial t} + \mathcal{J}^c + \mathcal{J}^i \\ \mathcal{M}^t &= \frac{\partial \mathcal{B}}{\partial t} + \mathcal{M}^i \end{aligned} \quad (1\text{-}17)$$

where the superscripts t, c, and i denote total, conduction, and impressed currents. The symbols i and k will be used to denote net electric and magnetic currents, and the same superscripts will indicate the type.

Thus, the circuit form corresponding to Eqs. (1-17) is

$$i^t = \frac{d\psi^e}{dt} + i^c + i^i$$
$$k^t = \frac{d\psi}{dt} + k^i \tag{1-18}$$

The i and k are, of course, related to the $\mathcal{g}$ and $\mathfrak{M}$ by

$$i = \iint \mathcal{g} \cdot d\mathbf{s} \qquad k = \iint \mathfrak{M} \cdot d\mathbf{s} \tag{1-19}$$

where these apply to any of the various types of current.

In terms of the generalized current concept, the basic equations of electromagnetism become, in the differential form,

$$\nabla \times \mathcal{E} = -\mathfrak{M}^t \qquad \nabla \times \mathcal{H} = \mathcal{g}^t \tag{1-20}$$

and in the integral form,

$$\oint \mathcal{E} \cdot d\mathbf{l} = -\iint \mathfrak{M}^t \cdot d\mathbf{s} \qquad \oint \mathcal{H} \cdot d\mathbf{l} = \iint \mathcal{g}^t \cdot d\mathbf{s} \tag{1-21}$$

Also, the mixed field-circuit form is

$$\oint \mathcal{E} \cdot d\mathbf{l} = -k^t \qquad \oint \mathcal{H} \cdot d\mathbf{l} = i^t \tag{1-22}$$

and the circuit form is

$$\Sigma v = -k^t \qquad \Sigma u = i^t \tag{1-23}$$

Note that these look simpler than the equations of Sec. 1-2. Actually, we have merely included many concepts in the functions $\mathfrak{M}^t$ and $\mathcal{g}^t$; so some of the information contained in the original Maxwell equations has become hidden. However, our study comprises only a small portion of the general theory of electromagnetism, and the forms of Eqs. (1-20) to (1-23) are well suited to our purposes.

Note that we have omitted the "divergence equations" of Maxwell from our above sets of equations. We have done so to emphasize that this information is included in the above sets. For example, taking the divergence of each of Eqs. (1-20), we obtain

$$\nabla \cdot \mathfrak{M}^t = 0 \qquad \nabla \cdot \mathcal{g}^t = 0 \tag{1-24}$$

for $\nabla \cdot \nabla \times \mathcal{C} = 0$ is an identity. Similarly, Eqs. (1-21) applied to closed surfaces became

$$\oiint \mathfrak{M}^t \cdot d\mathbf{s} = 0 \qquad \oiint \mathcal{g}^t \cdot d\mathbf{s} = 0 \tag{1-25}$$

Thus, the total currents are solenoidal. Lines of total current have no beginning or end but must be continuous.

As an illustration of the generalized current concept, consider the circuits of Figs. 1-5 and 1-6. In Fig. 1-5, the "current source" $\mathcal{J}^i$ produces a conduction current $\mathcal{J}^c$ through the resistor and a displacement current $\mathcal{J}^d = \partial\mathfrak{D}/\partial t$ through the capacitor. In Fig. 1-6, the "voltage source" $\mathfrak{M}^i$ produces an electric current in the wire which in turn causes the magnetic displacement current $\mathfrak{M}^d = \partial\mathfrak{B}/\partial t$ in the magnetic core. In these pictures we have used the convention that a single-headed arrow represents an electric current, a double-headed arrow represents a magnetic current.

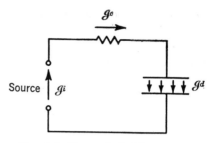

Fig. 1-5. Types of electric current.

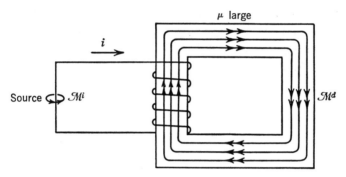

Fig. 1-6. Types of magnetic current.

It is not possible at this time to give the reader a complete picture of the usefulness of impressed currents. Figures 1-5 and 1-6 anticipate one application, namely, that of representing sources. More generally, the impressed currents are those currents we *view* as sources. In a sense, the impressed currents are those currents in terms of which the field is expressed. In one problem, a conduction current might be considered as the source, or impressed, current. In another problem, a polarization or magnetization current might be considered as the source current. Our understanding of the concept will grow as we learn to use it.

1-5. Energy and Power. Consider a region of electromagnetic field, as suggested by Fig. 1-7. The field obeys the Maxwell equations, which in generalized current

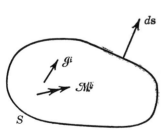

Fig. 1-7. A region containing sources.

notation are Eqs. (1-20). As an extension of circuit concepts, it can be shown that a product $\mathcal{E} \cdot \mathcal{J}$ is a power density. This suggests a scalar multiplication of the second of Eqs. (1-20) by $\mathcal{E}$. Also, in view of the vector identity

$$\nabla \cdot (\mathcal{E} \times \mathcal{H}) = \mathcal{H} \cdot \nabla \times \mathcal{E} - \mathcal{E} \cdot \nabla \times \mathcal{H}$$

a scalar multiplication of the first of Eqs. (1-20) by $\mathcal{H}$ is suggested. The difference of the resulting two equations is

$$\nabla \cdot (\mathcal{E} \times \mathcal{H}) + \mathcal{E} \cdot \mathcal{J}^t + \mathcal{H} \cdot \mathcal{M}^t = 0 \quad (1\text{-}26)$$

If this equation is integrated throughout a region, and the divergence theorem applied to the first term, there results

$$\oint \mathcal{E} \times \mathcal{H} \cdot d\mathbf{s} + \iiint (\mathcal{E} \cdot \mathcal{J}^t + \mathcal{H} \cdot \mathcal{M}^t)\, d\tau = 0 \quad (1\text{-}27)$$

We shall interpret these as equations for the *conservation of energy*, Eq. (1-26) being the differential form and Eq. (1-27) being the integral form.

The generally accepted interpretation of Eqs. (1-26) and (1-27) is as follows. The *Poynting vector*

$$\mathbf{S} = \mathcal{E} \times \mathcal{H} \quad (1\text{-}28)$$

is postulated to be a density-of-power flux. The point relationship

$$p_f = \nabla \cdot \mathbf{S} = \nabla \cdot (\mathcal{E} \times \mathcal{H}) \quad (1\text{-}29)$$

is then a volume density of power leaving the point, and the integral

$$\mathcal{P}_f = \oint \mathbf{S} \cdot d\mathbf{s} = \oint \mathcal{E} \times \mathcal{H} \cdot d\mathbf{s} \quad (1\text{-}30)$$

is the total power leaving the region bounded by the surface of integration. The other terms of Eq. (1-26) can then be interpreted as the rate of increase in energy density at a point. Similarly, the other terms of Eq. (1-27) can be interpreted as the rate of increase in energy within the region. Further identification of this energy can be made in particular cases.

For media linear in the simple sense, as defined by Eqs. (1-15), the last two terms of Eq. (1-26) become

$$\begin{aligned}
\mathcal{E} \cdot \mathcal{J}^t &= \frac{\partial}{\partial t}\left(\frac{1}{2}\epsilon \mathcal{E}^2\right) + \sigma \mathcal{E}^2 + \mathcal{E} \cdot \mathcal{J}^i \\
\mathcal{H} \cdot \mathcal{M}^t &= \frac{\partial}{\partial t}\left(\frac{1}{2}\mu \mathcal{H}^2\right) + \mathcal{H} \cdot \mathcal{M}^i
\end{aligned} \quad (1\text{-}31)$$

where $\mathcal{J}^i$ and $\mathcal{M}^i$ represent possible source currents. The terms

$$w_e = \tfrac{1}{2}\epsilon \mathcal{E}^2 \qquad w_m = \tfrac{1}{2}\mu \mathcal{H}^2 \quad (1\text{-}32)$$

are identified as the electric and magnetic energy densities of static fields, and this interpretation is retained for dynamic fields. The term

$$p_d = \sigma \mathcal{E}^2 \tag{1-33}$$

is identified as the density of power converted to heat energy, called *dissipated power*. Finally, the density of power supplied by the source currents is defined as

$$p_s = -(\mathcal{E} \cdot \mathcal{J}^i + \mathcal{K} \cdot \mathcal{M}^i) \tag{1-34}$$

The reference direction for source power is opposite to that for dissipated power, as evidenced by the minus sign of Eq. (1-34). In terms of the above-defined quantities, we can rewrite Eq. (1-26) as

$$p_s = p_f + p_d + \frac{\partial}{\partial t}(w_e + w_m) \tag{1-35}$$

A word statement of this equation is: At any point, the density of power supplied by the sources must equal that leaving the point plus that dissipated plus the rate of increase in stored electric and magnetic energy densities.

A more common statement of the conservation of energy is that which refers to an entire region. Corresponding to the densities of Eqs. (1-32), we define the net electric and magnetic energies within a region as

$$\mathcal{W}_e = \tfrac{1}{2} \iiint \epsilon \mathcal{E}^2 \, d\tau \qquad \mathcal{W}_m = \tfrac{1}{2} \iiint \mu \mathcal{K}^2 \, d\tau \tag{1-36}$$

Corresponding to Eq. (1-33), we define the net power converted to heat energy as

$$\mathcal{P}_d = \iiint \sigma \mathcal{E}^2 \, d\tau \tag{1-37}$$

Finally, corresponding to Eq. (1-34), we define the net power supplied by sources within the region as

$$\mathcal{P}_s = - \iiint (\mathcal{E} \cdot \mathcal{J}^i + \mathcal{K} \cdot \mathcal{M}^i) \, d\tau \tag{1-38}$$

In terms of these definitions, Eq. (1-27) can be written as

$$\mathcal{P}_s = \mathcal{P}_f + \mathcal{P}_d + \frac{d}{dt}(\mathcal{W}_e + \mathcal{W}_m) \tag{1-39}$$

Thus, the power supplied by the sources within a region must equal that leaving the region plus that dissipated within the region plus the rate of increase in electric and magnetic energies stored within the region.

If we proceed to the general definition of linearity, Eqs. (1-16), the separation of power into a reversible energy change (storage) and an irreversible energy change (dissipation) is no longer easy. Contributions to energy storage and to energy dissipation may originate from both conduction and displacement currents. However, Eqs. (1-35) and (1-39) still apply to media linear in the general sense. We merely cannot identify the various terms. In Sec. 1-10 we shall see that for a-c fields the division of energy into stored and dissipated components again assumes a simple form.

1-6. Circuit Concepts. The usual equations of circuit theory are specializations of the field equations. Our knowledge of circuit concepts can therefore be of help to us in understanding field concepts. In this section we shall quickly review this relationship of circuits to fields.

Kirchhoff's current law for circuits is an application of the equation of conservation of charge to surfaces enclosing wire junctions. To demonstrate, consider the parallel RLC circuit of Fig. 1-8. Let the letter o denote the junction, and the letters a, b, c, d denote the upper terminals of the elements. We apply Eq. (1-7) to a surface enclosing the junction, as represented by the dotted line in Fig. 1-8. The result is

$$i_{oa} + i_{ob} + i_{oc} + i_{od} + i_l + \frac{dq}{dt} = 0$$

where the i_{on} are the currents in the wires, i_l is the leakage current crossing the surface outside of the wires, and q is the charge on the junction. The term dq/dt can be thought of as the current through the stray capacitance between the top and bottom junctions. In most circuit applications both i_l and dq/dt are negligible, and the above equation reduces to

$$i_{oa} + i_{ob} + i_{oc} + i_{od} = 0$$

This is the usual expression of the Kirchhoff current law for the circuit of Fig. 1-8.

Kirchhoff's voltage law for circuits is an application of the first Maxwell equation to closed contours following the connecting wires of the circuit and closing across the terminals of the elements. To demonstrate, consider the series RLC circuit of Fig. 1-9. Let the letters a to h denote

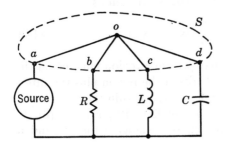

Fig. 1-8. A parallel RLC circuit.

FUNDAMENTAL CONCEPTS 13

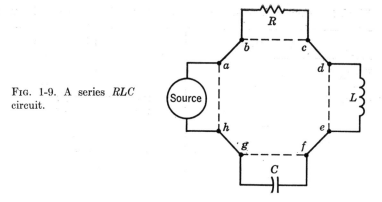

FIG. 1-9. A series RLC circuit.

the terminals of the elements as shown. We apply the first of Eqs. (1-6) to the contour $abcdefgha$, following the dotted lines between terminals. This gives

$$v_{ab} + v_{bc} + v_{cd} + v_{de} + v_{ef} + v_{fg} + v_{gh} + v_{ha} + \frac{d\psi}{dt} = 0$$

where the v_{mn} are the voltage drops along the contour and ψ is the magnetic flux enclosed. The voltages v_{ab}, v_{cd}, v_{ef}, and v_{gh} are due to the resistance of the wire. The term $d\psi/dt$ is the voltage of the stray inductance of the loop. When the wire resistance and the stray inductance can be neglected, the above equation reduces to

$$v_{bc} + v_{de} + v_{fg} + v_{ha} = 0$$

This is the usual form of Kirchhoff's voltage law for the circuit of Fig. 1-9.

In addition to Kirchhoff's laws, circuit theory uses a number of "element laws." Ohm's law for resistors, $v = Ri$, is a specialization of the constitutive relationship $\mathcal{J} = \sigma \mathcal{E}$. The law for capacitors, $q = Cv$, expresses the same concept as $\mathcal{D} = \epsilon \mathcal{E}$. We have from the equation of continuity $i = dq/dt$, so the capacitor law can also be written as $i = C\, dv/dt$. The law for inductors, $\psi = Li$, expresses the same concept as $\mathcal{B} = \mu \mathcal{H}$. From the first Maxwell equation we have $v = d\psi/dt$, so the inductor law can also be written as $v = L\, di/dt$. Finally, the various energy relationships for circuit theory can be considered as specializations of those for field theory. Detailed expositions of the various specializations mentioned above can be found in elementary textbooks. Table 1-1 summarizes the various correspondences between field concepts and circuit concepts.

1-7. Complex Quantities. When the fields are a-c, that is, when the time variation is harmonic, the mathematical analysis can be simplified

TABLE 1-1. CORRESPONDENCES BETWEEN CIRCUIT CONCEPTS AND FIELD CONCEPTS

Circuit concepts	Field concepts
Voltage v	Electric intensity $\mathcal{E}$
Current i	Electric current density $\mathcal{J}$ or magnetic intensity $\mathcal{H}$
Magnetic flux ψ	Magnetic flux density $\mathcal{B}$
Charge q	Charge density q_v or electric flux density $\mathcal{D}$
Kirchhoff's voltage law (generalized) $$\sum v_n = -\frac{d\psi}{dt}$$	Maxwell-Faraday equation $$\nabla \times \mathcal{E} = -\frac{\partial \mathcal{B}}{\partial t}$$
Kirchhoff's current law (generalized) $$\sum i_n = -\frac{dq}{dt}$$	Equation of continuity $$\nabla \cdot \mathcal{J}^c = -\frac{\partial q_v}{\partial t}$$
Element laws (linear) Resistors $i = \frac{1}{R} v$ Capacitors $q = Cv$ or $i = C \frac{dv}{dt}$ Inductors $\psi = Li$ or $v = L \frac{di}{dt}$	Constitutive relationships (linear in the simple sense) Conductors $\mathcal{J}^c = \sigma \mathcal{E}$ Dielectrics $\mathcal{D} = \epsilon \mathcal{E}$ or $\mathcal{J}^d = \epsilon \frac{\partial \mathcal{E}}{\partial t}$ Magnetic properties $\mathcal{B} = \mu \mathcal{H}$ or $\mathcal{M}^d = \mu \frac{\partial \mathcal{H}}{\partial t}$
Power flow $p_f = vi$	Power flow $\mathcal{S} = \mathcal{E} \times \mathcal{H}$
Power dissipation in resistors $$\mathcal{P}_d = vi = \frac{1}{R} v^2$$	Power dissipation $$p_d = \mathcal{E} \cdot \mathcal{J}^c = \sigma \mathcal{E}^2$$
Energy in capacitors $\mathcal{W}_e = \frac{1}{2} qv = \frac{1}{2} Cv^2$	Electric energy $w_e = \frac{1}{2} \mathcal{D} \cdot \mathcal{E} = \frac{1}{2} \epsilon \mathcal{E}^2$
Energy in inductors $\mathcal{W}_m = \frac{1}{2} \psi i = \frac{1}{2} Li^2$	Magnetic energy $w_m = \frac{1}{2} \mathcal{B} \cdot \mathcal{H} = \frac{1}{2} \mu \mathcal{H}^2$

FUNDAMENTAL CONCEPTS

by using complex quantities. The basis for this is Euler's identity

$$e^{j\alpha} = \cos \alpha + j \sin \alpha$$

where $j = \sqrt{-1}$. This gives us a relationship between real sinusoidal functions and the complex exponential function.

Any a-c quantity can be represented by a complex quantity. A scalar quantity is interpreted according to[1]

$$v = \sqrt{2}\,|V|\cos(\omega t + \alpha) = \sqrt{2}\,\text{Re}\,(Ve^{j\omega t}) \qquad (1\text{-}40)$$

where v is called the *instantaneous quantity* and $V = |V|e^{j\alpha}$ is called the *complex quantity*. The notation Re ( ) stands for "the real part of," that is, the part not associated with j. Other names for V are "phasor quantity" and "vector quantity," the last name causing confusion with space vectors. In our notation v represents a voltage, hence V is a *complex voltage*. Equation (1-40) with v replaced by i and V replaced by I would define a *complex current*, and so on. Note that the complex quantity is *not* a function of time but it may be a function of position. Note also that the magnitude of the complex quantity is the effective (root-mean-square) value of the instantaneous quantity. We have chosen it so because (1) a-c quantities are usually specified or measured in effective values in practice, and (2) equations for complex power and energy retain the same proportionality factors as do their instantaneous counterparts. For example, in circuit theory the instantaneous power is $p = vi$, and complex power is $P = VI^*$. A factor of $\tfrac{1}{2}$ appears in the equation for complex power if peak values of v and i are used for $|V|$ and $|I|$.

Complex notation can readily be extended to vectors having sinusoidal time variation. A *complex* **E** is defined as related to an *instantaneous* **ε** according to

$$\boldsymbol{\mathcal{E}} = \sqrt{2}\,\text{Re}\,(\mathbf{E}e^{j\omega t}) \qquad (1\text{-}41)$$

This means that the spatial components of **E** are related to the spatial components of **ε** by Eq. (1-40). For example, the x components of **E** and **ε** are related by

$$\mathcal{E}_x = \sqrt{2}\,\text{Re}\,(E_x e^{j\omega t}) = \sqrt{2}\,|E_x|\cos(\omega t + \alpha_x)$$

where $E_x = |E_x|e^{j\alpha_x}$. Similar equations relate the y and z components of **E** and **ε**. The phase of each component may be different from the phases of the other two components, that is, α_x, α_y, and α_z are not necessarily equal. In our notation **ε** is an electric intensity, hence **E** is called the *complex electric intensity*. Equation (1-41) with **E** replaced by **H** and **ε** by **ℋ**

---

[1] The convention $v = \sqrt{2}\,\text{Im}\,(Ve^{j\omega t})$ can also be used, where Im ( ) stands for "the imaginary part of." The factor $\sqrt{2}$ can be omitted if it is desired that $|V|$ be the peak value of v.

defines a *complex magnetic intensity* **H**, representing the instantaneous magnetic intensity ℋ, and so on. Note that the magnitude of a component of the complex vector is the effective value of the corresponding component of the instantaneous vector. This choice corresponds to that taken for complex scalars and has essentially the same advantages.

A real vector, such as 𝓔 or ℋ, can be thought of as a triplet of real scalar functions, namely, the x, y, and z components. At any instant of time, the vector has a definite magnitude and direction at every point in space and can be represented in three dimensions by arrows. A complex vector, such as **E** or **H**, is a group of six real scalar functions, namely, the real and imaginary parts of the x, y, and z components. It *cannot* be represented by arrows in three-dimensional space except in special cases. One such special case is that for which $\alpha_x = \alpha_y = \alpha_z$, so that the vector has a real direction in space. In this case the instantaneous vector always points in the same direction (or opposite direction), at a point in space, changing only in amplitude. We could define a "complex magnitude" and a "complex direction" for a complex vector as extensions of the corresponding definitions for real vectors, but these would have little use.

Throughout this book we shall use the following notation. Instantaneous quantities are denoted by script letters or lower-case letters. Complex quantities which represent the instantaneous quantities are denoted by the corresponding capital letter. Vectors are denoted by boldface type.

1-8. Complex Equations. The symbol Re ( ) can be considered as a mathematical operator which selects the real part of a complex quantity. A set of rules for manipulating the operator Re ( ) can be formulated from the properties of complex functions. The following are the rules we shall need. Let a capital letter denote a complex quantity and a lower-case letter denote a real quantity. Then

$$\begin{aligned}
\operatorname{Re}(A) + \operatorname{Re}(B) &= \operatorname{Re}(A + B) \\
\operatorname{Re}(aA) &= a \operatorname{Re}(A) \\
\frac{\partial}{\partial x} \operatorname{Re}(A) &= \operatorname{Re}\left(\frac{\partial A}{\partial x}\right) \\
\int \operatorname{Re}(A)\, dx &= \operatorname{Re}\left(\int A\, dx\right)
\end{aligned} \qquad (1\text{-}42)$$

The proof of these is left to the reader.

In addition to the above equations we shall need the following lemma. *If A and B are complex quantities, and $\operatorname{Re}(Ae^{j\omega t}) = \operatorname{Re}(Be^{j\omega t})$ for all t, then $A = B$.* We can readily show this by first taking $t = 0$, obtaining $\operatorname{Re}(A) = \operatorname{Re}(B)$, and then taking $\omega t = \pi/2$, obtaining $\operatorname{Im}(A) = \operatorname{Im}(B)$. Thus, $A = B$, for the above two equalities are the definition of this.

To illustrate the derivation of an equation for complex quantities from one for instantaneous quantities, consider

$$v = \int \mathcal{E} \cdot d\mathbf{l}$$

Expressing v and $\mathcal{E}$ in terms of their complex counterparts, we have

$$\sqrt{2}\,\text{Re}\,(Ve^{j\omega t}) = \int \sqrt{2}\,\text{Re}\,(\mathbf{E}e^{j\omega t}) \cdot d\mathbf{l}$$

By steps justifiable by Eqs. (1-42), this reduces to

$$\sqrt{2}\,\text{Re}\,(Ve^{j\omega t}) = \sqrt{2}\,\text{Re}\left(e^{j\omega t}\int \mathbf{E} \cdot d\mathbf{l}\right)$$

Cancellation of the $\sqrt{2}$'s and application of the above lemma then gives

$$V = \int \mathbf{E} \cdot d\mathbf{l}$$

Note that this is of the same form as the original instantaneous equation. We have illustrated the procedure with a scalar equation, but the same steps apply to the components of a vector equation.

From our rules for manipulation of the Re ( ) operator, it should be apparent that any equation linearly relating instantaneous quantities and *not* involving time differentiation takes the same form for complex quantities. Thus, the complex circuit quantities V, I, U, and K are related to the complex field quantities $\mathbf{E}$, $\mathbf{H}$, $\mathbf{J}$, and $\mathbf{M}$ according to

$$V = \int \mathbf{E} \cdot d\mathbf{l} \qquad U = \int \mathbf{H} \cdot d\mathbf{l}$$
$$I = \iint \mathbf{J} \cdot d\mathbf{s} \qquad K = \iint \mathbf{M} \cdot d\mathbf{s} \qquad (1\text{-}43)$$

There is no time differentiation explicit in the field equations written in generalized current notation. The complex forms of these must therefore also be the same as the instantaneous forms. For example, the complex form of Eqs. (1-20) is

$$\nabla \times \mathbf{E} = -\mathbf{M}^i \qquad \nabla \times \mathbf{H} = \mathbf{J}^i \qquad (1\text{-}44)$$

Even though these complex equations look the same as the corresponding instantaneous equations, we should always keep in mind the difference in meaning.

As an illustration of the procedure when the instantaneous equation exhibits a time differentiation, consider the equation

$$\nabla \times \mathcal{E} = -\frac{\partial \mathcal{B}}{\partial t}$$

Again we express the instantaneous quantities in terms of the complex quantities, and obtain

$$\nabla \times [\sqrt{2}\,\text{Re}\,(\mathbf{E}e^{j\omega t})] = -\frac{\partial}{\partial t}[\sqrt{2}\,\text{Re}\,(\mathbf{B}e^{j\omega t})]$$

The time variation is explicit, and the differentiation can be performed. By steps justifiable by Eqs. (1-42), the above equation becomes

$$\sqrt{2}\,\text{Re}\,(\nabla \times \mathbf{E}e^{j\omega t}) = -\sqrt{2}\,\text{Re}\,(j\omega \mathbf{B}e^{j\omega t})$$

By the foregoing lemma, this reduces to

$$\nabla \times \mathbf{E} = -j\omega \mathbf{B}$$

It should now be apparent that each time derivative in a linear instantaneous equation is replaced by a $j\omega$ multiplier in the corresponding complex equation. For example, the Maxwell equations in complex form corresponding to Eqs. (1-1) are

$$\begin{aligned}\nabla \times \mathbf{E} &= -j\omega \mathbf{B} & \nabla \cdot \mathbf{B} &= 0 \\ \nabla \times \mathbf{H} &= j\omega \mathbf{D} + \mathbf{J} & \nabla \cdot \mathbf{D} &= Q_v\end{aligned} \quad (1\text{-}45)$$

The other forms of these can be obtained in a similar fashion.

1-9. Complex Constitutive Parameters. The constitutive relationships for matter linear in the general sense can be specialized to the a-c case by the procedure of the preceding section. To illustrate, consider the first of Eqs. (1-16), which is

$$\mathfrak{D} = \left(\epsilon + \epsilon_1 \frac{\partial}{\partial t} + \epsilon_2 \frac{\partial^2}{\partial t^2} + \cdots\right)\mathcal{E}$$

The complex form of this equation is readily found as

$$\mathbf{D} = (\epsilon + j\omega\epsilon_1 - \omega^2\epsilon_2 + \cdots)\mathbf{E}$$

The quantity $(\epsilon + j\omega\epsilon_1 - \omega^2\epsilon_2 + \cdots)$ is just a complex function of ω, which we shall denote by $\hat{\epsilon}(\omega)$. Thus, the complex equation

$$\mathbf{D} = \hat{\epsilon}(\omega)\mathbf{E}$$

which looks like the form for simple media, is actually valid for media linear in the general sense.

The other two of Eqs. (1-16) simplify in a similar manner; so we have the *a-c constitutive relationships*

$$\begin{aligned}\mathbf{D} &= \hat{\epsilon}(\omega)\mathbf{E} \\ \mathbf{B} &= \hat{\mu}(\omega)\mathbf{H} \\ \mathbf{J}^c &= \hat{\sigma}(\omega)\mathbf{E}\end{aligned} \quad (1\text{-}46)$$

for linear media. We call $\hat{\epsilon}$ the *complex permittivity* of the medium, $\hat{\mu}$ the *complex permeability* of the medium, and $\hat{\sigma}$ the *complex conductivity* of the medium. Remember that these parameters are not necessarily the d-c parameters, but

$$\hat{\epsilon}(\omega), \hat{\mu}(\omega), \hat{\sigma}(\omega) \xrightarrow[\omega \to 0]{} \epsilon, \mu, \sigma$$

The d-c parameters may apply over a wide range of frequencies for some materials but never over all frequencies (vacuum excepted).

In terms of the generalized current concept, the induced currents (caused by the field) are

$$\begin{aligned} \mathbf{J} &= (\hat{\sigma} + j\omega\hat{\epsilon})\mathbf{E} = \hat{y}(\omega)\mathbf{E} \\ \mathbf{M} &= j\omega\hat{\mu}\mathbf{H} = \hat{z}(\omega)\mathbf{H} \end{aligned} \quad (1\text{-}47)$$

The parameter $\hat{y}(\omega)$ has the dimensions of admittance per length and will be called the *admittivity* of the medium. The parameter $\hat{z}(\omega)$ has the dimensions of impedance per length and will be called the *impedivity* of the medium. Note that $\hat{y}$ is a combination of the $\hat{\sigma}$ and $\hat{\epsilon}$ parameters. A measurement of $\hat{y}$ is relatively simple, but it is difficult to separate $\hat{\sigma}$ from $\hat{\epsilon}$. The distinction is primarily philosophical. If the current is due to free charge, we include its effect in $\hat{\sigma}$. If the current is due to bound charge, we include its effect in $\hat{\epsilon}$. Thus, when talking of conductors, the usual convention is to let $\hat{y} = \hat{\sigma} + j\omega\epsilon_0$. When discussing dielectrics, it is common to let $\hat{y} = j\omega\hat{\epsilon}$.

To represent sources, impressed currents are added to the induced currents of Eqs. (1-47). Thus, the general form of the a-c field equations is

$$\begin{aligned} -\nabla \times \mathbf{E} &= \hat{z}(\omega)\mathbf{H} + \mathbf{M}^i \\ \nabla \times \mathbf{H} &= \hat{y}(\omega)\mathbf{E} + \mathbf{J}^i \end{aligned} \quad (1\text{-}48)$$

The $\hat{z}(\omega)$ and $\hat{y}(\omega)$ specify the characteristics of the media. The $\mathbf{J}^i$ and $\mathbf{M}^i$ represent the sources. Equations (1-48) are therefore two equations for determining the complex field $\mathbf{E}, \mathbf{H}$. Solutions to these equations are the principal topic of this book.

1-10. Complex Power. In Sec. 1-5 we considered expressions for instantaneous power and energy in terms of the instantaneous field vectors. We shall show now that similar expressions in terms of the complex field vectors represent time-average power and energy in a-c fields. For this, we shall need the concept of complex conjugate quantities, denoted by *, and defined as follows. If $A = a' + ja'' = |A|e^{j\alpha}$, the conjugate of A is $A^* = a' - ja'' = |A|e^{-j\alpha}$. It follows from this that $AA^* = |A|^2$.

Let us first consider any two a-c quantities $\mathcal{A}$ and $\mathcal{B}$, which may be scalars or components of vectors. These are in general of the form

$$\begin{aligned} \mathcal{A} &= \sqrt{2}\,|A|\cos(\omega t + \alpha) = \sqrt{2}\,\text{Re}\,(Ae^{j\omega t}) \\ \mathcal{B} &= \sqrt{2}\,|B|\cos(\omega t + \beta) = \sqrt{2}\,\text{Re}\,(Be^{j\omega t}) \end{aligned}$$

where $A = |A|e^{j\alpha}$ and $B = |B|e^{j\beta}$. The product of two such quantities is

$$\mathcal{A}\mathcal{B} = \sqrt{2}\,|A|\cos(\omega t + \alpha)\,\sqrt{2}\,|B|\cos(\omega t + \beta)$$
$$= |A|\,|B|[\cos(\alpha - \beta) + \cos(2\omega t + \alpha + \beta)] \qquad (1\text{-}49)$$

We shall denote the time average of a quantity by a bar over that quantity. The time average of the above expression is

$$\overline{\mathcal{A}\mathcal{B}} = |A|\,|B|\cos(\alpha - \beta)$$

We also note that

$$AB^* = |A|\,|B|[\cos(\alpha - \beta) + j\sin(\alpha - \beta)]$$

so it is evident that

$$\overline{\mathcal{A}\mathcal{B}} = \mathrm{Re}\,(AB^*) \qquad (1\text{-}50)$$

This identity forms the basis of definitions of complex power.

The instantaneous Poynting vector [Eq. (1-28)] can be expanded in rectangular coordinates as

$$\mathbf{S} = \mathbf{u}_x(\mathcal{E}_y\mathcal{H}_z - \mathcal{E}_z\mathcal{H}_y) + \mathbf{u}_y(\mathcal{E}_z\mathcal{H}_x - \mathcal{E}_x\mathcal{H}_z) + \mathbf{u}_z(\mathcal{E}_x\mathcal{H}_y - \mathcal{E}_y\mathcal{H}_x)$$

This is a sum of terms, each of which is the form of Eq. (1-49). It therefore follows that

$$\overline{\mathbf{S}} = \overline{\mathcal{E} \times \mathcal{H}} = \mathrm{Re}\,(\mathbf{E} \times \mathbf{H}^*)$$

In view of this we define a *complex Poynting vector*

$$\mathbf{S} = \mathbf{E} \times \mathbf{H}^* \qquad (1\text{-}51)$$

whose real part is the time average of the instantaneous Poynting vector, or

$$\overline{\mathbf{S}} = \mathrm{Re}\,(\mathbf{S}) \qquad (1\text{-}52)$$

We shall interpret the imaginary part of **S** later.

We can obtain an equation in which **S** appears by operating on the complex field equations in a manner similar to that used in the instantaneous case. Starting from Eqs. (1-44), we scalarly multiply the first by **H*** and the conjugate of the second by **E**. The difference of the resulting two equations is

$$\mathbf{E}\cdot\nabla\times\mathbf{H}^* - \mathbf{H}^*\cdot\nabla\times\mathbf{E} = \mathbf{E}\cdot\mathbf{J}^{i*} + \mathbf{H}^*\cdot\mathbf{M}^i$$

The left-hand term is $-\nabla\cdot(\mathbf{E}\times\mathbf{H}^*)$ by a mathematical identity; so we have

$$\nabla\cdot(\mathbf{E}\times\mathbf{H}^*) + \mathbf{E}\cdot\mathbf{J}^{i*} + \mathbf{H}^*\cdot\mathbf{M}^i = 0 \qquad (1\text{-}53)$$

The integral form of this is obtained by integrating throughout a region and applying the divergence theorem. This results in

$$\oint \mathbf{E} \times \mathbf{H}^* \cdot d\mathbf{s} + \iiint (\mathbf{E} \cdot \mathbf{J}^{i*} + \mathbf{H}^* \cdot \mathbf{M}^i) \, d\tau = 0 \qquad (1\text{-}54)$$

Compare these with Eqs. (1-26) and (1-27). We shall call Eqs. (1-53) and (1-54) expressions for the *conservation of complex power*, the former applying at a point and the latter applying to an entire region.

The various terms of the above equations are interpreted as follows. As suggested by Eqs. (1-29) and (1-52), we define a *complex volume density of power leaving a point* as

$$\hat{p}_f = \nabla \cdot \mathbf{S} = \nabla \cdot (\mathbf{E} \times \mathbf{H}^*) \qquad (1\text{-}55)$$

The real part of this is a time-average volume density of power leaving a point, or

$$\operatorname{Re}(\hat{p}_f) = \bar{p}_f \qquad (1\text{-}56)$$

where p_f is defined by Eq. (1-29). Similarly, we define the *complex power leaving a region* as

$$P_f = \oint \mathbf{S} \cdot d\mathbf{s} = \oint \mathbf{E} \times \mathbf{H}^* \cdot d\mathbf{s} \qquad (1\text{-}57)$$

It is evident from Eqs. (1-30) and (1-52) that the real part of this is the time-average power flow, or

$$\operatorname{Re}(P_f) = \bar{\mathcal{O}}_f \qquad (1\text{-}58)$$

Note that these relationships are quite different from those used to interpret most complex quantities [Eqs. (1-40) and (1-41)]. This is because $\mathbf{S}$, p, and $\mathcal{O}$ are *not* sinusoidal quantities but are formed of products of sinusoidal quantities.

To interpret the other terms of Eq. (1-53), let us first specialize to the case of a source-free field in media linear in the simple sense. We then have

$$\mathbf{J}^i = \hat{y}\mathbf{E} = (\sigma + j\omega\epsilon)\mathbf{E}$$
$$\mathbf{M}^i = \hat{z}\mathbf{H} = j\omega\mu\mathbf{H}$$
so
$$\mathbf{E} \cdot \mathbf{J}^{i*} = \sigma|E|^2 - j\omega\epsilon|E|^2$$
$$\mathbf{H}^* \cdot \mathbf{M}^i = j\omega\mu|H|^2$$

where $|E|^2$ means $\mathbf{E} \cdot \mathbf{E}^*$ and $|H|^2$ means $\mathbf{H} \cdot \mathbf{H}^*$. In terms of the instantaneous energy and power definitions of Eqs. (1-32) and (1-33), we have

$$\left. \begin{array}{l} \bar{p}_d = \sigma|E|^2 \\ \bar{w}_e = \tfrac{1}{2}\epsilon|E|^2 \\ \bar{w}_m = \tfrac{1}{2}\mu|H|^2 \end{array} \right\} \quad \text{in simple media} \qquad (1\text{-}59)$$

We can now write Eq. (1-53) as

$$\nabla \cdot \mathbf{S} + \bar{p}_d + j2\omega(\bar{w}_m - \bar{w}_e) = 0 \qquad (1\text{-}60)$$

Thus, the imaginary part of $\hat{p}_f$ as defined by Eq. (1-55) is 2ω times the difference between the time-average electric and magnetic energy densities. The integral relationships corresponding to Eqs. (1-59) are

$$\left. \begin{aligned} \bar{\mathcal{P}}_d &= \iiint \sigma |E|^2 \, d\tau \\ \bar{\mathcal{W}}_e &= \frac{1}{2} \iiint \epsilon |E|^2 \, d\tau \\ \bar{\mathcal{W}}_m &= \frac{1}{2} \iiint \mu |H|^2 \, d\tau \end{aligned} \right\} \quad \text{in simple media} \qquad (1\text{-}61)$$

where $\mathcal{P}_d$, $\mathcal{W}_e$, and $\mathcal{W}_m$ are defined by Eqs. (1-36) and (1-37). The specialization of Eq. (1-54) to source-free simple media is therefore

$$\oiint \mathbf{S} \cdot d\mathbf{s} + \bar{\mathcal{P}}_d + j2\omega(\bar{\mathcal{W}}_m - \bar{\mathcal{W}}_e) = 0 \qquad (1\text{-}62)$$

corresponding to the point relationship of Eq. (1-60). Note that this interpretation of complex power is precisely that chosen in circuit theory.

If sources are present, a *complex power density supplied by the sources* can be defined as

$$\hat{p}_s = -(\mathbf{E} \cdot \mathbf{J}^{i*} + \mathbf{H}^* \cdot \mathbf{M}^i) \qquad (1\text{-}63)$$

The real part of this is the time-average power density supplied by the sources, or

$$\operatorname{Re}(\hat{p}_s) = \bar{p}_s \qquad (1\text{-}64)$$

where p_s is defined by Eq. (1-34). We can write Eq. (1-53) in general as

$$\hat{p}_s = \hat{p}_f + \bar{p}_d + j2\omega(\bar{w}_m - \bar{w}_e) \qquad (1\text{-}65)$$

where all terms have been identified for simple media. Similarly, the *total complex power supplied by sources within a region* can be defined as

$$P_s = -\iiint (\mathbf{E} \cdot \mathbf{J}^{i*} + \mathbf{H}^* \cdot \mathbf{M}^i) \, d\tau \qquad (1\text{-}66)$$

where, from Eq. (1-38), it is evident that

$$\operatorname{Re}(P_s) = \bar{\mathcal{P}}_s \qquad (1\text{-}67)$$

Then the form of Eq. (1-65) applicable to an entire region is

$$P_s = P_f + \bar{\mathcal{P}}_d + j2\omega(\bar{\mathcal{W}}_m - \bar{\mathcal{W}}_e) \qquad (1\text{-}68)$$

The real part of this represents a time-average power balance. The imaginary part is related to time-average energies, and, in conformity with circuit theory nomenclature, is called *reactive power*.

Note that we have never defined $\mathcal{P}_d$, $\mathcal{W}_m$, or $\mathcal{W}_e$ for media linear in the general sense. We can, however, continue to use Eq. (1-68) for the general case of linear media by extending our definitions. This is done as follows. The time-average power dissipation is defined in general as

$$\bar{\mathcal{P}}_d = \text{Re}\left[\iiint (\hat{y}|E|^2 + \hat{z}|H|^2)\, d\tau\right] \qquad (1\text{-}69)$$

which reduces to the first of Eqs. (1-61) in simple media. The first term of the integrand represents both conduction and dielectric losses, and the second term represents magnetic losses. The time-average electric and magnetic energies are defined in general as

$$\begin{aligned}\bar{\mathcal{W}}_e &= \frac{1}{2\omega}\text{Im}\left(\iiint \hat{y}|E|^2\, d\tau\right)\\ \bar{\mathcal{W}}_m &= \frac{1}{2\omega}\text{Im}\left(\iiint \hat{z}|H|^2\, d\tau\right)\end{aligned} \qquad (1\text{-}70)$$

which reduce to the last two of Eqs. (1-61) in simple media. The first of Eqs. (1-70) includes kinetic energy stored by free charges as well as the usual field and polarization energies. More discussion of this concept is given in the next section.

1-11. A-C Characteristics of Matter. In source-free regions, the complex field equations read

$$-\nabla \times \mathbf{E} = \hat{z}(\omega)\mathbf{H} \qquad \nabla \times \mathbf{H} = \hat{y}(\omega)\mathbf{E}$$

In free space, $\hat{z}$ and $\hat{y}$ assume their simplest forms, being

$$\left.\begin{aligned}\hat{y}(\omega) &= j\omega\epsilon_0\\ \hat{z}(\omega) &= j\omega\mu_0\end{aligned}\right\} \quad \text{in free space} \qquad (1\text{-}71)$$

These hold for all frequencies and all field intensities. In metals, the conductivity remains very close to the d-c value for all radio frequencies, that is, up to the infrared frequency spectrum. The permittivity of metals is hard to measure but appears to be approximately that of vacuum. Thus,

$$\left.\begin{aligned}\hat{y}(\omega) &= \sigma + j\omega\epsilon_0\\ \hat{z}(\omega) &= j\omega\mu_0\end{aligned}\right\} \quad \text{in nonmagnetic metals} \qquad (1\text{-}72)$$

In ferromagnetic metals, μ_0 would be replaced by $\hat{\mu}$. We shall consider this case later.

In good dielectrics, it is common practice to neglect $\hat{\sigma}$ and express $\hat{y}$ entirely in terms of $\hat{\epsilon}$. Thus,

$$\left.\begin{aligned}\hat{y}(\omega) &= j\omega\hat{\epsilon}\\ \hat{z}(\omega) &= j\omega\mu_0\end{aligned}\right\} \quad \text{in nonmagnetic dielectrics} \qquad (1\text{-}73)$$

Let us now consider $\hat{\epsilon}(\omega)$ in more detail.[1] We can express $\hat{\epsilon}$ in both rec-

---

[1] A. Von Hipple, "Dielectric Materials and Applications," John Wiley & Sons, Inc., New York, 1954.

tangular and polar form as

$$\hat{\epsilon}(\omega) = \epsilon' - j\epsilon'' = |\hat{\epsilon}|e^{-j\delta} \tag{1-74}$$

where ϵ', ϵ'', and δ are real quantities. We call ϵ' the *a-c capacitivity*, ϵ'' the *dielectric loss factor*, and δ the *dielectric loss angle*. In Sec. 1-13 we shall see that they are related to the capacitance, resistance, and loss angle, respectively, of an ideal circuit capacitor. In terms of power and energy, we have from Eqs. (1-69) and (1-70) that

$$\begin{aligned} \bar{\mathcal{W}}_e &= \frac{1}{2} \iiint \epsilon' |E|^2 \, d\tau \\ \bar{\mathcal{P}}_d &= \iiint \omega \epsilon'' |E|^2 \, d\tau \end{aligned} \tag{1-75}$$

Thus, ϵ' contributes to stored energy (acts like ϵ in simple matter), and $\omega \epsilon''$ contributes to power dissipation (acts like σ in simple matter). Measured values of $\hat{\epsilon}(\omega)$ are usually expressed in terms of ϵ' and tan δ, or in terms of ϵ' and ϵ''. We shall use the latter representation.

A "perfect dielectric" would be one for which $\epsilon'' = 0$. The only perfect dielectric is vacuum. A "good dielectric" is defined to be one for which ϵ' remains almost constant at all radio frequencies and for which ϵ'' is very small. Examples of good dielectrics are polystyrene, paraffin, and Teflon. Figure 1-10 shows ϵ' and ϵ'' versus frequency for polystyrene to illustrate the characteristics of a good dielectric. There is also a group of "lossy dielectrics," characterized by a varying ϵ' and a large ϵ'' in the radio-frequency range. Examples of lossy dielectrics are Plexiglas, porcelain, and Bakelite. Figure 1-11 shows ϵ' and ϵ'' versus frequency for Plexiglas to illustrate the characteristics of a lossy dielectric. There is a group of dielectrics which have unusually high dielectric constants. The titanate and ferrite ceramics fall into this

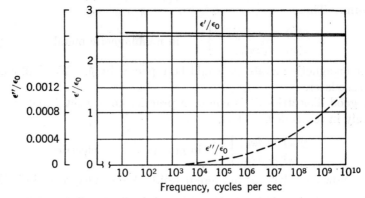

FIG. 1-10. $\hat{\epsilon}(\omega) = \epsilon' - j\epsilon''$ versus frequency for polystyrene at 25°C.

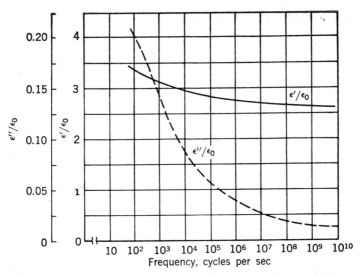

FIG. 1-11. $\hat{\epsilon}(\omega) = \epsilon' - j\epsilon''$ versus frequency for Plexiglas at 25°C.

class (the latter also being ferromagnetic). Such dielectrics are usually lossy. A qualitative explanation of the behavior of $\hat{\epsilon}$ can be made in terms of atomic concepts, but we shall view $\hat{\epsilon}$ as simply a measured parameter. A table of $\hat{\epsilon}$ for some common dielectrics is given in Appendix B.

In ferromagnetic matter, when it can be considered linear, both conduction and dielectric losses may be significant. In addition to these, magnetic losses become important. Thus,

$$\left.\begin{array}{l} \hat{y} = \sigma + j\omega\hat{\epsilon} \\ \hat{z} = j\omega\hat{\mu} \end{array}\right\} \quad \text{in ferromagnetic matter} \quad (1\text{-}76)$$

The parameter $\hat{\mu}(\omega)$ can be treated in a manner analogous to the treatment of $\hat{\epsilon}(\omega)$. Thus, we express $\hat{\mu}$ in both rectangular and polar form as

$$\hat{\mu}(\omega) = \mu' - j\mu'' = |\hat{\mu}|e^{-j\delta_m} \quad (1\text{-}77)$$

where μ', μ'', and δ_m are real quantities. We call μ' the *a-c inductivity*, μ'' the *magnetic loss factor*, and δ_m the *magnetic loss angle*. In Sec. 1-13 we shall see that they are related to the inductance, resistance, and loss angle, respectively, of an ideal circuit inductor. In terms of power and energy, we have from Eqs. (1-69) and (1-70) that

$$\begin{aligned} \bar{\mathcal{W}}_m &= \frac{1}{2} \iiint \mu' |H|^2 \, d\tau \\ \bar{\mathcal{P}}_d &= \iiint \omega\mu'' |H|^2 \, d\tau \end{aligned} \quad (1\text{-}78)$$

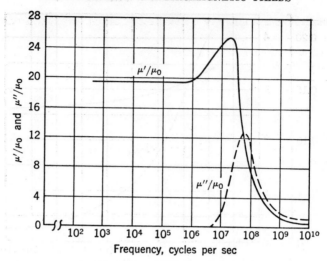

Fig. 1-12. $\hat{\mu}(\omega) = \mu' - j\mu''$ versus frequency for Ferramic A at 25°C.

where the above $\bar{\mathcal{P}}_d$ is only the time-average magnetic power loss, to which must be added the conduction and dielectric losses for the total power dissipation. Thus, μ' contributes to stored energy and μ'' to power dissipation. Measured values of $\hat{\mu}(\omega)$ are usually expressed in terms of μ' and tan δ_m, or in terms of μ' and μ''. We shall use the latter representation.

Ferromagnetic metals are extremely lossy materials (primarily due to σ), and also quite nonlinear with respect to $\hat{\mu}$. They are seldom intentionally used at radio frequencies. However, the ferromagnetic ceramics can be profitably used at radio frequencies to obtain high values of μ'. They are lossy in the magnetic sense, in that they also have appreciable μ''. Figure 1-12 shows μ' and μ'' versus frequency for Ferramic A, to illustrate the characteristics of ferrite ceramics. These materials become even more useful when magnetized by a d-c magnetic field, in which case $\hat{\mu}$ assumes the form of an asymmetrical tensor. Magnetized ferrites can be used to build "nonreciprocal" devices, such as "isolators" and "circulators."[1]

1-12. A Discussion of Current. The concept of current has broadened considerably since its inception. Originally, the term current meant the flow of free charges in conductors. This concept was extended to include displacement current, which was visualized as the displacement of bound charge in matter and in an "ether." The existence of an ether has been disproved, but the concept of displacement current has been retained,

---

[1] C. L. Hogan, The Ferromagnetic Effect at Microwave Frequencies, *Bell System Tech. J.*, vol. 31, no. 1, January, 1952.

even though it is not entirely a motion of charge. A further generalization was made to include magnetic displacement current as a "dual" concept of the electric displacement current. Finally, impressed currents, both electric and magnetic, have been introduced to represent sources. Because of the breadth of the concept of current, many different phenomena are included, and the nomenclature used is somewhat lengthy. We shall summarize the notation and concepts in complex form in this section.

Consider the complex electric current density. Internal to conductors, the current is, for all practical purposes, due entirely to the motion of free electrons. Such current is called the *conduction current* and is expressed mathematically by $\mathbf{J} = \sigma \mathbf{E}$. (We shall consider $\hat{\sigma} = \sigma$, a real quantity, for this discussion. This is usually true at radio frequencies.) Even in dielectrics there is some conduction current, but it is usually small. In free space there is no motion of charges at all, and we have only a *free-space displacement current*, given by $\mathbf{J} = j\omega\epsilon_0\mathbf{E}$. In matter, in addition to the conduction current and the free-space displacement current, we have a current due to the motion of bound charges. This is called the *polarization current* and is expressed mathematically by $\mathbf{J} = j\omega(\hat{\epsilon} - \epsilon_0)\mathbf{E}$. Because the term $\mathbf{J} = j\omega\hat{\epsilon}\mathbf{E}$ is of the same mathematical form as the free-space displacement current, it is called the *displacement current*. For our purposes, still another division of the electric current is convenient. This involves viewing the current in terms of a component in phase with $\mathbf{E}$, called the *dissipative current*, $\mathbf{J} = (\sigma + \omega\epsilon'')\mathbf{E}$, and a component out of phase with $\mathbf{E}$, called the *reactive current*, $\mathbf{J} = j\omega\epsilon'\mathbf{E}$. This is essentially a generalization of the circuit concept of current, where the dissipative current produces the power loss and the reactive current gives rise to the stored energy. All the currents mentioned are classified as *induced currents*, that is, are caused by the field. *Impressed currents* are used to represent sources or known quantities. In this sense, they are independent of the field and are said to cause the field. The total electric current is the sum of the induced currents plus the impressed currents. The nomenclature used for electric currents is summarized in the first column of Table 1-2.

Both the nomenclature and the concepts of complex magnetic currents are similar to those for electric currents. The one essential difference in the two concepts is the nonexistence of magnetic "charges" in nature. Thus, there is no free magnetic charge and no magnetic conduction current. In absence of matter, we have a *magnetic free-space displacement current*, $\mathbf{M} = j\omega\mu_0\mathbf{H}$, analogous to the electric case. When matter is present, we have magnetic effects due to the motion of the atomic particles, giving rise to an induced magnetic current in addition to the free-space displacement current. We call this the *magnetic polarization*

TABLE 1-2. CLASSIFICATION OF ELECTRIC AND MAGNETIC CURRENTS

Type	Complex electric current density	Complex magnetic current density
Conduction	$\sigma \mathbf{E}$	
Free-space displacement	$j\omega\epsilon_0 \mathbf{E}$	$j\omega\mu_0 \mathbf{H}$
Polarization	$j\omega(\hat{\epsilon} - \epsilon_0)\mathbf{E}$	$j\omega(\hat{\mu} - \mu_0)\mathbf{H}$
Displacement	$j\omega\hat{\epsilon}\mathbf{E}$	$j\omega\hat{\mu}\mathbf{H}$
Dissipative	$(\sigma + \omega\epsilon'')\mathbf{E}$	$\omega\mu''\mathbf{H}$
Reactive	$j\omega\epsilon'\mathbf{E}$	$j\omega\mu'\mathbf{H}$
Induced	$\hat{y}\mathbf{E} = (\sigma + j\omega\hat{\epsilon})\mathbf{E}$ $= (\sigma + \omega\epsilon'' + j\omega\epsilon')\mathbf{E}$	$\hat{z}\mathbf{H} = j\omega\hat{\mu}\mathbf{H}$ $= (\omega\mu'' + j\omega\mu')\mathbf{H}$
Impressed	$\mathbf{J}^i$	$\mathbf{M}^i$
Total	$\mathbf{J}^t = \hat{y}\mathbf{E} + \mathbf{J}^i$	$\mathbf{M}^t = \hat{z}\mathbf{H} + \mathbf{M}^i$

current, expressed by $\mathbf{M} = j\omega(\hat{\mu} - \mu_0)\mathbf{H}$. The term $\mathbf{M} = j\omega\hat{\mu}\mathbf{H}$ is called the *magnetic displacement current*, being the sum of the free-space displacement current and the polarization current. We find it convenient to divide the magnetic current into a component in phase with $\mathbf{H}$, called the *magnetic dissipative current*, $\mathbf{M} = \omega\mu''\mathbf{H}$, and a component out of phase with $\mathbf{H}$, called the *magnetic reactive current*, $\mathbf{M} = j\omega\mu'\mathbf{H}$. The dissipative magnetic current contributes to the power loss, and the reactive magnetic current contributes to the stored energy. All the aforementioned magnetic currents are *induced currents*, that is, caused by the field. In nonmagnetic matter, the induced magnetic current is simply the free-space displacement current, $\mathbf{M} = j\omega\mu_0\mathbf{H}$, a reactive current. To represent sources or known quantities, we use *impressed currents*. The nomenclature for magnetic currents is summarized in the second column of Table 1-2.

A convenient classification of matter from the electric current standpoint can be made in terms of a *quality factor Q*. This is defined as

$$Q = \frac{\text{magnitude of reactive current density}}{\text{magnitude of dissipative current density}}$$
$$= \frac{\omega\epsilon'}{\sigma + \omega\epsilon''} \tag{1-79}$$

In nonmagnetic matter, this involves a ratio of stored electric energy to power dissipated. In terms of the energy and power densities, Eq. (1-79) can be written as

$$Q = \frac{\omega \epsilon' |E|^2}{(\sigma + \omega \epsilon'')|E|^2}$$
$$= \omega \frac{\text{peak density of electric energy}}{\text{average density of power dissipated}}$$
$$= 2\pi \frac{\text{peak density of electric energy}}{\text{density of energy dissipated in one cycle}} \quad (1\text{-}80)$$

Thus, the concept of Q in nonmagnetic matter can be considered as an extension of the concept of Q for capacitors in circuit theory. A good dielectric is a high-Q material, while conductors have an extremely low Q.

When magnetic matter is considered, there is an additional power dissipation due to magnetic hysteresis loss. The interpretation given to Eq. (1-80) must be modified, since it includes only the power loss due to electric effects. In this case, the Q defined above would be called the electric Q, and an analogous magnetic quality factor Q_m could be defined. Since we deal principally with nonmagnetic materials, we shall not expand this concept further.

1-13. A-C Behavior of Circuit Elements. The complex notation used for a-c fields is the extension of the complex notation used for a-c circuits. The complex field equations bear a relationship to the complex circuit equations which is similar to that for the time-varying case, given in Sec. 1-6. Circuit elements (resistors, capacitors, and inductors) are merely configurations of matter and thus have characteristics which depend upon the properties of matter. Insight into the interpretation of the impedivity and admittivity functions of field theory can be gained by considering their relationship to the more familiar characteristics of impedance and admittance of circuit elements.

The basic elements of circuit theory are small[1] two-terminal structures whose fields are largely confined internal to the elements. According to the concepts of Sec. 1-10, the complex power supplied to a circuit element is

$$P = \bar{\mathcal{P}}_d + j2\omega(\bar{\mathcal{W}}_m - \bar{\mathcal{W}}_e) \quad (1\text{-}81)$$

In terms of circuit concepts, the power supplied to an element also can be written as

$$P = |I|^2 Z = |V|^2 Y^* \quad (1\text{-}82)$$

where Z and Y are the impedance and admittance of the element. In general, an element is called an impedor. When P is primarily real, the

---
[1] The smallness of an element depends upon the frequency, or wavelength, as we shall see in Chap. 2.

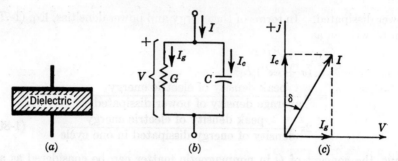

FIG. 1-13. A capacitor according to circuit concepts. (a) Physical capacitor; (b) equivalent circuit; (c) complex diagram.

element is called a resistor, and when P is primarily imaginary, the element is called a reactor. A reactor is called an inductor or capacitor according as Im (Z) is positive or negative, respectively. It should be noted that P, and hence Z, is a function of frequency. Thus, the designation of an element as a resistor, inductor, or capacitor is, to some degree, dependent upon frequency. We usually classify elements according to their low-frequency behavior.

For an explicit discussion, consider the parallel-plate capacitor of Fig. 1-13a. The low-frequency equivalent circuit of this element is shown in Fig. 1-13b, where the conductance G accounts for energy dissipation and the capacitance C accounts for energy storage. The relationship of complex terminal current I to complex terminal voltage V is

$$I = I_g + I_c = YV = (G + j\omega C)V \qquad (1\text{-}83)$$

Figure 1-13c shows the complex diagram representing this equation. The complex power to the element is[1]

$$P = |V|^2 (G - j\omega C)$$

For a "good" capacitor ($\omega C \gg G$) the current leads the voltage by almost 90°, and the power is principally reactive. For a "poor" capacitor ($G \gg \omega C$) the current and voltage are almost in phase, and the power is principally dissipative. The element in this case could be classified as a resistor. The angle between I_c and I is called the loss angle δ, as shown in Fig. 1-13c.

Let us idealize the problem to a capacitor with perfectly conducting plates. Furthermore, we shall approximate the field by

$$E = \frac{V}{d} \qquad J = \frac{I}{A}$$

---

[1] We are using the convention $P = VI^*$. Some authors define $P = IV^*$, in which case the sign of reactive power is opposite to that which we get.

where A is the area of the plates and d is their separation. The a-c constitutive relationship for the field between the capacitor plates is

$$J = \hat{y}E = (\sigma + \omega\epsilon'' + j\omega\epsilon')E$$

where we have taken $\hat{\sigma} = \sigma$. Substituting for E and J from the preceding equations, we have

$$I = \hat{y}\frac{A}{d}V = (\sigma + \omega\epsilon'' + j\omega\epsilon')\frac{A}{d}V$$

A comparison of this with Eqs. (1-83) shows that

$$Y = \hat{y}\frac{A}{d} \qquad G = (\sigma + \omega\epsilon'')\frac{A}{d} \qquad C = \epsilon'\frac{A}{d}$$

Thus, for our idealized circuit element, the admittance is proportional to the admittivity of the matter between the plates. The equivalency of "field power," Eq. (1-81), to "circuit power," Eq. (1-82), also can be demonstrated. For our idealized element

$$P = \iiint \hat{y}^*|E|^2\, d\tau = \hat{y}^*|E|^2 A d = |V|^2 Y^*$$

We can use this result to define the admittance of a cube and then view admittivity $\hat{y}$ as the admittance of a unit cube.

The magnetic properties of matter are similarly related to the circuit behavior of an inductor. To demonstrate this, consider the toroidal inductor of Fig. 1-14a. The low-frequency equivalent circuit of this element is shown in Fig. 1-14b, where the resistance R accounts for energy dissipation and the inductance L accounts for energy storage. The relationship of complex terminal voltage V to complex terminal current I is

$$V = V_r + V_l = ZI = (R + j\omega L)I \tag{1-84}$$

The complex diagram representing this equation is shown in Fig. 1-14c.

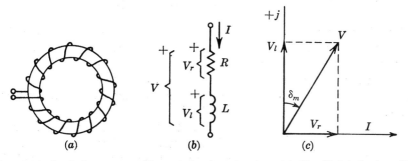

Fig. 1-14. An inductor according to circuit concepts. (a) Toroidal inductor; (b) equivalent circuit; (c) complex diagram.

The complex power to the element is

$$P = |I|^2(R + j\omega L)$$

For a good inductor ($\omega L \gg R$) the current lags the voltage by almost 90°, and the power is principally reactive. For a poor inductor ($R \gg \omega L$) the current and voltage are almost in phase, and the power is principally dissipative. The element in this case could be classified as a resistor. The angle between V_l and V is called the magnetic loss angle δ_m, as shown on Fig. 1-14c.

We now idealize the problem to an inductor of perfectly conducting wire and approximate the field by

$$H = \frac{NI}{l} \qquad M = \frac{V}{NA}$$

where N is the number of turns, l is the average circumference, and A is the cross-sectional area. The magnetic constitutive relationship for the field in the core is

$$M = \hat{z}H = (\omega\mu'' + j\omega\mu')H$$

A substitution for H and M from the preceding equations gives

$$V = \hat{z}\frac{N^2A}{l} I = (\omega\mu'' + j\omega\mu') \frac{N^2A}{l} I$$

Comparing this with Eq. (1-84), we see that

$$Z = \hat{z}\frac{N^2A}{l} \qquad R = \omega\mu'' \frac{N^2A}{l} \qquad L = \mu' \frac{N^2A}{l}$$

Thus, for the idealized inductor, the impedance is proportional to the impedivity of the matter. From Eq. (1-82), the power supplied to the inductor is

$$P = \iiint \hat{z}|H|^2 \, d\tau = \hat{z}|H|^2 Al = |I|^2 Z$$

which is consistent with Eq. (1-82). Using this result to define the impedance of a cube, we can think of impedivity as the impedance of a unit cube.

This development serves to illustrate the close correspondences between a-c circuit concepts and a-c field concepts. A summary of the various concepts is given in Table 1-3.

1-14. Singularities of the Field. A field is said to be *singular* at a point for which the function or its derivatives are discontinuous. Most of our discussion so far has been about well-behaved fields, but we have meant to include by implication certain types of allowable singularities.

TABLE 1-3. CORRESPONDENCES BETWEEN A-C CIRCUIT CONCEPTS AND A-C FIELD CONCEPTS

A-C circuit concepts	A-C field concepts
Complex voltage V	Complex electric intensity $\mathbf{E}$ Complex magnetic current density $\mathbf{M}$
Complex current I	Complex electric current density $\mathbf{J}$ Complex magnetic intensity $\mathbf{H}$
Complex power flow VI^*	Density of complex power flow $\mathbf{E} \times \mathbf{H}^*$
Impedance $Z(\omega)$	Impedivity $\hat{z}(\omega)$
Admittance $Y(\omega)$	Admittivity $\hat{y}(\omega)$

Resistors:
  Admittance, $Y(\omega) = \dfrac{1}{R}$
  Current, $I = \dfrac{1}{R} V$
  Power dissipation $\dfrac{1}{R} VV^*$
	Conductors ($\sigma \gg \omega\epsilon_0$):
	  Admittivity, $\hat{y}(\omega) \approx \sigma$
	  Current density, $\mathbf{J} \approx \sigma \mathbf{E}$
	  Density of power dissipation, $\sigma \mathbf{E} \cdot \mathbf{E}^*$

Capacitors:
  Admittance, $Y(\omega) = \dfrac{1}{R} + j\omega C$
  Current, $I = \left(\dfrac{1}{R} + j\omega C\right) V$
  Stored energy $\tfrac{1}{2} CVV^*$
  Power dissipation $\dfrac{1}{R} VV^*$
	Dielectrics ($\omega\epsilon'' \gg \sigma$):
	  Admittivity, $\hat{y}(\omega) \approx \omega\epsilon'' + j\omega\epsilon'$
	  Current density, $\mathbf{J} \approx (\omega\epsilon'' + j\omega\epsilon')\mathbf{E}$
	  Density of stored energy, $\tfrac{1}{2}\epsilon' \mathbf{E} \cdot \mathbf{E}^*$
	  Density of power dissipation, $\omega\epsilon'' \mathbf{E} \cdot \mathbf{E}^*$

Inductors:
  Impedance, $Z(\omega) = R + j\omega L$
  Voltage, $V = (R + j\omega L)I$
  Stored energy, $\tfrac{1}{2} LII^*$
  Power dissipation, RII^*
	Magnetic properties:
	  Impedivity, $\hat{z}(\omega) = \omega\mu'' + j\omega\mu'$
	  Magnetic current, $\mathbf{M} = (\omega\mu'' + j\omega\mu')\mathbf{H}$
	  Density of stored energy, $\tfrac{1}{2}\mu' \mathbf{H} \cdot \mathbf{H}^*$
	  Density of power dissipation, $\omega\mu'' \mathbf{H} \cdot \mathbf{H}^*$

These can occur at material boundaries (discontinuous $\hat{z}$ and $\hat{y}$) and at singular source distributions, such as sheets and filaments of currents.

As evidenced by Eqs. (1-44), the total electric and magnetic currents are vortices of $\mathbf{H}$ and $-\mathbf{E}$, respectively. Suppose we have a surface distribution of currents $\mathbf{J}_s$ and $\mathbf{M}_s$, as represented by Fig. 1-15. By applying

$$\oint \mathbf{H} \cdot d\mathbf{l} = I^t \qquad \oint \mathbf{E} \cdot d\mathbf{l} = -K^t \qquad (1\text{-}85)$$

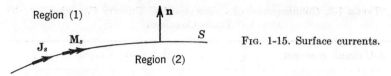

Fig. 1-15. Surface currents.

to rectangular paths enclosing a portion of the surface currents, we obtain[1]

$$\mathbf{n} \times [\mathbf{H}^{(1)} - \mathbf{H}^{(2)}] = \mathbf{J}_s \qquad [\mathbf{E}^{(1)} - \mathbf{E}^{(2)}] \times \mathbf{n} = \mathbf{M}_s \qquad (1\text{-}86)$$

where $\mathbf{n}$ is the unit vector normal to the surface and pointing into region (1). The superscripts (1) and (2) denote the side of S on which $\mathbf{E}$ or $\mathbf{H}$ is evaluated. Equations (1-86) are essentially the field equations at sheets of currents. They express at current sheets the same concept as Eqs. (1-44) express at volume distributions of currents. If $\mathbf{J}_s$ and $\mathbf{M}_s$ are impressed currents, Eqs. (1-86) are the "boundary conditions" to be satisfied at the source.

Equations (1-86) apply regardless of whether or not a discontinuity in media exists on S. Whenever $\mathbf{J}_s$ and $\mathbf{M}_s$ are zero, Eqs. (1-86) state that the tangential components of $\mathbf{E}$ and $\mathbf{H}$ are continuous across the surface. If $\hat{z}$ and $\hat{y}$ are finite in both regions 1 and 2, no induced surface current can result. Thus, *tangential components of* $\mathbf{E}$ *and* $\mathbf{H}$ *are continuous across any material boundary, perfect conductors excepted.* If one side of S is a perfect electric conductor, say region 2, a surface conduction current $\mathbf{J}_s$ can exist even though $\mathbf{E}$ is zero, since $\hat{y} = \sigma$ is infinite. In this case, Eqs. (1-86) reduce to

$$\left. \begin{array}{l} \mathbf{n} \times \mathbf{H} = \mathbf{J}_s \\ \mathbf{n} \times \mathbf{E} = 0 \end{array} \right\} \qquad \text{at a perfect conductor} \qquad (1\text{-}87)$$

where $\mathbf{n}$ points into the region of field. Thus, the "boundary condition" at a perfect electric conductor is vanishing tangential components of $\mathbf{E}$. The *perfect magnetic conductor* is defined to be a material for which the tangential components of $\mathbf{H}$ are zero at its surface. This is, however, purely a mathematical concept. The necessary "magnetic conduction current" on its surface has no physical significance.

Finally, at a filament of current, the field must be singular such that Eqs. (1-85) yield the current enclosed, no matter how small the contour. For example, at a filament of electric current I, the boundary condition for $\mathbf{H}$ is

$$\oint_C \mathbf{H} \cdot d\mathbf{l} \xrightarrow[\text{radius of } C \to 0]{} I \qquad (1\text{-}88)$$

A similar limit of the second of Eqs. (1-85) must be satisfied at a filament of magnetic current.

---

[1] R. F. Harrington, "Introduction to Electromagnetic Engineering," McGraw-Hill Book Company, Inc., p. 74, 1958.

FUNDAMENTAL CONCEPTS

It is often convenient for mathematical and discussional purposes to consider the various singular quantities as limits of nonsingular quantities. For example, we can think of an abrupt material boundary as the limit of a continuous, but rapid, change in $\hat{y}$ and $\hat{z}$. Similarly, a sheet of current can be thought of as a volume distribution of current having a large magnitude and confined to a thin shell. By such expediencies we can avoid much tedium in the exposition of the theory.

PROBLEMS

1-1. Using Stokes' theorem and the divergence theorem, show that Eqs. (1-1) are equivalent to Eqs. (1-3).

1-2. The conduction current in conductors is affected by the magnetic field as well as by the electric field (Hall effect). Using an atomic model, justify that

$$\mathcal{J} \approx \sigma \mathcal{E} + \sigma^2 h \mathcal{E} \times \mathcal{B}$$

where h is the Hall constant. For copper ($h = -5.5 \times 10^{-11}$), determine the $\mathcal{B}$ for which the second term of the above equation is 1 per cent of the first term.

1-3. Given $\mathcal{E} = \mathbf{u}_x y^2 \sin \omega t$ and $\mathcal{H} = \mathbf{u}_y x \cos \omega t$, determine $\mathcal{J}^t$ and $\mathcal{M}^t$. Determine i^t and k^t through the disk $z = 0$, $x^2 + y^2 = 1$.

1-4. For the field of Prob. 1-3, determine the Poynting vector. Show that Eq. (1-26) is satisfied for this field.

1-5. Starting from Maxwell's equations, derive the circuit law for capacitors, $i = C \, dv/dt$, and the circuit law for inductors, $v = L \, di/dt$.

1-6. Determine the instantaneous quantities corresponding to (a) $I = 10 + j5$, (b) $\mathbf{E} = \mathbf{u}_x(5 + j3) + \mathbf{u}_y(2 + j3)$, (c) $\mathbf{H} = (\mathbf{u}_x + \mathbf{u}_y)e^{j(x+y)}$.

1-7. Prove Eqs. (1-42).

1-8. Given $\mathbf{H} = \mathbf{u}_x \sin y$ in a source-free region of Plexiglas, determine $\mathbf{E}$ and $\mathcal{E}$ at a frequency of (a) 1 megacycle, (b) 100 megacycles.

1-9. Show that $Q_v = 0$ (complex charge density vanishes) in a source-free region of homogeneous matter, linear in the general sense.

1-10. Show that the instantaneous Poynting vector is given by

$$\mathcal{S} = \mathrm{Re} \, (\mathbf{S} + \mathbf{E} \times \mathbf{H} e^{j2\omega t})$$

Why is $\mathcal{S}$ not related to $\mathbf{S}$ by Eq. (1-41)?

1-11. Consider the unit cube shown in Fig. 1-16 which has all sides except the face $x = 0$ covered by perfect conductors. If $E_z = 100 \sin (\pi y)$ and $H_y = e^{j\pi/6} \sin (\pi y)$

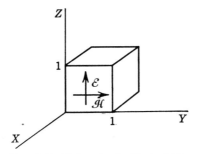

Fig. 1-16. Unit cube for Prob. 1-11.

over the open face and no sources exist within the cube, determine (a) the time-average power dissipated within the cube, (b) the difference between the time-average electric and magnetic energies within the cube.

1-12. Suppose a filament of z-directed electric current $I^i = 10$ is impressed along the z axis from $z = 0$ to $z = 1$. If $\mathbf{E} = \mathbf{u}_z(1 + j)$, determine the complex power and the time-average power supplied by this source.

1-13. Suppose we have a 10-megacycle field $\mathbf{E} = \mathbf{u}_x 5$, $\mathbf{H} = \mathbf{u}_y 2$, at some point in a material having $\sigma = 10^{-4}$, $\hat{\epsilon} = (8 - j10^{-2})\epsilon_0$, and $\hat{\mu} = (14 - j)\mu_0$ at the operating frequency. Determine each type of current (except impressed) listed in Table 1-2.

1-14. A small capacitor has a d-c capacitance of 300 micromicrofarads when air-filled. When it is oil-filled, it is found to have an impedance of $(500 - j) \times 10^3$ at $\omega = 10^6$. Determine $\hat{y}$, ϵ', and ϵ'' of the oil, neglecting conductor losses.

1-15. For a practical toroidal inductor of the type shown in Fig. 1-14a, show that the power loss in the wire will usually be much larger than that in a core of low-loss ferromagnetic material.

1-16. Assume that $\hat{\epsilon} = \epsilon' - j\epsilon''$ is an analytic function of ω and show that

$$\epsilon'(\omega) = \epsilon_0 + \frac{2}{\pi} \int_0^\infty \frac{w\epsilon''(w)\,dw}{w^2 - \omega^2}$$

$$\epsilon''(\omega) = -\frac{2}{\pi} \int_0^\infty \frac{w[\epsilon'(w) - \epsilon_0]\,dw}{w^2 - \omega^2}$$

(Equations of this type are valid for any analytic function regular in the lower half plane.)

1-17. Derive Eqs. (1-86).

CHAPTER 2

INTRODUCTION TO WAVES

2-1. The Wave Equation. A field that is a function of both time and space coordinates can be called a wave. We shall, however, be a bit more restrictive in our definition and use the term wave to denote a solution to a particular type of equation, called a wave equation. Electromagnetic fields obey wave equations, so the terms wave and field are synonymous for time-varying electromagnetism. In this chapter we shall consider a number of simple wave solutions to introduce and illustrate various a-c electromagnetic phenomena.

For the present, let us consider fields in regions which are source-free ($\mathbf{J}^i = \mathbf{M}^i = 0$), linear ($\hat{z}$ and $\hat{y}$ independent of $|E|$ and $|H|$), homogeneous ($\hat{z}$ and $\hat{y}$ independent of position), and isotropic ($\hat{z}$ and $\hat{y}$ are scalar). The complex field equations are then

$$\nabla \times \mathbf{E} = -\hat{z}\mathbf{H}$$
$$\nabla \times \mathbf{H} = \hat{y}\mathbf{E} \qquad (2\text{-}1)$$

The curl of the first equation is

$$\nabla \times \nabla \times \mathbf{E} = -\hat{z}\nabla \times \mathbf{H}$$

which, upon substitution for $\nabla \times \mathbf{H}$ from the second equation, becomes

$$\nabla \times \nabla \times \mathbf{E} = -\hat{z}\hat{y}\mathbf{E}$$

The frequently encountered parameter

$$k = \sqrt{-\hat{z}\hat{y}} \qquad (2\text{-}2)$$

is called the *wave number* of the medium. In terms of k, the preceding equation becomes

$$\nabla \times \nabla \times \mathbf{E} - k^2\mathbf{E} = 0 \qquad (2\text{-}3)$$

which we shall call the *complex vector wave equation*. If we return to Eqs. (2-1), take the curl of the second equation, and substitute from the first equation, we obtain

$$\nabla \times \nabla \times \mathbf{H} - k^2\mathbf{H} = 0 \qquad (2\text{-}4)$$

Thus, $\mathbf{H}$ is a solution to the same complex wave equation as is $\mathbf{E}$.

The wave equation is often written in another form by defining an operation
$$\nabla^2 \mathbf{A} = \nabla(\nabla \cdot \mathbf{A}) - \nabla \times \nabla \times \mathbf{A}$$
In rectangular components, this reduces to
$$\nabla^2 \mathbf{A} = \mathbf{u}_x \nabla^2 A_x + \mathbf{u}_y \nabla^2 A_y + \mathbf{u}_z \nabla^2 A_z$$
where $\mathbf{u}_x$, $\mathbf{u}_y$, and $\mathbf{u}_z$ are the rectangular-coordinate unit vectors and ∇^2 is the Laplacian operator. It is implicit in the wave equations that
$$\nabla \cdot \mathbf{E} = 0 \qquad \nabla \cdot \mathbf{H} = 0 \tag{2-5}$$
shown by taking the divergence of Eqs. (2-3) and (2-4). Using Eqs. (2-5) and the operation defined above, we can write Eqs. (2-3) and (2-4) as
$$\nabla^2 \mathbf{E} + k^2 \mathbf{E} = 0$$
$$\nabla^2 \mathbf{H} + k^2 \mathbf{H} = 0 \tag{2-6}$$
These we shall also call vector wave equations. They are not, however, so general as the previous forms, for they do not imply Eqs. (2-5). In other words, Eqs. (2-6) *and* Eqs. (2-5) are equivalent to Eqs. (2-3) and (2-4). Thus, the *rectangular components* of **E** and **H** satisfy the *complex scalar wave equation* or *Helmholtz equation*[1]
$$\nabla^2 \psi + k^2 \psi = 0 \tag{2-7}$$
We can construct electromagnetic fields by choosing solutions to Eq. (2-7) for E_x, E_y, and E_z or H_x, H_y, and H_z, such that Eqs. (2-5) are also satisfied.

To illustrate the wave behavior of electromagnetic fields, let us construct a simple solution. Take the medium to be a perfect dielectric, in which case $\hat{y} = j\omega\epsilon$, $\hat{z} = j\omega\mu$, and
$$k = \omega \sqrt{\epsilon\mu} \tag{2-8}$$
Also, take **E** to have only an x component independent of x and y. The first of Eqs. (2-6) then reduces to
$$\frac{d^2 E_x}{dz^2} + k^2 E_x = 0$$
which is the one-dimensional Helmholtz equation. Solutions to this are linear combinations of e^{jkz} and e^{-jkz}. In particular, let us consider a solution
$$E_x = E_0 e^{-jkz} \tag{2-9}$$
This satisfies $\nabla \cdot \mathbf{E} = 0$ and is therefore a possible electromagnetic field.

[1] We shall use the symbol ψ to denote "wave functions," that is, solutions to Eq. (2-7). Do not confuse these ψ's with magnetic flux.

The associated magnetic field is found according to

$$j\omega\mu \mathbf{H} = -\nabla \times \mathbf{E} = \mathbf{u}_y jk E_x$$

which, using Eq. (2-8), can be written as

$$E_x = \sqrt{\frac{\mu}{\epsilon}} H_y \qquad (2\text{-}10)$$

Ratios of components of **E** to components of **H** have the dimensions of impedance and are called *wave impedances*. The wave impedance associated with our present solution,

$$\eta = \frac{E_x}{H_y} = \sqrt{\frac{\mu}{\epsilon}} \qquad (2\text{-}11)$$

is called the *intrinsic impedance* of the medium. In vacuum,

$$\eta_0 = \sqrt{\frac{\mu_0}{\epsilon_0}} \approx 120\pi \approx 377 \text{ ohms} \qquad (2\text{-}12)$$

We shall see later that the intrinsic impedance of a medium enters into wave transmission and reflection problems in the same manner as the characteristic impedance of transmission lines.

To interpret this solution, let E_0 be real and determine $\mathcal{E}$ and $\mathcal{H}$ according to Eq. (1-41). The instantaneous fields are found as

$$\begin{aligned}\mathcal{E}_x &= \sqrt{2}\, E_0 \cos(\omega t - kz) \\ \mathcal{H}_y &= \frac{\sqrt{2}}{\eta} E_0 \cos(\omega t - kz)\end{aligned} \qquad (2\text{-}13)$$

This is called a *plane wave* because the phase (kz) of $\mathcal{E}$ and $\mathcal{H}$ is constant over a set of planes (defined by $z =$ constant) called *equiphase surfaces*. It is called a *uniform* plane wave because the amplitudes (E_0 and E_0/η) of $\mathcal{E}$ and $\mathcal{H}$ are constant over the equiphase planes. $\mathcal{E}$ and $\mathcal{H}$ are said to be *in phase* because they have the same phase at any point. At some specific time, $\mathcal{E}$ and $\mathcal{H}$ are sinusoidal functions of z. The vector picture of Fig. 2-1 illustrates $\mathcal{E}$ and $\mathcal{H}$ along the z axis at $t = 0$. The direction of an arrow represents the direction of a vector, and the length of an arrow represents the magnitude of a vector. If we take a slightly later instant of time, the picture of Fig. 2-1 will be shifted in the $+z$ direction. We say that the wave is traveling in the $+z$ direction and call it a *traveling wave*. The term *polarization* is used to specify the behavior of $\mathcal{E}$ lines. In this wave, the $\mathcal{E}$ lines are always parallel to the x axis, and the wave is said to be *linearly polarized* in the x direction.

The velocity at which an equiphase surface travels is called the *phase velocity* of the wave. An equiphase plane $z = z_p$ is defined by

$$\omega t - k z_p = \text{constant}$$

that is, the argument of the cosine functions of Eq. (2-13) is constant. As t increases, the value of z_p must also increase to maintain this constancy, and the plane $z = z_p$ will move in the $+z$ direction. This is illustrated by Fig. 2-2, which is a plot of $\mathcal{E}$ for several instants of time. To obtain the phase velocity dz_p/dt, differentiate the above equation. This gives

$$\omega - k \frac{dz_p}{dt} = 0$$

The phase velocity of this wave is called the *intrinsic phase velocity* v_p of the dielectric and is, according to the above equation,

$$v_p = \frac{dz_p}{dt} = \frac{\omega}{k} = \frac{1}{\sqrt{\epsilon\mu}} \qquad (2\text{-}14)$$

In vacuum, this is the velocity of light: 3×10^8 meters per second.

The *wavelength* of a wave is defined as the distance in which the phase increases by 2π at any instant. This distance is shown on Fig. 2-2. The wavelength of the particular wave of Eqs. (2-13) is called the *intrinsic wavelength* λ of the medium. It is given by $k\lambda = 2\pi$, or

$$\lambda = \frac{2\pi}{k} = \frac{2\pi v_p}{\omega} = \frac{v_p}{f} \qquad (2\text{-}15)$$

where f is the frequency in cycles per second. The wavelength is often used as a measure of whether a distance is long or short. The range of wavelengths encountered in electromagnetic engineering is large. For example, the free-space wavelength of a 60-cycle wave is 5000 kilometers, whereas the free-space wavelength of a 1000-megacycle wave is only 30 centimeters. Thus, a distance of 1 kilometer is very short at 60 cycles,

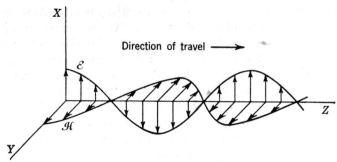

FIG. 2-1. A linearly polarized uniform plane traveling wave.

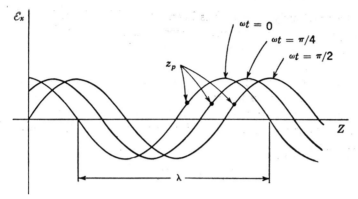

FIG. 2-2. $\mathcal{E}$ at several instants of time in a linearly polarized uniform plane traveling wave.

but very long at 1000 megacycles. The usual circuit theory is based on the assumption that distances are much shorter than a wavelength.

2.2. Waves in Perfect Dielectrics. In this section we shall consider the properties of uniform plane waves in perfect dielectrics, of which free space is the most common example. We have already given a special case of the uniform plane wave in the preceding section. To summarize,

where
$$E_x = E_0 e^{-jkz} \qquad H_y = \frac{E_0}{\eta} e^{-jkz}$$
$$k = \omega \sqrt{\mu \epsilon} = \frac{2\pi}{\lambda} = \frac{\omega}{v_p}$$
$$\eta = \sqrt{\frac{\mu}{\epsilon}}$$
(2-16)

It is an x-polarized, $+z$ traveling wave. Because of the symmetry of the rectangular coordinate system, other uniform plane-wave solutions can be obtained by rotations of the coordinate axes, corresponding to cyclic interchanges of coordinate variables. We wish to restrict consideration to $+z$ and $-z$ traveling waves; so we shall consider only the transformations (x,y,z) to $(-y,x,z)$, to $(x,-y,-z)$, and to $(y,x,-z)$. This procedure, together with our original solution, gives us the four waves

$$E_x^+ = A e^{-jkz} \qquad H_y^+ = \frac{A}{\eta} e^{-jkz}$$
$$E_y^+ = B e^{-jkz} \qquad H_x^+ = \frac{-B}{\eta} e^{-jkz}$$
$$E_x^- = C e^{jkz} \qquad H_y^- = \frac{-C}{\eta} e^{jkz}$$
$$E_y^- = D e^{jkz} \qquad H_x^- = \frac{D}{\eta} e^{jkz}$$
(2-17)

where the previously used E_0 has been replaced by A, B, C, or D. The superscript $+$ denotes a $+z$ traveling wave, and the superscript $-$ denotes a $-z$ traveling wave. The most general uniform plane wave is a superposition of Eqs. (2-17).

We have already interpreted the first wave of Eqs. (2-17) in Sec. 2-1. This also constitutes an interpretation of the other three waves if the appropriate interchanges of coordinates are made. We have not yet mentioned power and energy considerations, so let us do so now. Given the traveling wave

$$E_x = E_0 e^{-jkz} \qquad H_y = \frac{E_0}{\eta} e^{-jkz}$$

we evaluate the various energy and power quantities as

$$\begin{aligned} w_e &= \frac{\epsilon}{2} \mathcal{E}^2 = \epsilon E_0^2 \cos^2(\omega t - kz) \\ w_m &= \frac{\mu}{2} \mathcal{H}^2 = \epsilon E_0^2 \cos^2(\omega t - kz) \\ \mathbf{S} &= \mathcal{E} \times \mathcal{H} = \mathbf{u}_z \frac{2}{\eta} E_0^2 \cos^2(\omega t - kz) \\ \mathbf{S} &= \mathbf{E} \times \mathbf{H}^* = \mathbf{u}_z \frac{E_0^2}{\eta} \end{aligned} \qquad (2\text{-}18)$$

Thus, the electric and magnetic energy densities are equal, half of the energy of the wave being electric and half magnetic. We can define a *velocity of propagation of energy* v_e as

$$v_e = \frac{\text{power flow density}}{\text{energy density}} = \frac{\mathbf{S}}{w_e + w_m} \qquad (2\text{-}19)$$

For the uniform plane traveling wave, from Eqs. (2-18) and (2-19) we find

$$v_e = \frac{1}{\sqrt{\mu\epsilon}}$$

which is also the phase velocity [Eq. (2-14)]. These two velocities are not necessarily equal for other types of electromagnetic waves. In general, the phase velocity may be greater or less than the velocity of light, but the velocity of propagation of energy is never greater than the velocity of light.

Another property of waves can be illustrated by the *standing wave*

$$E_x = E_0 \sin kz \qquad H_y = j\frac{E_0}{\eta} \cos kz \qquad (2\text{-}20)$$

obtained by combining the first and third waves of Eqs. (2-17) with

$A = -C = jE_0/2$. The corresponding instantaneous fields are

$$\mathcal{E}_x = \sqrt{2}\, E_0 \sin kz \cos \omega t \qquad \mathcal{H}_y = -\sqrt{2}\, \frac{E_0}{\eta} \cos kz \sin \omega t$$

Note that the phase is now independent of z, there being no traveling motion; hence the name *standing wave*. A picture of $\mathcal{E}$ and $\mathcal{H}$ at some instant of time is shown in Fig. 2-3. The field oscillates in amplitude, with $\mathcal{E}$ reaching its peak value when $\mathcal{H}$ is zero, and vice versa. In other words, $\mathcal{E}$ and $\mathcal{H}$ are 90° out of phase. The planes of zero $\mathcal{E}$ and $\mathcal{H}$ are fixed in space, the zeros of $\mathcal{E}$ being displaced a quarter-wavelength from the zeros of $\mathcal{H}$. Successive zeros of $\mathcal{E}$ or of $\mathcal{H}$ are separated by a half-wavelength, as shown on Fig. 2-3. The wave is still a *plane* wave, for equiphase surfaces are planes. It is still a *uniform* wave, for its amplitude is constant over equiphase surfaces. It is still *linearly polarized*, for $\mathcal{E}$ always points in the same direction (or opposite direction when $\mathcal{E}$ is negative).

The energy and power quantities associated with this wave are

$$w_e = \frac{\epsilon}{2} \mathcal{E}^2 = \epsilon E_0^2 \sin^2 kz \cos^2 \omega t$$

$$w_m = \frac{\mu}{2} \mathcal{H}^2 = \epsilon E_0^2 \cos^2 kz \sin^2 \omega t$$

$$\mathbf{S} = \mathcal{E} \times \mathcal{H} = -\mathbf{u}_z \frac{E_0^2}{2\eta} \sin 2kz \sin 2\omega t \qquad (2\text{-}21)$$

$$\mathbf{S} = \mathbf{E} \times \mathbf{H}^* = -\mathbf{u}_z \frac{jE_0^2}{2\eta} \sin 2kz$$

The time-average Poynting vector $\bar{\mathbf{s}} = \mathrm{Re}\,(\mathbf{S})$ is zero, showing no power flow on the average. The electric energy density is a maximum when the magnetic energy density is zero, and vice versa. A picture of energy

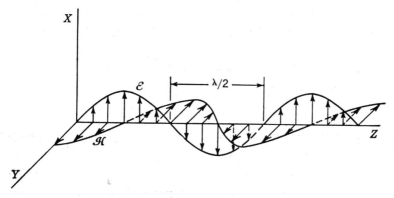

Fig. 2-3. A linearly polarized uniform plane standing wave.

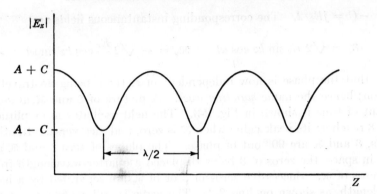

FIG. 2-4. Standing-wave pattern of two oppositely traveling waves of unequal amplitudes.

oscillating between the electric and magnetic forms can be used for this wave. Note that we have planes of zero electric intensity at $kz = n\pi$, n an integer. Thus, perfect electric conductors can be placed over one or more of these planes. If an electric conductor covers the plane $z = 0$, Eqs. (2-20) represent the solution to the problem of reflection of a uniform plane wave normally incident on this conductor. If two electric conductors cover the planes $kz = n_1\pi$ and $kz = n_2\pi$, Eqs. (2-20) represent the solution of a one-dimensional "resonator."

A more general x-polarized field is one consisting of waves traveling in opposite directions with unequal amplitudes. This is a superposition of the first and third of Eqs. (2-17), or

$$E_x = Ae^{-jkz} + Ce^{jkz}$$
$$H_y = \frac{1}{\eta}(Ae^{-jkz} - Ce^{jkz})$$
(2-22)

If $A = 0$ or $C = 0$, we have a pure traveling wave, and if $|A| = |C|$, we have a pure standing wave. For $A \neq C$, let us take A and C real[1] and express the field in terms of an amplitude and phase. This gives

$$E_x = \sqrt{A^2 + C^2 + 2AC\cos 2kz}\, e^{-j\tan^{-1}\left(\frac{A-C}{A+C}\tan kz\right)}$$
(2-23)

The rms amplitude of E is

$$\sqrt{A^2 + C^2 + 2AC\cos 2kz}$$

which is called the *standing-wave pattern* of the field. This is illustrated by Fig. 2-4. The voltage output of a small probe (receiving antenna) connected to a detector would essentially follow this standing-wave pat-

[1] This is actually no restriction on the generality of our interpretation, for it corresponds to a judicious choice of z and t origins.

tern. For a pure traveling wave, the standing-wave pattern is a constant, and for a pure standing wave, it is of the form $|\cos kz|$, that is, a "rectified" sine wave. The ratio of the maximum of the standing-wave pattern to the minimum is called the *standing-wave ratio* (SWR). From Fig. 2-4, it is evident that

$$\text{SWR} = \frac{A + C}{A - C} \qquad (2\text{-}24)$$

because the two traveling-wave components [Eqs. (2-22)] add in phase at some points and add 180° out of phase at other points. The distance between successive minima is $\lambda/2$. The standing-wave ratio of a pure traveling wave is unity, that of a pure standing wave is infinite. Plane traveling waves reflected by dielectric or imperfectly conducting boundaries will result in partial standing waves, with SWR's between one and infinity.

Let us now consider a traveling wave in which both E_x and E_y exist. This is a superposition of the first and second of Eqs. (2-17), that is,

$$\begin{aligned} \mathbf{E} &= (\mathbf{u}_x A + \mathbf{u}_y B)e^{-jkz} \\ \mathbf{H} &= (-\mathbf{u}_x B + \mathbf{u}_y A)\frac{1}{\eta} e^{-jkz} \end{aligned} \qquad (2\text{-}25)$$

If $B = 0$, the wave is linearly polarized in the x direction. If $A = 0$, the wave is linearly polarized in the y direction. If A and B are both real (or complex with equal phases), we again have a linearly polarized wave, with the axis of polarization inclined at an angle $\tan^{-1}(B/A)$ with respect to the x axis. This is illustrated by Fig. 2-5a. If A and B are complex with different phase angles, $\mathcal{E}$ will no longer point in a single spatial direction. Letting $A = |A|e^{ja}$ and $B = |B|e^{jb}$, we have the instan-

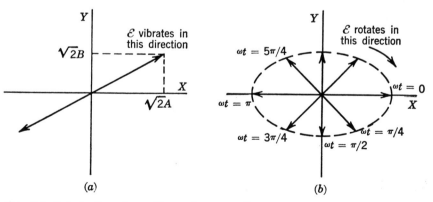

FIG. 2-5. Polarization of a uniform plane traveling wave. (a) Linear polarization; (b) elliptical polarization.

taneous electric intensity given by

$$\mathcal{E}_x = \sqrt{2}\,|A|\cos(\omega t - kz + a)$$
$$\mathcal{E}_y = \sqrt{2}\,|B|\cos(\omega t - kz + b)$$

A vector picture of $\mathcal{E}$ for various instants of time changes in both amplitude and direction, going through this variation once each cycle. For example, let $|A| = 2|B|$, $a = 0$, and $b = \pi/2$. A plot of $\mathcal{E}$ for various values of t in the plane $z = 0$ is shown in Fig. 2-5b. The tip of the arrow in the vector picture traces out an ellipse, and the field is said to be *elliptically polarized*. Depending upon A and B, this ellipse can be of arbitrary orientation in the xy plane and of arbitrary axial ratio. Linear polarization can be considered as the special case of elliptic polarization for which the axial ratio is infinite.

If the axial ratio is unity, the tip of the arrow traces out a circle, and the field is said to be *circularly polarized*. The polarization is said to be *right-handed* if $\mathcal{E}$ rotates in the direction of the fingers of the right hand when the thumb points in the direction of propagation. The polarization is said to be *left-handed* if $\mathcal{E}$ rotates in the opposite direction. The specialization of Eq. (2-25) to right-handed circular polarization is obtained by setting $A = jB = E_0$, giving

$$\mathbf{E} = (\mathbf{u}_x - j\mathbf{u}_y)E_0 e^{-jkz}$$
$$\mathbf{H} = (\mathbf{u}_x - j\mathbf{u}_y)j\frac{E_0}{\eta} e^{-jkz} \qquad (2\text{-}26)$$

A vector picture of the type of Fig. 2-1 for this wave would show $\mathcal{E}$ and $\mathcal{H}$ in the form of two corkscrews, with $\mathcal{E}$ perpendicular to $\mathcal{H}$ at each point. As time increases, this picture would rotate giving a corkscrew type of motion in the z direction. The various energy and power quantities associated with this wave are

$$w_e = \frac{\epsilon}{2}\mathcal{E}^2 = \epsilon E_0^2$$
$$w_m = \frac{\mu}{2}\mathcal{H}^2 = \epsilon E_0^2$$
$$\mathbf{S} = \mathcal{E} \times \mathcal{H} = \mathbf{u}_z \frac{2}{\eta} E_0^2 \qquad (2\text{-}27)$$
$$\mathbf{S} = \mathbf{E} \times \mathbf{H}^* = \mathbf{u}_z \frac{2}{\eta} E_0^2$$

Thus, there is no change in energy and power densities with time or space. Circular polarization gives a steady power flow, analogous to circuit-theory power transmission in a two-phase system.

INTRODUCTION TO WAVES

As a final example, consider the circularly polarized standing-wave field specified by

$$\mathbf{E} = (\mathbf{u}_x + j\mathbf{u}_y)E_0 \sin kz$$
$$\mathbf{H} = (\mathbf{u}_x + j\mathbf{u}_y)\frac{E_0}{\eta} \cos kz \qquad (2\text{-}28)$$

This is the superposition of Eqs. (2-17) for which $A = -C = jE_0/2$, $D = -B = E_0/2$. The corresponding instantaneous fields are

$$\boldsymbol{\mathcal{E}} = (\mathbf{u}_x \cos \omega t - \mathbf{u}_y \sin \omega t)\sqrt{2}\, E_0 \sin kz$$
$$\boldsymbol{\mathcal{H}} = (\mathbf{u}_x \cos \omega t - \mathbf{u}_y \sin \omega t)\sqrt{2}\, \frac{E_0}{\eta} \cos kz$$

Note that $\boldsymbol{\mathcal{E}}$ and $\boldsymbol{\mathcal{H}}$ are always *parallel* to each other. A vector picture of $\boldsymbol{\mathcal{E}}$ and $\boldsymbol{\mathcal{H}}$ at $t = 0$ is shown in Fig. 2-6. As time progresses, this picture rotates about the z axis, the amplitudes of $\boldsymbol{\mathcal{E}}$ and $\boldsymbol{\mathcal{H}}$ being independent of time. It is only the direction of $\boldsymbol{\mathcal{E}}$ and $\boldsymbol{\mathcal{H}}$ which changes with time. The amplitudes of $\boldsymbol{\mathcal{E}}$ and $\boldsymbol{\mathcal{H}}$ are, however, a function of z, giving a standing-wave pattern in the z direction. The energy and power densities associated with this wave are

$$w_e = \frac{\epsilon}{2} \boldsymbol{\mathcal{E}}^2 = \epsilon E_0^2 \sin^2 kz$$
$$w_m = \frac{\mu}{2} \boldsymbol{\mathcal{H}}^2 = \epsilon E_0^2 \cos^2 kz$$
$$\mathbf{s} = \boldsymbol{\mathcal{E}} \times \boldsymbol{\mathcal{H}} = 0 \qquad (2\text{-}29)$$
$$\mathbf{S} = -\mathbf{u}_z \frac{j}{\eta} E_0^2 \sin 2kz$$

It is interesting to note that the instantaneous energy and power densities are independent of time. This field can represent resonance between two perfectly conducting planes situated where E is zero. It thus seems that the picture of energy oscillating between the electric and magnetic forms

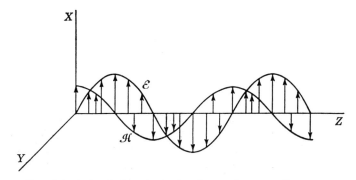

Fig. 2-6. A circularly polarized uniform plane standing wave.

is not generally valid for resonance. However, the circularly polarized standing wave is the sum of two linearly polarized waves which can exist independently of each other. We actually have two coincident resonances (called a *degenerate case*), and the picture of energy oscillating between electric and magnetic forms applies to each linearly polarized resonance.

2-3. Intrinsic Wave Constants. When the wave aspects of electromagnetism are emphasized, the wave number k and the intrinsic impedance η, given by

$$k = \sqrt{-\hat{z}\hat{y}} \qquad \eta = \sqrt{\frac{\hat{z}}{\hat{y}}} \qquad (2\text{-}30)$$

play an important role. The second equation is a generalization of Eq. (2-11), obtained in the same manner as Eq. (2-11) when $\hat{z}$ and $\hat{y}$ are not specialized to the case of a perfect dielectric. We can solve Eqs. (2-30) for $\hat{z}$ and $\hat{y}$, obtaining

$$\hat{z} = jk\eta \qquad \hat{y} = \frac{jk}{\eta} \qquad (2\text{-}31)$$

A knowledge of k and η is equivalent to a knowledge of $\hat{z}$ and $\hat{y}$, and hence specifies the characteristics of the medium.

The wave number is, in general, complex, and may be written as

$$k = k' - jk'' \qquad (2\text{-}32)$$

where k' is the *intrinsic phase constant* and k'' is the *intrinsic attenuation constant*. We have already seen that when $k = k'$, it enters into the phase function of the wave. We shall see in the next section that k'' causes an exponential attenuation of the wave amplitude. The behavior of k can be illustrated by a complex diagram relating k to $\hat{z}$ and $\hat{y}$. This is shown in Fig. 2-7. In the expressions

$$\hat{y} = \sigma + \omega\epsilon'' + j\omega\epsilon'$$
$$\hat{z} = \omega\mu'' + j\omega\mu'$$

σ, ϵ'', and μ'' are always positive in source-free media, for they account for energy dissipation. The parameters ϵ' and μ' are usually positive but may be negative for certain types of atomic resonance. Thus, $\hat{z}$ and $\hat{y}$ usually lie in the first quadrant of the complex plane, as shown in Fig. 2-7. The product $-\hat{z}\hat{y}$ then usually lies in the bottom half of the complex

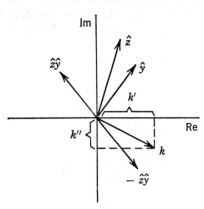

FIG. 2-7. Complex diagram relating k to $\hat{z}$ and $\hat{y}$.

FIG. 2-8. Complex diagram relating η to $\hat{z}$ and $\hat{y}$.

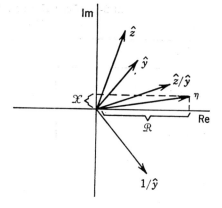

plane. The principal square root, $k = \sqrt{-\hat{z}\hat{y}}$, lies in the fourth quadrant, showing that k' and k'' are usually positive. Even when ϵ' or μ' is negative, k'' is positive; it is only k' that could conceivably be negative. In lossless media, $\hat{y} = j\omega\epsilon$, $\hat{z} = j\omega\mu$, and k is real.

The intrinsic wave impedance can be considered in an analogous manner. Expressing η in rectangular components, we have

$$\eta = \mathcal{R} + j\mathcal{X} \tag{2-33}$$

where $\mathcal{R}$ is the *intrinsic wave resistance* and $\mathcal{X}$ is the *intrinsic wave reactance*. For a wave in a perfect dielectric, η is purely resistive and is therefore the ratio of the amplitude of $\mathcal{E}$ to $\mathcal{H}$. We shall see in Sec. 2-4 that $\mathcal{X}$ introduces a phase difference between $\mathcal{E}$ and $\mathcal{H}$. The complex diagram relating η to $\hat{y}$ and $\hat{z}$ in general is shown in Fig. 2-8. In source-free regions, σ, ϵ'', and μ'' are always positive, and ϵ' and μ' are usually positive. Thus $\hat{z}$ usually lies in the first quadrant and $1/\hat{y}$ in the fourth quadrant. The ratio $\hat{z}/\hat{y}$ therefore usually lies in the right half plane and η in the sector $\pm 45°$ with respect to the positive real axis. When ϵ' or μ' is negative, η may lie anywhere in the right half plane, but $\mathcal{R}$ is never negative. In lossless media, the wave impedance is real.

There are several special cases of particular interest to us. First, consider the case of no magnetic losses. From the first of Eqs. (2-31), we have

$$\eta = \frac{\hat{z}}{jk} = \frac{\hat{z}k^*}{jkk^*} = -\frac{jk^*\hat{z}}{|\hat{z}||\hat{y}|}$$

the last equality following from Eqs. (2-30). Now for $\hat{z} = j\omega\mu = j|\hat{z}|$, we have

$$\eta = \frac{k^*}{|\hat{y}|} \qquad \text{no magnetic losses} \tag{2-34}$$

TABLE 2-1. WAVE NUMBER ($k = k' - jk''$) AND INTRINSIC IMPEDANCE ($\eta = \mathcal{R} + j\mathcal{X} = |\eta|e^{j\zeta}$)

	k'	k''	$\mathcal{R}$	$\mathcal{X}$
General	$\operatorname{Re}\sqrt{-\hat{z}\hat{y}}$	$-\operatorname{Im}\sqrt{-\hat{z}\hat{y}}$	$\operatorname{Re}\sqrt{\dfrac{\hat{z}}{\hat{y}}}$	$\operatorname{Im}\sqrt{\dfrac{\hat{z}}{\hat{y}}}$
No magnetic losses	$\operatorname{Im}\sqrt{j\omega\mu\hat{y}}$	$\operatorname{Re}\sqrt{j\omega\mu\hat{y}}$	$\dfrac{k'}{|\hat{y}|}$	$\dfrac{k''}{|\hat{y}|}$
Perfect dielectric	$\omega\sqrt{\mu\epsilon}$	0	$\sqrt{\dfrac{\mu}{\epsilon}}$	0
Good dielectric	$\omega\sqrt{\mu\epsilon'}$	$\dfrac{\omega\epsilon''}{2}\sqrt{\dfrac{\mu}{\epsilon'}}$	$\sqrt{\dfrac{\mu}{\epsilon'}}$	$\dfrac{\epsilon''}{2\epsilon'}\sqrt{\dfrac{\mu}{\epsilon'}}$
Good conductor	$\sqrt{\dfrac{\omega\mu\sigma}{2}}$	$\sqrt{\dfrac{\omega\mu\sigma}{2}}$	$\sqrt{\dfrac{\omega\mu}{2\sigma}}$	$\sqrt{\dfrac{\omega\mu}{2\sigma}}$

Separation into real and imaginary parts is shown explicitly in row 2 of Table 2-1. A similar simplification can be made for the case of no electric losses. (See Prob. 2-13.) Three special cases of materials with no magnetic losses are (1) perfect dielectrics, (2) good dielectrics, and (3) good conductors. The perfect dielectric case is that for which

$$k = \omega\sqrt{\mu\epsilon} \qquad \eta = \sqrt{\dfrac{\mu}{\epsilon}}$$

This is summarized in row 3 of Table 2-1. A good dielectric is characterized by $\hat{z} = j\omega\mu$, $\hat{y} = \omega\epsilon'' + j\omega\epsilon'$, with $\epsilon' \gg \epsilon''$. In this case, we have

$$k = \omega\sqrt{\mu\epsilon'\left(1 - j\dfrac{\epsilon''}{\epsilon'}\right)} \approx \omega\sqrt{\mu\epsilon'}\left(1 - j\dfrac{\epsilon''}{2\epsilon'}\right)$$

$$\eta = \dfrac{k^*}{|\hat{y}|} \approx \sqrt{\dfrac{\mu}{\epsilon'}}\left(1 + j\dfrac{\epsilon''}{2\epsilon'}\right)$$

which is summarized in row 4 of Table 2-1. Finally, a good conductor is characterized by $\hat{z} = j\omega\mu$, $\hat{y} = \sigma + j\omega\epsilon$, with $\sigma \gg \omega\epsilon$. In this case, we have

$$k = \sqrt{-j\omega\mu(\sigma + j\omega\epsilon)} \approx \sqrt{-j\omega\mu\sigma}$$

$$\eta = \dfrac{k^*}{|\hat{y}|} \approx \sqrt{\dfrac{j\omega\mu}{\sigma}}$$

The last row of Table 2-1 shows these parameters separated into real and imaginary parts.

2-4. Waves in Lossy Matter.

The only difference between the wave equation, Eq. (2-7), for lossy media and loss-free media is that k is complex in lossy media and real in loss-free media. Thus, Eq. (2-9) is still a solution in lossy media. In terms of the real and imaginary parts of k, it is

$$E_x = E_0 e^{-jkz} = E_0 e^{-k''z} e^{-jk'z} \tag{2-35}$$

Also, **H** is still given by Eq. (2-10), except that η is now complex. Thus, the **H** associated with the **E** of Eq. (2-35) is

$$H_y = \frac{E_0}{\eta} e^{-jkz} = \frac{E_0}{|\eta|} e^{-j\zeta} e^{-k''z} e^{-jk'z} \tag{2-36}$$

where $\eta = |\eta| e^{j\zeta}$. The instantaneous fields corresponding to Eqs. (2-35) and (2-36) are

$$\begin{aligned} \mathcal{E}_x &= \sqrt{2}\, E_0 e^{-k''z} \cos(\omega t - k'z) \\ \mathcal{H}_y &= \sqrt{2}\, \frac{E_0}{|\eta|} e^{-k''z} \cos(\omega t - k'z - \zeta) \end{aligned} \tag{2-37}$$

Thus, in lossy matter, a traveling wave is attenuated in the direction of travel according to $e^{-k''z}$, and $\mathcal{H}$ is no longer in phase with $\mathcal{E}$. A sketch of $\mathcal{E}$ and $\mathcal{H}$ versus z at some instant of time would be similar to Fig. 2-1 except that the amplitudes of $\mathcal{E}$ and $\mathcal{H}$ would decrease exponentially with z, and $\mathcal{H}$ would not be in phase with $\mathcal{E}$ ($\mathcal{H}$ usually lags $\mathcal{E}$). A sketch of $\mathcal{E}_x$ versus z for several instants of time is shown in Fig. 2-9 for a case of fairly large attenuation. A sketch of $\mathcal{H}_y$ versus z would be similar in form.

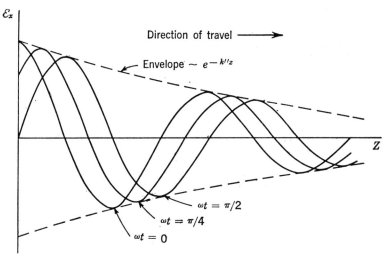

FIG. 2-9. $\mathcal{E}$ at several instants of time in a linearly polarized uniform plane traveling wave in dissipative matter.

The wave of Eq. (2-37) is still uniform, still plane, and still linearly polarized. So that our definitions of phase velocity and wavelength will be unchanged for lossy media, we should replace k and k' in the loss-free formulas, or

$$v_p = \frac{\omega}{k'} \qquad \lambda = \frac{2\pi}{k'} = \frac{v_p}{f} \qquad (2\text{-}38)$$

Then v_p is still the velocity of a plane of constant phase, and λ is still the distance in which the phase increases by 2π.

Two cases of particular interest are (1) good dielectrics (low-loss), and (2) good conductors (high-loss). For the first case, we have (see Table 2-1)

$$\left.\begin{array}{l} k' = \omega\sqrt{\mu\epsilon'} \\[4pt] k'' = \dfrac{\omega\epsilon''}{2}\sqrt{\dfrac{\mu}{\epsilon'}} \\[4pt] |\eta| = \sqrt{\dfrac{\mu}{\epsilon'}} \\[4pt] \zeta = \tan^{-1}\dfrac{\epsilon''}{2\epsilon'} \end{array}\right\} \text{ in good dielectrics } (\epsilon'' \ll \epsilon') \qquad (2\text{-}39)$$

Thus, the attenuation is very small, and $\mathcal{E}$ and $\mathcal{H}$ are nearly in phase. The wave is almost the same as in a loss-free dielectric. For example, in polystyrene (see Fig. 1-10), a 10-megacycle wave is attenuated only 0.5 per cent per kilometer, and the phase difference between $\mathcal{E}$ and $\mathcal{H}$ is only 0.003°. The intrinsic impedance of a dielectric is usually less than that of free space, since usually $\epsilon' > \epsilon_0$ and $\mu = \mu_0$. The intrinsic phase velocity and wavelength in a dielectric are also less than those of free space.

In the high-loss case (see Table 2-1), we have

$$\left.\begin{array}{l} k' = \sqrt{\dfrac{\omega\mu\sigma}{2}} \\[4pt] k'' = \sqrt{\dfrac{\omega\mu\sigma}{2}} \\[4pt] |\eta| = \sqrt{\dfrac{\omega\mu}{\sigma}} \\[4pt] \zeta = \dfrac{\pi}{4} \end{array}\right\} \text{ in good conductors } (\sigma \gg \omega\epsilon) \qquad (2\text{-}40)$$

Thus, the attenuation is very large, and $\mathcal{H}$ lags $\mathcal{E}$ by 45°. The intrinsic impedance of a good conductor is extremely small at radio frequencies, having a magnitude of 1.16×10^{-3} ohm for copper at 10 megacycles. The wavelength is also very small compared to the free-space wavelength. For example, at 10 megacycles the free-space wavelength is 30 meters, while in copper the wavelength is only 0.131 millimeter. The attenuation in a good conductor is very rapid. For the above-mentioned 10-megacycle wave in copper the attenuation is 99.81 per cent in 0.131 millimeter of travel. Thus, waves do not penetrate metals very deeply. A metal acts as a shield against electromagnetic waves.

A wave starting at the surface of a good conductor and propagating inward is very quickly damped to insignificant values. The field is localized in a thin surface layer, this phenomenon being known as *skin effect*. The distance in which a wave is attenuated to $1/e$ (36.8 per cent) of its initial value is called the *skin depth* or *depth of penetration* δ. This is defined by $k''\delta = 1$, or

$$\delta = \sqrt{\frac{2}{\omega\mu\sigma}} = \frac{1}{k''} = \frac{\lambda_m}{2\pi} \qquad (2\text{-}41)$$

where λ_m is the wavelength *in the metal*. The skin depth is very small for good conductors at radio frequencies, for λ_m is very small. For example, the depth of penetration into copper at 10 megacycles is only 0.021 millimeter. The density of power flow into the conductor, which must also be that dissipated within the conductor, is given by

$$\mathbf{S} = \mathbf{E} \times \mathbf{H}^* = \mathbf{u}_z |H_0|^2 \eta_m$$

where H_0 is the amplitude of $\mathbf{H}$ at the surface. The time-average power dissipation per unit area of surface cross section is the real part of the above power flow, or

$$\bar{\mathcal{P}}_d = |H_0|^2 \mathcal{R} \qquad \text{watts per square meter} \qquad (2\text{-}42)$$

where $\mathcal{R} = \text{Re}(\eta_m)$ is the intrinsic resistance of the metal. $\mathcal{R}$ is also called the *surface resistance* and η_m the *surface impedance* of the metal. Eq. (2-42) is strictly true only when the wave propagates normally into the conductor. In the next section we shall see that this is usually so. In most problems Eq. (2-42) can be used to calculate power losses in conducting boundaries. (An important exception to this occurs at sharp points and corners extending outward from conductors.)

More general waves can be constructed by superposition of waves of the above type with various polarizations and directions of propagation. For waves uniform in the xy plane, the four basic waves, corresponding to Eqs. (2-17), are

$$\begin{aligned}
E_x^+ &= A e^{-k''z} e^{-jk'z} & H_y^+ &= \frac{A}{\eta} e^{-k''z} e^{-jk'z} \\
E_y^+ &= B e^{-k''z} e^{-jk'z} & H_x^+ &= \frac{-B}{\eta} e^{-k''z} e^{-jk'z} \\
E_x^- &= C e^{k''z} e^{jk'z} & H_y^- &= \frac{-C}{\eta} e^{k''z} e^{jk'z} \\
E_y^- &= D e^{k''z} e^{jk'z} & H_x^- &= \frac{D}{\eta} e^{k''z} e^{jk'z}
\end{aligned} \qquad (2\text{-}43)$$

The preceding discussion of this section applies to each of these waves if the appropriate interchange of coordinates is made.

A superposition of waves traveling in opposite directions, for example

$$E_x = Ae^{-k''z}e^{-jk'z} + Ce^{k''z}e^{jk'z}$$
$$H_y = \frac{1}{\eta}(Ae^{-k''z}e^{-jk'z} - Ce^{k''z}e^{jk'z})$$ (2-44)

gives us standing-wave phenomena. However, it is no longer possible to have two "equal" waves traveling in opposite directions. One wave is attenuated in the $+z$ direction, the other in the $-z$ direction; hence they can be equal only at one plane. Suppose that the wave components are equal at $z = 0$, that is, $A = C$ in Eq. (2-44). There will then be standing waves in the vicinity of $z = 0$, which will die out in both the $+z$ and $-z$ directions. This is illustrated by Fig. 2-10 for a material having fairly large losses. Far in the $+z$ direction the $+z$ traveling wave has died out, leaving only the $-z$ traveling wave. Similarly, far in the $-z$ direction we have only the $+z$ traveling wave. The standing-wave ratio is now a function of z, being large in the vicinity of $z = 0$ and approaching unity as $|z|$ becomes large. For very small amounts of dissipation, say in a good dielectric, the attenuation of the wave is small, and standing-wave patterns are almost the same as for the dissipationless case.

Other superpositions of Eqs. (2-43) can be formed to give elliptically and circularly polarized waves. In a picture of a circularly polarized wave traveling in dissipative media, the "corkscrews" for $\mathcal{E}$ and $\mathcal{H}$ would be attenuated in the direction of propagation. Also, $\mathcal{E}$ would be somewhat out of phase with $\mathcal{H}$. A circularly polarized standing wave would be a localized phenomenon in dissipative media, just as a linearly polarized standing wave is localized.

2-5. Reflection of Waves. We saw in Sec. 1-14 that the tangential components of **E** and **H** must be continuous across a material boundary.

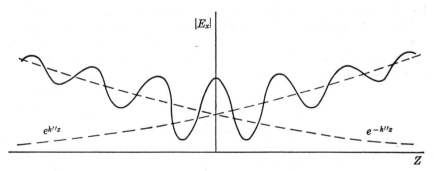

Fig. 2-10. Standing-wave pattern of two oppositely traveling waves in dissipative matter.

A ratio of a component of **E** to a component of **H** is called the wave impedance in the direction defined by the cross-product rule applied to the two components. Thus, continuity of tangential **E** and **H** requires that *wave impedances normal to a material boundary must be continuous.*

The simplest reflection problem is that of a uniform plane wave normally incident upon a plane boundary between two media. This is illustrated by Fig. 2-11. In region 1 the field will be the sum of an incident wave plus a reflected wave. The ratio of the reflected electric intensity to the incident electric intensity at the interface is defined to be the *reflection coefficient* Γ. Hence, for region 1

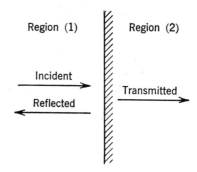

Fig. 2-11. Reflection at a plane dielectric interface, normal incidence.

$$E_x^{(1)} = E_0(e^{-jk_1z} + \Gamma e^{jk_1z})$$
$$H_y^{(1)} = \frac{E_0}{\eta_1}(e^{-jk_1z} - \Gamma e^{jk_1z})$$

In region 2 there will be a transmitted wave. The ratio of the transmitted electric intensity to the incident electric intensity at the interface is defined to be the transmission coefficient T. Hence, for region 2

$$E_x^{(2)} = E_0 T e^{-jk_2z}$$
$$H_y^{(2)} = \frac{E_0}{\eta_2} T e^{-jk_2z}$$

For continuity of wave impedance at the interface, we have

$$Z_z\bigg|_{z=0} = \frac{E_x^{(1)}}{H_y^{(1)}}\bigg|_{z=0} = \eta_1 \frac{1+\Gamma}{1-\Gamma} = \eta_2$$

where η_1 and η_2 are the intrinsic wave impedances of media 1 and 2. Solving for the reflection coefficient, we have

$$\Gamma = \frac{\eta_2 - \eta_1}{\eta_2 + \eta_1} \tag{2-45}$$

From the continuity of E_x at $z = 0$, we have the transmission coefficient given by

$$T = 1 + \Gamma = \frac{2\eta_2}{\eta_2 + \eta_1} \tag{2-46}$$

If region 1 is a perfect dielectric, the standing-wave ratio is

$$\text{SWR} = \frac{E_{\max}^{(1)}}{E_{\min}^{(1)}} = \frac{1+|\Gamma|}{1-|\Gamma|} \tag{2-47}$$

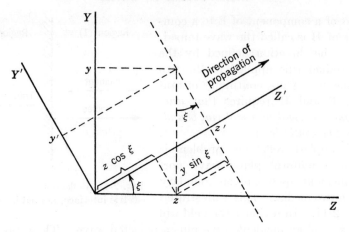

Fig. 2-12. A plane wave propagating at an angle ξ with respect to the x-z plane.

because the incident and reflected waves add in phase at some points and add 180° out of phase at other points. The density of power transmitted across the interface is

$$\mathcal{S}_{\text{trans}} = \text{Re } \mathbf{E} \times \mathbf{H}^* \cdot \mathbf{u}_z \Big|_{z=0} = \mathcal{S}_{\text{inc}}(1 - |\Gamma|^2) \qquad (2\text{-}48)$$

where $\mathcal{S}_{\text{inc}} = E_0^2/\eta_1$ is the incident power density. The difference between the incident and transmitted power must be that reflected, or

$$\mathcal{S}_{\text{refl}} = \mathcal{S}_{\text{inc}}|\Gamma|^2 \qquad (2\text{-}49)$$

We have used an x-polarized wave for the analysis, but the results are valid for arbitrary polarization, since the x axis may be in any direction tangential to the boundary. Those of us familiar with transmission-line theory should note the complete analogy between the above plane-wave problem and the transmission-line problem.

Another reflection problem of considerable interest is that of a plane wave incident at an angle upon a plane dielectric boundary. Before considering this problem, let us express the uniform plane wave in coordinates rotated with respect to the direction of propagation. Let Fig. 2-12 represent a plane wave propagating at an angle ξ with respect to the xz plane. An equiphase plane z' in terms of the unprimed coordinates is

$$z' = z \cos \xi + y \sin \xi$$

and the unit vector in the y' direction in terms of the unprimed coordinate unit vectors is

$$\mathbf{u}_{y'} = \mathbf{u}_y \cos \xi - \mathbf{u}_z \sin \xi$$

The expression for a uniform plane wave with **E** parallel to the $z = 0$ plane is the first of Eqs. (2-17) with all coordinates primed. Substituting from the above two equations, we have

$$E_x = E_0 e^{-jk(y \sin \xi + z \cos \xi)}$$
$$\mathbf{H} = (\mathbf{u}_y \cos \xi - \mathbf{u}_z \sin \xi) \frac{E_0}{\eta} e^{-jk(y \sin \xi + z \cos \xi)} \quad (2\text{-}50)$$

The wave impedance in the z direction for this wave is

$$Z_z = \frac{E_x}{H_y} = \frac{\eta}{\cos \xi} \quad (2\text{-}51)$$

In a similar manner, from the second of Eqs. (2-17), the expression for a uniform plane wave with **H** parallel to the $z = 0$ plane is found to be

$$\mathbf{E} = (\mathbf{u}_y \cos \xi - \mathbf{u}_z \sin \xi) E_0 e^{-jk(y \sin \xi + z \cos \xi)}$$
$$H_x = -\frac{E_0}{\eta} e^{-jk(y \sin \xi + z \cos \xi)} \quad (2\text{-}52)$$

The wave impedance in the z direction for this wave is

$$Z_z = -\frac{E_y}{H_x} = \eta \cos \xi \quad (2\text{-}53)$$

Thus, the z-directed wave impedance for **E** parallel to the $z = 0$ plane is always greater than the intrinsic impedance, and for **H** parallel to the $z = 0$ plane it is always less than the intrinsic impedance of the medium.

Now suppose that a uniform plane wave is incident at an angle $\xi = \theta_i$ upon a dielectric interface at $z = 0$, as shown in Fig. 2-13. Part of the wave will be reflected at an angle $\xi = \pi - \theta_r$, and part transmitted at an angle $\xi = \theta_t$. Each of these partial fields will be of the form of Eqs. (2-50) if **E** is parallel to the interface or of the form of Eqs. (2-52) if **H** is parallel to the interface. (Arbitrary polarization is a superposition of these two

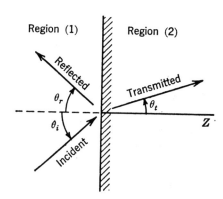

FIG. 2-13. Reflection at a plane dielectric interface, arbitrary angle of incidence.

cases.) For continuity of tangential **E** and **H** over the entire interface, the y variation of all three partial fields must be the same. This is so if

$$k_1 \sin \theta_i = k_1 \sin \theta_r = k_2 \sin \theta_t$$

From the first equality, we have

$$\theta_r = \theta_i \qquad (2\text{-}54)$$

that is, *the angle of reflection is equal to the angle of incidence.* From the second equality, we have

$$\frac{\sin \theta_t}{\sin \theta_i} = \frac{k_1}{k_2} = \frac{v_2}{v_1} = \sqrt{\frac{\epsilon_1 \mu_1}{\epsilon_2 \mu_2}} \qquad (2\text{-}55)$$

where v is the phase velocity. Equation (2-55) is known as *Snell's law of refraction*. The direction of propagation of the transmitted wave is thus different from that of the incident wave unless $\epsilon_1 \mu_1 = \epsilon_2 \mu_2$. In practically all low-loss dielectrics, $\mu_1 = \mu_2 = \mu_0$. If medium 2 is free space and medium 1 is a nonmagnetic dielectric, the right-hand side of Eq. (2-55) becomes $\sqrt{\epsilon_1/\epsilon_0} = \sqrt{\epsilon_r}$, which is called the *index of refraction* of the dielectric.

The magnitudes of the reflected and transmitted fields depend upon the polarization. For **E** parallel to the interface, we have in region 1

$$E_x^{(1)} = A(e^{-jk_1 z \cos \theta_i} + \Gamma e^{jk_1 z \cos \theta_r})$$
$$H_y^{(1)} = \frac{A}{\eta_1} \cos \theta_i (e^{-jk_1 z \cos \theta_i} - \Gamma e^{jk_1 z \cos \theta_r})$$

where A includes the y dependence. Thus, the z-directed wave impedance in region 1 at the interface is

$$Z_z^{(1)} = \frac{E_x^{(1)}}{H_y^{(1)}} = \frac{\eta_1}{\cos \theta_i} \frac{1 + \Gamma}{1 - \Gamma}$$

This must be equal to the z-directed wave impedance in region 2 at the interface, which is Eq. (2-51) with $\xi = \theta_t$. Thus,

$$\Gamma = \frac{\eta_2 \sec \theta_t - \eta_1 \sec \theta_i}{\eta_2 \sec \theta_t + \eta_1 \sec \theta_i} \qquad (2\text{-}56)$$

Note that this is of the same form as the corresponding equation for normal incidence, Eq. (2-45). The intrinsic impedances are merely replaced by the z-directed wave impedances of single traveling waves. It should be apparent from the form of the equations that, for **H** parallel to the interface, the reflection coefficient is given by

$$\Gamma = \frac{\eta_2 \cos \theta_t - \eta_1 \cos \theta_i}{\eta_2 \cos \theta_t + \eta_1 \cos \theta_i} \qquad (2\text{-}57)$$

In both cases we have standing waves in the z direction, the standing-wave ratio being given by Eq. (2-47).

Two cases of special interest are (1) that of total transmission and (2) that of total reflection. The first case occurs when $\Gamma = 0$. For **E** parallel to the interface, we see from Eq. (2-56) that $\Gamma = 0$ when

$$\frac{\eta_2}{\cos \theta_t} = \frac{\eta_1}{\cos \theta_i}$$

Substituting for θ_t from Eq. (2-55) and for the η's from Eq. (2-11) we obtain

$$\sin \theta_i = \sqrt{\frac{\epsilon_2/\epsilon_1 - \mu_2/\mu_1}{\mu_1/\mu_2 - \mu_2/\mu_1}} \qquad (2\text{-}58)$$

as the angle at which no reflection occurs. This does not always have a real solution for θ_i. In fact,

$$\sin \theta_i \xrightarrow[\mu_1 \to \mu_2]{} \infty$$

For nonmagnetic dielectrics ($\mu_1 = \mu_2 = \mu_0$) there is no angle of total transmission when **E** is parallel to the boundary. For the case of **H** parallel to the boundary, we find from Eq. (2-57) that $\Gamma = 0$ when

$$\sin \theta_i = \sqrt{\frac{\epsilon_2/\epsilon_1 - \mu_2/\mu_1}{\epsilon_2/\epsilon_1 - \epsilon_1/\epsilon_2}} \qquad (2\text{-}59)$$

Again this does not always have a real solution for arbitrary μ and ϵ. But in the nonmagnetic case

$$\theta_i = \sin^{-1} \sqrt{\frac{\epsilon_2}{\epsilon_1 + \epsilon_2}} = \tan^{-1} \sqrt{\frac{\epsilon_2}{\epsilon_1}} \qquad (2\text{-}60)$$

There is usually an angle of total transmission when **H** is parallel to the boundary. The angle specified by Eq. (2-60) is called the *polarizing angle* or *Brewster angle*. If an arbitrarily polarized wave is incident upon a nonmagnetic boundary at this angle, the reflected wave will be polarized with **E** parallel to the boundary.

The case of total reflection occurs when $|\Gamma| = 1$. We are considering lossless media; so the η's are real. It is apparent from Eqs. (2-56) and (2-57) that $|\Gamma| \neq 1$ for real values of θ_i and θ_t. However, when $\epsilon_1 \mu_1 > \epsilon_2 \mu_2$, Eq. (2-55) says that $\sin \theta_t$ can be greater than unity. What does this mean? Our initial assumption was that the transmitted wave was a uniform plane wave. But Eqs. (2-50) specify a solution to Maxwell's equations regardless of the value of $\sin \xi$. It can be real or complex. All that is changed is our interpretation of the field. To illustrate, suppose $\sin \xi > 1$ in Eqs. (2-50) and let

$$k \sin \xi = \beta$$
$$k \cos \xi = k \sqrt{1 - \sin^2 \xi} = \pm j\alpha \qquad (2\text{-}61)$$

If we choose the minus sign for α, Eqs. (2-50) become

$$E_x = E_0 e^{-j\beta y} e^{-\alpha z}$$
$$\mathbf{H} = -\left(\mathbf{u}_y \frac{j\alpha}{k} + \mathbf{u}_z \frac{\beta}{k}\right) \frac{E_0}{\eta} e^{-j\beta y} e^{-\alpha z} \qquad (2\text{-}62)$$

which is a field exponentially attenuated in the z direction. Note the 90° phase difference between E_x and H_y; so the wave impedance in the z direction is imaginary, and there is no power flow in the z direction. A similar interpretation applies to Eqs. (2-52) when $\sin \xi > 1$. Returning now to our reflection problem, from Eq. (2-55) it is evident that $\sin \theta_t$ is greater than unity when $\sin \theta_i > \sqrt{\epsilon_2 \mu_2 / \epsilon_1 \mu_1}$. Thus, the point of transition from real values of θ_t (wave impedance real in region 2) to imaginary values of θ_t (wave impedance imaginary in region 2) is

$$\sin \theta_i = \sqrt{\frac{\epsilon_2 \mu_2}{\epsilon_1 \mu_1}} \qquad (2\text{-}63)$$

The angle specified by Eq. (2-63) is called the *critical angle*. A wave incident upon the boundary at an angle equal to or greater than the critical angle will be totally reflected. Note that there is a real critical angle only if $\epsilon_1 \mu_1 > \epsilon_2 \mu_2$ or, in the nonmagnetic case, if $\epsilon_1 > \epsilon_2$. Thus, total reflection occurs only if the wave passes from a "dense" material into a "less dense" material. The reflection coefficient, Eq. (2-56) or Eq. (2-57), becomes of the form

$$\Gamma = \frac{R - jX}{R + jX}$$

when total reflection occurs. It is evident in this case that $|\Gamma|$ is unity. Remember that the field in region 2 is not zero when total reflection occurs. It is an exponentially decaying field, called a *reactive* field or an *evanescent* field. Optical prisms make use of the phenomenon of total reflection.

All the theory of this section can be applied to dissipative media if the η's and θ's are allowed to be complex. Of particular interest is the case of a plane wave incident upon a good conductor at an angle θ_i. When region 1 is a nonmagnetic dielectric and region 2 is a nonmagnetic conductor, Eq. (2-55) becomes

$$\frac{\sin \theta_t}{\sin \theta_i} = \frac{k_1}{k_2} \approx \sqrt{\frac{j\omega\epsilon}{\sigma}}$$

INTRODUCTION TO WAVES 61

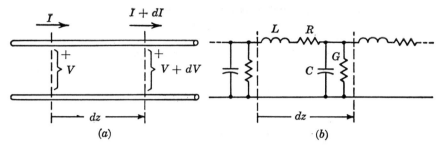

FIG. 2-14. A transmission line according to circuit concepts. (a) Physical line; (b) equivalent circuit.

This is an extremely small quantity for good conductors. For most practical purposes, the wave can be considered to propagate normally into the conductor regardless of the angle of incidence.

2-6. Transmission-line Concepts. Let us review the circuit concept of a transmission line and then show its relationship to the field concept. Let Fig. 2-14a represent a two-conductor transmission line. For each incremental length of line dz there is a series voltage drop dV and a shunt current dI. The circuit theory postulate is that the voltage drop is proportional to the line current I. Thus,

$$dV = -IZ\,dz$$

where Z is a series *impedance per unit length*. It is also postulated that the shunt current is proportional to the line voltage V. Thus,

$$dI = -VY\,dz$$

where Y is a *shunt admittance per unit length*. Dividing by dz, we have the *a-c transmission-line equations*

$$\frac{dV}{dz} = -IZ \qquad \frac{dI}{dz} = -VY \qquad (2\text{-}64)$$

Implicit in this development are the assumptions that (1) no mutual impedance exists between incremental sections of line and (2) the shunt current dI flows in planes transverse to z. The transmission line is said to be *uniform* if Z and Y are independent of z.

Taking the derivative of the first of Eqs. (2-64) and substituting from the second, we obtain

$$\frac{d^2V}{dz^2} - ZYV = 0 \qquad \frac{d^2I}{dz^2} - ZYI = 0 \qquad (2\text{-}65)$$

which are one-dimensional Helmholtz equations. The general solution

TABLE 2-2. COMPARISON OF TRANSMISSION-LINE WAVES
TO UNIFORM PLANE WAVES

Transmission line	Uniform plane wave
$\dfrac{d^2V}{dz^2} - \gamma^2 V = 0$	$\dfrac{d^2E_x}{dz^2} + k^2 E_x = 0$
$\dfrac{d^2I}{dz^2} - \gamma^2 I = 0$	$\dfrac{d^2H_y}{dz^2} + k^2 H_y = 0$
$\gamma = \sqrt{ZY}$	$jk = \sqrt{\hat{z}\hat{y}}$
$V = V_0^+ e^{-\gamma z} + V_0^- e^{\gamma z}$	$E_x = E_0^+ e^{-jkz} + E_0^- e^{jkz}$
$I = I_0^+ e^{-\gamma z} + I_0^- e^{\gamma z}$	$H_y = H_0^+ e^{-jkz} + H_0^- e^{jkz}$
$Z_0 = \dfrac{V_0^+}{I_0^+} = -\dfrac{V_0^-}{I_0^-} = \sqrt{\dfrac{Z}{Y}}$	$\eta = \dfrac{E_0^+}{H_0^+} = -\dfrac{E_0^-}{H_0^-} = \sqrt{\dfrac{\hat{z}}{\hat{y}}}$
$P = VI^*$	$S_z = E_x H_y^*$

is a sum of a $+z$ traveling wave and a $-z$ traveling wave, with propagation constant

$$\gamma = \sqrt{ZY} \tag{2-66}$$

Choosing the $+z$ traveling wave

$$V^+ = V_0 e^{-\gamma z} \qquad I^+ = I_0 e^{-\gamma z}$$

we have from Eqs. (2-64) that

$$\frac{V^+}{I^+} = \frac{Z}{\gamma} = \frac{\gamma}{Y}$$

Substituting for γ from Eq. (2-66), we have

$$Z_0 = \frac{V^+}{I^+} = \sqrt{\frac{Z}{Y}} \tag{2-67}$$

which is called the *characteristic impedance* of the transmission line. The imaginary parts of Z and Y are usually positive, and it is common practice to write

$$Z = R + j\omega L \qquad Y = G + j\omega C \tag{2-68}$$

The equivalent circuit of the transmission line is then as shown in Fig. 2-14b. The reader has probably already noted the complete analogy between the linearly polarized plane wave and the transmission line. This analogy is summarized by Table 2-2.

In the circuit theory development, we assumed no mutual coupling between adjacent elements of the transmission line. From the field theory point of view, this is equivalent to assuming that no E_z or H_z exists. Such a wave is called *transverse electromagnetic*, abbreviated TEM. This is not the only wave possible on a transmission line, for Maxwell's equations show that infinitely many wave types can exist. Each possible wave is called a *mode*, and a TEM wave is called a *transmission-line mode*. All other waves, which must have an E_z or an H_z or both, are called *higher-order modes*. The higher-order modes are usually important only in the vicinity of the feed point, or in the vicinity of a discontinuity on the line. In this section we shall restrict consideration to transmission-line, or TEM, modes.

For the TEM mode to exist exactly, the conductors must be perfect, or else an E_z is required to support the z-directed current. Let us therefore specialize the problem to that of perfect conductors immersed in a homogeneous medium. We assume $E_z = H_z = 0$ and z dependence of the form $e^{-\gamma z}$. Expansion of the field equations, Eqs. (2-1), then gives

$$\gamma E_y = -\hat{z} H_x \qquad \gamma H_y = \hat{y} E_x$$
$$\gamma E_x = \hat{z} H_y \qquad \gamma H_x = -\hat{y} E_y$$
$$\frac{\partial E_y}{\partial x} - \frac{\partial E_x}{\partial y} = 0 \qquad \frac{\partial H_y}{\partial x} - \frac{\partial H_x}{\partial y} = 0$$

It follows from these equations that

$$\gamma = jk \qquad (2\text{-}69)$$

The propagation constant of any TEM wave is the intrinsic propagation constant of the medium. The proportionality of components of **E** to those of **H** expressed by the above equations can be written concisely as

$$\mathbf{E} = \eta \mathbf{H} \times \mathbf{u}_z \qquad \mathbf{H} = \frac{1}{\eta} \mathbf{u}_z \times \mathbf{E} \qquad (2\text{-}70)$$

Thus, *the z-directed wave impedance of any TEM wave is the intrinsic wave impedance of the medium.* Finally, manipulation of the original six equations shows that each component of **E** and **H** satisfies the two-dimensional Laplace equation. We can summarize this by defining a *transverse Laplacian operator*

$$\nabla_t^2 = \frac{\partial^2}{\partial x^2} + \frac{\partial^2}{\partial y^2} \qquad (2\text{-}71)$$

and writing $\qquad \nabla_t^2 \mathbf{E} = 0 \qquad \nabla_t^2 \mathbf{H} = 0$

The boundary conditions for the problem are

$$\left. \begin{array}{l} E_t = 0 \\ H_n = 0 \end{array} \right\} \quad \text{at the conductors} \qquad (2\text{-}72)$$

Thus, the boundary-value problem for **E** is the same as the electrostatic

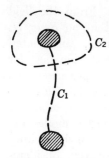

FIG. 2-15. Cross section of a transmission line.

problem having the same conducting boundaries. The boundary-value problem for **H** is the same as the magnetostatic problem having "anticonducting" (no H_n) boundaries. It is for this reason that "static" capacitances and inductances can be used for transmission lines even though the field is time-harmonic.

To show the relationship of the static L's and C's to the Z_0 of the transmission line, consider a cross section of the line as represented by Fig. 2-15. In the transmission-line problem, the line voltage and current are related to the fields by

$$V = \int_{C_1} \mathbf{E} \cdot d\mathbf{l} \qquad I = \int_{C_2} \mathbf{H} \cdot d\mathbf{l} \qquad (2\text{-}73)$$

where C_1 and C_2 are as shown on Fig. 2-15. From the second of these and the second of Eqs. (2-70) we have

$$I = \frac{1}{\eta} \int_{C_2} \mathbf{u}_z \times \mathbf{E} \cdot d\mathbf{l} = \frac{1}{\eta} \int_{C_2} E_n \, dl$$

But in the corresponding electrostatic problem the capacitance is

$$C = \frac{q}{V} = \frac{\epsilon}{V} \int_{C_2} E_n \, dl$$

Thus, the characteristic impedance of the transmission line is related to the electrostatic capacitance per unit length by

$$Z_0 = \frac{V}{I} = \eta \frac{\epsilon}{C} \qquad (2\text{-}74)$$

Similarly, from the first of Eqs. (2-73) and (2-70) we have

$$V = \eta \int_{C_1} \mathbf{H} \times \mathbf{u}_z \cdot d\mathbf{l} = \eta \int_{C_1} H_n \, dl$$

In the corresponding magnetostatic problem we have

$$L = \frac{\psi}{I} = \frac{\mu}{I} \int_{C_1} H_n \, dl$$

Therefore, the characteristic impedance of the line is related to the magnetostatic inductance per unit length by

$$Z_0 = \frac{V}{I} = \eta \frac{L}{\mu} \qquad (2\text{-}75)$$

Note also that L and C are related to each other through Eqs. (2-74) and (2-75). The electrostatic and magnetostatic problems have **E** and **H** everywhere orthogonal to each other and are called *conjugate problems*.

TABLE 2-3. CHARACTERISTIC IMPEDANCES OF SOME COMMON TRANSMISSION LINES

Line	Geometry	Characteristic impedance
Two wire		$Z_0 \approx \dfrac{\eta}{\pi} \log \dfrac{2D}{d}$ $\quad D \gg d$
Coaxial		$Z_0 = \dfrac{\eta}{2\pi} \log \dfrac{b}{a}$
Confocal elliptic		$Z_0 = \dfrac{\eta}{2\pi} \log \dfrac{b + \sqrt{b^2 - c^2}}{a + \sqrt{a^2 - c^2}}$
Parallel plate		$Z_0 \approx \eta \dfrac{b}{w}$ $\quad w \gg b$
Collinear plate		$Z_0 \approx \dfrac{\eta}{\pi} \log \dfrac{4D}{w}$ $\quad D \gg w$
Wire above ground plane		$Z_0 \approx \dfrac{\eta}{2\pi} \log \dfrac{4h}{d}$ $\quad h \gg d$
Shielded pair		$Z_0 \approx \dfrac{\eta}{\pi} \log \left(\dfrac{2s}{d} \dfrac{D^2 - s^2}{D^2 + s^2} \right)$ $\quad \begin{array}{l} D \gg d \\ s \gg d \end{array}$
Wire in trough		$Z_0 \approx \dfrac{\eta}{2\pi} \log \left(\dfrac{4w}{\pi d} \tanh \dfrac{\pi h}{w} \right)$ $\quad \begin{array}{l} h \gg d \\ w \gg d \end{array}$

Once the electrostatic C or the magnetostatic L is known, the Z_0 of the corresponding transmission line is given by Eq. (2-74) or Eq. (2-75). Table 2-3 lists the characteristic impedances of some common transmission lines.

When the dielectric is lossy but the conductors still assumed perfect, all of our equations still apply. Z_0 (proportional to η) and γ ($= jk$)

become complex. The most important effect of this is that the wave is attenuated in the direction of travel. The attenuation constant in this case is the intrinsic attenuation constant of the dielectric (Table 2-1, column 2, row 4). When the conductors are imperfect, the field is no longer exactly TEM, and exact solutions are usually impractical. However, the waves will still be characterized by a propagation constant $\gamma = \alpha + j\beta$. Hence a $+z$-traveling wave will be of the form

$$V = V_0 e^{-(\alpha+j\beta)z} \qquad I = \frac{V}{Z_0}$$

and the power flow is given by

$$P_f = VI^* = \frac{|V_0|^2}{Z_0^*} e^{-2\alpha z} = P_0 e^{-2\alpha z}$$

or, in terms of time-average powers,

$$\bar{\mathcal{P}}_f = \text{Re}\,(P_f) = \text{Re}\,(P_0) e^{-2\alpha z}$$

The rate of decrease in $\bar{\mathcal{P}}_f$ versus z equals the time-average power dissipated per unit length $\bar{\mathcal{P}}_d$, or

$$\bar{\mathcal{P}}_d = -\frac{d\bar{\mathcal{P}}_f}{dz} = 2\alpha \bar{\mathcal{P}}_f$$

Thus, the attenuation constant is given by

$$\alpha = \frac{\bar{\mathcal{P}}_d}{2\bar{\mathcal{P}}_f} \tag{2-76}$$

While this equation is exact if $\bar{\mathcal{P}}_d$ and $\bar{\mathcal{P}}_f$ are determined exactly, its greatest use lies in approximating α by approximating $\bar{\mathcal{P}}_d$. For example, attenuation due to losses in imperfect conductors can be approximated by assuming that Eq. (2-42) holds at their surface. We shall carry out such a calculation for the rectangular waveguide in the next section.

2-7. Waveguide Concepts. The waves on a transmission line can be viewed as being guided by the conductors. This concept of wave guidance is quite general and applies to many configurations of matter. In general, systems which guide waves are called *waveguides*. Apart from transmission lines, the most commonly used waveguide is the *rectangular waveguide*, illustrated by Fig. 2-16. It is a hollow conducting tube

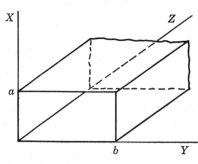

FIG. 2-16. The rectangular waveguide.

INTRODUCTION TO WAVES 67

of rectangular cross section. Fields existing within this tube must be characterized by zero tangential components of **E** at the conducting walls.

Consider two uniform plane waves traveling at the angles ξ and $-\xi$ with respect to the xz plane (see Fig. 2-12). If the waves are x-polarized, we use Eq. (2-50) and write

$$E_x = A(e^{-jky \sin \xi} - e^{jky \sin \xi})e^{-jkz \cos \xi}$$
$$= -2jA \sin (ky \sin \xi) e^{-jkz \cos \xi}$$

Let E_0 denote $(-2jA)$ and define

$$k_c = k \sin \xi \qquad \gamma = jk \cos \xi$$

In view of the trigonometric identity $\sin^2 \xi + \cos^2 \xi = 1$, the parameters γ and k_c are related by

$$\gamma^2 = k_c^2 - k^2 \qquad (2\text{-}77)$$

The above field can now be written as

$$E_x = E_0 \sin (k_c y) e^{-\gamma z} \qquad (2\text{-}78)$$

Let us see if this field can exist within the rectangular waveguide. There is only an E_x; so no component of **E** is tangential to the conductors $x = 0$ and $x = a$. Also, $E_x = 0$ at $y = 0$; so there is no tangential component of **E** at the wall $y = 0$. There remains the condition that $E_x = 0$ at $y = b$, which is satisfied if

$$k_c = \frac{n\pi}{b} \qquad n = 1, 2, 3, \ldots \qquad (2\text{-}79)$$

These permissible values of k_c are called *eigenvalues*, or *characteristic values* of the problem.

Each choice of n in Eq. (2-79) determines a possible field, or *mode*. The modes in a waveguide are usually classified according to the existence of z components of the field. A mode having no E_z is said to be a *transverse electric* (TE) mode. One having no H_z is said to be a *transverse magnetic* (TM) mode. All the modes in the rectangular waveguide fall into one of these two classes. The modes represented by Eqs. (2-78) and (2-79) have no E_z and are therefore TE modes. The particular modes that we are considering are TE$_{0n}$ modes, the subscript 0 denoting no variation with x, and the subscript n denoting the choice by Eq. (2-79). The complete system of modes will be considered in Sec. 4-3.

For k real (loss-free dielectric), the propagation constant γ can be expressed as

$$\gamma = \begin{cases} j\beta = j\sqrt{k^2 - \left(\frac{n\pi}{b}\right)^2} & k > \frac{n\pi}{b} \\ \alpha = \sqrt{\left(\frac{n\pi}{b}\right)^2 - k^2} & k < \frac{n\pi}{b} \end{cases} \qquad (2\text{-}80)$$

where α and β are real. This follows from Eqs. (2-77) and (2-79). When $\gamma = j\beta$, we have wave propagation in the z direction, and the mode is called a *propagating mode*. When $\gamma = \alpha$, the field decays exponentially with z, and there is no wave propagation. In this case, the mode is called a *nonpropagating mode*, or an *evanescent mode*. The transition from one type of behavior to the other occurs at $\alpha = 0$ or $k = n\pi/b$. Letting $k = 2\pi f \sqrt{\epsilon\mu}$, we can solve for the transition frequency, obtaining

$$f_c = \frac{n}{2b\sqrt{\epsilon\mu}} \quad (2\text{-}81)$$

This is called the *cutoff frequency* of the TE_{0n} mode. The corresponding intrinsic wavelength

$$\lambda_c = \frac{2b}{n} \quad (2\text{-}82)$$

is called the *cutoff wavelength* of the TE_{0n} mode. At frequencies greater than f_c (wavelengths less than λ_c), the mode propagates. At frequencies less than f_c (wavelengths greater than λ_c), the mode is nonpropagating.

A knowledge of f_c or λ_c is equivalent to a knowledge of k_c; so they also are eigenvalues. In particular, from Eqs. (2-79), (2-81), and (2-82), it is evident that

$$k_c = \frac{2\pi}{\lambda_c} = 2\pi f_c \sqrt{\epsilon\mu} \quad (2\text{-}83)$$

Using the last equality and $k = 2\pi f \sqrt{\epsilon\mu}$ in Eq. (2-80), we can express γ as

$$\gamma = \begin{cases} j\beta = jk\sqrt{1 - \left(\dfrac{f_c}{f}\right)^2} & f > f_c \\ \alpha = k_c\sqrt{1 - \left(\dfrac{f}{f_c}\right)^2} & f < f_c \end{cases} \quad (2\text{-}84)$$

Thus, the phase constant β of a propagating mode is always less than the intrinsic phase constant k of the dielectric, approaching k as $f \to \infty$. The attenuation constant of a nonpropagating mode is always less than k_c, approaching k_c as $f \to 0$. When a mode propagates, the concepts of wavelength and phase velocity can be applied to the mode field as a whole. Thus, the *guide wavelength* λ_g is defined as the distance in which the phase of E increases by 2π, that is, $\beta\lambda_g = 2\pi$. Using β from Eq. (2-84), we have

$$\lambda_g = \frac{\lambda}{\sqrt{1 - (f_c/f)^2}} \quad (2\text{-}85)$$

showing that the guide wavelength is always greater than the intrinsic wavelength of the dielectric. The *guide phase velocity* v_g is defined as the velocity at which a point of constant phase of $\mathcal{E}$ travels. Thus, in a manner analogous to that used to derive Eq. (2-14), we find

$$v_g = \frac{\omega}{\beta} = \frac{v_p}{\sqrt{1 - (f_c/f)^2}} \quad (2\text{-}86)$$

where v_p is the intrinsic phase velocity of the dielectric. The guide phase velocity is therefore greater than the intrinsic phase velocity.

Another important property of waveguide modes is the existence of a *characteristic wave impedance*. To show this, let us find **H** from the **E** of Eq. (2-78) according to $\nabla \times \mathbf{E} = -j\omega\mu\mathbf{H}$. The result is

$$\begin{aligned} E_x &= E_0 \sin(k_c y)\, e^{-\gamma z} \\ H_y &= \frac{\gamma}{j\omega\mu} E_0 \sin(k_c y)\, e^{-\gamma z} \\ H_z &= \frac{k_c}{j\omega\mu} E_0 \cos(k_c y)\, e^{-\gamma z} \end{aligned} \quad (2\text{-}87)$$

where E_x has been repeated for convenience. The wave impedance in the z direction is

$$Z_z = \frac{E_x}{H_y} = \frac{j\omega\mu}{\gamma} \quad (2\text{-}88)$$

This is called the characteristic impedance of the mode and plays the same role in reflection problems as does the Z_0 of transmission lines. If we substitute into the above equation for γ from Eq. (2-84), we find

$$Z_0 = Z_z = \begin{cases} \dfrac{\eta}{\sqrt{1 - (f_c/f)^2}} & f > f_c \\ \dfrac{j\eta}{\sqrt{(f_c/f)^2 - 1}} & f < f_c \end{cases} \quad (2\text{-}89)$$

Thus, the characteristic impedance of a TE$_{0n}$ propagating mode is always greater than the intrinsic impedance of the dielectric, approaching η as $f \to \infty$. The characteristic impedance of a nonpropagating mode is reactive and approaches zero as $f \to 0$.

All our discussion so far has dealt with waves traveling in the $+z$ direction. For each $+z$ traveling wave, a $-z$ traveling wave is possible, obtained by replacing γ by $-\gamma$ in Eqs. (2-87). The simultaneous existence of $+z$ and $-z$ traveling waves in the same mode gives rise to standing waves. The concepts of reflection coefficients, standing-wave ratios, etc., used in the case of uniform plane-wave reflection, also apply to waveguide problems.

The mode with the lowest cutoff frequency in a particular guide is called the *dominant mode*. The dominant mode in a rectangular waveguide, assuming $b > a$, is the TE$_{01}$ mode. (This we have not shown, for

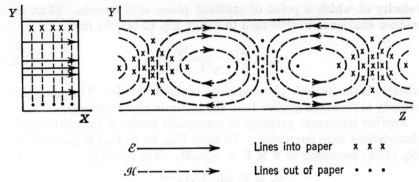

Fig. 2-17. Mode pattern for the TE_{01} waveguide mode.

we have not considered all modes.) From Eq. (2-82) with $n = 1$, we see that the cutoff wavelength of the TE_{01} mode is $\lambda_c = 2b$. Thus, wave propagation can take place in a rectangular waveguide only when its widest side is greater than a half-wavelength.[1] A sketch of the instantaneous field lines at some instant is called a *mode pattern*. The mode pattern of the TE_{01} mode in the propagating state is shown in Fig. 2-17. This figure is obtained by determining $\mathcal{E}$ and $\mathcal{H}$ from the **E** and **H** of Eqs. (2-87) and specializing the result to some instant of time. As time progresses, the mode pattern moves in the z direction.

It is admittedly confusing to learn that many modes exist on a given guiding system. It is not, however, so bad as it seems at first. If only one mode propagates in a waveguide, this will be the only mode of appreciable magnitude except near sources or discontinuities. The rectangular waveguide is usually operated so that only the TE_{01} mode propagates. This is therefore the only wave of significant amplitude along the guide except near sources and discontinuities.

Because of the importance of the TE_{01} mode, let us consider it in a little more detail. Table 2-4 specializes our preceding equations to this mode and includes some additional parameters which we shall now consider.

The power transmitted along the waveguide can be found by integrating the axial component of the Poynting vector over a guide cross section. This gives

$$P_f = \int_0^a \int_0^b E_x H_y^* \, dx \, dy = |E_0|^2 \frac{ab}{2Z_0^*}$$

which, above cutoff, is real and is therefore the time-average power transmitted. Below cutoff, the power is imaginary, indicating no time-average

---

[1] We are referring to the intrinsic wavelength of the dielectric filling the waveguide, which is usually free space.

TABLE 2-4. SUMMARY OF WAVEGUIDE PARAMETERS FOR THE DOMINANT MODE (TE_{01}) IN A RECTANGULAR WAVEGUIDE

Complex field	$E_x = E_0 \sin \frac{\pi y}{b} e^{-\gamma z}$ $H_y = \frac{E_0}{Z_0} \sin \frac{\pi y}{b} e^{-\gamma z}$ $H_z = \frac{E_0}{j\eta} \frac{f_c}{f} \cos \frac{\pi y}{b} e^{-\gamma z}$
Cutoff frequency	$f_c = \dfrac{1}{2b \sqrt{\epsilon \mu}}$
Cutoff wavelength	$\lambda_c = 2b$
Propagation constant	$\gamma = \begin{cases} j\beta = jk\sqrt{1 - (f_c/f)^2} & f > f_c \\ \alpha = \dfrac{2\pi}{\lambda_c}\sqrt{1 - (f/f_c)^2} & f < f_c \end{cases}$
Characteristic impedance	$Z_0 = \dfrac{j\omega\mu}{\gamma} = \begin{cases} \eta/\sqrt{1 - (f_c/f)^2} & f > f_c \\ j\eta/\sqrt{(f_c/f)^2 - 1)} & f < f_c \end{cases}$
Guide wavelength	$\lambda_g = \dfrac{\lambda}{\sqrt{1 - (f_c/f)^2}}$
Guide phase velocity	$v_g = \dfrac{v_p}{\sqrt{1 - (f_c/f)^2}}$
Power transmitted	$P = \dfrac{|E_0|^2 ab}{2Z_0}$
Attenuation due to lossy dielectric	$\alpha_d = \dfrac{\omega \epsilon''}{2} \eta \sqrt{1 - (f_c/f)^2}$
Attenuation due to imperfect conductor	$\alpha_c = \dfrac{\mathcal{R}}{a\eta \sqrt{1 - (f_c/f)^2}} \left[1 + \dfrac{2a}{b} \left(\dfrac{f_c}{f}\right)^2 \right]$

power transmitted. (The preceding equation applies only at $z = 0$ below cutoff unless the factor $e^{-2\alpha z}$ is added.) It is also interesting to note that the time-average electric and magnetic energies per unit length of guide are equal above cutoff (see Prob. 2-32).

In contrast to the transmission-line mode, there is no unique voltage and current associated with a waveguide mode. However, the amplitude of a modal traveling wave (E_0 in Table 2-4) enters into waveguide reflection problems in the same manner as V in transmission-line problems.

To emphasize this correspondence, it is common to define a *mode voltage V* and a *mode current I* such that

$$Z_0 = \frac{V}{I} \qquad P = VI^* \qquad (2\text{-}90)$$

From Table 2-4, it is evident that

$$V = E_0 \sqrt{\frac{ab}{2}}\, e^{-\gamma z} \qquad I = \frac{V}{Z_0} \qquad (2\text{-}91)$$

satisfy this definition. Remember that we are dealing with only a $+z$ traveling wave. In the $-z$ traveling wave, $I = -V/Z_0$. When waves in both directions are present, the ratio V/I is a function of z. Other definitions of mode voltage, mode current, and characteristic impedance can be found in the literature. These alternative definitions will always be proportional to our definitions (see Prob. 2-34).

Our treatment has so far been confined to the ideal loss-free guide. When losses are present in the dielectric but not in the conductor, all our equations still apply, except that most parameters become complex. There is no longer a real cutoff frequency, for γ never goes to zero. Also, the characteristic impedance is complex at all frequencies. The behavior of $\gamma = \alpha + j\beta$ in the low-loss case is sketched in Fig. 2-18. The behavior of γ for the loss-free case is shown dashed. The most important effect of dissipation is the existence of an attenuation constant at all frequencies. In the low-loss case, we can continue to use the relationship

$$\gamma = \alpha + j\beta \approx jk\sqrt{1 - \left(\frac{f_c}{f}\right)^2}$$

provided f is not too close to f_c. Letting $k = k' - jk''$ and referring to

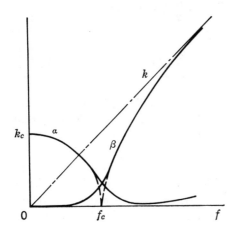

Fig. 2-18. Propagation constant for a lossy waveguide (loss-free case shown dashed).

Table 2-1, we find

$$\alpha_d \approx \frac{\omega\epsilon''}{2}\sqrt{\frac{\mu}{\epsilon'}}\sqrt{1 - \left(\frac{f_c}{f}\right)^2} \qquad (2\text{-}92)$$

This is the attenuation constant due to a lossy dielectric in the guide.

Even more important is the attenuation due to imperfectly conducting guide walls. Our solution is no longer exact in this case, because the boundary conditions are changed. The tangential component of **E** is now not quite zero at the conductor. However, for good conductors, the tangential component of **E** is very small, and the field is only slightly changed, or "perturbed," from the loss-free solution. The loss-free solution is used to approximate **H** at the conductor, and Eq. (2-42) is used to approximate the power dissipated in the conductor. Such a procedure is called a *perturbational method* (see Chap. 7). The power per unit length dissipated in the wall $y = 0$ is

$$\bar{\mathcal{P}}_d\bigg|_{y=0} = \mathcal{R}\int_0^a |H_z|^2\,dx = \mathcal{R}|E_0|^2\left(\frac{f_c}{\eta f}\right)^2\int_0^a dx$$
$$= \mathcal{R}|E_0|^2 a\left(\frac{f_c}{\eta f}\right)^2$$

and an equal amount is dissipated in the wall $y = b$. The power per unit length dissipated in the wall $x = 0$ is

$$\bar{\mathcal{P}}_d\bigg|_{x=0} = \mathcal{R}\int_0^b (|H_y|^2 + |H_z|^2)\,dy$$
$$= \mathcal{R}|E_0|^2\int_0^b \left[\frac{\sin^2(\pi y/b)}{Z_0^2} + \left(\frac{f_c}{\eta f}\right)^2 \cos^2\frac{\pi y}{b}\right] dy$$
$$= \mathcal{R}|E_0|^2 \left[\frac{b}{2Z_0^2} + \left(\frac{f_c}{\eta f}\right)^2 \frac{b}{2}\right]$$

and an equal amount is dissipated in the wall $x = a$. The total power dissipated per unit length is the sum of that for the four walls, or

$$\bar{\mathcal{P}}_d = \mathcal{R}|E_0|^2\left[\frac{b}{Z_0^2} + \left(\frac{f_c}{\eta f}\right)^2 (2a + b)\right]$$

Equation (2-76) is valid for any traveling wave; so using the above $\bar{\mathcal{P}}_d$, and $\bar{\mathcal{P}}_f = P$ of Table 2-4, we have

$$\alpha_c = \frac{\mathcal{R} Z_0}{ab}\left[\frac{b}{Z_0^2} + \left(\frac{f_c}{\eta f}\right)^2 (2a + b)\right]$$
$$= \frac{\mathcal{R}}{a\eta\sqrt{1 - (f_c/f)^2}}\left[1 + \frac{2a}{b}\left(\frac{f_c}{f}\right)^2\right] \qquad (2\text{-}93)$$

This is the attenuation constant due to conductor losses. When both

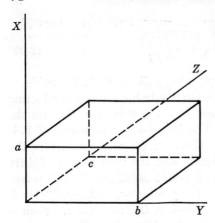

Fig. 2-19. The rectangular cavity.

dielectric losses and conductor losses need to be considered, the total attenuation constant is

$$\alpha = \alpha_d + \alpha_c \quad (2\text{-}94)$$

for by Eq. (2-76) we merely add the two losses.

2-8. Resonator Concepts. In Sec. 2-2 we noted a similarity between standing waves and circuit theory resonance. In the loss-free case, electromagnetic fields can exist within a source-free region enclosed by a perfect conductor. These fields can exist only at specific frequencies, called *resonant frequencies*. When losses are present, a source must exist to sustain oscillations. The input impedance seen by the source behaves, in the vicinity of a resonant frequency, like the impedance of an LC circuit. Resonators can therefore be used for the same purposes at high frequencies as LC resonators are used at lower frequencies.

To illustrate resonator concepts, consider the "rectangular cavity" of Fig. 2-19. This consists of a conductor enclosing a dielectric, both of which we will assume to be perfect at present. We desire to find solutions to the field equations having zero tangential components of **E** over the entire boundary. The TE_{01} waveguide mode already satisfies this condition over four of the walls. We recall that standing waves have planes of zero field, which suggests trying the standing-wave TE_{01} field. For E_x to be zero at $z = 0$, we choose

$$E_x = E_x{}^+ + E_x{}^- = A \sin \frac{\pi y}{b} (e^{-j\beta z} - e^{j\beta z})$$
$$= E_0 \sin \left(\frac{\pi y}{b}\right) \sin \beta z$$

For E_x to be zero at $z = c$, we choose $\beta c = \pi$, which, according to Table 2-4, is

$$\pi = ck \sqrt{1 - \left(\frac{f_c}{f}\right)^2} = c 2\pi f \sqrt{\epsilon\mu} \sqrt{1 - \frac{1}{(2b \sqrt{\epsilon\mu}\, f)^2}}$$

Solving for the resonant frequency $f = f_r$, we have

$$f_r = \frac{1}{2bc} \sqrt{\frac{b^2 + c^2}{\epsilon\mu}} \quad (2\text{-}95)$$

When a is the smallest cavity dimension, this is the resonant frequency of the dominant mode, called the TE_{011} mode. The additional subscript 1 indicates that we have chosen the first zero of $\sin \beta z$. The higher zeros give higher-order modes, that is, modes with higher resonant frequencies. Setting $\beta = \pi/c$ in the above expression for E_x and determining $\mathbf{H}$ from the Maxwell equations, we have for the TE_{011} mode

$$E_x = E_0 \sin \frac{\pi y}{b} \sin \frac{\pi z}{c}$$
$$H_y = \frac{jbE_0}{\eta \sqrt{b^2 + c^2}} \sin \frac{\pi y}{b} \cos \frac{\pi z}{c} \qquad (2\text{-}96)$$
$$H_z = -\frac{jcE_0}{\eta \sqrt{b^2 + c^2}} \cos \frac{\pi y}{b} \sin \frac{\pi z}{c}$$

Note that E and H are 90° out of phase; so $\mathcal{E}$ is maximum when $\mathcal{H}$ is minimum and vice versa. A sketch of the instantaneous field lines at some time when both $\mathcal{E}$ and $\mathcal{H}$ exist is given in Fig. 2-20. Also of interest is the energy stored within the cavity. From the conservation of complex power, Eq. (1-68), we know that $\overline{\mathcal{W}}_m = \overline{\mathcal{W}}_e$. Thus, the time-average electric and magnetic energies are

$$\overline{\mathcal{W}}_m = \overline{\mathcal{W}}_e = \frac{\epsilon}{2} \iiint_{\text{cavity}} |E|^2 \, d\tau = \frac{\epsilon}{8} |E_0|^2 abc \qquad (2\text{-}97)$$

We also know from conservation of energy, Eq. (1-39), that the total energy within the resonator is independent of time. If we choose a time for which $\mathcal{H}$ is zero, $\mathcal{W}_m$ will be zero, and $\mathcal{W}_e$ will be maximum and twice its average value. Therefore,

$$\mathcal{W} = 2\overline{\mathcal{W}}_e = \frac{\epsilon}{4} |E_0|^2 abc \qquad (2\text{-}98)$$

is the total energy stored within the cavity.

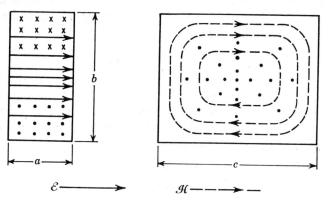

FIG. 2-20. Mode pattern for the TE_{011} cavity mode.

When the resonator has losses, we define its quality factor as

$$Q = \frac{\omega \times \text{energy stored}}{\text{average power dissipated}} = \frac{\omega \mathcal{W}}{\bar{\mathcal{P}}_d} \qquad (2\text{-}99)$$

by analogy to the Q of an LC circuit. If the losses are dielectric losses, we have

$$Q_d = \frac{\omega \epsilon' \iiint |E|^2 \, d\tau}{\omega \epsilon'' \iiint |E|^2 \, d\tau} = \frac{\epsilon'}{\epsilon''} \qquad (2\text{-}100)$$

so the Q of the resonator is that of the dielectric, Eq. (1-79). This is valid for any mode in a cavity of arbitrary shape. Usually more important in determining the Q is the loss due to imperfect conductors. This is determined to the same approximation as we used for waveguide attenuation. We assume H at walls to be that of the loss-free mode and calculate $\bar{\mathcal{P}}_d$ by Eq. (2-42). To summarize,

$$\bar{\mathcal{P}}_d = \mathcal{R} \oint_{\substack{\text{cavity} \\ \text{walls}}} |H|^2 \, ds = \frac{\mathcal{R}|E_0|^2}{2\eta^2(b^2+c^2)} [bc(b^2+c^2) + 2a(b^3+c^3)]$$

Substituting this, Eq. (2-98), and Eq. (2-95) into Eq. (2-99), we have

$$Q_c = \frac{\pi \eta}{2\mathcal{R}} \frac{a(b^2+c^2)^{3/2}}{bc(b^2+c^2) + 2a(b^3+c^3)} \qquad (2\text{-}101)$$

From the symmetry of Q_c in b and c, it is evident that $b = c$ for maximum Q. For a "square-base" cavity ($b = c$), we have

$$f_r = \frac{1}{b\sqrt{2\epsilon\mu}} \qquad Q_c = \frac{1.11\eta}{\mathcal{R}(1 + b/2a)} \qquad (2\text{-}102)$$

The Q also increases as a increases, but if $a > b$ we no longer have the dominant mode. As an example of the Q's obtainable, consider a cubic cavity constructed of copper. In this case we have

$$Q_c = 1.07 \times 10^9/\sqrt{f} \qquad (2\text{-}103)$$

which, at microwave frequencies, gives Q's of several thousand. This idealized Q will, however, be lowered in practice by the introduction of a feed system, by imperfections in the construction, and by corrosion of the metal. When both conductor losses and dielectric losses are considered, the Q of the cavity becomes

$$\frac{1}{Q} = \frac{1}{Q_d} + \frac{1}{Q_c} \qquad (2\text{-}104)$$

which is evident from Eq. (2-99).

2-9. Radiation.

We shall now show that a source in unbounded space is characterized by a radiation of energy. Consider the field equations

$$\nabla \times \mathbf{E} = -j\omega\mu\mathbf{H} \qquad \nabla \times \mathbf{H} = j\omega\epsilon\mathbf{E} + \mathbf{J} \qquad (2\text{-}105)$$

where $\mathbf{J}$ is the source, or impressed, current. These equations apply explicitly to a perfect dielectric, but the extension to lossy media is effected by replacing $j\omega\mu$ by $\hat{z}$ and $j\omega\epsilon$ by $\hat{y}$. In homogeneous media, the divergence of the first equation is

$$\nabla \cdot \mathbf{H} = 0$$

Any divergenceless vector is the curl of some other vector; so

$$\mathbf{H} = \nabla \times \mathbf{A} \qquad (2\text{-}106)$$

where $\mathbf{A}$ is called a *magnetic vector potential*.[1] Substituting Eq. (2-106) into the first of Eqs. (2-105), we have

$$\nabla \times (\mathbf{E} + j\omega\mu\mathbf{A}) = 0$$

Any curl-free vector is the gradient of some scalar. Hence,

$$\mathbf{E} + j\omega\mu\mathbf{A} = -\nabla\Phi \qquad (2\text{-}107)$$

where Φ is an *electric scalar potential*. To obtain the equation for $\mathbf{A}$, substitute Eqs. (2-106) and (2-107) into the second of Eqs. (2-105). This gives

$$\nabla \times \nabla \times \mathbf{A} - k^2\mathbf{A} = \mathbf{J} - j\omega\epsilon\nabla\Phi \qquad (2\text{-}108)$$

which, by a vector identity, becomes

$$\nabla(\nabla \cdot \mathbf{A}) - \nabla^2\mathbf{A} - k^2\mathbf{A} = \mathbf{J} - j\omega\epsilon\nabla\Phi$$

Only $\nabla \times \mathbf{A}$ was specified by Eq. (2-106). We are still free to choose $\nabla \cdot \mathbf{A}$. If we let

$$\nabla \cdot \mathbf{A} = -j\omega\epsilon\Phi \qquad (2\text{-}109)$$

the equation for $\mathbf{A}$ simplifies to

$$\nabla^2\mathbf{A} + k^2\mathbf{A} = -\mathbf{J} \qquad (2\text{-}110)$$

This is the Helmholtz equation, or complex wave equation. Solutions to Eq. (2-110) are called *wave potentials*. In terms of the magnetic wave potential, we have

$$\mathbf{E} = -j\omega\mu\mathbf{A} + \frac{1}{j\omega\epsilon}\nabla(\nabla \cdot \mathbf{A}) \qquad (2\text{-}111)$$
$$\mathbf{H} = \nabla \times \mathbf{A}$$

---

[1] In general electromagnetic theory it is more common to let $\mathbf{A}$ be the vector potential of $\mathbf{B}$. In homogeneous media the two potentials are in the ratio μ, a constant.

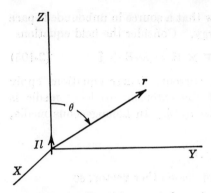

Fig. 2-21. A z-directed current element at the coordinate origin.

obtained from Eqs. (2-106), (2-107), and (2-109). The principal advantages of using $\mathbf{A}$ instead of $\mathbf{E}$ or $\mathbf{H}$ are (1) rectangular components of $\mathbf{A}$ have corresponding rectangular components of $\mathbf{J}$ as their sources and (2) $\mathbf{A}$ need not be divergenceless.

Let us first determine $\mathbf{A}$ for a current I extending over an incremental length l, forming a *current element* or *electric dipole* of moment Il. Take this current element to be z-directed and situated at the coordinate origin, as shown in Fig. 2-21. The current is z-directed; so we take $\mathbf{A}$ to have only a z component, satisfying

$$\nabla^2 A_z + k^2 A_z = 0$$

everywhere except at the origin. The scalar quantity A_z has a point source Il and should therefore be spherically symmetric. Thus, let $A_z = A_z(r)$, and the above equation reduces to

$$\frac{1}{r^2}\frac{d}{dr}\left(r^2 \frac{dA_z}{dr}\right) + k^2 A_z = 0$$

This has the two independent solutions

$$\frac{1}{r} e^{-jkr} \qquad \frac{1}{r} e^{jkr}$$

the first of which represents an outward-traveling wave, and the second an inward-traveling wave. (In dissipative media, $k = k' - jk''$, and the first solution vanishes as $r \to \infty$, and the second solution becomes infinite.) We therefore choose the first solution, and take

$$A_z = \frac{C}{r} e^{-jkr}$$

where C is a constant.[1] As $k \to 0$, Eq. (2-110) reduces to Poisson's equation, for which the solution is

$$A_z = \frac{Il}{4\pi r}$$

---

[1] To be precise, C might be a function of k, but the solution must also reduce to the static field as $r \to 0$. Hence, C is not a function of k.

INTRODUCTION TO WAVES

Our constant C must therefore be

$$C = \frac{Il}{4\pi}$$

and hence

$$A_z = \frac{Il}{4\pi r} e^{-jkr} \tag{2-112}$$

is the desired solution for the current element of Fig. 2-21. The outward-traveling wave represented by Eq. (2-112) is called a *spherical wave*, since surfaces of constant phase are spheres.

The electromagnetic field of the current element is obtained by substituting Eq. (2-112) into Eqs. (2-111). The result is

$$\begin{aligned} E_r &= \frac{Il}{2\pi} e^{-jkr} \left(\frac{\eta}{r^2} + \frac{1}{j\omega\epsilon r^3} \right) \cos\theta \\ E_\theta &= \frac{Il}{4\pi} e^{-jkr} \left(\frac{j\omega\mu}{r} + \frac{\eta}{r^2} + \frac{1}{j\omega\epsilon r^3} \right) \sin\theta \\ H_\phi &= \frac{Il}{4\pi} e^{-jkr} \left(\frac{jk}{r} + \frac{1}{r^2} \right) \sin\theta \end{aligned} \tag{2-113}$$

Very close to the current element, the **E** reduces to that of a static charge dipole, the **H** reduces to that of a constant current element, and the field is said to be *quasi-static*. Far from the current element, Eqs. (2-113) reduce to

$$\left. \begin{aligned} E_\theta &= \eta \frac{jIl}{2\lambda r} e^{-jkr} \sin\theta \\ H_\phi &= \frac{jIl}{2\lambda r} e^{-jkr} \sin\theta \end{aligned} \right\} \quad r \gg \lambda \tag{2-114}$$

which is called the *radiation field*. At intermediate values of r the field is called the *induction field*. The outward-directed complex power over a sphere of radius r is

$$\begin{aligned} P_f &= \oint \mathbf{E} \times \mathbf{H}^* \cdot d\mathbf{s} = \int_0^{2\pi} d\phi \int_0^{\pi} d\theta\, r^2 \sin\theta\, E_\theta H_\phi^* \\ &= \eta \frac{2\pi}{3} \left| \frac{Il}{\lambda} \right|^2 \left[1 - \frac{j}{(kr)^3} \right] \end{aligned} \tag{2-115}$$

The time-average power radiated is the real part of P_f, or

$$\bar{\mathcal{P}}_f = \eta \frac{2\pi}{3} \left| \frac{Il}{\lambda} \right|^2 \tag{2-116}$$

This is independent of r and can be most simply obtained from the radiation field, Eq. (2-114). The reactive power, which is negative, indicates that there is an excess of electric energy over magnetic energy in the near field.

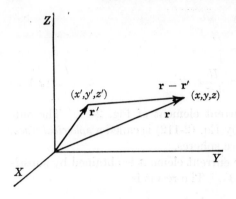

Fig. 2-22. Radius vector notation.

To obtain the field of an arbitrary distribution of electric currents, we need only superimpose the solutions for each element, for the equations are linear. A superposition of vector potentials is usually the most convenient one. For this purpose, we shall use the radius vector notation illustrated by Fig. 2-22. The "field coordinates" are specified by

$$\mathbf{r} = \mathbf{u}_x x + \mathbf{u}_y y + \mathbf{u}_z z$$

and the "source coordinates" by

$$\mathbf{r}' = \mathbf{u}_x x' + \mathbf{u}_y y' + \mathbf{u}_z z'$$

In Eq. (2-112), r is the distance from the source to the field point. For Il not at the coordinate origin, r should be replaced by

$$|\mathbf{r} - \mathbf{r}'| = \sqrt{(x - x')^2 + (y - y')^2 + (z - z')^2}$$

Note the direction of the vector potential is that of the current; so Eq. (2-112) can be generalized to a current element of arbitrary orientation by replacing Il by $I\mathbf{l}$ and A_z by $\mathbf{A}$. Thus, the vector potential from current element of arbitrary location and orientation is

$$\mathbf{A} = \frac{I\mathbf{l}\, e^{-jk|\mathbf{r}-\mathbf{r}'|}}{4\pi|\mathbf{r} - \mathbf{r}'|}$$

To emphasize that $\mathbf{A}$ is evaluated at the field point (x,y,z) and $I\mathbf{l}$ is situated at the source point (x',y',z'), we shall use the notation $\mathbf{A}(\mathbf{r})$ and $I\mathbf{l}(\mathbf{r}')$. The above equation then becomes

$$\mathbf{A}(\mathbf{r}) = \frac{I\mathbf{l}(\mathbf{r}')e^{-jk|\mathbf{r}-\mathbf{r}'|}}{4\pi|\mathbf{r} - \mathbf{r}'|} \tag{2-117}$$

Finally, for a current distribution $\mathbf{J}$, the current element contained in a volume element $d\tau$ is $\mathbf{J}\, d\tau$, and a superposition over all such elements is

$$\mathbf{A}(\mathbf{r}) = \frac{1}{4\pi} \iiint \frac{\mathbf{J}(\mathbf{r}')e^{-jk|\mathbf{r}-\mathbf{r}'|}}{|\mathbf{r} - \mathbf{r}'|}\, d\tau' \tag{2-118}$$

INTRODUCTION TO WAVES

The prime on $d\tau'$ emphasizes that the integration is over the source coordinates. Equation (2-118) is called the *magnetic vector potential integral*. It is intended to include the cases of surface currents and filamentary currents by implication. We therefore have a formal solution for any problem characterized by electric currents in an unbounded homogeneous medium. The medium may be dissipative if k is considered to be complex.

2-10. Antenna Concepts. A device whose primary purpose is to radiate or receive electromagnetic energy is called an antenna. To illustrate antenna concepts, we shall consider the *linear antenna* of Fig. 2-23. It consists of a straight wire carrying a current $I(z)$. When it is energized at the center, it is called a *dipole antenna*. The magnetic vector potential, Eq. (2-118), for this particular problem is

$$A_z = \frac{1}{4\pi} \int_{-L/2}^{L/2} \frac{I(z')e^{-jk|\mathbf{r}-\mathbf{r}'|}}{|\mathbf{r}-\mathbf{r}'|} dz' \qquad (2\text{-}119)$$

where
$$|\mathbf{r}-\mathbf{r}'| = \sqrt{r^2 + z'^2 - 2rz'\cos\theta} \qquad (2\text{-}120)$$

The radiation field (r large) is of primary interest, in which case

$$|\mathbf{r}-\mathbf{r}'| \approx r - z'\cos\theta \qquad r \gg z' \qquad (2\text{-}121)$$

and
$$A_z \approx \frac{e^{-jkr}}{4\pi r} \int_{-L/2}^{L/2} I(z')e^{jkz'\cos\theta}\,dz' \qquad r \gg L \qquad (2\text{-}122)$$

Note that the second term of Eq. (2-121) must be retained in the "phase term" $e^{-jk|\mathbf{r}-\mathbf{r}'|}$, but not in the "amplitude term" $|\mathbf{r}-\mathbf{r}'|^{-1}$. To obtain the field components, substitute Eq. (2-122) into Eqs. (2-111) and retain only the $1/r$ terms. This gives

$$\left.\begin{aligned} E_\theta &= j\omega\mu \sin\theta\, A_z \\ H_\phi &= \frac{1}{\eta} E_\theta \end{aligned}\right\} \quad r \text{ large} \qquad (2\text{-}123)$$

This result is equivalent to superimposing Eqs. (2-114) for all elements of current.

To evaluate the radiation field, we must know the current on the antenna. An exact determination of the current requires the solution to a boundary-value problem. Fortunately, the radiation field is relatively insensitive to minor changes in current distribution, and much use-

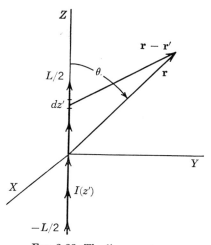

FIG. 2-23. The linear antenna.

ful information can be obtained from an approximate current distribution. We have already seen that on transmission lines the current is a harmonic function of kz. This is also true for the principal mode on a single thin wire. The current on the dipole antenna must be zero at the ends of the wire, symmetrical in z, and continuous at the source ($z = 0$). Thus, we choose

$$I(z) = I_m \sin\left[k\left(\frac{L}{2} - |z|\right)\right] \quad (2\text{-}124)$$

The vector potential in the radiation zone can now be evaluated as

$$A_z = \frac{I_m e^{-jkr}}{4\pi r} \int_{-L/2}^{L/2} \sin\left[k\left(\frac{L}{2} - |z'|\right)\right] e^{jkz'\cos\theta} \, dz'$$

$$= \frac{I_m e^{-jkr}}{4\pi r} \frac{2\left[\cos\left(k\frac{L}{2}\cos\theta\right) - \cos\left(k\frac{L}{2}\right)\right]}{k\sin^2\theta}$$

From Eq. (2-123), the radiation field is

$$E_\theta = \frac{j\eta I_m e^{-jkr}}{2\pi r} \left[\frac{\cos\left(k\frac{L}{2}\cos\theta\right) - \cos\left(k\frac{L}{2}\right)}{\sin\theta}\right] \quad (2\text{-}125)$$

with $H_\phi = E_\theta/\eta$. Note that the radiation field is linearly polarized, for there is only an E_θ. The density of power radiated is the r component of the Poynting vector

$$S_r = E_\theta H_\phi^* = \frac{\eta |I_m|^2}{(2\pi r)^2} \left[\frac{\cos\left(k\frac{L}{2}\cos\theta\right) - \cos\left(k\frac{L}{2}\right)}{\sin\theta}\right]^2 \quad (2\text{-}126)$$

The total power radiated is obtained by integrating S_r over a large sphere, or

$$\bar{\mathcal{P}}_f = \int_0^{2\pi} \int_0^\pi S_r r^2 \sin\theta \, d\theta \, d\phi$$

$$= \frac{\eta |I_m|^2}{2\pi} \int_0^\pi \frac{\left[\cos\left(k\frac{L}{2}\cos\theta\right) - \cos\left(k\frac{L}{2}\right)\right]^2}{\sin\theta} \, d\theta \quad (2\text{-}127)$$

The radiation resistance R_r of an antenna is defined as

$$R_r = \frac{\bar{\mathcal{P}}_f}{|I|^2} \quad (2\text{-}128)$$

where I is some arbitrary reference current. For the dipole antenna, the reference current is usually picked as I_m. Hence,

$$R_r = \frac{\eta}{2\pi} \int_0^\pi \frac{\left[\cos\left(k\frac{L}{2}\cos\theta\right) - \cos\left(k\frac{L}{2}\right)\right]^2}{\sin\theta} \, d\theta \quad (2\text{-}129)$$

Fig. 2-24. Radiation resistance of the dipole antenna.

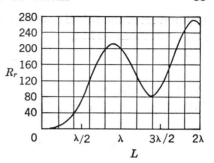

This integral can be evaluated in terms of tabulated functions (see Prob. 2-44). A graph of R_r versus L is given in Fig. 2-24.

The *radiation field pattern* of an antenna is a plot of $|E|$ at constant r in the radiation zone. For a dipole antenna, the radiation field pattern is essentially the bracketed term of Eq. (2-125). This is shown in Fig. 2-25 for kL small (short dipole), $kL = \pi$ (half-wavelength dipole), and $kL = 2\pi$ (full-wavelength dipole). The *radiation power pattern*, defined as a plot of $|S_r|$ at constant r, is an alternative method of showing radiation characteristics. When the radiation field is linearly polarized, as it is for the dipole antenna, the power pattern is the square of the field pattern. The *gain g* of an antenna in a given direction is defined as the ratio of the power required from an omnidirectional antenna to the power

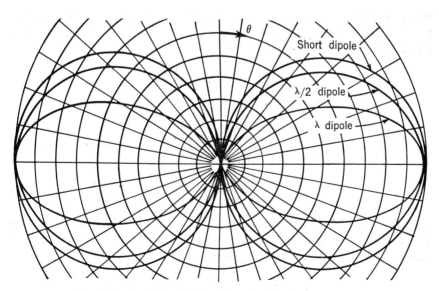

Fig. 2-25. Radiation field patterns for the dipole antenna.

required from the actual antenna, assuming equal power densities in the given direction. Thus,

$$g(\theta) = \frac{4\pi r^2 S_r(\theta)}{\bar{\mathcal{P}}_f} \tag{2-130}$$

For $L \leq \lambda$, the maximum gain of a dipole antenna occurs at $\theta = \pi/2$. From Eqs. (2-126) and (2-128), we have

$$g\left(\frac{\pi}{2}\right) = \frac{\eta |I_m|^2 \left(1 - \cos\frac{kL}{2}\right)^2}{\pi \bar{\mathcal{P}}_f} = \frac{\eta \left(1 - \cos\frac{kL}{2}\right)^2}{\pi R_r} \tag{2-131}$$

In the limit $kL \to 0$, we have $g(\pi/2) = 1.5$; so the maximum gain of a short dipole is 1.5. For a half-wave dipole, we can use Fig. 2.24 and calculate a maximum gain of 1.64. Similarly, for a full-wave dipole, the maximum gain is 2.41.

The *input impedance* of an antenna is the impedance seen by the source, that is, the ratio of the complex terminal voltage to the complex terminal current. A knowledge of the reactive power, which cannot be obtained from radiation zone fields, is needed to evaluate the input reactance. The input resistance accounts for the radiated power (and dissipated power if losses are present). We define the input resistance of a loss-free antenna as

$$R_i = \frac{\bar{\mathcal{P}}_f}{|I_i|^2} \tag{2-132}$$

where $\bar{\mathcal{P}}_f$ is the power radiated and I_i is the input current. If losses are present, a "loss resistance" must be added to Eq. (2-132) to obtain the input resistance. For the dipole antenna,

$$I_i = I_m \sin \frac{kL}{2}$$

and the input resistance is

$$R_i = \frac{R_r}{\sin^2 [k(L/2)]} \tag{2-133}$$

In the limit as kL is made small, we find

$$R_i = \frac{\eta (kL)^2}{24\pi} \qquad L \ll \lambda \tag{2-134}$$

The short dipole therefore has a very small input resistance. For example, if $L = \lambda/10$, the input resistance is about 2 ohms. For the half-wavelength dipole, we use Fig. 2-24 and Eq. (2-133) and find

$$R_i = R_r = 73.1 \text{ ohms} \qquad L = \frac{\lambda}{2} \tag{2-135}$$

For the full-wavelength dipole, Eq. (2-133) shows $R_i = \infty$. This incorrect result is due to our initial choice of current, which has a null at the source. The input resistance of the full-wavelength dipole is actually large, but not infinite, and depends markedly on the wire diameter (see Fig. 7-13).

2-11. On Waves in General. A complex function of coordinates representing an instantaneous function according to Eq. (1-40) is called a *wave function*. A wave function ψ, which may be either a scalar field or the component of a vector field, may be expressed as

$$\psi = A(x,y,z)e^{j\Phi(x,y,z)} \tag{2-136}$$

where A and Φ are real. The corresponding instantaneous function is

$$\sqrt{2}\, A(x,y,z) \cos[\omega t + \Phi(x,y,z)] \tag{2-137}$$

The *magnitude* A of the complex function is the rms amplitude of the instantaneous function. The *phase* Φ of the complex function is the initial phase of the instantaneous function. Surfaces over which the phase is constant (instantaneous function vibrates in phase) are called *equiphase surfaces*. These are defined by

$$\Phi(x,y,z) = \text{constant} \tag{2-138}$$

Waves are called *plane, cylindrical,* or *spherical* according as their equiphase surfaces are planes, cylinders, or spheres. Waves are called *uniform* when the amplitude A is constant over the equiphase surfaces. Perpendiculars to the equiphase surfaces are called *wave normals*. These are, of course, in the direction of $\nabla\Phi$ and are the curves along which the phase changes most rapidly.

The rate at which the phase decreases in some direction is called the *phase constant* in that direction. (The term phase constant is used even though it is not, in general, a constant.) For example, the phase constants in the cartesian coordinate directions are

$$\beta_x = -\frac{\partial \Phi}{\partial x} \qquad \beta_y = -\frac{\partial \Phi}{\partial y} \qquad \beta_z = -\frac{\partial \Phi}{\partial z} \tag{2-139}$$

These may be considered as components of a *vector phase constant* defined by

$$\boldsymbol{\beta} = -\nabla\Phi \tag{2-140}$$

The maximum phase constant is therefore along the wave normal and is of magnitude $|\nabla\Phi|$.

The *instantaneous phase* of a wave is the argument of the cosine function of Eq. (2-137). A *surface of constant phase* is defined as

$$\omega t + \Phi(x,y,z) = \text{constant} \tag{2-141}$$

that is, the instantaneous phase is constant. At any instant, the surfaces of constant phase coincide with the equiphase surfaces. As time increases, Φ must decrease to maintain the constancy of Eq. (2-141), and the surfaces of constant phase move in space. For any increment ds the change in Φ is

$$\nabla\Phi \cdot d\mathbf{s} = \frac{\partial \Phi}{\partial x} dx + \frac{\partial \Phi}{\partial y} dy + \frac{\partial \Phi}{\partial z} dz$$

To keep the instantaneous phase constant for an incremental increase in time, we must have

$$\omega\, dt + \nabla\Phi \cdot d\mathbf{s} = 0$$

That is, the total differential of Eq. (2-141) must vanish. The *phase velocity* of a wave in a given direction is defined as the velocity of surfaces of constant phase in that direction. For example, the phase velocities along cartesian coordinates are

$$\begin{aligned}
v_x &= -\frac{\omega}{\partial \Phi/\partial x} = \frac{\omega}{\beta_x} \\
v_y &= -\frac{\omega}{\partial \Phi/\partial y} = \frac{\omega}{\beta_y} \\
v_z &= -\frac{\omega}{\partial \Phi/\partial z} = \frac{\omega}{\beta_z}
\end{aligned} \qquad (2\text{-}142)$$

The phase velocity along a wave normal ($d\mathbf{s}$ in the direction of $-\nabla\Phi$) is

$$v_p = -\frac{\omega}{|\nabla\Phi|} = \frac{\omega}{\beta} \qquad (2\text{-}143)$$

which is the *smallest* phase velocity for the wave. Phase velocity is *not* a vector quantity.

We can also express the wave function, Eq. (2-136), as

$$\psi = e^{\Theta(x,y,z)} \qquad (2\text{-}144)$$

where Θ is a complex function whose imaginary part is the phase Φ. A *vector propagation constant* can be defined in terms of the rate of change of Θ as

$$\boldsymbol{\gamma} = -\nabla\Theta = \boldsymbol{\alpha} + j\boldsymbol{\beta} \qquad (2\text{-}145)$$

where $\boldsymbol{\beta}$ is the phase constant of Eq. (2-140) and $\boldsymbol{\alpha}$ is the *vector attenuation constant*. The components of $\boldsymbol{\alpha}$ are the logarithmic rates of change of the magnitude of ψ in the various directions.

In the electromagnetic field, ratios of components of **E** to components of **H** are called *wave impedances*. The direction of a wave impedance is defined according to the right-hand "cross-product" rule of component **E**

INTRODUCTION TO WAVES 87

rotated into component **H**. For example,

$$\frac{E_x}{H_y} = Z_{xy}{}^+ = Z_z \tag{2-146}$$

is a wave impedance in the $+z$ direction, while

$$-\frac{E_x}{H_y} = Z_{xy}{}^- = Z_{-z} \tag{2-147}$$

is a wave impedance in the $-z$ direction. The wave impedance in the $+z$ direction involving E_y and H_x is

$$\frac{-E_y}{H_x} = Z_{yx}{}^+ = -Z_{yx}{}^- \tag{2-148}$$

The Poynting vector can be expressed in terms of wave impedances. For example, the z component is

$$S_z = (\mathbf{E} \times \mathbf{H}^*)_z = E_x H_y^* - E_y H_x^*$$
$$= Z_{xy}{}^+ |H_y|^2 + Z_{yx}{}^+ |H_x|^2 \tag{2-149}$$

The concept of wave impedance is most useful when the wave impedances are constant over equiphase surfaces.

Let us illustrate the various concepts by specializing them to the uniform plane wave. Consider the x-polarized z-traveling wave in lossy matter,

$$E_x = E_0 e^{-k''z} e^{-jk'z}$$
$$H_y = \frac{E_0}{\eta} e^{-k''z} e^{-jk'z}$$

The amplitude of E_x is $E_0 e^{-k''z}$ and its phase is $-k'z$. Equiphase surfaces are defined by $-k'z = $ constant, or, since k' is constant, by $z = $ constant. These are planes; so the wave is a plane wave. The amplitude of E_x is constant over each equiphase surface; so the wave is uniform. The wave normals all point in the z direction. The cartesian components of the phase constant are $\beta_x = \beta_y = 0$, $\beta_z = k'$; so the vector phase constant is $\boldsymbol{\beta} = \mathbf{u}_z k'$. The phase velocity in the direction of the wave normals is $v_p = \omega/k'$. The cartesian components of the attenuation constant are $\alpha_x = \alpha_y = 0$, $\alpha_z = k''$; so the vector attenuation constant is $\boldsymbol{\alpha} = \mathbf{u}_z k''$. The vector propagation constant is

$$\boldsymbol{\gamma} = \boldsymbol{\alpha} + j\boldsymbol{\beta} = \mathbf{u}_z(k'' + jk') = \mathbf{u}_z jk$$

The wave impedance in the z direction is $Z_z = Z_{xy}{}^+ = E_x/H_y = \eta$. Note that the various parameters specialized to the uniform plane traveling wave are all intrinsic parameters. This is, by definition, the meaning of the word "intrinsic."

PROBLEMS

2-1. Show that $E_z = E_0 e^{-jkz}$ satisfies Eq. (2-6) but not Eq. (2-5). Show that it does not satisfy Eq. (2-3). This is *not* a possible electromagnetic field.

2-2. Derive the "wave equations" for inhomogeneous media

$$\nabla \times (\hat{z}^{-1} \nabla \times \mathbf{E}) + \hat{y}\mathbf{E} = 0$$
$$\nabla \times (\hat{y}^{-1} \nabla \times \mathbf{H}) + \hat{z}\mathbf{H} = 0$$

Are these valid for nonisotropic media? Do Eqs. (2-5) hold for inhomogeneous media?

2-3. Show that for any lossless nonmagnetic dielectric

$$k = k_0 \sqrt{\epsilon_r} \qquad \eta = \frac{\eta_0}{\sqrt{\epsilon_r}}$$

$$\lambda = \frac{\lambda_0}{\sqrt{\epsilon_r}} \qquad v_p = \frac{c}{\sqrt{\epsilon_r}}$$

where ϵ_r is the dielectric constant and k_0, η_0, λ_0, and c are the intrinsic parameters of vacuum.

2-4. Show that the quantities of Eqs. (2-18) satisfy Eq. (1-35). Repeat for Eqs. (2-21), (2-27), and (2-29).

2-5. For the field of Eqs. (2-20), show that the velocity of propagation of energy as defined by Eq. (2-19) is

$$v_e = \frac{1}{\sqrt{\epsilon\mu}} \frac{\sin 2kz \sin 2\omega t}{1 - \cos 2kz \cos 2\omega t} \leq \frac{1}{\sqrt{\epsilon\mu}}$$

2-6. For the field of Eqs. (2-22), show that the phase velocity is

$$v_p = \frac{1}{\sqrt{\epsilon\mu}} \left(\frac{A+C}{A-C} \cos^2 kz + \frac{A-C}{A+C} \sin^2 kz \right)$$

2-7. For the field of Eqs. (2-28), show that the z-directed wave impedances are

$$Z_{xy}^+ = \frac{E_x}{H_y} = -j\eta \tan kz$$

$$Z_{yx}^+ = \frac{-E_y}{H_x} = -j\eta \tan kz$$

Would you expect $Z_{xy}^+ = Z_{yx}^+$ to be true for all a-c fields?

2-8. Given a uniform plane wave traveling in the $+z$ direction, show that the wave is circularly polarized if

$$\frac{E_x}{E_y} = \pm j$$

being right-handed if the ratio is $+j$ and left-handed if the ratio is $-j$.

2-9. Show that the uniform plane traveling wave of Eq. (2-25) can be expressed as the sum of a right-hand circularly polarized wave and a left-hand circularly polarized wave.

2-10. Show that the uniform plane traveling wave of Eq. (2-25) can be expressed as

$$\mathbf{E} = (\mathbf{E}_1 + j\mathbf{E}_2)e^{-jkz}$$

where $\mathbf{E}_1$ and $\mathbf{E}_2$ are real vectors lying in the xy plane. Relate $\mathbf{E}_1$ and $\mathbf{E}_2$ to A and B.

2-11. Show that the tip of the arrow representing $\mathbf{\mathcal{E}}$ for an arbitrary complex $\mathbf{E}$ traces out an ellipse in space. [*Hint:* let $\mathbf{E} = \text{Re}(\mathbf{E}) + j\,\text{Im}(\mathbf{E})$ and use the results of Prob. 2-10.]

2-12. For the frequencies 10, 100, and 1000 megacycles, determine $k = k' - jk''$ and $\eta = \mathcal{R} + j\mathcal{X}$ for (*a*) polystyrene, Fig. 1-10, (*b*) Plexiglas, Fig. 1-11, (*c*) Ferramic A, Fig. 1-12, $\epsilon_r = 10$, and (*d*) copper, $\sigma = 5.8 \times 10^7$.

2-13. Show that when all losses are of the magnetic type ($\sigma = \epsilon'' = 0$),

$$\eta = \frac{k}{|y|} = \frac{k'}{\omega\epsilon} - j\frac{k''}{\omega\epsilon}$$

2-14. Show that for nonmagnetic dielectrics

$$\left.\begin{aligned}
k' &\approx \omega\sqrt{\mu\epsilon'}\left(1 + \frac{1}{8Q^2}\right) \\
k'' &\approx \frac{\omega\epsilon''}{2}\sqrt{\frac{\mu}{\epsilon'}}\left(1 - \frac{1}{8Q^2}\right) \\
\mathcal{R} &\approx \sqrt{\frac{\mu}{\epsilon'}}\left(1 - \frac{3}{8Q^2}\right) \\
\mathcal{X} &\approx \frac{\epsilon''}{2\epsilon'}\sqrt{\frac{\mu}{\epsilon'}}\left(1 - \frac{5}{8Q^2}\right)
\end{aligned}\right\} Q \gg 1$$

where Q is defined by Eq. (1-79).

2-15. Show that for nonmagnetic conductors

$$\left.\begin{aligned}
k' &\approx \sqrt{\frac{\omega\mu\sigma}{2}}\left(1 + \frac{Q}{2}\right) \\
k'' &\approx \sqrt{\frac{\omega\mu\sigma}{2}}\left(1 - \frac{Q}{2}\right) \\
\mathcal{R} &\approx \sqrt{\frac{\omega\mu}{2\sigma}}\left(1 + \frac{Q}{2}\right) \\
\mathcal{X} &\approx \sqrt{\frac{\omega\mu}{2\sigma}}\left(1 - \frac{Q}{2}\right)
\end{aligned}\right\} Q \ll 1$$

where Q is defined by Eq. (1-79).

2-16. Show that for metals

$$\eta = \mathcal{R}(1 + j) \qquad k = \frac{1}{\delta}(1 - j) \qquad \mathcal{R} = \frac{1}{\sigma\delta}$$

where $\mathcal{R}$ is the surface resistance, δ is the skin depth, and σ is the conductivity.

2-17. Derive the following formulas

$$\begin{aligned}
\mathcal{R}\text{ (silver)} &= 2.52 \times 10^{-7}\sqrt{f} \\
\mathcal{R}\text{ (copper)} &= 2.61 \times 10^{-7}\sqrt{f} \\
\mathcal{R}\text{ (gold)} &= 3.12 \times 10^{-7}\sqrt{f} \\
\mathcal{R}\text{ (aluminum)} &= 3.26 \times 10^{-7}\sqrt{f} \\
\mathcal{R}\text{ (brass)} &= 5.01 \times 10^{-7}\sqrt{f}
\end{aligned}$$

where f is the frequency in cycles per second.

2-18. Find the power per square meter dissipated in a copper sheet if the rms magnetic intensity at its surface is 1 ampere per meter at (*a*) 60 cycles, (*b*) 1 megacycle, (*c*) 1000 megacycles.

2-19. Make a sketch similar to Fig. 2-6 for a circularly polarized standing wave in dissipative media. Give a verbal description of $\mathcal{E}$ and $\mathcal{H}$.

2-20. Given a uniform plane wave normally incident upon a plane air-to-dielectric interface, show that the standing-wave ratio is

$$\text{SWR} = \sqrt{\epsilon_r} = \text{index of refraction}$$

where ϵ_r is the dielectric constant of the dielectric (assumed nonmagnetic and loss-free).

2-21. Take the index of refraction of water to be 9, and calculate the percentage of power reflected and transmitted when a plane wave is normally incident on a calm lake.

2-22. Calculate the two polarizing angles (internal and external) and the critical angle for a plane interface between air and (*a*) water, $\epsilon_r = 81$, (*b*) high-density glass, $\epsilon_r = 9$, and (*c*) polystyrene, $\epsilon_r = 2.56$.

2-23. Suppose a uniform plane wave in a dielectric just grazes a plane dielectric-to-air interface. Calculate the attenuation constant in the air [α as defined by Eq. (2-61)] for the three cases of Prob. 2-22. Calculate the distance from the boundary in which the field is attenuated to $1/e$ (36.8 per cent) of its value at the boundary. What is the value of α at the critical angle?

2-24. From Eqs. (2-66) and (2-68), show that when $R \ll \omega L$ and $G \ll \omega C$

$$\alpha \approx \frac{R}{2\sqrt{L/C}} + \frac{G\sqrt{L/C}}{2}$$
$$\beta \approx \omega\sqrt{LC}$$

where $\gamma = \alpha + j\beta$.

2-25. Show that G and C of a transmission line are related by

$$G = \frac{\omega \epsilon''}{\epsilon'} C = \frac{\omega \epsilon'' \eta}{Z_0}$$

when the dielectric is homogeneous. Show that R of a transmission line is approximately equal to the d-c resistance per unit length of hollow conductors having thickness δ (skin depth) provided H is approximately constant over each conductor and the radius of curvature of the conductors is large compared to δ.

2-26. Using results of Prob. 2-25, show that for the two-wire line of Table 2-3

$$R \approx \frac{2\mathcal{R}}{\pi d} \quad \begin{array}{c} d \gg \delta \\ D \gg d \end{array}$$

and that for the coaxial line

$$R \approx \frac{\mathcal{R}}{2\pi} \frac{a+b}{ab} \quad a \gg \delta$$

and that for the parallel-plate line

$$R \approx \frac{2\mathcal{R}}{w} \quad w \gg b$$

2-27. Verify Eqs. (2-70).

2-28. Consider a parallel-plate waveguide formed by conductors covering the planes $y = 0$ and $y = b$. Show that the field

$$E_x = E_0 \sin\frac{n\pi y}{b} e^{-\gamma z} \quad n = 1, 2, 3, \ldots$$

defines a set of TE$_n$ modes and the field

$$H_x = H_0 \cos\frac{n\pi y}{b} e^{-\gamma z} \qquad n = 0, 1, 2, \ldots$$

defines a set of TM$_n$ modes, where

$$\gamma = \sqrt{\left(\frac{n\pi}{b}\right)^2 - k^2}$$

in both cases. Show that the cutoff frequencies of the TE$_n$ and TM$_n$ modes are

$$f_c = \frac{n}{2b\sqrt{\epsilon\mu}}$$

Show that Eqs. (2-83) to (2-86) apply to the parallel-plate waveguide modes.

2-29. Show that the power transmitted per unit width (x direction) of the parallel-plate waveguide of Prob. 2-28 is

$$P = \frac{b|E_0|^2}{2\eta}\sqrt{1 - \left(\frac{f_c}{f}\right)^2}$$

for the TE$_n$ modes, and

$$P = \frac{b|H_0|^2\eta}{2}\sqrt{1 - \left(\frac{f_c}{f}\right)^2}$$

for the TM$_n$ modes ($n \neq 0$).

2-30. For the parallel-plate waveguide of Prob. 2-28, show that the attenuation due to conductor losses is

$$\alpha_c = \frac{2\mathcal{R}(f_c/f)^2}{b\eta\sqrt{1 - (f_c/f)^2}}$$

for the TE$_n$ modes, and

$$\alpha_c = \frac{2\mathcal{R}}{b\eta\sqrt{1 - (f_c/f)^2}}$$

for the TM$_n$ modes ($n \neq 0$).

2-31. Show that the TM$_0$ mode of the parallel-plate waveguide as defined in Prob. 2-28 is actually a TEM mode. Show that for this mode the attenuation due to conductor losses is

$$\alpha_c = \frac{\mathcal{R}}{b\eta}$$

Compare this with α obtained by using the results of Probs. 2-26 and 2-24.

2-32. For the TE$_{01}$ rectangular waveguide mode, show that the time-average electric and magnetic energies per unit length are

$$\bar{\mathcal{W}}_e = \bar{\mathcal{W}}_m = \frac{\epsilon_0}{4}|E_0|^2 ab$$

Can this equality of $\bar{\mathcal{W}}_e$ and $\bar{\mathcal{W}}_m$ be predicted from Eq. (1-62)?

2-33. Show that the time-average velocity of propagation of energy down a rectangular waveguide is

$$\bar{v}_e = \frac{\bar{\mathcal{S}}_z}{\bar{\mathcal{W}}} = \frac{1}{\sqrt{\epsilon\mu}}\sqrt{1 - \left(\frac{f_c}{f}\right)^2}$$

for the TE$_{01}$ mode.

2-34. For the TE_{01} rectangular waveguide mode, define a voltage V as $\int \mathbf{E} \cdot d\mathbf{l}$ across the center of the guide and a current I as the total z-directed current in the guide wall $x = 0$. Show that these are

$$V = aE_0 e^{-j\beta z} \qquad I = \frac{2bE_0}{\pi Z_0} e^{-j\beta z}$$

Show that $P \neq VI^*$. Why? Define a characteristic impedance $Z_{VI} = V/I$ and show that it is proportional to Z_0 of Table 2-4.

2-35. Let a rectangular waveguide have a discontinuity in dielectric at $z = 0$, that is, ϵ_1, μ_1 for $z < 0$ and ϵ_2, μ_2 for $z > 0$. Show that the reflection and transmission coefficients for a TE_{01} wave incident from $z < 0$ are

$$\Gamma = \frac{Z_{02} - Z_{01}}{Z_{02} + Z_{01}} \qquad T = \frac{2Z_{02}}{Z_{02} + Z_{01}}$$

where Z_{01} and Z_{02} are the characteristic impedances $z < 0$ and $z > 0$, respectively. These results are valid for any waveguide mode.

2-36. Show that there is no reflected wave for the TE_{01} mode in Prob. 2-35 when

$$\frac{f}{f_{c1}} = \sqrt{\frac{\epsilon_1(\mu_1^2 - \mu_2^2)}{\mu_2(\mu_1\epsilon_2 - \mu_2\epsilon_1)}}$$

where f_{c1} is the cutoff frequency $z < 0$. Note that we cannot have a reflectionless interface when both dielectrics are nonmagnetic. This result is valid for any TE mode.

2-37. Take a parallel-plate waveguide with ϵ_1, μ_1 for $z < 0$ and ϵ_2, μ_2 for $z > 0$. Show that there is no reflected wave for a TM mode incident from $z < 0$ when

$$\frac{f}{f_{c1}} = \sqrt{\frac{\mu_1(\epsilon_1^2 - \epsilon_2^2)}{\epsilon_2(\epsilon_1\mu_2 - \epsilon_2\mu_1)}}$$

For nonmagnetic dielectrics, this reduces to

$$\frac{f}{f_{c1}} = \sqrt{\frac{\epsilon_1 + \epsilon_2}{\epsilon_2}}$$

Compare this to Eq. (2-60). These results are valid for any TM mode.

2-38. Design a square-base cavity with height one-half the width of the base to resonate at 1000 megacycles (a) when it is air-filled and (b) when it is polystyrene-filled. Calculate the Q in each case.

2-39. For the rectangular cavity of Fig. 2-19, define a voltage V as that between mid-points of the top and bottom walls and a current I as the total x-directed current in the side walls. Show that

$$V = E_0 a \qquad I = \frac{j4E_0}{\pi \eta} \sqrt{b^2 + c^2}$$

Define a mode conductance G as $G = \bar{\mathcal{P}}_d/|V|^2$ and show that

$$G = \frac{\mathcal{R}[bc(b^2 + c^2) + 2a(b^3 + c^3)]}{2\eta^2 a^2(b^2 + c^2)}$$

Define a mode resistance R as $R = \bar{\mathcal{P}}_d/|I|^2$ and show that

$$R = \frac{\pi^2 \mathcal{R}[bc(b^2 + c^2) + 2a(b^3 + c^3)]}{32(b^2 + c^2)^2}$$

INTRODUCTION TO WAVES 93

2-40. Derive Eqs. (2-123).

2-41. Consider the small loop of constant current I as shown in Fig. 2-26. Show that the magnetic vector potential is

$$A_\phi = A_y \bigg|_{\phi=0} = \frac{Ia}{4\pi} \int_0^{2\pi} f \cos \phi' \, d\phi'$$

where

$$f = \frac{\exp(-jk\sqrt{r^2 + a^2 - 2ra \sin\theta \cos\phi'})}{\sqrt{r^2 + a^2 - 2ra \sin\theta \cos\phi'}}$$

Expand f in a Maclaurin series about $a = 0$ and show that

$$A_\phi \xrightarrow[a \to 0]{} \frac{I\pi a^2}{4\pi} e^{-jkr} \left(\frac{jk}{r} + \frac{1}{r^2}\right) \sin\theta$$

The quantity $I\pi a^2 = IS$ is called the magnetic moment of the loop.

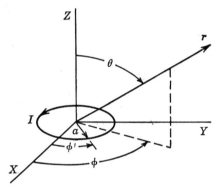

Fig. 2-26. A circular loop of current.

2-42. Show that the field of the small current loop of Prob. 2-41 is

$$H_r = \frac{IS}{2\pi} e^{-jkr} \left(\frac{jk}{r^2} + \frac{1}{r^3}\right) \cos\theta$$

$$H_\theta = \frac{IS}{4\pi} e^{-jkr} \left(-\frac{k^2}{r} + \frac{jk}{r^2} + \frac{1}{r^3}\right) \sin\theta$$

$$E_\phi = \frac{\eta IS}{4\pi} e^{-jkr} \left(\frac{k^2}{r} - \frac{jk}{r^2}\right) \sin\theta$$

Show that the radiation resistance of the small loop referred to I is

$$R_r = \eta \frac{2\pi}{3} \left(\frac{kS}{\lambda}\right)^2$$

2-43. Consider the current element of Fig. 2-21 and the current loop of Fig. 2-26 to exist simultaneously. Show that the radiation field is everywhere circularly polarized if

$$Il = kIS$$

2-44. In terms of the tabulated functions

$$\text{Si}(x) = \int_0^x \frac{\sin x}{x} \, dx \qquad \text{Ci}(x) = -\int_x^\infty \frac{\cos x}{x} \, dx$$

show that Eq. (2-129) can be expressed as

$$R_r = \frac{\eta}{2\pi}\left[C + \log kL - \operatorname{Ci} kL + \sin kL \,(\tfrac{1}{2}\operatorname{Si} 2kL - \operatorname{Si} kL) \right.$$
$$\left. + \tfrac{1}{2}\cos kL \left(C + \log k\frac{L}{2} + \operatorname{Ci} 2kL - 2\operatorname{Ci} kL\right)\right]$$

where $C = 0.5772 \cdots$ is Euler's constant.

2-45. If the linear antenna of Fig. 2-23 is an integral number of half-wavelengths long, the current will assume the form

$$I(z) = I_m \sin k\left(z + \frac{L}{2}\right)$$

regardless of the position of the feed as long as it is not near a current null. Such an antenna is said to be of *resonant length*. Show that the radiation field of the antenna is

$$E_\theta = \frac{j\eta I_m}{2\pi r} e^{-jkr} \frac{\cos\left(\dfrac{n\pi}{2}\cos\theta\right)}{\sin\theta} \qquad n \text{ odd}$$

$$E_\theta = \frac{\eta I_m}{2\pi r} e^{-jkr} \frac{\sin\left(\dfrac{n\pi}{2}\cos\theta\right)}{\sin\theta} \qquad n \text{ even}$$

where $n = 2L/\lambda$ is an integer.

2-46. For an antenna of resonant length (Prob. 2-45), show that the radiation resistance referred to I_m is

$$R_r = \frac{\eta}{4\pi}[C + \log 2n\pi - \operatorname{Ci}(2n\pi)]$$

where $n = 2L/\lambda$, $C = 0.5772$, and Ci is as defined in Prob. 2-44. Show that the input resistance for a loss-free antenna with feed point at $z = a\lambda$ is

$$R_i = \frac{R_r}{\sin 2\pi(a + n/4)}$$

Specialize this result to $L = \lambda/2$, $a = 0$ (the half-wave dipole) and show that $R_i = 73$ ohms.

CHAPTER 3

SOME THEOREMS AND CONCEPTS

3-1. The Source Concept. The complex field equations for linear media are

$$-\nabla \times \mathbf{E} = \hat{z}\mathbf{H} + \mathbf{M} \qquad \nabla \times \mathbf{H} = \hat{y}\mathbf{E} + \mathbf{J} \qquad (3\text{-}1)$$

where **J** and **M** are sources in the most general sense. We have purposely omitted superscripts on **J** and **M** because their interpretations vary from problem to problem. In one problem, they might represent actual sources, in which case we would call them impressed currents. In another problem, **J** might represent a conduction current that we wish to keep separate from the $\hat{y}\mathbf{E}$ term. In still another problem, **M** might represent a magnetic polarization current that we wish to keep separate from the $\hat{z}\mathbf{H}$ term, and so on. We can think of **J** and **M** as "mathematical sources," regardless of their physical interpretation.

For our first illustration, let us show how to represent "circuit sources" in terms of the "field sources" **J** and **M**. The *current source* of circuit theory is defined as one whose current is independent of the load. In terms of field concepts it can be pictured as a short filament of impressed electric current in series with a perfectly conducting wire. This is shown in Fig. 3-1a. That it has the characteristics of the current source of circuit theory can be demonstrated as follows. We make the usual circuit assumption that the displacement current through the surrounding medium is negligible. It then follows from the conservation of charge that the current in the leads is equal to the impressed current, independent of the load. The field formula for power, Eq. (1-66), reduces to

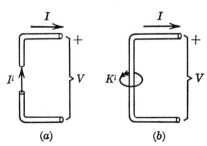

FIG. 3-1. Circuit sources in terms of impressed currents. (a) Current source; (b) voltage source.

the circuit formula for this source. We have only electric currents; hence

$$P_s = - \iiint \mathbf{E} \cdot \mathbf{J}^{i*} \, d\tau = -I^{i*} \int \mathbf{E} \cdot d\mathbf{l} = VI^*$$

The "internal impedance" of the source is infinite, since a removal of the impressed current leaves an open circuit.

The *voltage source* of circuit theory is defined as one whose voltage is independent of the load. In terms of field concepts it can be pictured as a small loop of impressed magnetic current encircling a perfectly conducting wire. This is illustrated by Fig. 3-1b. To show that it has the characteristics of the voltage source of circuit theory, we neglect displacement current and apply the field equation $K = -\oint \mathbf{E} \cdot d\mathbf{l}$ to a path coincident with the wire and closing across the terminals. The $\mathbf{E}$ is zero in the wire; so the line integral is merely the terminal voltage, that is, $K^i = -V$. The impressed current, and therefore the terminal voltage, is independent of load. The field formula for power, Eq. (1-66), reduces in this case to

$$P_s = - \iiint \mathbf{H}^* \cdot \mathbf{M}^i \, d\tau = -K^i \oint \mathbf{H}^* \cdot d\mathbf{l} = VI^*$$

which is the usual circuit formula. The internal impedance of the source is zero, since a removal of the impressed current leaves a short circuit.

We can use the circuit sources in field problems when the source and input region are of "circuit dimensions," that is, of dimensions small compared to a wavelength. Given a pair of terminals close together, we can apply the current source of Fig. 3-1a, that is, a short filament of impressed electric current. Given a conductor of small cross section, we can apply the voltage source of Fig. 3-1b, that is, a small loop of impressed magnetic current. As an example of the use of a circuit source, consider the linear antenna of Fig. 2-23. The geometry of the physical antenna is two sections of wire separated by a small gap at the input. To excite the antenna, we can place a current source (a short filament of electric current) across the gap, which causes a current in the antenna wire. An exact solution to the problem involves a determination of the resulting current in the wire. This is difficult to do. Instead, we approximate the current in the wire, drawing on qualitative and experimental knowledge. We then use this current, plus the current source across the gap, in the potential integral formula to give us an approximation to the field.

We shall find much use for the concept of current sheets, considered in Sec. 1-14. As an example, suppose we have a $\mathbf{J}_s$ over the cross section of a rectangular waveguide, as shown in Fig. 3-2. Furthermore, we postulate that this current should produce only the TE_{01} waveguide mode,

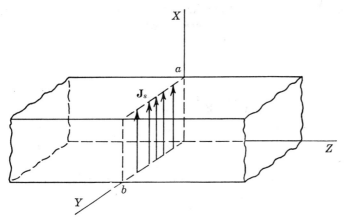

Fig. 3-2. A sheet of current in a rectangular waveguide.

which propagates outward from the current sheet. Abstracting from Table 2-4, we have the wave

$$\left.\begin{array}{l} E_x^+ = A \sin \dfrac{\pi y}{b} e^{-j\beta z} \\[4pt] H_y^+ = \dfrac{A}{Z_0} \sin \dfrac{\pi y}{b} e^{-j\beta z} \\[4pt] H_z^+ = \dfrac{A}{j\eta} \dfrac{f_c}{f} \cos \dfrac{\pi y}{b} e^{-j\beta z} \end{array}\right\} \quad z > 0$$

where the constant A specifies the mode amplitude. The $-z$ traveling wave is of the same form with β replaced by $-\beta$ and Z_0 by $-Z_0$. Thus,

$$\left.\begin{array}{l} E_x^- = B \sin \dfrac{\pi y}{b} e^{j\beta z} \\[4pt] H_y^- = -\dfrac{B}{Z_0} \sin \dfrac{\pi y}{b} e^{j\beta z} \\[4pt] H_z^- = \dfrac{B}{j\eta} \dfrac{f_c}{f} \cos \dfrac{\pi y}{b} e^{j\beta z} \end{array}\right\} \quad z < 0$$

where B is the mode amplitude of the $-z$ traveling wave. At $z = 0$, Eqs. (1-86) must be satisfied. Take the (1) side to be $z > 0$, so that $\mathbf{n} = \mathbf{u}_z$, and obtain

$$-\mathbf{u}_z[H_y^+ - H_y^-]_{z=0} = \mathbf{J}_s \qquad [E_x^+ - E_x^-]_{z=0} = 0$$

Substitution for H_y and E_x from above reduces these equations to

$$-\mathbf{u}_x \frac{A + B}{Z_0} \sin \frac{\pi y}{b} = \mathbf{J}_s \qquad A - B = 0$$

Let

$$\mathbf{J}_s = \mathbf{u}_x J_0 \sin \frac{\pi y}{b} \qquad (3\text{-}2)$$

The preceding equations then have the solution $A = B = -J_0Z_0/2$. Thus, if the current of Eq. (3-2) exists over the guide cross section $z = 0$, then

$$E_x = \begin{cases} -\dfrac{J_0Z_0}{2} \sin \dfrac{\pi y}{b} e^{-j\beta z} & z > 0 \\ -\dfrac{J_0Z_0}{2} \sin \dfrac{\pi y}{b} e^{j\beta z} & z < 0 \end{cases} \qquad (3\text{-}3)$$

It would admittedly be difficult to obtain the current of Eq. (3-2) in practice, but this is not of concern at present. We shall learn how to treat more practical problems later. Note that our approach in this problem was to assume the field and find the current. This we shall find to be a very powerful concept.

3-2. Duality. If the equations describing two different phenomena are of the same mathematical form, solutions to them will take the same mathematical form. The formal recognition of this is called the *concept of duality*. Two equations of the same mathematical form are called *dual equations*. Quantities occupying the same position in dual equations are called *dual quantities*. Note that the field equations, Eqs. (3-1), are duals of each other. A systematic interchange of symbols changes the first equation into the second, and vice-versa.

A duality of importance to us is that between a problem for which all sources are of the electric type and a problem for which all sources are of the magnetic type. The first two rows of Table 3-1 give the field equations in each case. The last two formulas of column (1) were derived in Sec. 2-9 for homogeneous space. The corresponding equations for the magnetic source case are evidently the last two formulas of column (2), obtained by systematically interchanging symbols. The particular interchange of symbols is summarized by Table 3-2. The reader should check for himself that a replacement of the symbols of

TABLE 3-1. DUAL EQUATIONS FOR PROBLEMS IN WHICH (1) ONLY ELECTRIC SOURCES EXIST AND (2) ONLY MAGNETIC SOURCES EXIST

(1) Electric sources	(2) Magnetic sources
$\nabla \times \mathbf{H} = \hat{y}\mathbf{E} + \mathbf{J}$	$-\nabla \times \mathbf{E} = \hat{z}\mathbf{H} + \mathbf{M}$
$-\nabla \times \mathbf{E} = \hat{z}\mathbf{H}$	$\nabla \times \mathbf{H} = \hat{y}\mathbf{E}$
$\mathbf{H} = \nabla \times \mathbf{A}$	$\mathbf{E} = -\nabla \times \mathbf{F}$
$\mathbf{A} = \dfrac{1}{4\pi} \iiint \dfrac{\mathbf{J}e^{-jk|\mathbf{r}-\mathbf{r}'|}}{|\mathbf{r}-\mathbf{r}'|} d\tau'$	$\mathbf{F} = \dfrac{1}{4\pi} \iiint \dfrac{\mathbf{M}e^{-jk|\mathbf{r}-\mathbf{r}'|}}{|\mathbf{r}-\mathbf{r}'|} d\tau'$

TABLE 3-2. DUAL QUANTITIES FOR PROBLEMS IN WHICH (1) ONLY ELECTRIC SOURCES EXIST, AND (2) ONLY MAGNETIC SOURCES EXIST

(1) Electric sources	(2) Magnetic sources
E	**H**
H	**−E**
J	**M**
A	**F**
$\hat{y}$	$\hat{z}$
$\hat{z}$	$\hat{y}$
k	k
η	$1/\eta$

column (1) of Table 3-2 by those of column (2) in the equations of column (1) of Table 3-1 results in the equations of column (2). The quantity **F** of these tables is called an *electric vector potential,* in analogy to **A**, a magnetic vector potential.

The concept of duality is important for several reasons. It is an aid to remembering equations, since almost half of them are duals of other equations. It shows us how to take the solution to one type of problem, interchange symbols, and obtain the solution to another type of problem. We can also use a physical or intuitive picture that applies to one type of problem and carry it over to the dual problem. For example, the picture of electric charge in motion giving rise to an electric current can also be used for magnetic case. That is, we can picture magnetic charge in motion as giving rise to magnetic current. Such a picture can serve as a guide to the mathematical development but cannot, of course, serve to argue for the existence of magnetic charges in nature. The concept of duality is based wholly on the mathematical symmetry of equations.

It is often convenient to divide a single problem into dual parts, thus cutting the mathematical labor in half. For example, suppose we have both electric and magnetic sources in a homogeneous medium of infinite extent. The field equations, Eqs. (3-1), are linear; so the total field can be considered as the sum of two parts, one produced by **J** and the other by **M**. To be explicit, let

$$\mathbf{E} = \mathbf{E}' + \mathbf{E}'' \qquad \mathbf{H} = \mathbf{H}' + \mathbf{H}''$$

where
$$\nabla \times \mathbf{H}' = \hat{y}\mathbf{E}' + \mathbf{J} \qquad -\nabla \times \mathbf{E}' = \hat{z}\mathbf{H}'$$
and
$$\nabla \times \mathbf{H}'' = \hat{y}\mathbf{E}'' \qquad -\nabla \times \mathbf{E}'' = \hat{z}\mathbf{H}'' + \mathbf{M}$$

We have the solution for each of these partial problems in Table 3-1. The complete solution is therefore just the superposition of the two partial solutions, or

$$\mathbf{E} = -\nabla \times \mathbf{F} + \hat{y}^{-1}(\nabla \times \nabla \times \mathbf{A} - \mathbf{J})$$
$$\mathbf{H} = \nabla \times \mathbf{A} + \hat{z}^{-1}(\nabla \times \nabla \times \mathbf{F} - \mathbf{M}) \qquad (3\text{-}4)$$

where
$$A(\mathbf{r}) = \frac{1}{4\pi} \iiint \frac{\mathbf{J}(\mathbf{r}')e^{-jk|\mathbf{r}-\mathbf{r}'|}}{|\mathbf{r}-\mathbf{r}'|} d\tau'$$
$$F(\mathbf{r}) = \frac{1}{4\pi} \iiint \frac{\mathbf{M}(\mathbf{r}')e^{-jk|\mathbf{r}-\mathbf{r}'|}}{|\mathbf{r}-\mathbf{r}'|} d\tau'$$
(3-5)

We thus have the formal solution for any problem consisting of electric and magnetic currents in an unbounded homogeneous region. The above formulas are meant to include by implication sheets and filaments of currents.

It is instructive to show that *an infinitesimal dipole of magnetic current is indistinguishable from an infinitesimal loop of electric current*. We might suspect this from the circuit source representations of Fig. 3-1. However, rather than rely on this argument, let us consider the fields explicitly. A z-directed magnetic current dipole of moment Kl at the coordinate origin is the dual problem to the electric current dipole (Fig. 2-21). An interchange of symbols, according to Table 3-2, in Eqs. (2-113) will give us the field of the magnetic current element. For example, the electric intensity is

$$E_\phi = \frac{-Kl}{4\pi} e^{-jkr} \left(\frac{jk}{r} + \frac{1}{r^2} \right) \sin \theta$$

The small loop of electric current is considered in Probs. 2-41 and 2-42 and is pictured in Fig. 2-26. Abstracting from Prob. 2-42, we have the electric intensity given by

$$E_\phi = \frac{\eta IS}{4\pi} e^{-jkr} \left(\frac{k^2}{r} - \frac{jk}{r^2} \right) \sin \theta$$

A comparison of the above two equations shows that they are identical if

$$Kl = j\omega\mu IS \tag{3-6}$$

This equality is illustrated by Fig. 3-3. Thus, effect of an element of magnetic current can be realized in practice by a loop of electric current.

3-3. Uniqueness. A solution is said to be unique when it is the only one possible among a given class of solutions. It is important to have

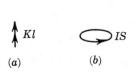

FIG. 3-3. These two sources radiate the same field if $Kl = j\omega\mu IS$. (a) Magnetic current element; (b) electric current loop.

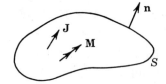

FIG. 3-4. S encloses linear matter and sources **J**, **M**.

precise theorems on uniqueness for several reasons. First of all, they tell us what information is needed to obtain the solution. Secondly, it is comforting to know that a solution is the only solution. Finally, uniqueness theorems establish conditions for a one-to-one correspondence of a field to its sources. This allows us to calculate the sources from a field, as well as the more usual reverse procedure.

Suppose we have a set of sources $\mathbf{J}$ and $\mathbf{M}$ acting in a region of linear matter bounded by the surface S, as suggested by Fig. 3-4. Any field within S must satisfy the complex field equations, Eqs. (3-1). Consider two possible solutions, $\mathbf{E}^a$, $\mathbf{H}^a$ and $\mathbf{E}^b$, $\mathbf{H}^b$. (These can be thought of as the fields when the sources *outside* of S are different.) We form the difference field $\delta \mathbf{E}$, $\delta \mathbf{H}$ according to

$$\delta \mathbf{E} = \mathbf{E}^a - \mathbf{E}^b \qquad \delta \mathbf{H} = \mathbf{H}^a - \mathbf{H}^b$$

Subtracting Eqs. (3-1) for the a field from those for the b field, we obtain

$$\left. \begin{array}{l} -\nabla \times \delta \mathbf{E} = \hat{z}\, \delta \mathbf{H} \\ \nabla \times \delta \mathbf{H} = \hat{y}\, \delta \mathbf{E} \end{array} \right\} \quad \text{within } S$$

Thus, the difference field satisfies the source-free field equations within S. The conditions for uniqueness are those for which $\delta \mathbf{E} = \delta \mathbf{H} = 0$ everywhere within S, for then $\mathbf{E}^a = \mathbf{E}^b$ and $\mathbf{H}^a = \mathbf{H}^b$.

We now apply Eq. (1-54) to the difference field and obtain

$$\oint (\delta \mathbf{E} \times \delta \mathbf{H}^*) \cdot d\mathbf{s} + \iiint (\hat{z}|\delta H|^2 + \hat{y}^*|\delta E|^2)\, d\tau = 0$$

Whenever
$$\oint (\delta \mathbf{E} \times \delta \mathbf{H}^*) \cdot d\mathbf{s} = 0 \qquad (3\text{-}7)$$

over S, the volume integral must also vanish. Thus, if Eq. (3-7) is true, then

$$\begin{aligned} \iiint [\operatorname{Re}(\hat{z})|\delta H|^2 + \operatorname{Re}(\hat{y})|\delta E|^2]\, d\tau &= 0 \\ \iiint [\operatorname{Im}(\hat{z})|\delta H|^2 - \operatorname{Im}(\hat{y})|\delta E|^2]\, d\tau &= 0 \end{aligned} \qquad (3\text{-}8)$$

For dissipative media, $\operatorname{Re}(\hat{z})$ and $\operatorname{Re}(\hat{y})$ are always positive. If we assume some dissipation everywhere, however slight, then Eqs. (3-8) are satisfied only if $\delta \mathbf{E} = \delta \mathbf{H} = 0$ everywhere within S.

Some of the more important cases for which Eq. (3-7) is satisfied, and therefore uniqueness is obtained in lossy regions, are as follows. (1) The field is unique among a class $\mathbf{E}$, $\mathbf{H}$ having $\mathbf{n} \times \mathbf{E}$ specified on S, for then $\mathbf{n} \times \delta \mathbf{E} = 0$ over S. (2) The field is unique among a class $\mathbf{E}$, $\mathbf{H}$ having $\mathbf{n} \times \mathbf{H}$ specified on S, for then $\mathbf{n} \times \delta \mathbf{H} = 0$ over S. (3) The field is unique among a class $\mathbf{E}$, $\mathbf{H}$ having $\mathbf{n} \times \mathbf{E}$ specified over part of S and $\mathbf{n} \times \mathbf{H}$ specified over the rest of S. These possibilities can be summarized by the following uniqueness theorem. *A field in a lossy region is uniquely specified by the sources within the region plus the tangential components of* **E** *over the boundary, or the tangential components of* **H** *over the boundary, or the former over part of the boundary and the latter over the rest of the boundary.* Note that our uniqueness proof breaks down for dissipationless media. To obtain uniqueness in this case, *we consider the field in a dissipationless medium to be the limit of the corresponding field in a lossy medium as the dissipation goes to zero.*

We have explicitly considered only volume distributions of sources and closed surfaces in our development, but the results are much more general than this. Singular sources, such as current sheets and current filaments, can be thought of as limiting cases of volume distributions and therefore are included by implication. Surfaces of infinite extent can be thought of as closed at infinity and can be included by appropriate limiting procedures. Of particular importance is the case for which the bounding surface is a sphere of radius $r \to \infty$, so that all space is included. If the sources are of finite extent, the vector potential solution of Eqs. (3-4) and (3-5) vanishes exponentially as $e^{-k''r}$, $r \to \infty$. We therefore have

$$\lim_{r \to \infty} \oiint \mathbf{E} \times \mathbf{H}^* \cdot d\mathbf{s} = 0 \qquad (3\text{-}9)$$

for this solution (in lossy media). According to our uniqueness proof this must be the only solution for a class **E**, **H** satisfying Eq. (3-9). Thus, *given sources of finite extent in an unbounded lossy region, any solution satisfying Eq. (3-9) must be identically equal to the potential integral solution.* The loss-free case can be treated as the limit of the lossy case as dissipation vanishes.

To illustrate the above concepts, consider the current element of Fig. 2-21. Our solution at large r is Eq. (2-114). Let this be the a solution of our uniqueness proof, or

$$H_\phi{}^a = \frac{jIl}{2\lambda r} e^{-jkr} \sin \theta \qquad E_\theta{}^a = \eta H_\phi{}^a$$

It can be shown that the inward-traveling wave

$$H_\phi{}^b = \frac{-jIl}{2\lambda r} e^{jkr} \sin \theta \qquad E_\theta{}^b = -\eta H_\phi{}^b$$

is also a solution to the equations at large r. In Sec. 2-9, we threw out this second solution by reasoning that waves must travel outward from the source, not inward. Let us now consider these two solutions in the light of the uniqueness theorem. The difference field in this case is

$$\delta H_\phi = H_\phi{}^a - H_\phi{}^b = j \frac{Il}{\lambda r} \cos kr \sin \theta$$

$$\delta E_\theta = E_\theta{}^a - E_\theta{}^b = \eta \frac{Il}{\lambda r} \sin kr \sin \theta$$

In dissipationless media (k real), we can pick a sphere $r =$ constant such that either δH_ϕ or δE_θ vanishes. Thus, Eq. (3-7) can be satisfied without obtaining uniqueness of the solution. However, in lossy media, $\sin kr$ and $\cos kr$ have no zeros $r > 0$, and Eq. (3-7) cannot be satisfied for any r. In this case, only the a solution vanishes as $r \to \infty$. It is therefore the desired solution in loss-free media.

3-4. Image Theory. Problems for which the field in a given region of space is determined from a knowledge of the field over the boundary of the region are called *boundary-value problems*. The rectangular waveguide of Sec. 2-7 is an example of a boundary-value problem. We shall now consider a class of boundary-value problems for which the boundary surface is a perfectly conducting plane. The procedure is known as image theory.

The boundary conditions at a perfect electric conductor are vanishing tangential components of **E**. An element of source plus an "image" element of source, radiating in free space, produce zero tangential components of **E** over the plane bisecting the line joining the two elements. According to uniqueness concepts, the solution to this problem is also the solution for a current element adjacent to a plane conductor. The necessary orientation and excitation of image elements is summarized by Fig. 3-5. Matter also can be imaged. For example, if a conducting sphere is adjacent to the plane conductor in the original problem, then two conducting spheres at image points are necessary in the image problem. In other words, we must maintain symmetry in the image problem. The procedure also applies to magnetic conductors in a dual sense. The application of image theory in a-c fields is much more restricted than in d-c fields. It is exact only when the plane conductor is perfect.

As an example of image theory, consider a current element normal to the ground (conducting) plane, as shown in Fig. 3-6a. This must produce the same field above the ground plane as do the two elements of Fig. 3-6b. Let us determine the radiation field. The radius vector from each current element is then parallel to that from the origin and given by

$$r_o = r - d \cos \theta \brace r_i = r + d \cos \theta \qquad r \gg d$$

where subscripts o and i refer to original and image elements, respectively. The radiation field of a single element is given by Eq. (2-114); so the radiation field of the two elements of Fig. 3-6b is the superposition

$$H_\phi = \frac{jIl}{2\lambda}\left(\frac{e^{-jkr_o}}{r_o} + \frac{e^{-jkr_i}}{r_i}\right)\sin\theta$$

$$\approx \frac{jIl}{\lambda r}e^{-jkr}\cos(kd\cos\theta)\sin\theta \qquad (3\text{-}10)$$

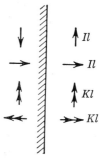

Fig. 3-5. A summary of image theory.

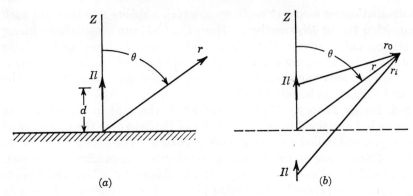

FIG. 3-6. A current element adjacent to a ground plane. (a) Original problem; (b) image problem.

and $E_\theta = \eta H_\phi$. According to image theory, this must also be the solution to Fig. 3-6a above the ground plane.

The problem of Fig. 3-6a represents the antenna system of a short dipole antenna adjacent to a ground plane. The total power radiated by the system is

$$\bar{\mathcal{P}}_f = \iint_{\substack{\text{hemi-}\\\text{sphere}}} E_\theta H_\phi^* \, ds = 2\pi\eta \int_0^{\pi/2} |H_\phi|^2 r^2 \sin\theta \, d\theta$$

where integration is over the large hemisphere $z > 0$, $r \to \infty$. Substituting from Eq. (3-10) and integrating, we have

$$\bar{\mathcal{P}}_f = 2\pi\eta \left|\frac{Il}{\lambda}\right|^2 \left[\frac{1}{3} - \frac{\cos 2kd}{(2kd)^2} + \frac{\sin 2kd}{(2kd)^3}\right] \quad (3\text{-}11)$$

As $kd \to \infty$, the power radiated is equal to that radiated by an isolated element [Eq. (2-116)]. As $kd \to 0$, the power radiated is double that radiated by an isolated element. The gain of the antenna system over an omnidirectional radiator, according to Eq. (2-130), is

$$g = \frac{4\pi r^2 \eta |H_\phi|^2}{\bar{\mathcal{P}}_f}$$

$$= \frac{2}{\dfrac{1}{3} - \dfrac{\cos 2kd}{(2kd)^2} + \dfrac{\sin 2kd}{(2kd)^3}} \quad (3\text{-}12)$$

along the ground plane. This is $g = 3$ at $kd = 0$, and $g = 6$ as $kd \to \infty$. The maximum gain occurs at $kd = 2.88$, for which $g = 6.57$. Thus, a gain of more than four times that of the isolated element (1.5) can be achieved. Figure 3-7 shows the radiation field patterns for the cases

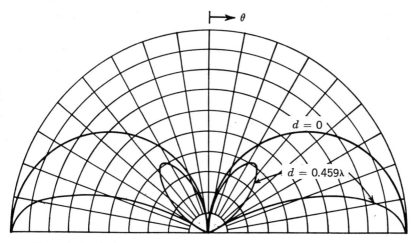

Fig. 3-7. Radiation field patterns for the current element of Fig. 3-6a.

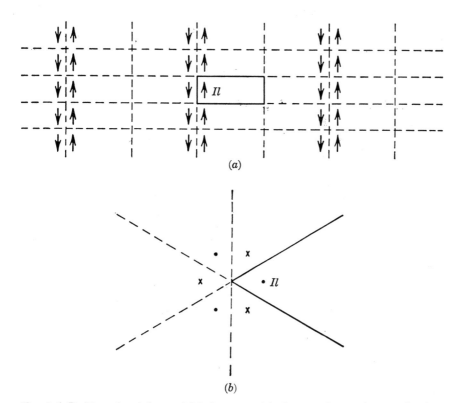

Fig. 3-8. Problems involving multiple images. (a) Current element in a conducting tube; (b) current element in a conducting wedge.

$d = 0$ (element at the gound plane surface) and $d = 0.459\lambda$ (maximum gain).

Image theory also can be applied in certain problems involving more than one conducting plane. Two such cases are illustrated by Fig. 3-8. In the case of a conducting tube (Fig. 3-8a), an infinite lattice of images is needed. In the case of a conducting wedge (Fig. 3-8b), a finite set of images results. Image theory can be used for conducting wedges when the wedge angle is $180°/n$ (n an integer).

3-5. The Equivalence Principle. Many source distributions outside a given region can produce the same field inside the region. For example, the image current element of Fig. 3-6b produces the same field above the plane $z = 0$ as do the currents on the conductor of Fig. 3-6a. Two sources producing the same field within a region of space are said to be *equivalent within that region*. When we are interested in the field in a given region of space, we do not need to know the actual sources. Equivalent sources will serve as well.

A simple application of the equivalence principle is illustrated by Fig. 3-9. Let Fig. 3-9a represent a source (perhaps a transmitter and antenna) internal to S and free space external to S. We can set up a problem equivalent to the original problem external to S as follows. Let the original field exist external to S, and the null field internal to S, with free space everywhere. This is shown in Fig. 3-9b. To support this field, there must exist surface currents $\mathbf{J}_s$, $\mathbf{M}_s$ on S according to Eqs. (1-86). These currents are therefore

$$\mathbf{J}_s = \mathbf{n} \times \mathbf{H} \qquad \mathbf{M}_s = \mathbf{E} \times \mathbf{n} \qquad (3\text{-}13)$$

where $\mathbf{n}$ points outward and $\mathbf{E}$, $\mathbf{H}$ are the original fields over S. Since the currents act in unbounded free space, we can determine the field from them by Eqs. (3-4) and (3-5). From the uniqueness theorem, we know that the field so calculated will be the originally postulated field, that is, $\mathbf{E}$, $\mathbf{H}$ external to S and zero internal to S. The final result of this procedure is a formula for $\mathbf{E}$ and $\mathbf{H}$ everywhere external to S in terms of the tangential components of $\mathbf{E}$ and $\mathbf{H}$ on S.

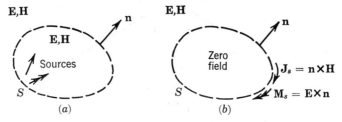

Fig. 3-9. The equivalent currents produce the same field external to S as do the original sources.

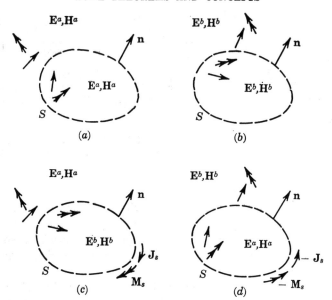

FIG. 3-10. A general formulation of the equivalence principle. (a) Original a problem; (b) original b problem; (c) equivalent to a external to S and to b internal to S; (d) equivalent to b external to S and to a internal to S.

We were overly restrictive in specifying the null field internal to S in the preceding example. Any other field would serve as well, giving us infinitely many equivalent currents as far as the external region is concerned. This general formulation of the equivalence principle is represented by Fig. 3-10. We have two original problems consisting of currents in linear media, as shown in Fig. 3-10a and b. We can set up a problem equivalent to a external to S and equivalent to b internal to S as follows. External to S, we specify that the field, medium, and sources remain the same as in the a problem. Internal to S, we specify that the field, medium, and sources remain the same as in the b problem. To support this field, there must be surface currents $\mathbf{J}_s$ and $\mathbf{M}_s$ on S. According to Eqs. (1-86), these are given by

$$\mathbf{J}_s = \mathbf{n} \times (\mathbf{H}^a - \mathbf{H}^b) \qquad \mathbf{M}_s = (\mathbf{E}^a - \mathbf{E}^b) \times \mathbf{n} \qquad (3\text{-}14)$$

where $\mathbf{E}^a$, $\mathbf{H}^a$ is the field of the a problem and $\mathbf{E}^b$, $\mathbf{H}^b$ is the field of the b problem. This equivalent problem is shown in Fig. 3-10c. We can also set up a problem equivalent to b external to S and to a internal to S in an analogous manner, as shown in Fig. 3-10d. In this case the necessary surface currents are the negative of Eqs. (3-14). Note that in each case we must keep the original sources and media in the region for which we keep the field. Note also that we cannot use Eqs. (3-4) and (3-5) to

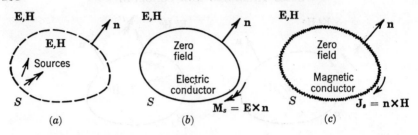

FIG. 3-11. The field external to S is the same in (a), (b), and (c). (a) Original problem; (b) magnetic current backed by an electric conductor; (c) electric current backed by a magnetic conductor.

determine the field of the currents unless the equivalent currents radiate into an unbounded homogeneous region. Finally, note that the restricted form of the equivalence principle (Fig. 3-9) is the special case of the general form for which all a sources and matter lie inside S and all b sources are zero.

So far, we have used the tangential components of *both* **E** and **H** in setting up our equivalent problems. From uniqueness concepts, we know that the tangential components of only **E** *or* **H** are needed to determine the field. We shall now show that equivalent problems can be found in terms of only magnetic currents (tangential **E**) or only electric currents (tangential **H**).

Consider a problem for which all sources lie within S, as shown in Fig. 3-11a. We set up the equivalent problem of Fig. 3-11b as follows. Over S we place a perfect electric conductor, and on top of this we place a sheet of magnetic current $\mathbf{M}_s$. External to S we specify the same field and medium as in the original problem. Since the tangential components of **E** are zero on the conductor (just behind $\mathbf{M}_s$), and equal to the original field components just in front of $\mathbf{M}_s$, it follows from Eqs. (1-86) that

$$\mathbf{M}_s = \mathbf{E} \times \mathbf{n} \qquad (3\text{-}15)$$

We now have the same tangential components of **E** over S in both Fig. 3-11a and b; so according to our uniqueness theorem the field outside of S must be the same in both cases. We can derive the alternative equivalent problem of Fig. 3-11c in an analogous manner. For this we need the perfect magnetic conductor, that is, a boundary of zero tangential components of **H**. We then find that the electric current sheet

$$\mathbf{J}_s = \mathbf{n} \times \mathbf{H} \qquad (3\text{-}16)$$

over a perfect magnetic conductor covering S produces the same field external to S as do the original sources.

By now, the general philosophy of the equivalence principle should be

SOME THEOREMS AND CONCEPTS 109

apparent. It is based upon the one-to-one correspondence between fields and sources when uniqueness conditions are met. If we specify the field and matter everywhere in space, we can determine all sources. We derived our various equivalences in this manner.

Considerable physical interpretation can be given to the equivalence principle. For example, in the problem of Fig. 3-9b, the field internal to S is zero. It therefore makes no difference what matter is within S as far as the field external to S is concerned. We have previously assumed that free space existed within S, so that the potential integral solution could be applied. We could just as well introduce a perfect electric conductor to back the current sheets of Fig. 3-9b. It can be shown by reciprocity (Sec. 3-8) that an electric current just in front of an electric current conductor produces no field. (We can think of the conductor as shorting out the current.) Therefore, the field is produced by the magnetic currents alone, in the presence of the electric conductor, which is Fig. 3-11b. Alternatively, we could back the equivalent currents of Fig. 3-9b with a perfect magnetic conductor and obtain the equivalent problem of Fig. 3-11c. When matter is placed within S in Fig. 3-9b, the partial fields produced by $\mathbf{J}_s$ alone and $\mathbf{M}_s$ alone will change external to S, but the total field must remain unchanged.

Perhaps it would help us to understand the equivalence principle if we considered the analogous concept in circuit theory. Consider a source (active network) connected to a passive network, as shown in Fig. 3-12a. We can set up a problem equivalent to this as far as the passive network is concerned, as follows. The original source is switched off, leaving the source impedance connected. A current source I, equal to the terminal current in the original problem, is placed across the terminals. A voltage

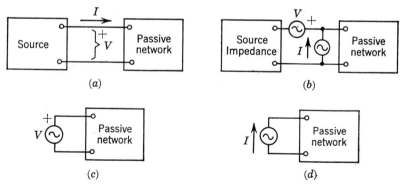

FIG. 3-12. A circuit theory analogue to the equivalence principle. (a) Original problem; (b) equivalent sources; (c) source impedance replaced by a short circuit; (d) source impedance replaced by an open circuit.

source V, equal to the terminal voltage in the original problem, is placed in series with the interconnection. This is illustrated by Fig. 3-12b. It is evident from the usual circuit concepts that there is no excitation of the source impedance from these equivalent sources, whereas the excitation of the passive network is unchanged. Thus, Fig. 3-12b is the circuit analogue to Fig. 3-9b.

Since there is no excitation of the source impedance in Fig. 3-12b, we may replace it by an arbitrary impedance without affecting the excitation of the passive network. This is analogous to the arbitrary placement of matter within S in the field equivalence of Fig. 3-9b. In particular, let the source impedance be replaced by a short circuit. This short-circuits the current source and leaves only the voltage source exciting the network (recall circuit theory superposition). Thus, the voltage source alone, as illustrated by Fig. 3-12c, produces the same excitation of the passive network as does the original source. This is analogous to the field problem of Fig. 3-11b. Now consider the source impedance of Fig. 3-12b replaced by an open circuit. This leaves only the current source exciting the network, as shown in Fig. 3-12d. This is analogous to the field problem of Fig. 3-11c.

3-6. Fields in Half-space. A combination of the equivalence principle and image theory can be used to obtain solutions to boundary-value problems for which the field in half-space is to be determined from its tangential components over the bounding plane. To illustrate, let the original problem consist of matter and sources $z < 0$, and free space $z > 0$, as shown in Fig. 3-13a. An application of the equivalence concepts of Fig. 3-11b yields the equivalent problem of Fig. 3-13b. This consists of the magnetic currents of Eq. (3-15) adjacent to an infinite

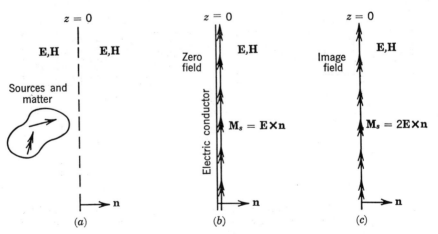

Fig. 3-13. Illustration of the steps used to establish Eq. (3-17).

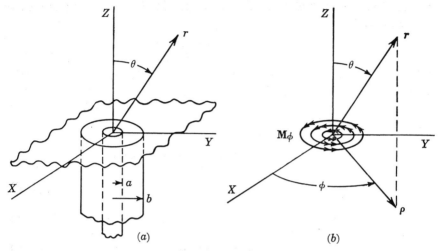

FIG. 3-14. A coaxial line opening onto a ground plane. (a) Original problem; (b) equivalent problem.

ground plane. We now image the magnetic currents in the ground plane, according to Fig. 3-5. The images are equal in magnitude to, and essentially coincident with, the $\mathbf{M}_s$ of Fig. 3-13b. Thus, as pictured in Fig. 3-13c, the magnetic currents $2\mathbf{M}_s$ radiating into unbounded space produce the same field $z > 0$ as do the original sources. They produce an image field $z < 0$, which is of no interest to us. The field of Fig. 3-13c is then calculated according to Eqs. (3-4) and (3-5) with $\mathbf{A} = 0$. This can be summarized mathematically by

$$\mathbf{E}(\mathbf{r}) = -\nabla \times \iint_{\text{plane}} \frac{e^{-jk|\mathbf{r}-\mathbf{r}'|}}{2\pi|\mathbf{r}-\mathbf{r}'|} \mathbf{E}(\mathbf{r}') \times d\mathbf{s}' \qquad (3\text{-}17)$$

This is a mathematical identity valid for any field $\mathbf{E}$ satisfying Eq. (2-3). The $\mathbf{H}$ field satisfies Eq. (2-4), which is identical to Eq. (2-3); so the above identity must also be valid for $\mathbf{E}$ replaced by $\mathbf{H}$. We can show this by reasoning dual to that used to establish Eq. (3-17).

The above result is particularly useful for problems involving apertures in conducting ground planes. As an example, suppose we have a coaxial transmission line opening into a ground plane (Fig. 3-14a). According to the above discussion, the field must be the same as that produced by Fig. 3-14b. Note that $\mathbf{M}_s$ exists only over the aperture (coax opening), for tangential $\mathbf{E}$ is zero over the ground plane. Let us asume that the field over the aperture is the transmission-line mode of the coax, that is

$$E_\rho = \frac{-V}{\rho \log (b/a)}$$

where V is the line voltage. To this approximation, the magnetic current in Fig. 3-14b is

$$M_\phi = \frac{V}{\rho \log (b/a)}$$

This is a loop of magnetic current which, if $b \ll \lambda$, acts as an electric dipole (dual to Fig. 3-3). Visualize this current as a continuous distribution of magnetic current filaments of strength $dK = M_\phi \, d\rho$. The total moment of the source is then

$$\begin{aligned} KS &= \int \pi \rho^2 \, dK = \frac{\pi V}{\log (b/a)} \int_a^b \rho \, d\rho \\ &= \frac{\pi V (b^2 - a^2)}{2 \log (b/a)} \end{aligned} \tag{3-18}$$

The equivalent electric current element must satisfy the equation dual to Eq. (3-6), or

$$Il = -j\omega\epsilon KS \tag{3-19}$$

We have now reduced the problem to that of Fig. 3-6a with $kd = 0$. From Eq. (3-10) and the above equalities, we have the radiation field given by

$$H_\phi = \frac{\omega\epsilon\pi V(b^2 - a^2)}{2\lambda r \log (b/a)} e^{-jkr} \sin \theta \tag{3-20}$$

and $E_\theta = \eta H_\phi$. Thus, the radiation field pattern is the $d = 0$ curve of Fig. 3-7. The gain of the antenna system is $g = 3$.

The power radiated is Eq. (3-11) with $kd = 0$ and Il given by Eqs. (3-18) and (3-19), or

$$\begin{aligned} \bar{\mathcal{P}}_f &= 2\pi\eta \left| \frac{\omega\epsilon\pi V(b^2 - a^2)}{2\lambda \log (b/a)} \right|^2 \frac{2}{3} \\ &= \frac{4\pi}{3\eta} \left| \frac{\pi^2(b^2 - a^2)V}{\lambda^2 \log (b/a)} \right|^2 \end{aligned} \tag{3-21}$$

Note that the power radiated varies inversely as λ^4. Note also that our answers are referred to a voltage, characteristic of aperture antennas. This is in contrast to answers referred to current for wire antennas. For aperture antennas we define a *radiation conductance* according to

$$G_r = \frac{\bar{\mathcal{P}}_f}{|V|^2} \tag{3-22}$$

where V is an arbitrary reference voltage. In the coaxial radiator of Fig. 3-14 it is logical to pick this V to be the coaxial V at the aperture. Hence, the radiation conductance is

$$G_r = \frac{4\pi^5}{3\eta} \left[\frac{b^2 - a^2}{\lambda^2 \log (b/a)} \right]^2 \tag{3-23}$$

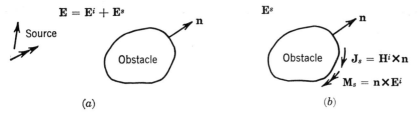

Fig. 3-15. Illustration of the induction theorem. (a) Original problem; (b) induction equivalent.

For the usual coaxial line, G_r is small, and the coaxial line sees nearly an open circuit. As a and b are made larger, the radiation becomes more pronounced, but our formulas must then be modified.[1]

3-7. The Induction Theorem. We now consider a theorem closely related in concept to the equivalence principle. Consider a problem in which a set of sources are radiating in the presence of an obstacle (material body). This is illustrated by Fig. 3-15a. Define the *incident field* $\mathbf{E}^i$, $\mathbf{H}^i$ as the field of the sources with the obstacle absent. Define the *scattered field* $\mathbf{E}^s$, $\mathbf{H}^s$ as the difference between the field with the obstacle present ($\mathbf{E}, \mathbf{H}$) and the incident field, that is,

$$\mathbf{E}^s = \mathbf{E} - \mathbf{E}^i \qquad \mathbf{H}^s = \mathbf{H} - \mathbf{H}^i \qquad (3\text{-}24)$$

This scattered field can be thought of as the field produced by the currents (conduction and polarization) on the obstacle. External to the obstacle, both $\mathbf{E}, \mathbf{H}$ and $\mathbf{E}^i, \mathbf{H}^i$ have the same sources. The scattered field $\mathbf{E}^s, \mathbf{H}^s$ is therefore a source-free field external to the obstacle.

We now construct a second problem as follows. Retain the obstacle, and postulate that the original field $\mathbf{E}, \mathbf{H}$ exists internal to it and that the scattered field $\mathbf{E}^s, \mathbf{H}^s$ exists external to it. Both these fields are source-free in their respective regions. To support these fields, there must be surface currents on S according to Eqs. (1-86), that is,

$$\mathbf{J}_s = \mathbf{n} \times (\mathbf{H}^s - \mathbf{H}) \qquad \mathbf{M}_s = (\mathbf{E}^s - \mathbf{E}) \times \mathbf{n}$$

where $\mathbf{n}$ points outward from S. According to Eqs. (3-24), these reduce to

$$\mathbf{J}_s = \mathbf{H}^i \times \mathbf{n} \qquad \mathbf{M}_s = \mathbf{n} \times \mathbf{E}^i \qquad (3\text{-}25)$$

It follows from the uniqueness theorem that these currents, radiating in the presence of the obstacle, produce the postulated field ($\mathbf{E}, \mathbf{H}$ internal to S, and $\mathbf{E}^s, \mathbf{H}^s$ external to S). This is the *induction theorem*, illustrated by Fig. 3-15b.

It is instructive to compare the induction theorem with the equiva-

---

[1] H. Levine and C. H. Papas, Theory of the Circular Diffraction Antenna, *J. Appl. Phy.*, vol. 22, no. 1, pp. 29–43, January, 1951.

lence theorem. The latter postulates **E**, **H** internal to S and zero field external to S, which must be supported by currents

$$\mathbf{J}_s = \mathbf{H} \times \mathbf{n} \qquad \mathbf{M}_s = \mathbf{n} \times \mathbf{E}$$

on S. These currents can be considered as radiating into an unbounded medium having constitutive parameters equal to those of the obstacle. Thus, we can use Eqs. (3-4) and (3-5) to calculate the field of the above currents. However, we do not know $\mathbf{J}_s$ and $\mathbf{M}_s$ until we know **E**, **H** on S, that is, until we have the solution to the problem of Fig. 3-15a. We can, however, approximate $\mathbf{J}_s$ and $\mathbf{M}_s$ and from these calculate an approximation to **E**, **H** within S.

In contrast to the above, the induction theorem yields known currents [Eqs. (3-25)]. (This assumes that $\mathbf{E}^i$, $\mathbf{H}^i$ is known.) We cannot, however, use Eqs. (3-4) and (3-5) to calculate the field from $\mathbf{J}_s$, $\mathbf{M}_s$, for they radiate in the presence of the obstacle. A determination of this field is a boundary-value problem of the same order of complexity as the original problem (Fig. 3-15a). We can, however, approximate the field of $\mathbf{J}_s$, $\mathbf{M}_s$ and thereby obtain an approximate formula for **E**, **H** internal to S and $\mathbf{E}^s$, $\mathbf{H}^s$ external to S.

A simplification of the induction theorem occurs when the obstacle is a perfect conductor. This situation is represented by Fig. 3-16a. The solution **E** must satisfy the boundary condition $\mathbf{n} \times \mathbf{E} = 0$ on S (zero tangential **E**). It then follows from the first of Eqs. (3-24) that

$$\mathbf{n} \times \mathbf{E}^s = -\mathbf{n} \times \mathbf{E}^i \qquad \text{on } S \qquad (3\text{-}26)$$

We now know the tangential components of $\mathbf{E}^s$ over S; so we can construct the induction representation of Fig. 3-16b as follows. We keep the perfectly conducting obstacle and specify that external to S the field $\mathbf{E}^s$, $\mathbf{H}^s$ exists. To support this field, there must be magnetic currents on S given by

$$\mathbf{M}_s = \mathbf{E}^s \times \mathbf{n} = \mathbf{n} \times \mathbf{E}^i \qquad (3\text{-}27)$$

We can visualize this current as causing the tangential components of **E** to jump from zero at the conductor to those of $\mathbf{E}^s$ just outside $\mathbf{M}_s$. The

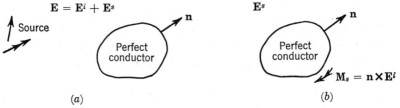

Fig. 3-16. The induction theorem as applied to a perfectly conducting obstacle. (a) Original problem; (b) induction equivalent.

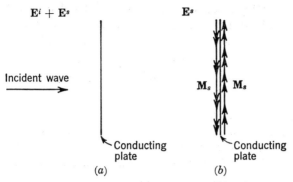

FIG. 3-17. Scattering by a conducting plate. (a) Original problem; (b) induction equivalent.

tangential components of **E** in Fig. 3-16b therefore have been forced to be **E**^{s}. Thus, according to uniqueness concepts, the currents of Eq. (3-27) radiating in the presence of the conducting obstacle must produce **E**^{s}, **H**^{s} external to S.

It is interesting to compare this result with the previous one (Fig. 3-15b). We found that, in general, both electric and magnetic currents exist on S in the induction representation. How, then, can both Fig. 3-15b and Fig. 3-16b be correct for a perfectly conducting obstacle? The answer must be that an electric current impressed along a perfect electric conductor produces no field. If the conductor is plane, this is evident from image theory. We can prove it, in general, by using the reciprocity concepts of the next section.

To illustrate an application of the induction theorem, consider the problem of determining the back scattering, or radar echo, from a large conducting plate. This problem is suggested by Fig. 3-17a. For normal incidence, let the plate lie in the $z = 0$ plane and let the incident field be specified by

$$E_x^i = E_0 e^{-jkz} \tag{3-28}$$

According to the induction theorem, the scattered field is produced by the currents $M_y = E_0$ on the side facing the source and $M_y = -E_0$ on the side away from the source. These currents radiate in the presence of the original conducting plate, as represented by Fig. 3-17b. Let the field from each element of current be approximated by the field from an element adjacent to a ground plane. According to image theory, this means that each element of M_y seen by the receiver radiates as $2M_y = 2E_0$ in free space. Hence, far from the plate, it contributes

$$dE_x^s = \frac{-jkE_0\,ds}{2\pi r} e^{-jkr}$$

in the back-scatter direction. Each element not seen by the receiver contributes nothing to the back-scattered field. Summing over the entire plate, we have the distant back-scattered field given by

$$E_x^s = \iint_{\text{plate}} dE_x^s = \frac{-jkE_0 A}{2\pi r} e^{-jkr} \tag{3-29}$$

where A is the area of the plate.

The *echo area* or *radar cross section* of an obstacle is defined as the area for which the incident wave contains sufficient power to produce, by omnidirectional radiation, the same back-scattered power density. In mathematical form, the echo area is

$$A_e = \lim_{r \to \infty} \left(4\pi r^2 \frac{\bar{\mathbb{S}}^s}{\bar{\mathbb{S}}^i} \right) \tag{3-30}$$

where $\bar{\mathbb{S}}^i$ is the incident power density and $\bar{\mathbb{S}}^s$ is the scattered power density. For our problem, $\bar{\mathbb{S}}^i = |E_0|^2/\eta$ and, from Eq. (3-29),

$$\bar{\mathbb{S}}^s = \frac{1}{\eta} \left| \frac{kE_0 A}{2\pi r} \right|^2$$

The echo area of a conducting plate is therefore

$$A_e \approx \frac{k^2 A^2}{\pi} = \frac{4\pi A^2}{\lambda^2} \tag{3-31}$$

valid for large plates and normal incidence.

3-8. Reciprocity. In its simplest sense, a reciprocity theorem states that a response of a system to a source is unchanged when source and measurer are interchanged. In a more general sense, reciprocity theorems relate a response at one source due to a second source to the response at the second source due to the first source. We shall establish this type of reciprocity relationship for a-c fields. The reciprocity theorem of circuit theory is a special case of this reciprocity theorem for fields.

Consider two sets of a-c sources, $\mathbf{J}^a$, $\mathbf{M}^a$ and $\mathbf{J}^b$, $\mathbf{M}^b$, of the same frequency, existing in the same linear medium. Denote the field produced by the a sources alone by $\mathbf{E}^a$, $\mathbf{H}^a$, and the field produced by the b sources alone by $\mathbf{E}^b$, $\mathbf{H}^b$. The field equations are then

$$\nabla \times \mathbf{H}^a = \hat{y}\mathbf{E}^a + \mathbf{J}^a \qquad \nabla \times \mathbf{H}^b = \hat{y}\mathbf{E}^b + \mathbf{J}^b$$
$$-\nabla \times \mathbf{E}^a = \hat{z}\mathbf{H}^a + \mathbf{M}^a \qquad -\nabla \times \mathbf{E}^b = \hat{z}\mathbf{H}^b + \mathbf{M}^b$$

We multiply the first equation scalarly by $\mathbf{E}^b$ and the last equation by $\mathbf{H}^a$ and add the resulting equations. This gives

$$-\nabla \cdot (\mathbf{E}^b \times \mathbf{H}^a) = \hat{y}\mathbf{E}^a \cdot \mathbf{E}^b + \hat{z}\mathbf{H}^a \cdot \mathbf{H}^b + \mathbf{E}^b \cdot \mathbf{J}^a + \mathbf{H}^a \cdot \mathbf{M}^b$$

where the left-hand term has been simplified by the identity

$$\nabla \cdot (\mathbf{A} \times \mathbf{B}) = \mathbf{B} \cdot \nabla \times \mathbf{A} - \mathbf{A} \cdot \nabla \times \mathbf{B}$$

An interchange of a and b in this result gives

$$-\nabla \cdot (\mathbf{E}^a \times \mathbf{H}^b) = \hat{y}\mathbf{E}^a \cdot \mathbf{E}^b + \hat{z}\mathbf{H}^a \cdot \mathbf{H}^b + \mathbf{E}^a \cdot \mathbf{J}^b + \mathbf{H}^b \cdot \mathbf{M}^a$$

A subtraction of the former equation from the latter yields

$$-\nabla \cdot (\mathbf{E}^a \times \mathbf{H}^b - \mathbf{E}^b \times \mathbf{H}^a) = \mathbf{E}^a \cdot \mathbf{J}^b + \mathbf{H}^b \cdot \mathbf{M}^a - \mathbf{E}^b \cdot \mathbf{J}^a - \mathbf{H}^a \cdot \mathbf{M}^b \tag{3-32}$$

At any point for which the fields are source-free ($\mathbf{J} = \mathbf{M} = 0$), this reduces to

$$\nabla \cdot (\mathbf{E}^a \times \mathbf{H}^b - \mathbf{E}^b \times \mathbf{H}^a) = 0 \tag{3-33}$$

which is called the *Lorentz reciprocity theorem*. If Eq. (3-33) is integrated throughout a source-free region and the divergence theorem applied, we have

$$\oint (\mathbf{E}^a \times \mathbf{H}^b - \mathbf{E}^b \times \mathbf{H}^a) \cdot d\mathbf{s} = 0 \tag{3-34}$$

which is the integral form of the Lorentz reciprocity theorem for a source-free region.

For a region containing sources, integration of Eq. (3-32) throughout the region gives

$$-\oint (\mathbf{E}^a \times \mathbf{H}^b - \mathbf{E}^b \times \mathbf{H}^a) \cdot d\mathbf{s}$$
$$= \iiint (\mathbf{E}^a \cdot \mathbf{J}^b - \mathbf{H}^a \cdot \mathbf{M}^b - \mathbf{E}^b \cdot \mathbf{J}^a + \mathbf{H}^b \cdot \mathbf{M}^a) \, d\tau \tag{3-35}$$

Let us now postulate that all sources and matter are of finite extent. Distant from the sources and matter, we have (see Sec. 3-13)

$$E_\theta = \eta H_\phi \qquad E_\phi = -\eta H_\theta$$

The left-hand term of Eq. (3-35), integrated over a sphere of radius $r \to \infty$, is then

$$-\eta \oint (H_\theta^a H_\theta^b + H_\phi^a H_\phi^b - H_\theta^b H_\theta^a - H_\phi^b H_\phi^a) \, ds = 0$$

Equation (3-35) now reduces to

$$\iiint (\mathbf{E}^a \cdot \mathbf{J}^b - \mathbf{H}^a \cdot \mathbf{M}^b) \, d\tau = \iiint (\mathbf{E}^b \cdot \mathbf{J}^a - \mathbf{H}^b \cdot \mathbf{M}^a) \, d\tau \tag{3-36}$$

where the integration extends over all space. This is the most useful form of the reciprocity theorem for our purposes. Equation (3-36) also applies to regions of finite extent whenever Eq. (3-34) is satisfied. For example, fields in a region bounded by a perfect electric conductor satisfy Eq. (3-34); hence Eq. (3-36) applies in this case.

The integrals appearing in Eq. (3-36) *do not* in general represent power, since no conjugates appear. They have been given the name *reaction*.[1] By definition, the reaction of field a on source b is

$$\langle a,b \rangle = \iiint (\mathbf{E}^a \cdot \mathbf{J}^b - \mathbf{H}^a \cdot \mathbf{M}^b)\, d\tau \qquad (3\text{-}37)$$

In this notation, the reciprocity theorem is

$$\langle a,b \rangle = \langle b,a \rangle \qquad (3\text{-}38)$$

that is, the reaction of field a on source b is equal to the reaction of field b on source a. Reaction is a useful quantity primarily because of this conservative property. For example, reaction can be used as a measure of equivalency, since a source must have the same reaction with all fields equivalent over its extent. This equality of reaction is a necessary, but not sufficient, test of equivalence as defined in Sec. 3-5. We shall use the term *self-reaction* to denote the reaction of a field on its own sources, that is, $\langle a,a \rangle$.

A valuable tool for expositional purposes can be obtained by using the circuit sources of Fig. 3-1 in the reaction concept. For a current source (Fig. 3-1a), we have

$$\langle a,b \rangle = \int \mathbf{E}^a \cdot \mathbf{I}^b\, dl = I^b \int \mathbf{E}^a \cdot dl = -V^a I^b$$

where V^a is the voltage across the b source due to some (as yet unspecified) a source. For a voltage source (Fig. 3-1b), we have $K^b = -V^b$, and

$$\langle a,b \rangle = -\oint \mathbf{H}^a \cdot \mathbf{K}^b\, dl = -K^b \oint \mathbf{H}^a \cdot dl = V^b I^a$$

where I^a is the current through the b source due to some a source. To summarize, the "circuit reactions" are

$$\langle a,b \rangle = \begin{cases} -V^a I^b & b \text{ a current source} \\ +V^b I^a & b \text{ a voltage source} \end{cases} \qquad (3\text{-}39)$$

If we use a unit current source ($I^b = 1$), then $\langle a,b \rangle$ is a measure of V^a (the voltage at b due to another source a). If we use a unit voltage source ($V^b = 1$), then $\langle a,b \rangle$ is a measure of I^a (the current at b due to another source a).

To relate our reciprocity theorem to the usual circuit theory statement of reciprocity, consider the two-port (four-terminal) network of

---

[1] V. H. Rumsey, The Reaction Concept in Electromagnetic Theory, *Phys. Rev.*, ser. 2, vol. 94, no. 6, pp. 1483–1491, June 15, 1954.

Fig. 3-18. The characteristics of a linear network can be described by the impedance matrix $[z]$ defined by

$$\begin{bmatrix} V_1 \\ V_2 \end{bmatrix} = \begin{bmatrix} z_{11} & z_{12} \\ z_{21} & z_{22} \end{bmatrix} \begin{bmatrix} I_1 \\ I_2 \end{bmatrix} \quad (3\text{-}40)$$

Suppose we apply a current source I_1 at port 1 and a current source I_2 at port 2. Let the partial response V_{ij} be the voltage at port i due to source I_j at port j. Each current source sees the other port open-circuited (see Fig. 3-1a); hence

$$z_{ij} = \frac{V_{ij}}{I_j}$$

In terms of the circuit reactions [Eq. (3-39)], $\langle j,i \rangle = -V_{ij}I_i$; hence

$$z_{ij} = -\frac{\langle j,i \rangle}{I_i I_j} \quad (3\text{-}41)$$

Thus, the elements of the impedance matrix are the various reactions among two unit current sources. The reciprocity theorem [Eq. (3-38)], applied to Eq. (3-41), shows that

$$z_{ij} = z_{ji} \quad (3\text{-}42)$$

which is the usual statement of reciprocity in circuit theory. Equations (3-41) and (3-42) also apply to an N-port network. The use of voltage sources instead of current sources gives reactions proportional to the elements of the admittance matrix $[y]$, and reciprocity then states that $y_{ij} = y_{ji}$.

The proofs of many other theorems can be based on the reciprocity theorem. For example, the preceding paragraph is a proof that *any network constructed of linear isotropic matter has a symmetrical impedance matrix*. This "network" might be the two antennas of Fig. 3-19. Reciprocity in this case can be stated as: The voltage at b due to a current source at a is equal to the voltage at a due to the same current source at b. If the b antenna is infinitely remote from the a antenna, its field will be a plane wave in the vicinity of a, and vice versa. The *receiving pattern* of an antenna is defined as the voltage at the antenna

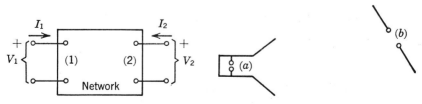

FIG. 3-18. A two-port network.   FIG. 3-19. Two antennas.

terminals due to a plane wave incident upon the antenna. The reciprocity theorem for antennas can thus be stated as: *The receiving pattern of any antenna constructed of linear isotropic matter is identical to its transmitting pattern.*

In Secs. 3-5 and 3-7, we used the fact that an electric current impressed along the surface of a perfect electric conductor radiated no field. The reciprocity theorem proves this, in general, as follows. Visualize a set of terminals a on the conductor and another set of terminals b in space away from the conductor. A current element at b produces no tangential component of **E** along the conductor; so V_{ab} (V at a due to I_b) is zero. By reciprocity, V_{ba} (V at b due to I_a) is zero. The terminals b are arbitrary; so the current element along the conductor (at a) produces no V between any two points in space; hence it produces no **E**. We can think of I_a as inducing currents on the conductor such that these currents produce a free-space field equal and opposite to the free-space field of I_a.

3-9. Green's Functions. Our reciprocity relationships are formulas symmetrical in two field-source pairs. Mathematical statements of reciprocity (symmetrical in two functions) are called *Green's theorems*. The difference between a Green's theorem and a reciprocity theorem is that no physical interpretation is given to the functions in the former.

The scalar Green's theorem is based on the identity

$$\nabla \cdot (\psi \nabla \phi) = \psi \nabla^2 \phi + \nabla \psi \cdot \nabla \phi$$

When this is integrated throughout a region and the divergence theorem applied to the left-hand term, we obtain *Green's first identity*

$$\oint \psi \frac{\partial \phi}{\partial n} ds = \iiint (\psi \nabla^2 \phi + \nabla \psi \cdot \nabla \phi) \, d\tau \qquad (3\text{-}43)$$

Interchanging ψ and ϕ in this identity and subtracting the interchanged equation from the original equation, we obtain *Green's second identity* or *Green's theorem*

$$\oint \left(\psi \frac{\partial \phi}{\partial n} - \phi \frac{\partial \psi}{\partial n} \right) ds = \iiint (\psi \nabla^2 \phi - \phi \nabla^2 \psi) \, d\tau \qquad (3\text{-}44)$$

This is a statement of reciprocity for scalar fields ψ and ϕ.

The vector analogue to Green's theorem is based on the identity

$$\nabla \cdot (\mathbf{A} \times \nabla \times \mathbf{B}) = \nabla \times \mathbf{A} \cdot \nabla \times \mathbf{B} - \mathbf{A} \cdot \nabla \times \nabla \times \mathbf{B}$$

An integration of this throughout a region and an application of the divergence theorem yields the vector analogue to Green's first identity

$$\oint (\mathbf{A} \times \nabla \times \mathbf{B}) \cdot d\mathbf{s} = \iiint (\nabla \times \mathbf{A} \cdot \nabla \times \mathbf{B} - \mathbf{A} \cdot \nabla \times \nabla \times \mathbf{B}) \, d\tau \qquad (3\text{-}45)$$

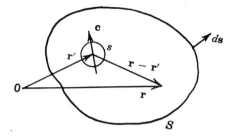

Fig. 3-20. Region to which Green's theorem is applied.

We can interchange **A** and **B** and subtract the resulting equation from the original equation. This gives the vector analogue to Green's second identity, or the vector Green's theorem,

$$\oiint (\mathbf{A} \times \nabla \times \mathbf{B} - \mathbf{B} \times \nabla \times \mathbf{A}) \cdot d\mathbf{s}$$
$$= \iiint (\mathbf{B} \cdot \nabla \times \nabla \times \mathbf{A} - \mathbf{A} \cdot \nabla \times \nabla \times \mathbf{B})\, d\tau \quad (3\text{-}46)$$

Our reciprocity theorem [Eq. (3-35)], for a *homogeneous* medium, is essentially Eq. (3-46) with $\mathbf{A} = \mathbf{E}^a$ and $\mathbf{B} = \mathbf{E}^b$. For an inhomogeneous medium, still another vector Green's theorem corresponds to our reciprocity theorem (see Prob. 3-28).

Green's theorems have been used extensively in the literature as follows. Suppose we desire the field **E** at a point $\mathbf{r}'$ in a region. Instead of solving this problem directly, a point source is placed at $\mathbf{r}'$, and its field is called a *Green's function* **G**. We then substitute $\mathbf{E} = \mathbf{A}$ and $\mathbf{G} = \mathbf{B}$ in Eq. (3-46). This gives a formula for **E** at $\mathbf{r}'$, as we shall discuss below. What we have done is solve the reciprocal problem (source at the field point of the original problem) and then apply reciprocity. The equivalence principle gives the solution more directly.

Let us summarize the various Green's functions used in the literature. Stratton chooses[1]

$$\mathbf{G}_1 = \mathbf{c}\phi \quad (3\text{-}47)$$

where

$$\phi = \frac{e^{-jk|\mathbf{r}-\mathbf{r}'|}}{|\mathbf{r}-\mathbf{r}'|} \quad (3\text{-}48)$$

and **c** is a constant vector. A comparison of Eq. (3-47) with Eq. (2-117) shows that $\mathbf{G}_1$ is the vector potential of a current element $Il = 4\pi\mathbf{c}$. Hence, $\mathbf{G}_1$ is a solution to Eq. (2-108), or

$$\nabla \times \nabla \times \mathbf{G}_1 - k^2 \mathbf{G}_1 = \nabla(\nabla \cdot \mathbf{G}_1) \quad \mathbf{r} \neq \mathbf{r}' \quad (3\text{-}49)$$

Now suppose we wish to find **E** at $\mathbf{r}'$ in a source-free region enclosed by S. The source of $\mathbf{G}_1$ is placed at $\mathbf{r}'$ and surrounded by an infinitesimal sphere s, as shown in Fig. 3-20. Equation (3-46) with $\mathbf{A} = \mathbf{E}$ and $\mathbf{B} = \mathbf{G}_1$ is now

---

[1] J. A. Stratton, "Electromagnetic Theory," p. 464, McGraw-Hill Book Company, Inc., New York, 1941.

applied to the region enclosed by S and s. The result is

$$-4\pi \mathbf{c} \cdot \mathbf{E} = \oint_S (\mathbf{E} \times \nabla \times \mathbf{G}_1 - \mathbf{G}_1 \times \nabla \times \mathbf{E} + \mathbf{E} \nabla \cdot \mathbf{G}_1) \cdot d\mathbf{s} \quad (3\text{-}50)$$

which is a formula for calculating $\mathbf{E}$ at $\mathbf{r}'$ in terms of $\mathbf{n} \times \mathbf{E}$, $\mathbf{n} \times \nabla \times \mathbf{E}$, and $\mathbf{n} \cdot \mathbf{E}$ on S. Furthermore, it is required that $\mathbf{E}$ be continuous and have continuous first derivatives on S. This is a severe restriction on the usefulness of Eq. (3-50), although it can be amended to admit singular $\mathbf{E}$'s on S.

A choice of Green's function which overcomes some of the disadvantages of Eq. (3-50) is[1]

$$\mathbf{G}_2 = \nabla \times \mathbf{c}\phi \quad (3\text{-}51)$$

where ϕ is given by Eq. (3-48). This is evidently the magnetic field of a current element $Il = 4\pi\mathbf{c}$. Hence, $\mathbf{G}_2$ is a solution to

$$\nabla \times \nabla \times \mathbf{G}_2 - k^2 \mathbf{G}_2 = 0 \quad \mathbf{r} \neq \mathbf{r}' \quad (3\text{-}52)$$

We now apply Eq. (3-46) with $\mathbf{A} = \mathbf{E}$ and $\mathbf{B} = \mathbf{G}_2$ to the region enclosed by S and s in Fig. 3-20. The result is[2]

$$4\pi \mathbf{c} \cdot \nabla' \times \mathbf{E} = \oint_S (\mathbf{G}_2 \times \nabla \times \mathbf{E} - \mathbf{E} \times \nabla \times \mathbf{G}_2) \cdot d\mathbf{s} \quad (3\text{-}53)$$

This is a formula for $\nabla' \times \mathbf{E}$ (hence for $\mathbf{H}$) at $\mathbf{r}'$ in terms of $\mathbf{n} \times \mathbf{E}$ and $\mathbf{n} \times \nabla \times \mathbf{E}$ on S. Equation (3-53) does not require $\mathbf{E}$ to be continuous on S, nor do we need to know $\mathbf{n} \cdot \mathbf{E}$ on S. Thus, Eq. (3-53) is a substantial improvement over Eq. (3-50). In fact, Eq. (3-53) can be shown to be identical to the formula obtained from the equivalence principle of Fig. 3-9, applied to a homogeneous medium.

Another useful Green's function is

$$\mathbf{G}_3 = \nabla \times \nabla \times \mathbf{c}\phi \quad (3\text{-}54)$$

where ϕ is given by Eq. (3-48). This is proportional to the electric field of an electric current element; so $\mathbf{G}_3$ also satisfies Eq. (3-52). An application of Eq. (3-46) would yield a formula for $\mathbf{E}$ at $\mathbf{r}'$, similar in form to Eq. (3-53).

All of the $\mathbf{G}$'s considered so far are "free-space" Green's functions, that is, they are fields of sources radiating into unbounded space. We can choose other $\mathbf{G}$'s such that they satisfy boundary conditions on S.

---

[1] J. R. Mentzer, "Scattering and Diffraction of Radio Waves," p. 14, Pergamon Press, New York, 1955.

[2] The left-hand side of this equation is a function only of the primed coordinates. Hence, a prime is placed on ∇' to indicate operation on $\mathbf{r}'$ instead of $\mathbf{r}$.

For example, let
$$\mathbf{G}_4 = \mathbf{G}_2 + \mathbf{G}_4{}^s \tag{3-55}$$
such that $\mathbf{G}_4$ satisfies Eq. (3-52) and
$$\mathbf{n} \times \nabla \times \mathbf{G}_4 = 0 \quad \text{on } S \tag{3-56}$$
The physical interpretation of $\mathbf{G}_4$ is that it is the magnetic field of a current element $Il = 4\pi\mathbf{c}$ radiating in the presence of a perfect electric conductor over S. The $\mathbf{G}_2$ is the incident field, and the $\mathbf{G}_4{}^s$ is the scattered field. Application of Eq. (3-46) with $\mathbf{A} = \mathbf{E}$ and $\mathbf{B} = \mathbf{G}_4$ results in Eq. (3-53) with the last term zero, because of Eq. (3-56). Thus,
$$4\pi\mathbf{c} \cdot \nabla' \times \mathbf{E} = \oint_S (\mathbf{G}_4 \times \nabla \times \mathbf{E}) \cdot d\mathbf{s} \tag{3-57}$$
which is a formula for $\nabla' \times \mathbf{E}$ in terms of only $\mathbf{n} \times \nabla \times \mathbf{E}$ over S. The same formula can be obtained from the equivalence principle of Fig. 3-11, as it applies to a homogeneous region.

Similarly, defining a $\mathbf{G}_5$ such that
$$\mathbf{n} \times \mathbf{G}_5 = 0 \quad \text{on } S \tag{3-58}$$
we can obtain a formula
$$4\pi\mathbf{c} \cdot \nabla' \times \mathbf{E} = -\oint_S (\mathbf{E} \times \nabla \times \mathbf{G}_5) \cdot d\mathbf{s} \tag{3-59}$$
and so on. All these various formulas, and many more, can be directly obtained from the equivalence principle. We have discussed the Green's function approach merely because it has been used extensively in the literature.

3-10. Tensor Green's Functions. We shall henceforth use the term "Green's function" to mean "field of a point source." Suppose we have a current element Il at $\mathbf{r}'$ and we wish to evaluate the field $\mathbf{E}$ at $\mathbf{r}$. The most general linear relationship between two vector quantities can be represented by a tensor. Hence, the field $\mathbf{E}$ is related to the source Il by
$$\mathbf{E} = [\Gamma]Il \tag{3-60}$$
where $[\Gamma]$ is called a *tensor Green's function*. In rectangular components and matrix notation, Eq. (3-60) becomes
$$\begin{bmatrix} E_x \\ E_y \\ E_z \end{bmatrix} = \begin{bmatrix} \Gamma_{xx} & \Gamma_{xy} & \Gamma_{xz} \\ \Gamma_{yx} & \Gamma_{yy} & \Gamma_{yz} \\ \Gamma_{zx} & \Gamma_{zy} & \Gamma_{zz} \end{bmatrix} \begin{bmatrix} Il_x \\ Il_y \\ Il_z \end{bmatrix} \tag{3-61}$$
Thus, Γ_{ij} is the ith component of $\mathbf{E}$ due to a unit j-directed electric current element. The $\mathbf{E}$ might be the free-space field of Il, in which case

[Γ] would be the "free-space Green's function." Alternatively, **E** might be the field of *I*l radiating in the presence of some matter, and [Γ] would then be called the "Green's function subject to boundary conditions." Still other Green's functions are those relating **H** to *I*l, those relating **E** to *K*l, and so on.

Our principal use of tensor Green's functions will be for concise mathematical expression. For example, the equation

$$\mathbf{E} = \iiint [\Gamma] \mathbf{J} \, d\tau' \qquad (3\text{-}62)$$

where [Γ] is the free-space Green's function defined by Eq. (3-60), represents the solution of Eq. (2-111), which is

$$\mathbf{E} = -j\omega\mu\mathbf{A} + \frac{1}{j\omega\epsilon} \nabla(\nabla \cdot \mathbf{A})$$
$$\mathbf{A} = \iiint \frac{\mathbf{J}e^{-jk|\mathbf{r}-\mathbf{r'}|}}{4\pi|\mathbf{r}-\mathbf{r'}|} \, d\tau' \qquad (3\text{-}63)$$

Equation (3-62) also represents the field of currents in the vicinity of a material body if [Γ] represents the appropriate Green's function, and so on. In other words, Eq. (3-62) is symbolic of the solution, regardless of whether or not we can find [Γ].

Even though we shall not use tensor Green's functions to find explicit solutions, it should prove instructive to find an explicit [Γ]. Let us take [Γ] to be the free-space Green's function defined by Eq. (3-60). If *I*l is x-directed,

$$A_x = \frac{Ile^{-jk|\mathbf{r}-\mathbf{r'}|}}{4\pi|\mathbf{r}-\mathbf{r'}|}$$

and

$$E_x = -j\omega\mu A_x + \frac{1}{j\omega\epsilon} \frac{\partial^2 A_x}{\partial x^2}$$
$$E_y = \frac{1}{j\omega\epsilon} \frac{\partial^2 A_x}{\partial y \, \partial x}$$
$$E_z = \frac{1}{j\omega\epsilon} \frac{\partial^2 A_x}{\partial z \, \partial x}$$

Comparing this with Eq. (3-61) for $Il_y = Il_z = 0$, we see that

$$\Gamma_{xx} = \left(-j\omega\mu + \frac{1}{j\omega\epsilon} \frac{\partial^2}{\partial x^2}\right)\psi$$
$$\Gamma_{yx} = \frac{1}{j\omega\epsilon} \frac{\partial^2 \psi}{\partial y \, \partial x}$$
$$\Gamma_{zx} = \frac{1}{j\omega\epsilon} \frac{\partial^2 \psi}{\partial z \, \partial x}$$

where

$$\psi = \frac{e^{-jk|\mathbf{r}-\mathbf{r'}|}}{4\pi|\mathbf{r}-\mathbf{r'}|} \qquad (3\text{-}64)$$

The other elements of $[\Gamma]$ are found by taking Il to be y-directed and then z-directed. From symmetry considerations, the other Γ_{ij}'s will differ only by a cyclic interchange of (x,y,z). The result is therefore

$$\Gamma_{ii} = \left(-j\omega\mu + \frac{1}{j\omega\epsilon}\frac{\partial^2}{\partial i^2}\right)\psi$$

$$\Gamma_{ij} = \frac{1}{j\omega\epsilon}\frac{\partial^2\psi}{\partial i\,\partial j} \qquad i \neq j \tag{3-65}$$

with ψ given by Eq. (3-64). The reciprocity theorem is reflected in the symmetry

$$\Gamma_{ij}(\mathbf{r},\mathbf{r}') = \Gamma_{ji}(\mathbf{r}',\mathbf{r}) \tag{3-66}$$

which can be proved for Γ's subject to boundary conditions as well.

3-11. Integral Equations. An integral equation is one for which the unknown quantity appears in an integrand. We already have the concepts needed to construct integral equations. For example, the potential integral of Eq. (2-118) is essentially an integral equation when $\mathbf{J}$ is unknown. Most problems can be formulated either in terms of integral equations or in terms of differential equations. When *exact* solutions are desired, the differential equation approach is usually the simpler one. An important use of integral equations is to obtain *approximate* solutions. There is good reason for this. Integration is a summation process, and it is not necessary that each element of the summation be correct. Errors in some elements of the summation may be compensated for by errors in other elements. Also, all elements do not contribute equally to a summation. It is much more important that the elements contributing most to the summation be correct than that the elements of minor contribution be correct. This is why we were able to obtain useful results by assuming the current on the linear antenna of Fig. 2-23, by assuming the field of each element of magnetic current in Fig. 3-17b, and so on.

To illustrate the formulation of an integral equation, consider the induction theorem of Fig. 3-16. Let $[\Gamma(\mathbf{r},\mathbf{r}')]$ be the tensor relating the $\mathbf{E}$ field at $\mathbf{r}$ due to an element of $\mathbf{M}$ at $\mathbf{r}'$ radiating in the presence of the conductor over S. In equation form, this is

$$d\mathbf{E}(\mathbf{r}) = [\Gamma(\mathbf{r},\mathbf{r}')]\,d\mathbf{M}(\mathbf{r}')$$

The total scattered field for the problem is then the summation

$$\mathbf{E}^s(\mathbf{r}) = \oiint_S [\Gamma(\mathbf{r},\mathbf{r}')]\mathbf{M}_s(\mathbf{r}')\,ds'$$

where $\mathbf{M}_s$ is given by Eq. (3-27). When $\mathbf{r}$ is on S, Eq. (3-26) must also be true; hence

$$\mathbf{n} \times \mathbf{E}^i(\mathbf{r}) = \mathbf{n} \times \oint_S [\Gamma(\mathbf{r},\mathbf{r}')]\mathbf{E}^i(\mathbf{r}') \times d\mathbf{s}' \qquad \mathbf{r} \text{ on } S \qquad (3\text{-}67)$$

The incident field $\mathbf{E}^i$ is assumed to be known; so Eq. (3-67) is an integral equation for determining $[\Gamma]$. As we mentioned earlier, an exact solution to Eq. (3-67) would be difficult even for the simplest specialization.

Problems involving a region homogeneous except for small "islands" of matter are commonly encountered. Examples of such problems are the linear antenna of Fig. 2-23 and the obstacle of Fig. 3-15a. To illustrate the general concepts involved, suppose we have an inhomogeneous region, possibly containing sources $\mathbf{J}^i$ and $\mathbf{M}^i$. Within this region, the field satisfies

$$-\nabla \times \mathbf{E} = \hat{z}\mathbf{H} + \mathbf{M}^i \qquad \nabla \times \mathbf{H} = \hat{y}\mathbf{E} + \mathbf{J}^i$$

where $\hat{z}$ and $\hat{y}$ are functions of position. We can define *normal* values of impedivity and admittivity, $\hat{z}_1$ and $\hat{y}_1$, which may be any convenient constants (usually the most common $\hat{z}$ and $\hat{y}$ in the region). We can now rewrite the field equations as

$$-\nabla \times \mathbf{E} = \hat{z}_1 \mathbf{H} + \mathbf{M} \qquad \nabla \times \mathbf{H} = \hat{y}_1 \mathbf{E} + \mathbf{J}$$

where the *effective* currents are

$$\begin{aligned}\mathbf{M} &= (\hat{z} - \hat{z}_1)\mathbf{H} + \mathbf{M}^i \\ \mathbf{J} &= (\hat{y} - \hat{y}_1)\mathbf{E} + \mathbf{J}^i\end{aligned} \qquad (3\text{-}68)$$

These effective currents can then be treated as source currents in a homogeneous region. Since $\mathbf{J}$ and $\mathbf{M}$ are functions of $\mathbf{E}$ and $\mathbf{H}$, a solution in terms of them will lead to an integral equation. However, if $\hat{z} = \hat{z}_1$ and $\hat{y} = \hat{y}_1$ except in small subregions, we can assume $\mathbf{J}$ and $\mathbf{M}$ in the subregions and obtain approximate expressions for $\mathbf{E}$ and $\mathbf{H}$ elsewhere. (Recall the linear antenna problem, where we assumed I on the antenna wire.) Note that, when the normal $\hat{z}$ and $\hat{y}$ are taken as the free-space parameters, Eqs. (3-68) reduce to

$$\begin{aligned}\mathbf{M} &= j\omega(\hat{\mu} - \mu_0)\mathbf{H} + \mathbf{M}^i \\ \mathbf{J} &= j\omega(\hat{\epsilon} - \epsilon_0)\mathbf{E} + \sigma\mathbf{E} + \mathbf{J}^i\end{aligned} \qquad (3\text{-}69)$$

The effective currents in excess of the true sources ($\mathbf{M}^i$ and $\mathbf{J}^i$) are now just those due to the motion of atomic particles in vacuum.

Let us reconsider the problem of scattering by an obstacle in the light of the above concepts. Given the problem of Fig. 3-15a, we can consider the total field to be the potential integral solution of Eqs. (3-4) and (3-5), with $\mathbf{J}$ and $\mathbf{M}$ given by Eqs. (3-69). The incident field is that produced by $\mathbf{J}^i$ and $\mathbf{M}^i$ outside of the obstacle, and the scattered field is that produced by

$$\mathbf{M} = j\omega(\hat{\mu} - \mu_0)\mathbf{H}$$
$$\mathbf{J} = j\omega(\hat{\epsilon} - \epsilon_0)\mathbf{E} + \sigma\mathbf{E} \qquad (3\text{-}70)$$

throughout the obstacle. To be explicit, outside of the obstacle

$$\mathbf{E}^s = -\nabla \times \mathbf{F} + \frac{1}{j\omega\epsilon_0}\nabla \times \nabla \times \mathbf{A} \qquad (3\text{-}71)$$

where

$$\mathbf{A} = \frac{1}{4\pi}\iiint_{\text{obstacle}} \frac{\mathbf{J}e^{-jk|\mathbf{r}-\mathbf{r}'|}}{|\mathbf{r}-\mathbf{r}'|}\,d\tau'$$

$$\mathbf{F} = \frac{1}{4\pi}\iiint_{\text{obstacle}} \frac{\mathbf{M}e^{-jk|\mathbf{r}-\mathbf{r}'|}}{|\mathbf{r}-\mathbf{r}'|}\,d\tau' \qquad (3\text{-}72)$$

with $\mathbf{J}$ and $\mathbf{M}$ given by Eq. (3-70). If we can guess $\mathbf{J}$ and $\mathbf{M}$ with reasonable accuracy, then Eqs. (3-71) and (3-72) will give us an approximate solution. For a nonmagnetic obstacle, $\mathbf{M}$, and consequently $\mathbf{F}$, will be zero. For a good conductor, $\mathbf{J}$ reduces to $\sigma\mathbf{E}$, and this current resides primarily on the surface of the obstacle. If we assume the obstacle perfectly conducting, then $\mathbf{J}$ becomes a true surface current. The solution in this case reduces to

$$\mathbf{E}^s = \frac{1}{4\pi j\omega\epsilon_0}\nabla \times \nabla \times \oint_S \frac{\mathbf{J}_s e^{-jk|\mathbf{r}-\mathbf{r}'|}}{|\mathbf{r}-\mathbf{r}'|}\,ds' \qquad (3\text{-}73)$$

If we specialize this equation to S, then Eq. (3-26) must be met, and we have an integral equation for determining $\mathbf{J}_s$.

An approximation to $\mathbf{J}_s$, known as the *physical optics approximation*, is as follows. Let Fig. 3-21a represent a perfectly conducting obstacle illuminated by some source. In terms of the total field, the surface current on the conductor is given by

$$\mathbf{J}_s = \mathbf{n} \times \mathbf{H}$$

When the obstacle is large, we assume that the total field is negligible in

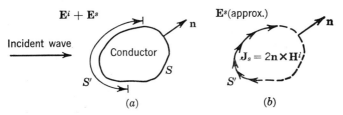

Fig. 3-21. The physical optics approximation. (a) Original problem; (b) the approximation.

the "shadow" region. Furthermore, if the obstacle is smooth and gently curved, each element of surface behaves similarly to an element of a ground plane. According to image theory, the tangential components of **H** at a ground plane are just twice those from the same source in unbounded space. We therefore approximate the current on the obstacle by

$$\mathbf{J}_s \approx 2\mathbf{n} \times \mathbf{H}^i \qquad \text{over } S' \tag{3-74}$$

where S' is the illuminated portion of S. The physical optics approximation to the scattered field is therefore

$$\mathbf{E}^s \approx \frac{1}{2\pi j\omega\epsilon_0} \nabla \times \nabla \times \iint_{S'} \frac{(\mathbf{n} \times \mathbf{H}^i)e^{-jk|\mathbf{r}-\mathbf{r}'|}}{|\mathbf{r}-\mathbf{r}'|} ds' \tag{3-75}$$

This approximation is illustrated by Fig. 3-21b.

As an explicit application of the physical optics approximation, again consider the large conducting plate of Fig. 3-17a. The incident **E** is given by Eq. (3-28); hence

$$H_y^i = \frac{E_0}{\eta} e^{-jkz}$$

The physical optics approximation to the obstacle current [Eq. (3-74)] is therefore

$$J_x = \frac{2E_0}{\eta}$$

Each element of this radiates as a current element in free space, as analyzed in Sec. 2-9. The contribution to the radiation field in the back-scatter direction from each $J_x \, ds$ is

$$dE_x^s = \frac{-jkE_0 \, ds}{2\pi r} e^{-jkr}$$

The total distant back-scattered field is therefore

$$E_x^s = \iint_{\text{plate}} dE_x^s = -\frac{jkE_0 A}{2\pi r} e^{-jkr} \tag{3-76}$$

which is identical to Eq. (3-29), the approximation obtained from the induction theorem. The physical optics approximation to the echo area of the plate is therefore that of Eq. (3-31). This equality of the two approximations to back scattering [Eqs. (3-29) and (3-76)] is no coincidence. It can be shown that the two approaches *always* give the same back scattering but *do not* give the same scattering in other directions.[1]

[1] R. F. Harrington, On Scattering by Large Conducting Bodies, *IRE Trans.*, vol. AP-7, no. 2, pp. 150–153, April, 1959.

3-12. Construction of Solutions. So far, we have explicitly considered only two types of solutions to the field equations, namely, uniform plane waves and the potential integrals. In the next three chapters, we shall learn how to construct many other solutions. A general method of obtaining these solutions is considered here.

In a homogeneous source-free region, the field satisfies

$$-\nabla \times \mathbf{E} = \hat{z}\mathbf{H} \qquad \nabla \cdot \mathbf{H} = 0 \\ \nabla \times \mathbf{H} = \hat{y}\mathbf{E} \qquad \nabla \cdot \mathbf{E} = 0 \tag{3-77}$$

In view of the divergenceless character of **E** and **H**, we can express the field in terms of a magnetic vector potential **A** or in terms of an electric vector potential **F**. More important, we can employ superposition and express part of the field in terms of **A** and part in terms of **F**. The **A** must be a solution to Eq. (2-108) with $\mathbf{J} = 0$, and the **F** a solution to the dual equation. The general equations for vector potentials are therefore

$$\nabla \times \nabla \times \mathbf{A} - k^2\mathbf{A} = -\hat{y}\nabla\Phi^a \\ \nabla \times \nabla \times \mathbf{F} - k^2\mathbf{F} = -\hat{z}\nabla\Phi^f \tag{3-78}$$

where Φ^a and Φ^f are arbitrary scalars. The electromagnetic field in terms of **A** and **F** is given by Eqs. (3-4) with $\mathbf{J} = \mathbf{M} = 0$, or

$$\mathbf{E} = -\nabla \times \mathbf{F} + \frac{1}{\hat{y}} \nabla \times \nabla \times \mathbf{A} \\ \mathbf{H} = \nabla \times \mathbf{A} + \frac{1}{\hat{z}} \nabla \times \nabla \times \mathbf{F} \tag{3-79}$$

Equations (3-78) and (3-79) are the general form for fields and potentials in homogeneous source-free regions.

There is a great deal of arbitrariness in the choice of vector potentials. For instance, we can choose the arbitrary Φ's according to

$$\nabla \cdot \mathbf{A} = -\hat{y}\Phi^a \qquad \nabla \cdot \mathbf{F} = -\hat{z}\Phi^f \tag{3-80}$$

This reduces Eqs. (3-78) to

$$\nabla^2\mathbf{A} + k^2\mathbf{A} = 0 \\ \nabla^2\mathbf{F} + k^2\mathbf{F} = 0 \tag{3-81}$$

Solutions to these equations are called *wave potentials*. Note that the rectangular components of the wave potentials satisfy the scalar wave equation, or Helmholtz equation,

$$\nabla^2\psi + k^2\psi = 0 \tag{3-82}$$

Also, when Eqs. (3-80) are satisfied, we can alternatively write Eqs.

(3-79) as

$$\mathbf{E} = -\nabla \times \mathbf{F} - \hat{z}\mathbf{A} + \frac{1}{\hat{y}}\nabla(\nabla \cdot \mathbf{A})$$

$$\mathbf{H} = \nabla \times \mathbf{A} - \hat{y}\mathbf{F} + \frac{1}{\hat{z}}\nabla(\nabla \cdot \mathbf{F}) \tag{3-83}$$

We have yet to decide how to divide the field between **A** and **F**. As a word of caution, *do not* make the mistake of thinking of **A** as due to **J** and **F** as due to **M**. This happened to be our choice for the potential integral solution, where we considered the sources everywhere. We are now concerned with regions of finite extent, and we can represent a field in terms of **A** or **F** or both, regardless of its actual source.

Let us now consider some particular choices of potentials. If we take **F** = 0 and

$$\mathbf{A} = \mathbf{u}_z \psi \tag{3-84}$$

then
$$\mathbf{E} = -\hat{z}\mathbf{A} + \frac{1}{\hat{y}}\nabla(\nabla \cdot \mathbf{A}) \qquad \mathbf{H} = \nabla \times \mathbf{A} \tag{3-85}$$

This can be expanded in rectangular coordinates as

$$\begin{aligned} E_x &= \frac{1}{\hat{y}}\frac{\partial^2 \psi}{\partial x\, \partial z} & H_x &= \frac{\partial \psi}{\partial y} \\ E_y &= \frac{1}{\hat{y}}\frac{\partial^2 \psi}{\partial y\, \partial z} & H_y &= -\frac{\partial \psi}{\partial x} \\ E_z &= \frac{1}{\hat{y}}\left(\frac{\partial^2}{\partial z^2} + k^2\right)\psi & H_z &= 0 \end{aligned} \tag{3-86}$$

A field with no H_z is called *transverse magnetic to z* (TM). We shall find it possible to choose ψ sufficiently general to express an arbitrary TM field in a homogeneous source-free region according to the above formulas.

In the dual sense, if we choose **A** = 0 and

$$\mathbf{F} = \mathbf{u}_z \psi \tag{3-87}$$

then
$$\mathbf{E} = -\nabla \times \mathbf{F} \qquad \mathbf{H} = -\hat{y}\mathbf{F} + \frac{1}{\hat{z}}\nabla(\nabla \cdot \mathbf{F}) \tag{3-88}$$

Expanded in rectangular coordinates, this is

$$\begin{aligned} E_x &= -\frac{\partial \psi}{\partial y} & H_x &= \frac{1}{\hat{z}}\frac{\partial^2 \psi}{\partial x\, \partial z} \\ E_y &= \frac{\partial \psi}{\partial x} & H_y &= \frac{1}{\hat{z}}\frac{\partial^2 \psi}{\partial y\, \partial z} \\ E_z &= 0 & H_z &= \frac{1}{\hat{z}}\left(\frac{\partial^2}{\partial z^2} + k^2\right)\psi \end{aligned} \tag{3-89}$$

A field with no E_z is called *transverse electric to z* (TE). We shall find it possible to choose ψ sufficiently general to express any TE field in a homogeneous source-free region according to the above formulas.

Now suppose we have a field neither TE nor TM. We can determine a ψ according to

$$\frac{\partial^2 \psi^a}{\partial z^2} + k^2 \psi^a = \hat{y} E_z$$

which will generate a field TM to z according to Eqs. (3-86). This TM field will have the same E_z as does the original field; so the difference between the two will be a TE field. We can therefore determine this difference field according to Eqs. (3-89), where the ψ is found from

$$\frac{\partial^2 \psi^f}{\partial z^2} + k^2 \psi^f = \hat{z} H_z$$

Thus, *an arbitrary field in a homogeneous source-free region can be expressed as the sum of a TM field and a TE field.* Explicit expressions for the field would be superposition of Eqs. (3-86) and (3-89), with superscripts a and f added to the ψ's to distinguish between them. Since the z direction is arbitrary, we can express this independent of the coordinate system by defining

$$\mathbf{A} = \mathbf{c}\psi^a \qquad \mathbf{F} = \mathbf{c}\psi^f \qquad (3\text{-}90)$$

where $\mathbf{c}$ is a *constant* vector. The field is then given by Eqs. (3-79), which become

$$\mathbf{E} = -\nabla \times (\mathbf{c}\psi^f) + \frac{1}{\hat{y}} \nabla \times \nabla \times (\mathbf{c}\psi^a)$$
$$\mathbf{H} = \nabla \times (\mathbf{c}\psi^a) + \frac{1}{\hat{z}} \nabla \times \nabla \times (\mathbf{c}\psi^f) \qquad (3\text{-}91)$$

where the ψ's are solutions to Eq. (3-82). We must therefore study solutions to the scalar Helmholtz equation to learn how to pick the ψ's.

If the region is not source-free but is still homogeneous, our starting equations are

$$-\nabla \times \mathbf{E} = \hat{z} \mathbf{H} + \mathbf{M}$$
$$\nabla \times \mathbf{H} = \hat{y} \mathbf{E} + \mathbf{J} \qquad (3\text{-}92)$$

instead of Eqs. (3-77). General solutions to Eqs. (3-92) can be constructed as the sum of any possible solution, called a *particular solution*, plus a solution to the source-free equations, called a *complementary solution*. We already have a particular solution, namely, the potential integral solution of Sec. 3-2. Therefore, solutions in a homogeneous region containing sources are given by

$$\mathbf{E} = \mathbf{E}_{ps} + \mathbf{E}_{cs} \qquad \mathbf{H} = \mathbf{H}_{ps} + \mathbf{H}_{cs} \qquad (3\text{-}93)$$

where the particular solution (ps) is formed according to Eqs. (3-4) and (3-5), and the complementary solution (cs) is constructed according to Eqs. (3-91). We can think of the particular solution as the field due to sources inside the region and the complementary solution as the field due to sources outside the region.

3-13. The Radiation Field. It is easier to evaluate the radiation (distant) field from sources of finite extent than to evaluate the near field. (See, for example, Secs. 2-9 and 2-10.) In this section, we shall formalize the procedure for specializing solutions to the radiation zone.

Consider a distribution of currents in the vicinity of the coordinate origin, immersed in a homogeneous region of infinite extent. The complete solution to the problem is represented by Eqs. (3-4) and (3-5). If we specialize to the radiation zone ($r \gg r'_{\max}$), as suggested by Fig. 3-22, we have

$$|\mathbf{r} - \mathbf{r}'| \to r - r' \cos \xi \tag{3-94}$$

where ξ is the angle between $\mathbf{r}$ and $\mathbf{r}'$. Furthermore, the second term of Eq. (3-94) can be neglected in the "magnitude factors," $|\mathbf{r} - \mathbf{r}'|^{-1}$, of Eqs. (3-5). It cannot, however, be neglected in the "phase factors," $\exp(-jk|\mathbf{r} - \mathbf{r}'|)$, unless $r'_{\max} \ll \lambda$. Thus, Eqs. (3-5) reduce to

$$\mathbf{A} = \frac{e^{-jkr}}{4\pi r} \iiint \mathbf{J}(\mathbf{r}') e^{jkr' \cos \xi} \, d\tau'$$
$$\mathbf{F} = \frac{e^{-jkr}}{4\pi r} \iiint \mathbf{M}(\mathbf{r}') e^{jkr' \cos \xi} \, d\tau' \tag{3-95}$$

in the radiation zone. Note that we now have the r dependence shown explicitly. Many of the operations of Eqs. (3-4) can therefore be performed.

Rather than blindly expanding Eqs. (3-4), let us draw upon some previous conclusions. In Sec. 2-9 it was shown that the distant field of an electric current element was essentially outward-traveling plane waves. The same is true of a magnetic current element, by duality. Hence, the

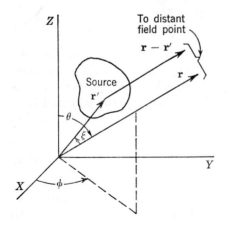

Fig. 3-22. Geometry for evaluating the radiation field.

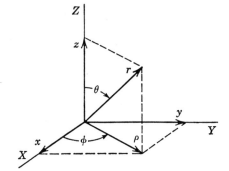

Fig. 3-23. Conventional coordinate orientation.

radiation zone must be characterized by

$$E_\theta = \eta H_\phi \qquad E_\phi = -\eta H_\theta \tag{3-96}$$

since it is a superposition of the fields from many current elements. We can evaluate the partial **H** field due to **J** according to $\mathbf{H}' = \nabla \times \mathbf{A}$ (see Sec. 3-2). Retaining only the dominant terms (r^{-1} variation), we have

$$H'_\theta = (\nabla \times \mathbf{A})_\theta = jkA_\phi$$
$$H'_\phi = (\nabla \times \mathbf{A})_\phi = -jkA_\theta$$

with $\mathbf{E}'$ given by Eqs. (3-96). Similarly, for the partial **E** field due to **M**, we have, in the radiation zone,

$$E''_\theta = -(\nabla \times \mathbf{F})_\theta = -jkF_\phi$$
$$E''_\phi = -(\nabla \times \mathbf{F})_\phi = jkF_\theta$$

with $\mathbf{H}''$ given by Eqs. (3-96). The total field is the sum of these partial fields, or

$$\begin{aligned} E_\theta &= -j\omega\mu A_\theta - jkF_\phi \\ E_\phi &= -j\omega\mu A_\phi + jkF_\theta \end{aligned} \tag{3-97}$$

in the radiation zone, with **H** given by Eqs. (3-96). Thus, no differentiation of the vector potentials is necessary to obtain the radiation field.

Also, for future reference, let us determine $r' \cos \xi$ as a function of the source coordinates. The three coordinate systems of primary interest are the rectangular, cylindrical, and spherical, as illustrated by Fig. 3-23. For the conventional orientation shown, we have the transformations

$$\begin{array}{ll} x = r \sin \theta \cos \phi & x = \rho \cos \phi \\ y = r \sin \theta \sin \phi & y = \rho \sin \phi \\ z = r \cos \theta & z = z \end{array} \tag{3-98}$$

To obtain $r' \cos \xi$, we form

$$rr' \cos \xi = \mathbf{r} \cdot \mathbf{r}' = xx' + yy' + zz' \tag{3-99}$$

Substituting for x, y, z from the first set of Eqs. (3-98), we obtain

$$r' \cos \xi = (x' \cos \phi + y' \sin \phi) \sin \theta + z' \cos \theta \qquad (3\text{-}100)$$

which is the desired form when rectangular coordinates are chosen for the source. Substituting into Eq. (3-100) for x', y', z' from the second set of Eqs. (3-98), we obtain

$$r' \cos \xi = \rho' \sin \theta \cos (\phi - \phi') + z' \cos \theta \qquad (3\text{-}101)$$

which is the desired form when cylindrical coordinates are chosen for the source. Finally, substituting into Eq. (3-100) for x', y', z' from the first set of Eqs. (3-98), we have

$$r' \cos \xi = r'[\cos \theta \cos \theta' + \sin \theta \sin \theta' \cos (\phi - \phi')] \qquad (3\text{-}102)$$

which is the desired form when spherical coordinates are chosen for the source.

PROBLEMS

3-1. Show that a current sheet

$$\mathbf{J} = \mathbf{u}_x J_0$$

over the $z = 0$ plane produces the outward-traveling plane waves

$$E_x = \begin{cases} -\dfrac{\eta J_0}{2} e^{-ikz} & z > 0 \\ -\dfrac{\eta J_0}{2} e^{ikz} & z < 0 \end{cases}$$

in an infinite homogeneous medium.

3-2. Instead of the electric current sheet, suppose that the magnetic current sheet

$$\mathbf{M}_s = \mathbf{u}_y M_0 \sin \frac{\pi y}{b}$$

exists over the cross section $z = 0$ in the waveguide of Fig. 3-2. Show that this magnetic current produces a field

$$E_x = \begin{cases} -\dfrac{M_0}{2} \sin \dfrac{\pi y}{b} e^{-i\beta z} & z > 0 \\ \dfrac{M_0}{2} \sin \dfrac{\pi y}{b} e^{i\beta z} & z < 0 \end{cases}$$

3-3. Suppose now that the two current sheets

$$\mathbf{J}_s = \mathbf{u}_x \frac{A}{Z_0} \sin \frac{\pi y}{b}$$

$$\mathbf{M}_s = \mathbf{u}_y A \sin \frac{\pi y}{b}$$

exist simultaneously over the cross section $z = 0$ of Fig. 3-2. Show that these produce a field

$$E_x = \begin{cases} -A \sin \dfrac{\pi y}{b} e^{-j\beta z} & z > 0 \\ 0 & z < 0 \end{cases}$$

This source is a "directional coupler."

3-4. In Fig. 3-2, suppose that a "shorting plate" (conductor) is placed over the cross section $z = -d$. Show that the current sheet of Eq. (3-2) now produces a field

$$E_x = \begin{cases} -\dfrac{J_0 Z_0}{2}(1 - e^{-j2\beta d}) \sin \dfrac{\pi y}{b} e^{-j\beta z} & z > 0 \\ -jJ_0 Z_0 e^{-j\beta d} \sin \dfrac{\pi y}{b} \sin [\beta(d + z)] & -d < z < 0 \end{cases}$$

Note that when d is an odd number of guide quarter-wavelengths, E_x for $z > 0$ is twice that for the current sheet alone [see Eq. (3-3)], but when d is an integral number of guide half-wavelengths, no E_x exists for $z > 0$.

3-5. The TE and TM modes of a parallel-plate waveguide (Prob. 2-28) are almost dual to each other. Show that the field dual to the TE_n mode of Prob. 2-28 is the TM_n mode for the parallel-plate guide having conductors over the planes $y = b/2$ and $y = -b/2$. Show that the field dual to the TM_n mode of Prob. 2-28 is the TE_n mode of this new waveguide.

3-6. Obtain the field of an infinitesimal loop of magnetic current having z-directed moment KS. Show that this produces the same field as the electric current element of Fig. 2-21 if

$$Il = -j\omega\epsilon KS$$

3-7. Figure 3-24a shows the cross section of a "twin-slot" transmission line. Show that the field distribution is dual to that of the collinear plate line of Fig. 3-24b. By integrating along the contours shown in Fig. 3-24c, determine the line voltages and

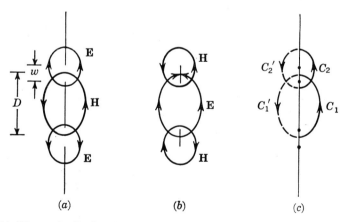

FIG. 3-24. Figures for Prob. 3-7. (a) Twin-slot line; (b) collinear plate line; (c) integration contours.

currents of both the slot line and the plate line. Show that

$$(Z_0)_{\text{slot line}} = \frac{\eta^2}{4(Z_0)_{\text{plate line}}}$$

From Table 2-3, it follows that

$$(Z_0)_{\text{slot line}} \approx \frac{\eta\pi}{4 \log (4D/w)} \qquad D \gg w$$

The two transmission lines are said to be complementary structures (see Babinet's principle, Sec. 7-12).

3-8. Show that the field

$$E_x = \begin{cases} \dfrac{J_0 Z_0}{2} \sin \dfrac{\pi y}{b} e^{j\beta z} & z > 0 \\ \dfrac{J_0 Z_0}{2} \sin \dfrac{\pi y}{b} e^{-j\beta z} & z < 0 \end{cases}$$

is also a mathematical solution to the problem of Fig. 3-2 with J_s given by Eq. (3-2). What do our uniqueness theorems say about this second solution? What can we say about it on physical grounds? Give a couple of other possible solutions to the problem, and interpret them physically.

3-9. Show that the current sheets

$$\mathbf{J}_s = -\mathbf{u}_\theta \frac{Il}{4\pi} e^{-jka} \left(\frac{jk}{a} + \frac{1}{a^2} \right) \sin \theta$$

$$\mathbf{M}_s = -\mathbf{u}_\phi \frac{Il}{4\pi} e^{-jka} \left(\frac{j\omega\mu}{a} + \frac{\eta}{a^2} + \frac{1}{j\omega\epsilon a^3} \right) \sin \theta$$

over the sphere $r = a$ produce the field of Eqs. (2-113) $r > a$ and zero field $r < a$.

3-10. If $\mathbf{E}$ is well-behaved in a homogeneous region bounded by S, and if $\hat{z}\mathbf{H} = -\nabla \times \mathbf{E}$, show that the currents

$$\mathbf{J} = -\hat{y}\mathbf{E} - \frac{1}{\hat{z}} \nabla \times \nabla \times \mathbf{E}$$

will support this and only this field among a class $\mathbf{E}$, $\mathbf{H}$ having identical tangential components of $\mathbf{E}$ on S. Show that the same $\mathbf{E}$, but different $\mathbf{H}$, can be obtained within this class if magnetic sources $\mathbf{K}$ are allowed in addition to $\mathbf{J}$.

3-11. Suppose there exists within the rectangular cavity of Fig. 2-19 a field

$$E_x = E_0 \sin \frac{\pi y}{b} \sinh \gamma z$$

where $\gamma = \sqrt{(\pi/b)^2 - k^2}$ and k is complex (lossy dielectric). Show that this field can be supported by the source

$$\mathbf{M}_s = -\mathbf{u}_y E_0 \sin \frac{\pi y}{b} \sinh \gamma c$$

at the wall $z = c$. Show that for a low-loss dielectric, $\mathbf{M}_s$ almost vanishes at the resonant frequency [Eq. (2-95)], that is, a small $\mathbf{M}_s$ produces a large $\mathbf{E}$.

3-12. Consider a z-directed current element Il a distance d in front of a ground plane covering the $y = 0$ plane, as shown in Fig. 3-25. Show that the radiation field is given by

$$E_\theta = \frac{-\eta Il}{\lambda r} e^{-jkr} \sin \theta \sin (kd \sin \phi \sin \theta)$$

and $\eta H_\phi = E_\theta$. Find the power radiated and show that the radiation resistance referred to I is

$$R_r = \frac{\eta \pi l^2}{\lambda^2} \left[\frac{2}{3} - \frac{\sin 2kd}{2kd} - \frac{\cos 2kd}{(2kd)^2} + \frac{\sin 2kd}{(2kd)^3} \right]$$

For $d \leq \lambda/4$, the maximum radiation is in the y direction. Show that

$$R_r \xrightarrow[kd \to 0]{} \eta \frac{32\pi^3 l^2 d^2}{15\lambda^4}$$

and that the gain is 7.5 for d small, 4.15 for $d = \lambda/4$, and approximately 6 for d large.

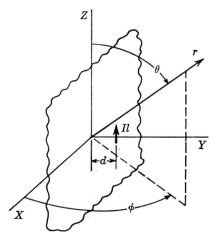

FIG. 3-25. Current element parallel to a ground plane.

3-13. In Fig. 3-6a, suppose we have a small loop of electric current with z-directed moment IS, instead of the current element. Show that the radiation field is given by

$$E_\phi = \frac{j\eta 2\pi IS}{\lambda^2 r} e^{-jkr} \sin(kd \cos \theta) \sin \theta$$

and $\eta H_\theta = -E_\phi$. Find the power radiated and show that the radiation resistance referred to I is

$$R_r = 2\pi\eta \left(\frac{kS}{\lambda}\right)^2 \left[\frac{1}{3} + \frac{\cos 2kd}{(2kd)^2} - \frac{\sin 2kd}{(2kd)^3} \right]$$

For small d,

$$E_\phi \xrightarrow[kd \to 0]{} \frac{j\eta\pi ISkd}{\lambda^2 r} e^{-jkr} \sin 2\theta$$

$$R_r \xrightarrow[kd \to 0]{} \frac{\pi\eta}{15} \left(\frac{kSkd}{\lambda}\right)^2$$

Thus, maximum radiation is at $\theta = 45°$ for small d. The gain at small d is 15. For large d, the maximum radiation lies close to the ground plane, and the gain is 6.

3-14. In Fig. 3-25, suppose we have a small loop of electric current with z-directed moment IS, instead of the current element. Show that the radiation field is given by

$$E_\phi = \frac{\eta k^2 IS}{2\pi r} e^{-jkr} \sin \theta \cos(kd \sin \phi \sin \theta)$$

and $\eta H_\theta = -E_\phi$. Show that the radiation resistance referred to I is

$$R_r = \pi\eta \left(\frac{kS}{\lambda}\right)^2 \left[\frac{2}{3} + \frac{\sin 2kd}{2kd} + \frac{\cos 2kd}{(2kd)^2} - \frac{\sin 2kd}{(2kd)^3}\right]$$

The maximum radiation is along the ground plane, in the x direction. For small kd,

$$R_r \xrightarrow[kd\to 0]{} \frac{4\pi\eta}{3}\left(\frac{kS}{\lambda}\right)^2$$

which is twice that for the isolated loop. For $d = 0$, the gain is 3; for $d = \lambda/4$, it is 7.1; and for $d \to \infty$, it is 6.

3-15. The monopole antenna consists of a straight wire perpendicular to a ground plane, fed at the ground plane, as shown in Fig. 3-26. Show that the field is the same as that from the dipole antenna (Fig. 2-23), fed at the center. Show that the gain of the monopole is twice that of the corresponding dipole and that the radiation resistance is one-half. For example, the radiation resistance of the $\lambda/4$ monopole is 36.6 ohms.

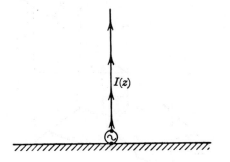

Fig. 3-26. The monopole antenna.

3-16. Consider an open-ended coaxial line (Fig. 3-14a without the ground plane) of small radii a and b. Treat the problem according to the equivalence principle as applied to a surface just enclosing the coax. Assume $\mathbf{n} \times \mathbf{H}$ is essentially zero over the entire surface and that tangential $\mathbf{E}$ is that of the transmission-line mode over the open end. Show that to this approximation the radiated field is one-half that of Eq. (3-20) and that the radiation conductance is one-half that of Eq. (3-23).

3-17. A slot antenna consists of a slot in a conducting ground plane, as shown in Fig. 3-27. It is called a dipole slot antenna when fed by a voltage impressed across the center of the slot. The slot and ground plane can be viewed as a transmission line, and the field in the slot will be essentially a harmonic function of kz. Assume

$$E_z = \frac{V_m}{w}\sin\left[k\left(\frac{L}{2} - |z|\right)\right]$$

in the slot, and obtain the magnetic current equivalent of the form of Fig. 3-13c. For w small, show that this equivalent representation is the dual problem to the dipole antenna of Sec. 2-10. Using duality, show that the radiation field is

$$\frac{jV_m e^{-jkr}}{\eta\pi r} \frac{\cos\left(k\frac{L}{2}\cos\theta\right) - \cos\left(k\frac{L}{2}\right)}{\sin\theta} = \begin{cases} H_\theta & y > 0 \\ -H_\theta & y < 0 \end{cases}$$

Define the radiation conductance of this antenna as $G_r = \bar{\mathcal{P}}_f/|V_m|^2$, and show that

$$(G_r)_{\text{slot dipole}} = \frac{4(R_r)_{\text{wire dipole}}}{\eta^2}$$

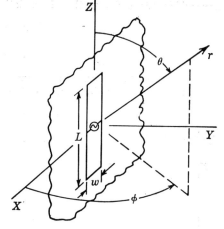

FIG. 3-27. A slot antenna.

where R_r is as plotted in Fig. 2-24. The input voltage V_i is related to V_m by $V_i = V_m \sin(kL/2)$; so the input conductance is given by

$$G_i = \frac{G_r}{\sin^2\left(k\dfrac{L}{2}\right)}$$

3-18. For the antenna of Fig. 3-27, assume E_z in the slot the same as in Prob. 3-17, and show that for arbitrary w

$$\frac{jV_m e^{-jkr}}{\eta \pi r} f(\theta,\phi) = \begin{cases} H_\theta & y > 0 \\ -H_\theta & y < 0 \end{cases}$$

where $$f(\theta,\phi) = \frac{\sin\left(k\dfrac{w}{2}\cos\phi\sin\theta\right)}{k\dfrac{w}{2}\cos\phi\sin\theta}\left[\frac{\cos\left(k\dfrac{L}{2}\cos\theta\right) - \cos\left(k\dfrac{L}{2}\right)}{\sin\theta}\right]$$

3-19. Figure 3-28 shows an aperture antenna consisting of a rectangular waveguide opening onto a ground plane. Assume that E_z in the aperture is that of the TE_{01}

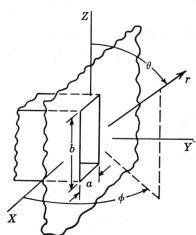

FIG. 3-28. A rectangular waveguide opening onto a ground plane.

waveguide mode, and show that the radiation field is

$$H_\theta = \frac{2jbE_0 e^{-jkr}}{\eta r} \frac{\sin\left(k\frac{a}{2}\cos\phi\sin\theta\right)\cos\left(k\frac{b}{2}\cos\theta\right)}{\cos\phi[\pi^2 - (kb\cos\theta)^2]}$$

3-20. Figure 3-29 represents a rectangular conducting plate of width a in the y direction and b in the z direction. Let the incident plane wave be specified by

$$E_z^i = E_0 e^{jk(x\cos\phi_0 + y\sin\phi_0)}$$

Use the induction theorem with the same approximation as was used in the problem

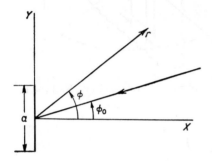

Fig. 3-29. Scattering by a rectangular plate.

of Fig. 3-17, and show that at large r the scattered field in the xy plane is

$$E_z^s \approx \frac{kE_0 abe^{-jkr}}{j2\pi r} \frac{\sin[k(a/2)(\sin\phi + \sin\phi_0)]}{k(a/2)(\sin\phi + \sin\phi_0)} \cos\phi$$

Show that the echo area is

$$A_e \approx 4\pi\left[\frac{ab\cos\phi_0\sin(ka\sin\phi_0)}{\lambda ka\sin\phi_0}\right]^2$$

3-21. Repeat Prob. 3-20 for the orthogonal polarization, that is,

$$H_z^i = H_0 e^{jk(x\cos\phi_0 + y\sin\phi_0)}$$

and show that at large r the scattered field in the xy plane is

$$H_z^s \approx \frac{jkH_0 abe^{-jkr}}{2\pi r} \frac{\sin[k(a/2)(\sin\phi + \sin\phi_0)]}{k(a/2)(\sin\phi + \sin\phi_0)} \cos\phi_0$$

Show that the echo area is the same as obtained in Prob. 3-20.

3-22. Use reciprocity to evaluate the radiation field of the dipole antenna of Sec. 2-10. To do this, place a θ-directed current element at large r, and apply Eq. (3-36), obtaining Eq. (2-125).

3-23. By applying voltage sources to the network of Fig. 3-18, show that the admittance matrix $[y]$ defined by

$$\begin{bmatrix} I_1 \\ I_2 \end{bmatrix} = \begin{bmatrix} y_{11} & y_{12} \\ y_{21} & y_{22} \end{bmatrix} \begin{bmatrix} V_1 \\ V_2 \end{bmatrix}$$

satisfies the reciprocity relationship $y_{12} = y_{21}$ when Eq. (3-38) is valid.

SOME THEOREMS AND CONCEPTS

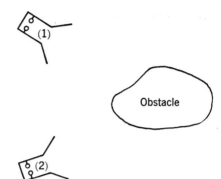

Fig. 3-30. Differential scattering.

3-24. Let Fig. 3-30 represent two antennas in the presence of an obstacle. Let V_1 be the voltage received at antenna 1 when a unit current source is applied at antenna 2 and V_2 be the voltage received at antenna 2 when a unit current source is applied at antenna 1. Let V_1^i and V_2^i be the corresponding voltages when the obstacle is absent. Define the scattered voltages as

$$V_1^s = V_1 - V_1^i \qquad V_2^s = V_2 - V_2^i$$

and show that $V_1^s = V_2^s$.

3-25. For the problem of Fig. 3-2, define the input impedance of the sheet of current as

$$Z = -\frac{\langle a,a \rangle}{I^2}$$

where $\langle a,a \rangle$ is the self-reaction of the currents and I is the total current of the sheet. Evaluate Z when the field is given by Eqs. (3-3).

3-26. Repeat Prob. 3-25 for the current sheet and field of Prob. 3-4.

3-27. In the vector Green's theorem [Eq. (3-46)], let $\mathbf{A} = \mathbf{E}^a$ and $\mathbf{B} = \mathbf{E}^b$ in a homogeneous isotropic region, and show that it reduces to Eq. (3-35).

3-28. Use the vector identity

$$\nabla \cdot (\mathbf{A} \times \phi \nabla \times \mathbf{B}) = \phi \nabla \times \mathbf{A} \cdot \nabla \times \mathbf{B} - \mathbf{A} \cdot \nabla \times \phi \nabla \times \mathbf{B}$$

and derive the modified vector Green's theorem

$$\oiint \phi(\mathbf{A} \times \nabla \times \mathbf{B} - \mathbf{B} \times \nabla \times \mathbf{A}) \cdot d\mathbf{s}$$
$$= \iiint (\mathbf{B} \cdot \nabla \times \phi \nabla \times \mathbf{A} - \mathbf{A} \cdot \nabla \times \phi \nabla \times \mathbf{B}) \, d\tau$$

Let $\mathbf{A} = \mathbf{E}^a$, $\mathbf{B} = \mathbf{E}^b$, $\phi = \hat{z}^{-1}$ in an inhomogeneous region, and show that the above theorem reduces to Eq. (3-35).

3-29. Derive the left-hand term of Eq. (3-50), that is, show

$$\oiint_s (\mathbf{E} \times \nabla \times \mathbf{G}_1 - \mathbf{G}_1 \times \nabla \times \mathbf{E} + \mathbf{E} \nabla \cdot \mathbf{G}_1) \cdot d\mathbf{s} \xrightarrow[|\mathbf{r}-\mathbf{r}'|\to 0]{} 4\pi c \cdot \mathbf{E}$$

3-30. Let G_4 be the magnetic field of a z-directed current element situated $y > 0$ and radiating in the presence of a perfect electric conductor covering the $y = 0$ plane. In other words, let $\mathbf{c} = \mathbf{u}_z$ and S be the $y = 0$ plane. Show that

$$\mathbf{G}_4 = \nabla \times \mathbf{u}_z \left(\frac{e^{-jkr_1}}{r_1} - \frac{e^{-jkr_2}}{r_2} \right)$$

where
$$r_1 = \sqrt{(x - x')^2 + (y - y')^2 + (z - z')^2}$$
$$r_2 = \sqrt{(x - x')^2 + (y + y')^2 + (z - z')^2}$$

3-31. Specialize the $\mathbf{G}_4$ of Prob. 3-30 to $r_1 \to \infty$, and apply Eq. (3-57) to the problem of Fig. 3-28. Show that this gives the same answer as obtained in Prob. 3-19.

3-32. Apply duality to Eqs. (3-65), and evaluate the magnetic tensor Green's function $[\Gamma]$ defined by

$$\mathbf{H} = [\Gamma] K \mathbf{l}$$

in free space.

3-33. Evaluate the Γ_{ij} for the free-space tensor Green's function defined by

$$\mathbf{H} = [\Gamma] I \mathbf{l}$$

3-34. Repeat Prob. 3-20 using the physical optics approximation, and show that the answer for $E_z{}^s$ differs from that of Prob. 3-20 by an interchange of ϕ and ϕ_0. Show that the echo area is identical to that of Prob. 3-20.

3-35. Repeat Prob. 3-21 using the physical optics approximation, and show that the answer for $H_z{}^s$ differs from that of Prob. 3-21 by an interchange of ϕ and ϕ_0. Show that the echo area is identical to that of Prob. 3-21.

3-36. Let $\psi = e^{-jky}$ in Eqs. (3-86), and evaluate the electromagnetic field. Classify this field in as many ways as you can (wave-type, polarization, etc.).

3-37. Let $\psi = e^{-jkz}$ in Eqs. (3-89), and evaluate the electromagnetic field. Classify this field in as many ways as you can.

3-38. Let $\mathbf{c} = \mathbf{u}_x$, $\psi^a = e^{-jkz}$, $\psi^f = je^{-jkz}$, and evaluate Eqs. (3-91). Classify this field in as many ways as you can.

3-39. Derive Eqs. (3-97) by expanding Eqs. (3-4) with $\mathbf{A}$ and $\mathbf{F}$ as given by Eqs. (3-95).

CHAPTER 4

PLANE WAVE FUNCTIONS

4-1. The Wave Functions. The problems that we have considered so far are of two types: (1) those reducible to sources in an unbounded homogeneous region, and (2) those solvable by using one or more uniform plane waves. Equations (3-91) show us how to construct general solutions to the field equations in homogeneous regions once we have general solutions to the scalar Helmholtz equation. By a method called *separation of variables*, general solutions to the Helmholtz equation can be constructed in certain coordinate systems.[1] In this section, we use the method of separation of variables to obtain solutions for the rectangular coordinate system.

The Helmholtz equation in rectangular coordinates is

$$\frac{\partial^2 \psi}{\partial x^2} + \frac{\partial^2 \psi}{\partial y^2} + \frac{\partial^2 \psi}{\partial z^2} + k^2 \psi = 0 \tag{4-1}$$

The method of separation of variables seeks to find solutions of the form

$$\psi = X(x)Y(y)Z(z) \tag{4-2}$$

that is, solutions which are the product of three functions of one coordinate each. Substitution of Eq. (4-2) into Eq. (4-1), and division by ψ, yields

$$\frac{1}{X}\frac{d^2X}{dx^2} + \frac{1}{Y}\frac{d^2Y}{dy^2} + \frac{1}{Z}\frac{d^2Z}{dz^2} + k^2 = 0 \tag{4-3}$$

Each term can depend, at most, on only one coordinate. Since each coordinate can be varied independently, Eq. (4-3) can sum to zero for all coordinate values only if each term is independent of x, y, and z. Thus, let

$$\frac{1}{X}\frac{d^2X}{dx^2} = -k_x{}^2 \qquad \frac{1}{Y}\frac{d^2Y}{dy^2} = -k_y{}^2 \qquad \frac{1}{Z}\frac{d^2Z}{dz^2} = -k_z{}^2$$

where k_x, k_y, and k_z are constants, that is, are independent of x, y, and z. (The choice of minus a constant squared is taken for later convenience.)

[1] It has been shown by Eisenhart (*Ann. Math.*, vol. 35, p. 284, 1934) that the Helmholtz equation is separable in 11 three-dimensional orthogonal coordinate systems.

We now have Eq. (4-1) separated into the trio of equations

$$\frac{d^2X}{dx^2} + k_x^2 X = 0$$
$$\frac{d^2Y}{dy^2} + k_y^2 Y = 0 \qquad (4\text{-}4)$$
$$\frac{d^2Z}{dz^2} + k_z^2 Z = 0$$

where, by Eq. (4-3), the separation parameters must satisfy

$$k_x^2 + k_y^2 + k_z^2 = k^2 \qquad (4\text{-}5)$$

This last equation is called the *separation equation*.

Equations (4-4) are all of the same form. They will be called *harmonic equations*. Any solution to the harmonic equation we shall call a *harmonic function*,[1] and denote it, in general, by $h(k_x x)$. Commonly used harmonic functions are

$$h(k_x x) \sim \sin k_x x, \ \cos k_x x, \ e^{jk_x x}, \ e^{-jk_x x} \qquad (4\text{-}6)$$

Any two of these are linearly independent. A constant times a harmonic function is still a harmonic function. A sum of harmonic functions is still a harmonic function. From Eqs. (4-2) and (4-4) it is evident that

$$\psi_{k_x k_y k_z} = h(k_x x) h(k_y y) h(k_z z) \qquad (4\text{-}7)$$

are solutions to the Helmholtz equation when the k_i satisfy Eq. (4-5). These solutions are called *elementary wave functions*.

Linear combinations of the elementary wave functions must also be solutions to the Helmholtz equation. As evidenced by Eq. (4-5), only two of the k_i may be chosen independently. We can therefore construct more general wave functions by summing over possible choices for one or two separation parameters. For example,

$$\psi = \sum_{k_x} \sum_{k_y} B_{k_x k_y} \psi_{k_x k_y k_z}$$
$$= \sum_{k_x} \sum_{k_y} B_{k_x k_y} h(k_x x) h(k_y y) h(k_z z) \qquad (4\text{-}8)$$

where the B_{ij} are constants, is a solution to the Helmholtz equation. The values of the k_i needed for any particular problem are determined by the boundary conditions of the problem and are called *eigenvalues* or *characteristic values*. The elementary wave functions corresponding to specific eigenvalues are called *eigenfunctions*.

---

[1] The term *harmonic function* also is used to denote a solution to Laplace's equation. This is not the present meaning of the term.

PLANE WAVE FUNCTIONS 145

Still more general wave functions can be constructed by integrating over one or two of the k_i. For example, a solution to the Helmholtz equation is

$$\psi = \int_{k_x}\int_{k_y} f(k_x,k_y)\psi_{k_xk_yk_z}\,dk_x\,dk_y$$
$$= \int_{k_x}\int_{k_y} f(k_x,k_y)h(k_xx)h(k_yy)h(k_zz)\,dk_x\,dk_y \quad (4\text{-}9)$$

where $f(k_x,k_y)$ is an analytic function, and the integration is over any path in the complex k_x and k_y domains. Equation (4-9) exhibits a continuous variation of the separation parameters, and we say that there exists a continuous spectrum of eigenvalues. We shall see that solutions for finite regions (waveguides and cavities) are characterized by discrete spectra of eigenvalues, while solutions for unbounded regions (antennas) often require continuous spectra. Wave functions of the form of Eq. (4-9) are most commonly used to construct Fourier integrals.

We should be familiar with the mathematical properties and with the physical interpretations of the various harmonic functions so that we can properly choose them for particular problems. Keep in mind that wave functions represent instantaneous quantities, according to Eq. (1-40). Solutions of the form $h(kx) = e^{-jkx}$ (k positive real) represent waves traveling unattenuated in the $+x$ direction. If k is complex and Re $(k) > 0$, we have $+x$ traveling waves which are attenuated or augmented according as Im (k) is negative or positive. Similarly, solutions of the form $h(kx) = e^{jkx}$, [Re $(k) > 0$] represent $-x$ traveling waves, attenuated or augmented if k is complex. If k is purely imaginary, the above two harmonic functions represent evanescent fields. Solutions of the form $h(kx) = \sin kx$ and $h(kx) = \cos kx$ with k real represent pure standing waves. If k is complex, they represent localized standing waves. If k is purely imaginary, say $k = -j\alpha$ with α real, then the "trigonometric functions" $\sin kx$ and $\cos kx$ can be expressed as "hyperbolic functions" $\sinh \alpha x$ and $\cosh \alpha x$. We should get used to thinking of the various functions as defined over the entire complex kx plane. The trigonometric and hyperbolic functions are then just specializations of the complex harmonic functions. Table 4-1 summarizes the above discussion. (The convention $k = \beta - j\alpha$ with α and β real is used.) Note that the degenerate case $k = 0$ has the harmonic functions $h(0x) = 1,x$. The choice of the proper harmonic functions in any particular case is largely a matter of experience, and facility in this respect will be gained as we use them.

4-2. Plane Waves. Consider an elementary wave function of the form

$$\psi = e^{-jk_xx}e^{-jk_yy}e^{-jk_zz} \quad (4\text{-}10)$$

TABLE 4-1. PROPERTIES OF THE HARMONIC FUNCTIONS*

$h(kx)$	Zeros†	Infinities†	Specializations of $k = \beta - j\alpha$	Special representations	Physical interpretation
e^{-jkx}	$kx \to -j\infty$	$kx \to j\infty$	k real k imaginary k complex	$e^{-j\beta x}$ $e^{-\alpha x}$ $e^{-\alpha x}e^{-j\beta x}$	$+x$ traveling wave Evanescent field Attenuated traveling wave
e^{jkx}	$kx \to j\infty$	$kx \to -j\infty$	k real k imaginary k complex	$e^{j\beta x}$ $e^{\alpha x}$ $e^{\alpha x}e^{j\beta x}$	$-x$ traveling wave Evanescent field Attenuated traveling wave
$\sin kx$	$kx = n\pi$	$kx \to \pm j\infty$	k real k imaginary k complex	$\sin \beta x$ $-j \sinh \alpha x$ $\sin \beta x \cosh \alpha x$ $-j \cos \beta x \sinh \alpha x$	Standing wave Two evanescent fields Localized standing waves
$\cos kx$	$kx = (n + \frac{1}{2})\pi$	$kx \to \pm j\infty$	k real k imaginary k complex	$\cos \beta x$ $\cosh \alpha x$ $\cos \beta x \cosh \alpha x$ $+j \sin \beta x \sinh \alpha x$	Standing wave Two evanescent fields Localized standing waves

* For $k = 0$, the harmonic functions are $h(0x) = 1, x$.
† For an essential singularity, this column gives the asymptotic behavior.

The k_i must satisfy Eq. (4-5), which is of the form of the scalar product of a vector

$$\mathbf{k} = \mathbf{u}_x k_x + \mathbf{u}_y k_y + \mathbf{u}_z k_z \qquad (4\text{-}11)$$

with itself. Note that in terms of $\mathbf{k}$ and the radius vector

$$\mathbf{r} = \mathbf{u}_x x + \mathbf{u}_y y + \mathbf{u}_z z \qquad (4\text{-}12)$$

we can express Eq. (4-10) as

$$\psi = e^{-j\mathbf{k} \cdot \mathbf{r}} \qquad (4\text{-}13)$$

For $\mathbf{k}$ real, we apply Eq. (2-140) and determine the vector phase constant

$$\boldsymbol{\beta} = -\nabla(-\mathbf{k} \cdot \mathbf{r}) = \mathbf{k}$$

Hence, the equiphase surfaces are planes perpendicular to $\mathbf{k}$. The amplitude of the wave is constant (unity). Equation (4-13) therefore represents a scalar uniform plane wave propagating in the direction of $\mathbf{k}$. Figure 4-1 illustrates this interpretation.

For $\mathbf{k}$ complex, we define two real vectors

$$\mathbf{k} = \boldsymbol{\beta} - j\boldsymbol{\alpha} \qquad (4\text{-}14)$$

and determine the vector propagation constant according to Eq. (2-145). This gives

$$\boldsymbol{\gamma} = -\nabla(-j\mathbf{k} \cdot \mathbf{r}) = j\mathbf{k} = \boldsymbol{\alpha} + j\boldsymbol{\beta}$$

We now have equiphase surfaces perpendicular to $\boldsymbol{\beta}$ and equiamplitude

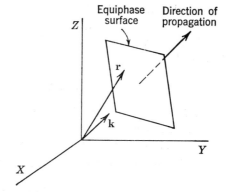

Fig. 4-1. A uniform plane wave.

surfaces perpendicular to α. Thus, when $\mathbf{k}$ is complex, Eq. (4-13) represents a plane wave propagating in the direction of β and attenuating in the direction of α. It is a uniform plane wave only if β and α are in the same direction. Note that definitions $\mathbf{k} = \beta - j\alpha$ and $k = k' - jk''$ do *not* imply that β equals k' or that α equals k'' in general. In fact, for loss-free media,

$$k^2 = \mathbf{k} \cdot \mathbf{k} = \beta^2 - \alpha^2 - j2\alpha \cdot \beta$$

must be positive real. Hence, either $\alpha = 0$ or $\alpha \cdot \beta = 0$. When $\alpha = 0$ we have the uniform plane wave discussed above. When α and β are mutually orthogonal we have an evanescent field, such as was encountered in total reflection [Eq. (2-62)].

The elementary wave functions of Eq. (4-10) or Eq. (4-13) are quite general, since sinusoidal wave functions are linear combinations of the exponential wave functions. Wave functions of the type of Eqs. (4-8) and (4-9) are linear combinations of the elementary wave functions. We therefore conjecture that all wave functions can be expressed as superpositions of plane waves.

Let us now consider the electromagnetic fields that we can construct from the wave functions of Eq. (4-10). Fields TM to z are obtained if ψ is interpreted according to $\mathbf{A} = \mathbf{u}_z \psi$. This choice results in Eqs. (3-86), which, for the ψ of Eq. (4-10), become

$$\mathbf{H} = -\mathbf{u}_x jk_y \psi + \mathbf{u}_y jk_x \psi$$
$$= \nabla \psi \times \mathbf{u}_z = j\psi \mathbf{u}_z \times \mathbf{k} \qquad (4\text{-}15)$$

and
$$\hat{y}\mathbf{E} = jk_z(\mathbf{u}_x jk_x + \mathbf{u}_y jk_y + \mathbf{u}_z jk_z)\psi + \mathbf{u}_z k^2 \psi$$
$$= (-k_z \mathbf{k} + \mathbf{u}_z k^2)\psi \qquad (4\text{-}16)$$

For $\mathbf{k}$ real, $\mathbf{H}$ is perpendicular to $\mathbf{k}$ by Eq. (4-15), and $\mathbf{E}$ is perpendicular to $\mathbf{k}$, since

$$\hat{y}\mathbf{k} \cdot \mathbf{E} = (-k_z k^2 + k_z k^2)\psi = 0$$

Thus, the wave is TEM to the direction of propagation (as well as TM to z). For **k** complex, define **α** and **β** by Eq. (4-14). It then follows that the wave is not necessarily TEM to the direction of propagation (that of **β**). It will be TEM to **β** only if **α** and **β** are in the same direction, that is, if

$$\mathbf{k} = \boldsymbol{\beta} - j\boldsymbol{\alpha} = (\mathbf{u}_x l + \mathbf{u}_y m + \mathbf{u}_z n)k$$

with l, m, n real. In this case, $\beta = k'$, $\alpha = k''$, and l, m, n are the direction cosines.

The dual procedure applies when ψ is interpreted according to $\mathbf{F} = \mathbf{u}_z \psi$. In this case, Eqs. (3-89) apply, giving

$$\begin{aligned} \mathbf{E} &= j\psi \mathbf{k} \times \mathbf{u}_z \\ \hat{z}\mathbf{H} &= (-k_z \mathbf{k} + \mathbf{u}_z k^2)\psi \end{aligned} \quad (4\text{-}17)$$

which are dual to Eqs. (4-15) and (4-16). For k real, this is a wave TEM to **k** and TE to z. Its polarization is orthogonal to the corresponding TM-to-z wave. For **k** complex, the wave is not necessarily TEM to the direction of propagation. All these fields are plane waves. An arbitrary electromagnetic field in a homogeneous region can be considered as a superposition of these plane waves.

4-3. The Rectangular Waveguide. The problem of determining modes in a rectangular waveguide provides a good illustration of the use of elementary wave functions. In Sec. 2-7 we considered only the dominant mode. In this section we shall consider the complete mode spectrum. The geometry of the rectangular waveguide is illustrated by Fig. 2-16.

It is conventional to classify the modes in a rectangular waveguide as TM to z (no H_z) and TE to z (no E_z). Modes TM to z are expressible in terms of an **A** having only a z component ψ. We wish to consider traveling waves; hence we consider wave functions of the form

$$\psi = h(k_x x)h(k_y y)e^{-jk_z z} \quad (4\text{-}18)$$

The electromagnetic field is given by Eqs. (3-86). In particular,

$$E_z = \frac{1}{\hat{y}}(k^2 - k_z^2)\psi$$

The boundary conditions on the problem are that tangential components of **E** vanish at the conducting walls. Hence, E_z must be zero at $x = 0$, $x = a$, $y = 0$, and $y = b$. The only harmonic functions having two or more zeros are the sinusoidal functions with k_i real. Thus, choose

$$h(k_x x) = \sin k_x x \qquad k_x = \frac{m\pi}{a} \qquad m = 1, 2, 3, \ldots$$

$$h(k_y y) = \sin k_y y \qquad k_y = \frac{n\pi}{b} \qquad n = 1, 2, 3, \ldots$$

so that the boundary conditions on E_z are satisfied. Each integer m and n specifies a possible field, or mode. The TM$_{mn}$ *mode functions* are therefore

$$\psi_{mn}{}^{\text{TM}} = \sin \frac{m\pi x}{a} \sin \frac{n\pi y}{b} e^{-jk_z z} \qquad (4\text{-}19)$$

with $m = 1, 2, 3, \ldots$, and $n = 1, 2, 3, \ldots$, and the separation parameter equation [Eq. (4-5)] becomes

$$\left(\frac{m\pi}{a}\right)^2 + \left(\frac{n\pi}{b}\right)^2 + k_z^2 = k^2 \qquad (4\text{-}20)$$

The TM$_{mn}$ mode fields are obtained by substituting the $\psi_{mn}{}^{\text{TM}}$ into Eqs. (3-86).

Modes TE to z are expressible in terms of an **F** having only a z component ψ. Again, we wish to find traveling waves; so the ψ must be of the form of Eqs. (4-18). The electromagnetic field this time will be given by Eqs. (3-89). In particular,

$$E_x = -\frac{\partial \psi}{\partial y} \qquad E_y = \frac{\partial \psi}{\partial x}$$

the first of which must vanish at $y = 0$, $y = b$, and the second at $x = 0$, $x = a$. Harmonic functions satisfying these boundary conditions are

$$h(k_x x) = \cos k_x x \qquad k_x = \frac{m\pi}{a} \qquad m = 0, 1, 2, \ldots$$

$$h(k_y y) = \cos k_y y \qquad k_y = \frac{n\pi}{b} \qquad n = 0, 1, 2, \ldots$$

Each integer m and n, except $m = n = 0$ (in which case **E** vanishes identically), specifies a mode. Hence, the TE$_{mn}$ mode functions are

$$\psi_{mn}{}^{\text{TE}} = \cos \frac{m\pi x}{a} \cos \frac{n\pi y}{b} e^{-jk_z z} \qquad (4\text{-}21)$$

with $m = 0, 1, 2, \ldots$; $n = 0, 1, 2, \ldots$; $m = n = 0$ excepted. The separation parameter equation remains the same as in the TM case [Eq. (4-20)]. The TE$_{mn}$ mode fields are obtained by substituting the $\psi_{mn}{}^{\text{TE}}$ into Eqs. (3-89).

Interpretation of each mode is similar to that of the dominant TE$_{01}$ mode, considered in Sec. 2-7. Equation (4-20) determines the mode propagation constant $\gamma = jk_z$. For k real, the propagation constant vanishes when k is

$$\sqrt{\left(\frac{m\pi}{a}\right)^2 + \left(\frac{n\pi}{b}\right)^2} = (k_c)_{mn} \qquad (4\text{-}22)$$

The $(k_c)_{mn}$ is called the *cutoff wave number* of the mn mode. For other values of k, we have

$$\gamma_{mn} = jk_z = \begin{cases} j\beta = j\sqrt{k^2 - (k_c)_{mn}^2} & k > k_c \\ \alpha = \sqrt{(k_c)_{mn}^2 - k^2} & k < k_c \end{cases} \quad (4\text{-}23)$$

Thus, for $k > k_c$ the mode is propagating, and for $k < k_c$ the mode is nonpropagating (evanescent). From Eq. (4-22) we determine the cutoff frequencies

$$(f_c)_{mn} = \frac{k_c}{2\pi\sqrt{\epsilon\mu}} = \frac{1}{2\sqrt{\epsilon\mu}}\sqrt{\left(\frac{m}{a}\right)^2 + \left(\frac{n}{b}\right)^2} \quad (4\text{-}24)$$

and the cutoff wavelengths

$$(\lambda_c)_{mn} = \frac{2\pi}{k_c} = \frac{2}{\sqrt{(m/a)^2 + (n/b)^2}} \quad (4\text{-}25)$$

In terms of the cutoff frequencies, we can re-express the mode propagation constants as

$$\gamma = jk_z = \begin{cases} j\beta = jk\sqrt{1 - \left(\frac{f_c}{f}\right)^2} & f > f_c \\ \alpha = k_c\sqrt{1 - \left(\frac{f}{f_c}\right)^2} & f < f_c \end{cases} \quad (4\text{-}26)$$

where mode indices mn are implied. We can also define mode wavelengths for each mode by Eq. (2-85) and mode phase velocities by Eq. (2-86), where mode indices are again implied.

It is apparent that $\gamma = jk_z$ for each mode has the same interpretation as γ for the TE_{01} mode. It is the physical size (compared to wavelength) of the waveguide that determines which modes propagate. Table 4-2 gives a tabulation of some of the smaller eigenvalues for various ratios b/a. Whenever two or more modes have the same cutoff frequency, they are said to be *degenerate modes*. The corresponding TE_{mn} and TM_{mn} modes are always degenerate in the rectangular guide (but not in other-shaped guides). In the square guide ($b/a = 1$), the TE_{mn}, TE_{nm}, TM_{mn}, and TM_{nm} modes form a foursome of degeneracy. Waveguides are usually constructed so that only one mode propagates, hence $b/a > 1$ usually. For $b/a = 2$, we have a 2:1 frequency range of single-mode operation, and this is the most common practical geometry. It is undesirable to make b/a greater than 2 for high-power operation, since, if the guide is too thin, arcing may occur. (The breakdown power is proportional to $\sqrt{a}$ for fixed b.) To illustrate the use of Table 4-2, suppose we wish to design an air-filled waveguide to propagate the TE_{01} mode at 10,000 megacycles ($\lambda = 3$ centimeters). We do not wish to operate too close to f_c, since the conductor losses are then large (see Table 2-4). If we take

TABLE 4-2. $\dfrac{(k_c)_{mn}}{(k_c)_{01}} = \dfrac{(f_c)_{mn}}{(f_c)_{01}} = \dfrac{(\lambda_c)_{01}}{(\lambda_c)_{mn}}$ FOR THE RECTANGULAR WAVEGUIDE, $b \geq a$

$\dfrac{b}{a}$	TE_{01}	TE_{10}	TE_{11} TM_{11}	TE_{02}	TE_{20}	TE_{12} TM_{12}	TE_{21} TM_{21}	TE_{22} TM_{22}	TE_{03}
1	1	1	1.414	2	2	2.236	2.236	2.828	3
1.5	1	1.5	1.803	2	3	2.500	3.162	3.606	3
2	1	2	2.236	2	4	2.828	4.123	4.472	3
3	1	3	3.162	2	6	3.606	6.083	6.325	3
∞	1	∞	∞	2	∞	∞	∞	∞	3

$b = 2$ centimeters, then $\lambda_c = 4$ centimeters for the TE_{01} mode, and we are operating well above cutoff. The next modes to become propagating are the TE_{10} and TE_{02} modes, at a frequency of 15,000 megacycles. The TE_{11} and TM_{11} modes become propagating at 16,770 megacycles, and so on.

The mode patterns (field lines) are also of interest. For this, we determine **E** and **H** from Eqs. (3-86) and (4-19) or Eqs. (3-89) and (4-21), and then determine $\mathcal{E}, \mathcal{H}$ from Eq. (1-41). The mode pattern is a plot of lines of $\mathcal{E}$ and $\mathcal{H}$ at some instant. (A more direct procedure for obtaining the mode patterns is considered in Sec. 8-1.) Figure 4-2 shows sketches of cross-sectional mode patterns for some of the lower-order modes. When a line appears to end in space in these patterns, it actually loops down the guide. A more complete picture is shown for the TE_{01} mode in Fig. 2-17.

In addition, each mode is characterized by a constant (with respect to

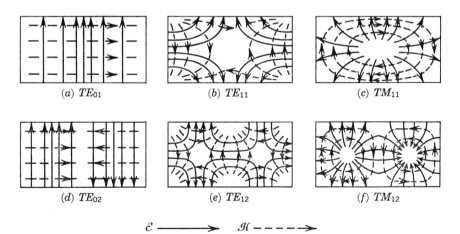

(a) TE_{01} (b) TE_{11} (c) TM_{11}
(d) TE_{02} (e) TE_{12} (f) TM_{12}

$\mathcal{E} \longrightarrow$ $\mathcal{H} \dashrightarrow$

FIG. 4-2. Rectangular waveguide mode patterns.

x, y) z-directed wave impedance. For the TE$_{mn}$ modes in loss-free media, we have from Eqs. (3-89) and (4-21)

$$j\omega\mu H_x = -jk_z \frac{\partial \psi}{\partial x} = -jk_z E_y$$

$$j\omega\mu H_y = -jk_z \frac{\partial \psi}{\partial y} = jk_z E_x$$

The TE$_{mn}$ characteristic wave impedances are therefore

$$(Z_0)_{mn}^{\text{TE}} = \frac{E_x}{H_y} = -\frac{E_y}{H_x} = \frac{\omega\mu}{k_z} = \begin{cases} \dfrac{\omega\mu}{\beta} & f > f_c \\ \dfrac{j\omega\mu}{\alpha} & f < f_c \end{cases} \quad (4\text{-}27)$$

Similarly, for the TM$_{mn}$ modes, we have from Eqs. (3-86) and (4-19)

$$j\omega\epsilon E_x = -jk_z \frac{\partial \psi}{\partial x} = jk_z H_y$$

$$j\omega\epsilon E_y = -jk_z \frac{\partial \psi}{\partial y} = -jk_z H_x$$

Thus, the TM$_{mn}$ characteristic wave impedances are

$$(Z_0)_{mn}^{\text{TM}} = \frac{E_x}{H_y} = \frac{-E_y}{H_x} = \frac{k_z}{\omega\epsilon} = \begin{cases} \dfrac{\beta}{\omega\epsilon} & f > f_c \\ \dfrac{\alpha}{j\omega\epsilon} & f < f_c \end{cases} \quad (4\text{-}28)$$

It is interesting to note that the product $(Z_0)_{mn}^{\text{TE}}(Z_0)_{mn}^{\text{TM}} = \eta^2$ at all frequencies. By Eq. (4-26), $\beta < k$ for propagating modes; so the TE characteristic wave impedances are always greater than η, and the TM characteristic wave impedances are always less than η. For nonpropagating modes, the TE characteristic impedances are inductive, and the TM characteristic impedances are capacitive. Figure 4-3 illustrates this behavior.

Attenuation of the higher-order modes due to dielectric losses is given by the same formula as for the dominant mode (see Table 2-4). Attenuation due to conductor losses is given in Prob. 4-4.

4-4. Alternative Mode Sets. The classification of waveguide modes into sets TE or TM to z is important because it applies also to guides of nonrectangular cross section. However, for many rectangular waveguide problems, more convenient classifications can be made. We now consider these alternative sets of modes.

If, instead of Eq. (3-84), we choose

$$\mathbf{A} = \mathbf{u}_x \psi \quad (4\text{-}29)$$

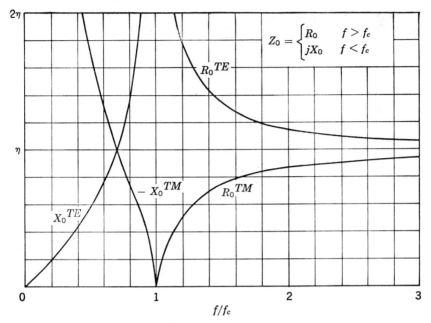

Fig. 4-3. Characteristic impedance of waveguide modes.

we have an electromagnetic field given by a set of equations differing from Eqs. (3-86) by a cyclic interchange of x, y, z. To be specific, the field is given by

$$E_x = \frac{1}{\hat{y}}\left(\frac{\partial^2}{\partial x^2} + k^2\right)\psi \qquad H_x = 0$$
$$E_y = \frac{1}{\hat{y}}\frac{\partial^2 \psi}{\partial x\, \partial y} \qquad H_y = \frac{\partial \psi}{\partial z} \qquad (4\text{-}30)$$
$$E_z = \frac{1}{\hat{y}}\frac{\partial^2 \psi}{\partial x\, \partial z} \qquad H_z = -\frac{\partial \psi}{\partial y}$$

This field is TM to x. Similarly, if, instead of Eq. (3-87), we choose

$$\mathbf{F} = \mathbf{u}_x \psi \qquad (4\text{-}31)$$

we have an electromagnetic field given by

$$E_x = 0 \qquad H_x = \frac{1}{\hat{z}}\left(\frac{\partial^2}{\partial x^2} + k^2\right)\psi$$
$$E_y = -\frac{\partial \psi}{\partial z} \qquad H_y = \frac{1}{\hat{z}}\frac{\partial^2 \psi}{\partial x\, \partial y} \qquad (4\text{-}32)$$
$$E_z = \frac{\partial \psi}{\partial y} \qquad H_z = \frac{1}{\hat{z}}\frac{\partial^2 \psi}{\partial x\, \partial z}$$

This field is TE to x. According to the concepts of Sec. 3-12, an arbitrary field can be constructed as a superposition of Eqs. (4-30) and (4-32).

The choice of ψ's to satisfy the boundary conditions for the rectangular waveguide (Fig. 2-16) is relatively simple. For modes TM to x (TMx_{mn} modes) we have

$$\psi_{mn}{}^{\text{TM}x} = \cos\frac{m\pi x}{a} \sin\frac{n\pi y}{b} e^{-jk_z z} \tag{4-33}$$

where $m = 0, 1, 2, \ldots$; $n = 1, 2, 3, \ldots$; and k_z is given by Eq. (4-26). The electromagnetic field is found by substituting Eq. (4-33) into Eqs. (4-30). For modes TE to x (TEx_{mn} modes) we have

$$\psi_{mn}{}^{\text{TE}x} = \sin\frac{m\pi x}{a} \cos\frac{n\pi y}{b} e^{-jk_z z} \tag{4-34}$$

where $m = 1, 2, 3, \ldots$; $n = 0, 1, 2, \ldots$; and k_z is again given by Eq. (4-26). The field is obtained by substituting Eq. (4-34) into Eqs. (4-32). Note that the TMx_{0n} modes are the TE$_{0n}$ modes of Sec. 4-3, and the TEx_{m0} modes are the TE$_{m0}$ modes. All other modes of Eqs. (4-33) and (4-34) are linear combinations of the degenerate sets of TE and TM modes. Note that our present set of modes have both an E_z and H_z (except for the 0-order modes). Such modes are called *hybrid*.

The mode patterns of these hybrid modes can be determined in the usual manner. (Determine **E**, **H**, then **ε**, **ℋ**, and specialize to some instant of time.) The TEx_{m0} mode patterns are those of the TE$_{m0}$ modes, and the TMx_{0n} mode patterns are those of the TE$_{0n}$ modes. Figure 4-4 shows the mode patterns for the TEx_{11} and TMx_{11} modes, to illustrate the character of the higher-order mode patterns.

The characteristic impedances of the hybrid modes are also of interest. For the TMx modes, we have from Eqs. (4-30) and (4-33)

$$j\omega\epsilon E_x = \left[k^2 - \left(\frac{m\pi}{a}\right)^2\right]\psi \qquad H_y = -jk_z\psi$$

Hence, the z-directed wave impedances are

$$(Z_0)_{mn}{}^{\text{TM}x} = \frac{E_x}{H_y} = \frac{k^2 - (m\pi/a)^2}{\omega\epsilon k_z} = \begin{cases} \dfrac{k^2 - (m\pi/a)^2}{\omega\epsilon\beta} & f > f_c \\ \dfrac{k^2 - (m\pi/a)^2}{-j\omega\epsilon\alpha} & f < f_c \end{cases} \tag{4-35}$$

Note that for a small, the cutoff TMx_{mn} modes, $m \neq 0$, have capacitive Z_0's, while the cutoff TMx_{0n} modes have inductive Z_0's. Similarly, from

PLANE WAVE FUNCTIONS 155

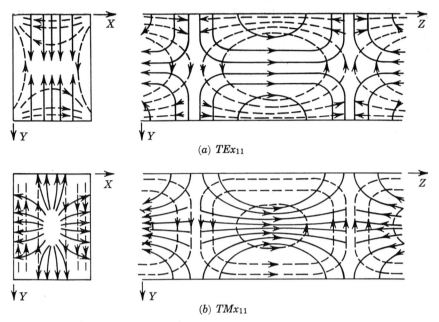

(a) TEx_{11}

(b) TMx_{11}

FIG. 4-4. Hybrid mode patterns.

Eqs. (4-32) and (4-34) we find

$$(Z_0)_{mn}^{TEx} = \frac{-E_y}{H_z} = \frac{\omega\mu k_z}{k^2 - (m\pi/a)^2} = \begin{cases} \dfrac{\omega\mu\beta}{k^2 - (m\pi/a)^2} & f > f_c \\ \dfrac{-j\omega\mu\alpha}{k^2 - (m\pi/a)^2} & f < f_c \end{cases} \quad (4\text{-}36)$$

Note that for a small, the cutoff TEx_{mn} modes all have inductive characteristic impedances.

Sets of modes TM and TE to y can be determined by letting $\mathbf{A} = \mathbf{u}_y\psi$ and $\mathbf{F} = \mathbf{u}_y\psi$, respectively. The fields would be given by equations similar to Eqs. (4-30) and (4-32) with x, y, z properly interchanged. The TMy and TEy mode functions would be given by Eqs. (4-33) and (4-34) with mx/a and ny/b interchanged.

4-5. The Rectangular Cavity. We considered the dominant mode of the rectangular cavity in Sec. 2-8. We shall now consider the complete mode spectrum. The geometry of the rectangular cavity is illustrated by Fig. 2-19.

The problem is symmetrical in x, y, z; so we can express the fields as TE or TM to any one of these coordinates. It is conventional to choose the z coordinate, and then the cavity modes are standing waves of the usual TE and TM waveguide modes. The wave functions of Eq. (4-19) satisfy the boundary condition of zero tangential **E** at four of the walls. It is merely necessary to repick $h(k_z z)$ to satisfy this condition at the remaining two walls. This is evidently accomplished if

$$\psi_{mnp}^{TM} = \sin\frac{m\pi x}{a} \sin\frac{n\pi y}{b} \cos\frac{p\pi z}{c} \qquad (4\text{-}37)$$

with $m = 1, 2, 3, \ldots; n = 1, 2, 3, \ldots; p = 0, 1, 2, \ldots$; and Eq. (4-20) becomes

$$\left(\frac{m\pi}{a}\right)^2 + \left(\frac{n\pi}{b}\right)^2 + \left(\frac{p\pi}{c}\right)^2 = k^2 \qquad (4\text{-}38)$$

The field of the TM_{mnp} mode is given by substitution of Eq. (4-37) into Eqs. (3-86). Similarly, the TE_{mnp} mode functions are given by

$$\psi_{mnp}^{TE} = \cos\frac{m\pi x}{a} \cos\frac{n\pi y}{b} \sin\frac{p\pi z}{c} \qquad (4\text{-}39)$$

with $m = 0, 1, 2, \ldots; n = 0, 1, 2, \ldots; p = 1, 2, 3, \ldots; m = n = 0$ excepted. The separation equation remains Eq. (4-38). The TE_{mnp} mode field is given by substitution of Eq. (4-39) into Eqs. (3-89).

As indicated by Eq. (4-38), each mode can exist at only a single k, given a, b, c. Setting $k = 2\pi f \sqrt{\epsilon\mu}$, we solve Eq. (4-38) for the resonant frequencies

$$(f_r)_{mnp} = \frac{1}{2\sqrt{\epsilon\mu}} \sqrt{\left(\frac{m}{a}\right)^2 + \left(\frac{n}{b}\right)^2 + \left(\frac{p}{c}\right)^2} \qquad (4\text{-}40)$$

For $a < b < c$, the dominant mode is the TE_{011} mode. Table 4-3 gives the ratio $(f_r)_{mnp}/(f_r)_{011}$ for cavities of various side lengths. Note that

TABLE 4-3. $\dfrac{(f_r)_{mnp}}{(f_r)_{011}}$ FOR THE RECTANGULAR CAVITY, $a \leq b \leq c$

$\dfrac{b}{a}$	$\dfrac{c}{a}$	TE_{011}	TE_{101}	TM_{110}	TM_{111} TE_{111}	TE_{012}	TE_{021}	TE_{201}	TE_{102}	TM_{120}	TM_{210}	TM_{112} TE_{112}
1	1	1	1	1	1.22	1.58	1.58	1.58	1.58	1.58	1.58	1.73
1	2	1	1	1.26	1.34	1.26	1.84	1.84	1.26	2.00	2.00	1.55
2	2	1	1.58	1.58	1.73	1.58	1.58	2.91	2.00	2.00	2.91	2.12
2	4	1	1.84	2.00	2.05	1.26	1.84	3.60	2.00	2.53	3.68	2.19
4	4	1	2.91	2.91	3.00	1.58	1.58	5.71	3.16	3.16	5.71	3.24
4	8	1	3.62	3.65	3.66	1.26	1.84	7.20	3.65	4.03	7.25	3.82
4	16	1	3.88	4.00	4.01	1.08	1.96	7.76	3.91	4.35	7.83	4.13

the TE_{mnp} and TM_{mnp} modes, mnp all nonzero, are always degenerate. When two or more sides of the cavity are of equal length, still other degeneracies occur. The greatest separation between the dominant mode and the next lowest-order mode is obtained for a square-base cavity ($b = c$) with height one-half or less of the base length ($b/a \geq 2$). In this case, the second resonance is $\sqrt{5/2} = 1.58$ times the first resonance.

The mode patterns of the rectangular cavity are similar to those of the TE or TM waveguide modes in a $z =$ constant plane, and similar to the hybrid mode patterns in the other two cross sections. The most significant difference between the waveguide patterns and the cavity patterns is that $\mathcal{E}$ is shifted from $\mathcal{H}$ by $\lambda_g/4$ in the latter case. Also, $\mathcal{E}$ and $\mathcal{H}$ are 90° out of phase in a cavity; so $\mathcal{E}$ is zero when $\mathcal{H}$ is maximum, and vice versa. The TE_{011} mode pattern is shown in Fig. 2-20. To illustrate higher-order mode patterns, Fig. 4-5 shows the TE_{123} mode pattern.

The quality factor Q of each cavity mode can be determined by the method used in Sec. 2-8 for the dominant mode. The Q due to dielectric losses is the same for all modes, given by Eq. (2-100). The Q's due to conductor losses for the various modes are given in Prob. 4-10. Note that the Q increases as the mode order increases. The Q varies roughly as the ratio of volume to surface area of the cavity, since the energy is

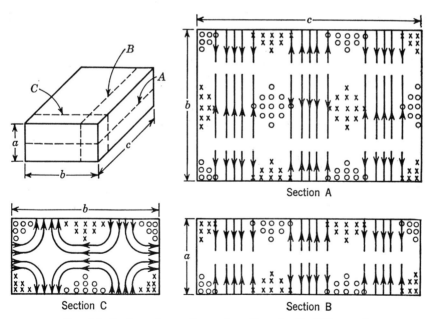

FIG. 4-5. Rectangular cavity mode pattern for the TE_{123} mode.

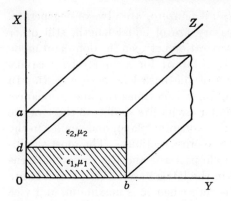

Fig. 4-6. A partially dielectric-filled rectangular waveguide.

stored in the dielectric and the losses are dissipated in the conducting walls.

4-6. Partially Filled Waveguide.[1] Consider a waveguide that is dielectric filled between $x = 0$ and $x = d$ (or has two dielectrics). This is illustrated by Fig. 4-6. The problem contains two homogeneous regions, $0 < x < d$ and $d < x < a$. Such problems are solved by finding solutions in each region such that tangential components of **E** and **H** are continuous across the common boundary. An attempt to find modes either TE to z or TM to z will prove unsuccessful, except for the TE_{m0} case. Most modes are therefore hybrid, having both E_z and H_z. An attempt to find modes TE or TM to x will prove successful, as we now show.

For fields TM to x, we choose ψ's in each region (region 1 is $x < d$, region 2 is $x > d$) to represent the x component of **A**, as in Eq. (4-29). The field in terms of the ψ's is then given by Eqs. (4-30). To satisfy the boundary conditions at the conducting walls, we take

$$\psi_1 = C_1 \cos k_{x1}x \sin \frac{n\pi y}{b} e^{-jk_z z}$$
$$\psi_2 = C_2 \cos [k_{x2}(a - x)] \sin \frac{n\pi y}{b} e^{-jk_z z} \qquad (4\text{-}41)$$

with $n = 1, 2, 3, \ldots$. It has been anticipated that $k_y = n\pi/b$ and k_z must be the same in each region for matching tangential **E** and **H** at $x = d$. The separation parameter equations in the two regions are

$$k_{x1}^2 + \left(\frac{n\pi}{b}\right)^2 + k_z^2 = k_1^2 = \omega^2 \epsilon_1 \mu_1$$
$$k_{x2}^2 + \left(\frac{n\pi}{b}\right)^2 + k_z^2 = k_2^2 = \omega^2 \epsilon_2 \mu_2 \qquad (4\text{-}42)$$

---

[1] L. Pincherle, Electromagnetic Waves in Metal Tubes Filled Longitudinally with Two Dielectrics, *Phys. Rev.*, vol. 66, no. 5, pp. 118–130, 1944.

From Eqs. (4-30) and (4-41) we calculate

$$E_{y1} = -\frac{1}{j\omega\epsilon_1} C_1 k_{x1} \frac{n\pi}{b} \sin k_{x1}x \cos \frac{n\pi y}{b} e^{-jk_z z}$$

$$E_{y2} = \frac{1}{j\omega\epsilon_2} C_2 k_{x2} \frac{n\pi}{b} \sin [k_{x2}(a - x)] \cos \frac{n\pi y}{b} e^{-jk_z z}$$

$$E_{z1} = \frac{1}{\omega\epsilon_1} C_1 k_{x1} k_z \sin k_{x1}x \sin \frac{n\pi y}{b} e^{-jk_z z}$$

$$E_{z2} = -\frac{1}{\omega\epsilon_2} C_2 k_{x2} k_z \sin [k_{x2}(a - x)] \sin \frac{n\pi y}{b} e^{-jk_z z}$$

Continuity of E_y and E_z at $x = d$ requires that

$$\frac{1}{\epsilon_1} C_1 k_{x1} \sin k_{x1}d = -\frac{1}{\epsilon_2} C_2 k_{x2} \sin [k_{x2}(a - d)] \qquad (4\text{-}43)$$

Similarly, from Eqs. (4-30) and (4-41) we calculate

$$H_{y1} = -jk_z C_1 \cos k_{x1}x \sin \frac{n\pi y}{b} e^{-jk_z z}$$

$$H_{y2} = -jk_z C_2 \cos [k_{x2}(a - x)] \sin \frac{n\pi y}{b} e^{-jk_z z}$$

$$H_{z1} = \frac{n\pi}{b} C_1 \cos k_{x1}x \cos \frac{n\pi y}{b} e^{-jk_z z}$$

$$H_{z2} = \frac{n\pi}{b} C_2 \cos [k_{x2}(a - x)] \cos \frac{n\pi y}{b} e^{-jk_z z}$$

Continuity of H_y and H_z at $x = d$ requires that

$$C_1 \cos k_{x1}d = C_2 \cos [k_{x2}(a - d)] \qquad (4\text{-}44)$$

Division of Eq. (4-43) by Eq. (4-44) gives

$$\frac{k_{x1}}{\epsilon_1} \tan k_{x1}d = -\frac{k_{x2}}{\epsilon_2} \tan [k_{x2}(a - d)] \qquad (4\text{-}45)$$

Both k_{x1} and k_{x2} are functions of k_z by Eqs. (4-42); so the above is a transcendental equation for determining possible k_z's (mode-propagation constants). Once the desired k_z is found, k_{x1} and k_{x2} are given by Eqs. (4-42), and the ratio C_2/C_1 is given by Eq. (4-43) or Eq. (4-44).

For fields TE to x, we choose ψ's in each region to represent the x component of **F**. To satisfy the boundary conditions at the conducting walls, we take

$$\psi_1 = C_1 \sin k_{x1}x \cos \frac{n\pi y}{b} e^{-jk_z z}$$

$$\psi_2 = C_2 \sin [k_{x2}(a - x)] \cos \frac{n\pi y}{b} e^{-jk_z z} \qquad (4\text{-}46)$$

with $n = 0, 1, 2, \ldots$. The separation parameter equations are again Eqs. (4-42). The field is calculated from the ψ's by Eqs. (4-32). A matching of tangential **E** and **H** at $x = d$ yields the characteristic equation

$$\frac{k_{x1}}{\mu_1} \cot k_{x1} d = -\frac{k_{x2}}{\mu_2} \cot [k_{x2}(a - d)] \qquad (4\text{-}47)$$

The k_{x1} and k_{x2} are functions of k_z by Eqs. (4-42); so the above is a transcendental equation for determining k_z's for the modes TE to x.

The modes of the partially filled rectangular waveguide are distorted versions of the TEx and TMx modes of Sec. 4-4. The mode patterns are similar to those of Fig. 4-4, except that the field tends to concentrate in the material of higher ϵ and μ. In the lossless case, the cutoff frequencies ($k_z = 0$) of the various modes will always lie between those for the corresponding modes of a guide filled with a material ϵ_1, μ_1, and those of a guide filled with a material ϵ_2, μ_2. (This can be shown by the perturbational procedure of Sec. 7-4.) In contrast to the filled guide, the cutoff frequencies of the corresponding TEx and TMx modes will be different. Also, a knowledge of the cutoff frequencies of the partially filled guide is *not* sufficient to determine k_z at other frequencies by Eq. (4-26). We have to solve Eqs. (4-45) and (4-47) at each frequency.

Of special interest is the dominant mode of a partially filled guide. For $b > a$, this is the mode corresponding to the TMx_{01} mode of the empty guide, which is also the TE$_{01}$ mode of the empty guide. For a given n, Eq. (4-45) has a denumerably infinite set of solutions. We shall let m denote the order of these solutions, as follows. The mode with the lowest cutoff frequency is denoted by $m = 0$, the next mode by $m = 1$, and so on. This numbering system is chosen so that the TMx_{mn} partially filled waveguide modes correspond to the TMx_{mn} empty-guide modes. The dominant mode of the partially filled guide is then the TMx_{01} mode when $b > a$. Hence, the propagation constant of the dominant mode is given by the lowest-order solution to Eq. (4-45) when the k_x's are given by Eqs. (4-42) with $n = 1$. Figure 4-7 shows some calculations for the case $\epsilon = 2.45\epsilon_0$.

When k_1 is not very different from k_2, we should expect k_{x1} and k_{x2} to be small (k_x is zero in an empty guide). If this is so, then Eq. (4-45) can be approximated by

$$\frac{k_{x1}^2 d}{\epsilon_1} \approx \frac{-k_{x2}^2(a - d)}{\epsilon_2} \qquad (4\text{-}48)$$

With this explicit relationship between k_{x1} and k_{x2}, we can solve Eqs. (4-42) simultaneously for k_{x1} and k_z (given ω). Note that when k_{x1} is real, k_{x2} is imaginary, and vice versa. The cutoff frequency is obtained by setting $k_z = 0$ in Eqs. (4-42). Using Eq. (4-48), we have for the

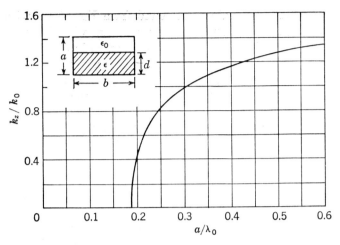

Fig. 4-7. Propagation constant for a rectangular waveguide partially filled with dielectric, $\epsilon = 2.45\epsilon_0$, $a/b = 0.45$, $d/a = 0.50$. (*After Frank.*)

dominant mode

$$k_{x1}^2 + \left(\frac{\pi}{b}\right)^2 = \omega^2 \epsilon_1 \mu_1$$

$$\frac{-\epsilon_2 d}{\epsilon_1(a-d)} k_{x1}^2 + \left(\frac{\pi}{b}\right)^2 = \omega^2 \epsilon_2 \mu_2$$

These we solve for the cutoff frequency $\omega = \omega_c$, obtaining

$$\omega_c \approx \frac{\pi}{b} \sqrt{\frac{\epsilon_1(a-d) + \epsilon_2 d}{\epsilon_1(a-d)\epsilon_2\mu_2 + \epsilon_2 d \epsilon_1 \mu_1}} \qquad (4\text{-}49)$$

valid when Eq. (4-48) applies. When $\mu_1 = \mu_2 = \mu$, this reduces to

$$\omega_c \approx \frac{\pi}{b} \sqrt{\frac{\epsilon_1(a-d) + \epsilon_2 d}{\mu \epsilon_1 \epsilon_2 a}} \qquad (4\text{-}50)$$

Note that this is the equation for resonance of a parallel-plate transmission line, shorted at each end, and having

$$L = \mu a \qquad C = \frac{\epsilon_1 \epsilon_2}{\epsilon_1(a-d) + \epsilon_2 d}$$

per unit width. All cylindrical (cross section independent of z) waveguides at cutoff are two-dimensional resonators.

A waveguide partially filled in the opposite manner (dielectric boundary parallel to the narrow side of the guide) is the same problem with $a > b$. The dominant mode of the empty guide is then the $\text{TE}x_{10}$ mode, or TE_{10} mode. The dominant mode of the partially filled guide will also be a

TEx mode; so the eigenvalues are found from Eq. (4-47) with $n = 0$. We shall order the modes by m as follows. That with the lowest cutoff frequency is denoted by $m = 1$, that with the next lowest by $m = 2$, and so on. This numbering system corresponds to that for the empty guide, the dominant mode being the TEx_{10} mode. When k_1 is not too different from k_2, we might expect k_{x1} and k_{x2} to be close to the empty-guide value $k_x = \pi/a$. An approximate solution to Eq. (4-47) could then be found by perturbing k_{x1} and k_{x2} about π/a. For the cutoff frequency of the

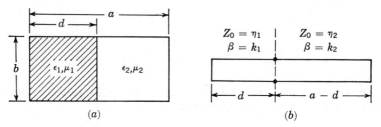

FIG. 4-8. (a) Partially filled waveguide; (b) transmission-line resonator. The cutoff frequency of the dominant mode of (a) is the resonant frequency of (b).

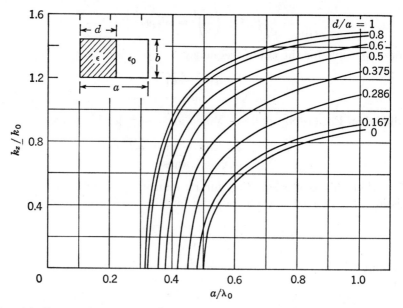

FIG. 4-9. Propagation constant for a rectangular waveguide partially filled with dielectric, $\epsilon = 2.45\epsilon_0$. (*After Frank.*)

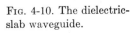

FIG. 4-10. The dielectric-slab waveguide.

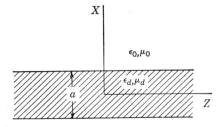

dominant mode, Eqs. (4-42) become

$$k_{x1}^2 = k_{1c}^2 = \omega_c^2 \epsilon_1 \mu_1$$
$$k_{x2}^2 = k_{2c}^2 = \omega_c^2 \epsilon_2 \mu_2$$

and Eq. (4-47) becomes

$$\frac{1}{\eta_1} \cot k_{1c} d = -\frac{1}{\eta_2} \cot [k_{2c}(a - d)] \qquad (4\text{-}51)$$

It is interesting to note that this is the equation for resonance of two short-circuited transmission lines having Z_0's of η_1 and η_2, and β's of k_{1c} and k_{2c}, as illustrated by Fig. 4-8. The reason for this is, at cutoff, the TEx_{10} mode reduces to the parallel-plate transmission-line mode that propagates in the x direction. This viewpoint has been used extensively by Frank.[1]

Some calculated propagation constants for the dominant mode are shown in Fig. 4-9 for the case $\epsilon = 2.45\epsilon_0$. Similar results for a centered dielectric slab are shown in Fig. 7-10, and the characteristic equation for that case is given in Prob. 4-19.

4-7. The Dielectric-slab Guide. It is not necessary to have conductors for the guidance or localization of waves. Such phenomena also occur in inhomogeneous dielectrics. The simplest illustration of this is the guidance of waves by a dielectric slab. The so-called slab waveguide is illustrated by Fig. 4-10.

We shall consider the problem to be two-dimensional, allowing no variation with the y coordinate. It is desired to find z-traveling waves, that is, $e^{-jk_z z}$ variation. Modes TE and TM to either x or z can be found, and we shall choose the latter representation. For modes TM to z, Eqs. (3-86) reduce to

$$E_x = \frac{-k_z}{\omega\epsilon}\frac{\partial \psi}{\partial x} \qquad E_z = \frac{1}{j\omega\epsilon}(k^2 - k_z^2)\psi \qquad H_y = -\frac{\partial \psi}{\partial x} \qquad (4\text{-}52)$$

We shall consider separately the two cases: (1) ψ an odd function of x, denoted by ψ^o, and (2) ψ an even function of x, denoted by ψ^e. For case

[1] N. H. Frank, Wave Guide Handbook, *MIT Rad. Lab. Rept.* 9, 1942.

(1), we choose in the dielectric region

$$\psi_d^o = A \sin ux \, e^{-jk_z z} \qquad |x| < \frac{a}{2} \tag{4-53}$$

and in the air region

$$\psi_a^o = B e^{-vx} e^{-jk_z z} \qquad x > \frac{a}{2}$$
$$\psi_a^o = -B e^{vx} e^{-jk_z z} \qquad x < -\frac{a}{2} \tag{4-54}$$

We have chosen $k_{xd} = u$ and $k_{x0} = jv$ for simplicity of notation. (It will be seen later that u and v are real for unattenuated wave propagation.) The separation parameter equations in each region become

$$u^2 + k_z^2 = k_d^2 = \omega^2 \epsilon_d \mu_d$$
$$-v^2 + k_z^2 = k_0^2 = \omega^2 \epsilon_0 \mu_0 \tag{4-55}$$

Evaluating the field components tangential to the air-dielectric interface, we have

$$\left. \begin{array}{l} E_z = \dfrac{A}{j\omega\epsilon_d} u^2 \sin ux \, e^{-jk_z z} \\ H_y = -Au \cos ux \, e^{-jk_z z} \end{array} \right\} \qquad |x| < \frac{a}{2}$$

$$H_y = Bv e^{-v|x|} e^{-jk_z z} \qquad |x| > \frac{a}{2}$$

$$E_z = \frac{-B}{j\omega\epsilon_0} v^2 e^{-vx} e^{-jk_z z} \qquad x > \frac{a}{2}$$

$$E_z = \frac{B}{j\omega\epsilon_0} v^2 e^{vx} e^{-jk_z z} \qquad x < -\frac{a}{2}$$

Continuity of E_z and H_y at $x = \pm a/2$ requires that

$$\frac{A}{\epsilon_d} u^2 \sin \frac{ua}{2} = \frac{-B}{\epsilon_0} v^2 e^{-va/2}$$

$$Au \cos \frac{ua}{2} = -Bv e^{-va/2}$$

The ratio of the first equation to the second gives

$$\frac{ua}{2} \tan \frac{ua}{2} = \frac{\epsilon_d}{\epsilon_0} \frac{va}{2} \tag{4-56}$$

This, coupled with Eqs. (4-55), is the characteristic equation for determining k_z's and cutoff frequencies of the odd TM modes.

For TM modes which are even functions of x, we choose

$$\psi_d{}^e = A \cos ux\, e^{-jk_z z} \qquad |x| < \frac{a}{2}$$
$$\psi_a{}^e = B e^{-v|x|} e^{-jk_z z} \qquad |x| > \frac{a}{2} \tag{4-57}$$

The separation parameter equations are still Eqs. (4-55). The field components are still given by Eqs. (4-52). In this case, matching E_z and H_y at $x = \pm a/2$ yields

$$-\frac{ua}{2} \cot \frac{ua}{2} = \frac{\epsilon_d}{\epsilon_0} \frac{va}{2} \tag{4-58}$$

This is the characteristic equation for determining the k_z's and cutoff frequencies of the even TM modes.

There is complete duality between the TM and TE modes of the slab waveguide; so the characteristic equations must be dual. For the TE modes with odd ψ we have

$$\frac{ua}{2} \tan \frac{ua}{2} = \frac{\mu_d}{\mu_0} \frac{va}{2} \tag{4-59}$$

as the characteristic equation, and for the TE modes with even ψ we have

$$-\frac{ua}{2} \cot \frac{ua}{2} = \frac{\mu_d}{\mu_0} \frac{va}{2} \tag{4-60}$$

as the characteristic equation. The u's and v's still satisfy Eqs. (4-55). The odd wave functions generating the TE modes are those of Eqs. (4-53) and (4-54), and the even wave functions generating the TE modes are those of Eqs. (4-57). The fields are, of course, obtained from the ψ's by equations dual to Eqs. (4-52), which are, explicitly,

$$H_x = -\frac{k_z}{\omega\mu} \frac{\partial \psi}{\partial x} \qquad H_z = \frac{1}{j\omega\mu}(k^2 - k_z{}^2)\psi \qquad E_y = \frac{\partial \psi}{\partial x} \tag{4-61}$$

These are specializations of Eqs. (3-89).

The concept of cutoff frequency for dielectric waveguides is given a somewhat different interpretation than for metal guides. Above the cutoff frequency, as we define it, the dielectric guide propagates a mode unattenuated (k_z is real). Below the cutoff frequency, there is attenuated propagation ($k_z = \beta - j\alpha$). Since the dielectric is loss free, this attenuation must be accounted for by radiation of energy as the wave progresses. Dielectric guides operated in a radiating mode (below cutoff) are used as antennas. The phase constant of an unattenuated mode lies between the intrinsic phase constant of the dielectric and that of air; that is,

$$k_0 < k_z < k_d$$

This can be shown as follows. Equations (4-55) require that u and v be either real or imaginary when k_z is real. The characteristic equations have solutions only when v is real. Furthermore, v must be positive, else the field will increase with distance from the slab [see Eqs. (4-54) or (4-57)]. When v is real and positive the characteristic equations have solutions only when u is also real. Hence, both u and v are real, and it follows from Eqs. (4-55) that $k_0 < k_z < k_d$. This result is a property of cylindrical dielectric waveguides in general.

The lowest frequency for which unattenuated propagation exists is called the *cutoff frequency*. From the above discussion, it is evident that cutoff occurs as $k_z \to k_0$, in which case $v \to 0$. The cutoff frequencies are therefore obtained from the characteristic equations by setting $u = \sqrt{k_d^2 - k_0^2}$ and $v = 0$. The result is

$$\tan\left(\frac{a}{2}\sqrt{k_d^2 - k_0^2}\right) = 0 \qquad \cot\left(\frac{a}{2}\sqrt{k_d^2 - k_0^2}\right) = 0$$

which apply to both TE and TM modes. These equations are satisfied when

$$\frac{a}{2}\sqrt{k_d^2 - k_0^2} = \frac{n\pi}{2} \qquad n = 0, 1, 2, \ldots$$

This we solve for the cutoff wavelengths

$$\lambda_c = \frac{2a}{n}\sqrt{\frac{\epsilon_d \mu_d}{\epsilon_0 \mu_0} - 1} \qquad n = 0, 1, 2, \ldots \tag{4-62}$$

and the cutoff frequencies

$$f_c = \frac{n}{2a\sqrt{\epsilon_d \mu_d - \epsilon_0 \mu_0}} \qquad n = 0, 1, 2, \ldots \tag{4-63}$$

The modes are ordered as TM_n and TE_n according to the choice of n in Eqs. (4-62) and (4-63). Note that f_c for the TE_0 and TM_0 modes is zero. In other words, *the lowest-order TE and TM modes propagate unattenuated no matter how thin the slab.* This is a general property of cylindrical dielectric waveguides; the cutoff frequency of the dominant mode (or modes) is zero. However, as the slab becomes very thin, $k_z \to k_0$ and $v \to 0$, so the field extends great distances from the slab. This characteristic is considered further in the next section. Finally, observe from Eq. (4-62) that when $\epsilon_d \mu_d \gg \epsilon_0 \mu_0$, the cutoffs occur when the guide width is approximately an integral number of half-wavelengths *in the dielectric*, zero half-wavelength included.

Simple graphical solutions of the characteristic equations exist to determine k_z at any frequency above cutoff. Let us demonstrate this for the TE modes. Elimination of k_z from Eqs. (4-55) gives

$$u^2 + v^2 = k_d{}^2 - k_0{}^2 = \omega^2(\epsilon_d\mu_d - \epsilon_0\mu_0)$$

Using this relationship, we can write the TE characteristic equations as

$$\left.\begin{aligned} \frac{\mu_0}{\mu_d}\frac{ua}{2}\tan\frac{ua}{2} \\ -\frac{\mu_0}{\mu_d}\frac{ua}{2}\cot\frac{ua}{2} \end{aligned}\right\} = \sqrt{\left(\frac{\omega a}{2}\right)^2(\epsilon_d\mu_d - \epsilon_0\mu_0) - \left(\frac{ua}{2}\right)^2}$$

Values of $ua/2$ for the various modes are the intersections of the plot of the left-hand terms with the circle specified by the right-hand term. Figure 4-11 shows a plot of the left-hand terms for $\mu_d = \mu_0$. A representative plot of the right-hand term is shown dashed. As ω or ϵ_d is varied, only the radius of the circle changes. (For the case shown, only three TE modes are above cutoff.) If $\mu_d \neq \mu_0$, the solid curves must be redrawn. The graphical solution for the TM mode eigenvalues is similar.

Sketches of the mode patterns are also of interest. Figure 4-12 shows the patterns of the TE_0 and TM_1 modes. These can also be interpreted as the mode patterns of the TM_0 and TE_1 modes if $\mathcal{E}$ and $\mathcal{H}$ are interchanged, for there is complete duality between the TE and TM cases.

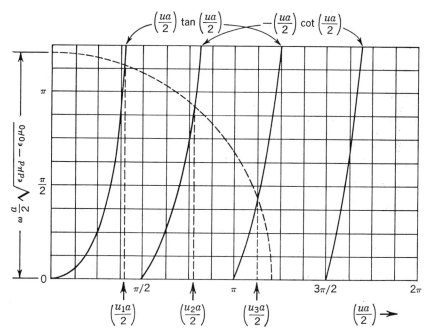

FIG. 4-11. Graphical solution of the characteristic equation for the slab waveguide.

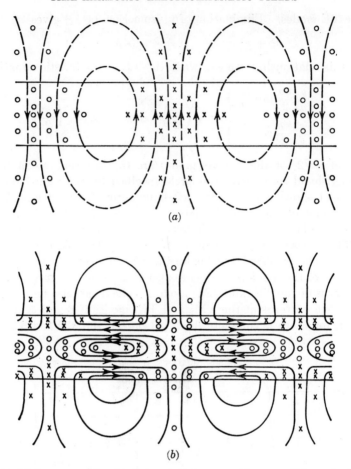

Fig. 4-12. Mode patterns for the dielectric-slab waveguide. (a) TE$_0$ mode ($\mathcal{H}$ lines dashed); (b) TM$_1$ mode ($\mathcal{E}$ lines solid).

As the mode number increases, more loops appear within the dielectric, but not in the air region.

4-8. Surface-guided Waves. We shall show that any "reactive boundary" will tend to produce wave guidance along that boundary. The wave impedances normal to the dielectric-to-air interfaces of the slab guide of Fig. 4-10 can be shown to be reactive. A simple way of obtaining a single reactive surface is to coat a conductor with a dielectric layer. This is shown in Fig. 4-13.

The modes of the dielectric-coated conductor are those of the dielectric slab having zero tangential **E** over the $x = 0$ plane. These are the TM$_n$, $n = 0, 2, 4, \ldots$, modes (odd ψ) and the TE$_n$, $n = 1, 3, 5, \ldots$, modes (even ψ) of the slab. We shall retain the same mode designations for the coated conductor. The characteristic equations for the TM modes of the coated conductor are therefore Eq. (4-56) with $a/2$ replaced by t (coating thickness). The characteristic equation for the TE modes is Eq. (4-60) with $a/2$ replaced by t. The cutoff frequencies are specified by Eq. (4-63), which, for the coated conductor, becomes

$$f_c = \frac{n}{4t \sqrt{\epsilon_d \mu_d - \epsilon_0 \mu_0}} \qquad (4\text{-}64)$$

where for TM modes $n = 0, 2, 4, \ldots$, and for TE modes $n = 1, 3, 5, \ldots$. The dominant mode is the TM_0 mode, which propagates unattenuated at all frequencies. The mode pattern of the TM_0 mode is sketched in Fig. 4-14.

Let us consider in more detail the manner in which the dominant mode decays with distance from the boundary. In the air space, the field attenuates as e^{-vx}. For thick coatings, $k_z \to k_d$, and, from Eq. (4-55),

$$v \xrightarrow[t \text{ large}]{} k_0 \sqrt{\frac{\epsilon_d \mu_d}{\epsilon_0 \mu_0} - 1} \qquad (4\text{-}65)$$

This attenuation is quite large for most dielectrics. For example, if the coating is polystyrene ($\epsilon_d = 2.56\epsilon_0$, $\mu_d = \mu_0$), the field in 0.12λ has decayed to 36.8 per cent of its value at the surface. However, for thin coatings,

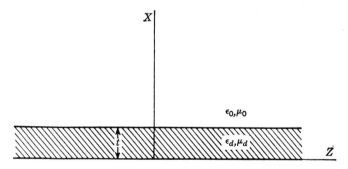

FIG. 4-13. A dielectric-coated conductor.

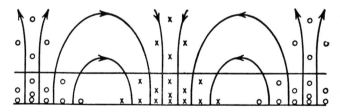

FIG. 4-14. The TM_0 mode pattern for the coated conductor ($\mathcal{E}$ lines solid.)

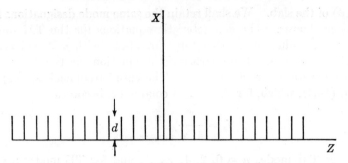

FIG. 4-15. A corrugated conductor.

the field decays slowly. In this case, $k_z \to k_0$, and

$$v \xrightarrow[t \text{ small}]{} 2\pi k_0 \left(\frac{\mu_d}{\mu_0} - \frac{\epsilon_0}{\epsilon_d} \right) \frac{t}{\lambda} \tag{4-66}$$

If the polystyrene coating were 0.0001 wavelength thick, we would have to go 40 wavelengths from the surface before the field decays to 36.8 per cent of its value at the surface. We say that the field is "tightly bound" to a thick dielectric coating and "loosely bound" to a thin dielectric coating.

Another way of obtaining a reactive surface is to "corrugate" a conducting surface, as suggested by Fig. 4-15. For a simple treatment of the problem, let us assume that the "teeth" are infinitely thin, and that there are many slots per wavelength. The teeth will essentially short out any E_y, permitting only E_z and E_x at the surface. The TM fields of the dielectric-slab guide are of this type; hence we shall assume that this field exists in the air region. Extracting from Sec. 4-7, we have

$$\left. \begin{array}{l} E_x = \dfrac{k_z}{\omega \epsilon_0} H_y \\[4pt] E_z = \dfrac{-B}{j \omega \epsilon_0} v^2 e^{-vx} e^{-jk_z z} \\[4pt] H_y = B v e^{-vx} e^{-jk_z z} \end{array} \right\} \quad x > d$$

where
$$-v^2 + k_z^2 = k_0^2 = \omega^2 \epsilon_0 \mu_0 \tag{4-67}$$

The wave impedance looking into the corrugated surface is

$$Z_{-x} = \frac{E_z}{H_y} = \frac{jv}{\omega \epsilon_0} \tag{4-68}$$

Note that this is inductively reactive; so to support such a field, the interface must be an inductively reactive surface. (The TE fields of Sec. 4-7 require a capacitively reactive surface.) In the slots of the corrugation, we assume that the parallel-plate transmission-line mode exists. These are then short-circuited transmission lines, of characteristic wave impedance η_0. Hence, the input wave impedance is

$$Z_{-x} = j\eta_0 \tan k_0 d \qquad (4\text{-}69)$$

For $k_0 d < \pi/2$, this is inductively reactive. Equating Eqs. (4-68) and (4-69), we have

$$v = k_0 \tan k_0 d \qquad (4\text{-}70)$$

and, from Eq. (4-67), we have

$$k_z = k_0 \sqrt{1 + \tan^2 k_0 d} \qquad (4\text{-}71)$$

It should be pointed out that this solution is approximate, for we have only approximated the wave impedance at $x = d$. In the true solution, the fields must differ from those assumed in the vicinity of $x = d$. (We should expect E_x to terminate on the edges of the teeth.)

When the teeth are considered to be of finite width, an approximate solution can be obtained by replacing Eq. (4-69) by the average wave impedance. This is found by assuming Eq. (4-69) to hold over the gaps, and by assuming zero impedance over the region occupied by the teeth. The result is[1]

$$k_z \approx k_0 \sqrt{1 + \left(\frac{g}{g+t}\right)^2 \tan^2 k_0 d}$$

where g = width of gaps and t = width of teeth.

While at this time we lack the concepts for estimating the accuracy of the above solution, it has been found to be satisfactory for small $k_0 d$. Note that, from Eq. (4-70), the wave is loosely bound for very small $k_0 d$, becoming more tightly bound as $k_0 d$ becomes larger (but still less than $\pi/2$). The mode pattern of the wave is similar to that for the TM$_0$ coated-conductor mode (Fig. 4-14), except in the vicinity of the corrugations.

4-9. Modal Expansions of Fields. The modes existing in a waveguide depend upon the excitation of the guide. The nonpropagating modes are of appreciable magnitude only in the vicinity of sources or discontinuities. Given the tangential components of **E** (or of **H**) over a waveguide cross section, we can determine the amplitudes of the various waveguide modes. This we shall illustrate for the rectangular waveguide.

Consider the rectangular waveguide of Fig. 2-16. Let $E_x = 0$ and $E_y = f(x,y)$ be known over the $z = 0$ cross section. We wish to determine the field $z > 0$, assuming that the guide is matched (only outward-traveling waves exist). The TEx modes of Sec. 4-4 have no E_x; so let us

---

[1] C. C. Cutler, Electromagnetic Waves Guided by Corrugated Conducting Surfaces, *Bell Telephone Lab. Rept.* MM-44-160-218, October, 1944.

take a superposition of these modes. This is

$$\psi = \sum_{m=1}^{\infty} \sum_{n=0}^{\infty} A_{mn} \sin \frac{m\pi x}{a} \cos \frac{n\pi y}{b} e^{-\gamma_{mn} z} \qquad (4\text{-}72)$$

where A_{mn} are mode amplitudes and the γ_{mn} are the mode-propagation constants, given by Eq. (4-23). In terms of ψ, the field is given by Eqs. (4-32). In particular, E_y at $z = 0$ is given by

$$E_y \bigg|_{z=0} = \sum_{m=1}^{\infty} \sum_{n=0}^{\infty} \gamma_{mn} A_{mn} \sin \frac{m\pi x}{a} \cos \frac{n\pi y}{b}$$

Note that this is in the form of a double Fourier series: a sine series in x and a cosine series in y (see Appendix C). It is thus evident that $\gamma_{mn} A_{mn}$ are the Fourier coefficients of E_y, or

$$\gamma_{mn} A_{mn} = E_{mn} = \frac{2\epsilon_n}{ab} \int_0^a dx \int_0^b dy \, E_y \bigg|_{z=0} \sin \frac{m\pi x}{a} \cos \frac{n\pi y}{b} \qquad (4\text{-}73)$$

where $\epsilon_n = 1$ for $n = 0$ and $\epsilon_n = 2$ for $n > 0$ (*Neumann's number*). The A_{mn}, and hence the field, are now evaluated. The solution for $E_x = f(x,y)$ and $E_y = 0$ given over the $z = 0$ cross section can be obtained from the above solution by a rotation of axes. The general case for which both E_x and E_y are given over the $z = 0$ cross section is a superposition of the two cases $E_x = 0$ and $E_y = 0$. The solution for the case H_x and H_y given over the $z = 0$ cross section can be obtained in a dual manner.

For a large class of waveguides, when many modes exist simultaneously, each mode transmits energy as if it existed alone. We shall show that the rectangular waveguide has this property. Given the wave function of Eq. (4-72), specifying a field according to Eqs. (4-32), the z-directed complex power at $z = 0$ is

$$P = \iint_{z=0} \mathbf{E} \times \mathbf{H}^* \cdot \mathbf{u}_z \, ds = -\int_0^a dx \int_0^b dy \, [E_y H_x^*]_{z=0}$$

$$= \int_0^a dx \int_0^b dy \left[\sum_{m,n} E_{mn} \sin \frac{m\pi x}{a} \cos \frac{n\pi y}{b} \right]$$

$$\times \left[\sum_{p,q} \frac{k^2 - \left(\frac{p\pi}{a}\right)^2}{j\omega\mu\gamma_{pq}^*} E_{pq}^* \sin \frac{p\pi x}{a} \cos \frac{q\pi y}{b} \right]$$

Because of the orthogonality relationships for the sinusoidal functions,

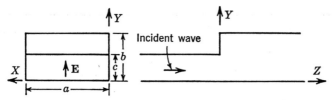

FIG. 4-16. A capacitive waveguide junction.

this reduces to

$$P = \sum_{m=1}^{\infty} \sum_{n=0}^{\infty} (Y_0)^*_{mn} |E_{mn}|^2 \frac{ab}{2\epsilon_n} \tag{4-74}$$

where $(Y_0)_{mn}$ are the TEx wave admittances, given by the reciprocal of Eqs. (4-36). The above equation is simply a summation of the powers for the individual modes. In a lossless guide, the power for a propagating mode is real and that for a nonpropagating mode is imaginary.

To illustrate the above theory, consider the waveguide junction of Fig. 4-16. The dimensions are such that only the dominant mode (TE$_{10}$) propagates in each section. Let there be a wave incident on the junction from the smaller guide, and let the larger guide be matched. For an approximate solution, assume that E_y at the junction is that of the incident wave

$$E_y \bigg|_{z=0} \approx \begin{cases} \sin \dfrac{\pi x}{a} & y < c \\ 0 & y > c \end{cases} \tag{4-75}$$

From Eq. (4-73), the only nonzero mode amplitudes are

$$\begin{aligned} E_{10} &= \gamma_{10} A_{10} = \frac{c}{b} \\ E_{1n} &= \gamma_{1n} A_{1n} = \frac{2}{n\pi} \sin \frac{n\pi c}{b} \end{aligned} \tag{4-76}$$

Thus, only the $m = 1$ term of the m summation remains in Eq. (4-72).

Let us use this solution to obtain an "aperture admittance" for the junction. From Eqs. (4-74) and (4-76), the complex power at $z = 0$ is

$$P = \frac{ac^2}{2b} \left\{ (Y_0)^*_{10} + 2 \sum_{n=1}^{\infty} (Y_0)^*_{1n} \left[\frac{\sin (n\pi c/b)}{n\pi c/b} \right]^2 \right\}$$

where, from Eqs. (4-36),

$$\begin{aligned} (Y_0)_{10} &= \frac{k^2 - (\pi/a)^2}{\omega \mu \beta} = \frac{\sqrt{1 - (f_c/f)^2}}{\eta} \\ (Y_0)_{1n} &= \frac{k^2 - (\pi/a)^2}{-j\omega\mu\alpha} = \frac{j2b(Y_0)_{10}}{\lambda_g \sqrt{n^2 - (2b/\lambda_g)^2}} \end{aligned}$$

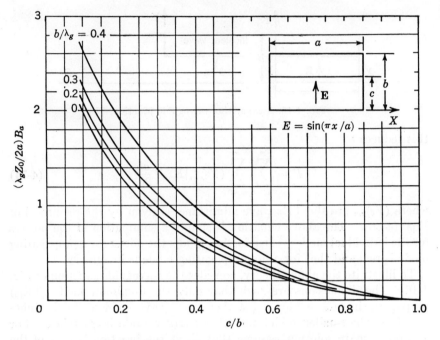

Fig. 4-17. Susceptance of a capacitive aperture.

The f_c and λ_g are those of the TE_{10} mode. We shall refer the aperture admittance to the voltage across the center of the aperture, which is $V = c$. The aperture admittance is then

$$Y_a = \frac{P^*}{|V|^2} = (Y_0)_{10} \left[\frac{a}{2b} + j \frac{2a}{\lambda_g} \sum_{n=1}^{\infty} \frac{\sin^2(n\pi c/b)}{(n\pi c/b)^2 \sqrt{n^2 - (2b/\lambda_g)^2}} \right] \quad (4\text{-}77)$$

The imaginary part of this is the aperture susceptance

$$B_a = \frac{2a}{\lambda_g Z_0} \sum_{n=1}^{\infty} \frac{\sin^2(n\pi c/b)}{(n\pi c/b)^2 \sqrt{n^2 - (2b/\lambda_g)^2}} \quad (4\text{-}78)$$

where λ_g and Z_0 are those of the dominant mode. Calculated values for B_a are shown in Fig. 4-17. For small c/b, we have[1]

$$\frac{\lambda_g Z_0}{2a} B_a \approx -\log\left\{ 0.656 \frac{c}{b} \left[1 + \sqrt{1 - \left(\frac{2b}{\lambda_g}\right)^2} \right] \right\} \quad (4\text{-}79)$$

[1] This equation is a quasi-static result. The direct specialization of Eq. (4-78) to small c/b yields a numerical factor of 0.379 instead of 0.656.

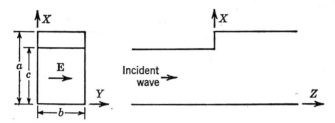

Fig. 4-18. An inductive waveguide junction.

The aperture susceptance is a quantity that will be useful for the treatment of microwave networks in Chap. 8. Note that the susceptance is capacitive (positive); so the original junction is called a *capacitive waveguide junction*. Remember that our solution is only approximate, since we assumed E in the aperture. (We shall see in Sec. 8-9 that the true susceptance cannot be greater than our present solution.) We have assumed that only one mode propagates in the guide; hence our solution is explicit only for

$$1 < \frac{f}{f_c} < \sqrt{1 + \left(\frac{a}{b}\right)^2}$$

When a second mode propagates, it contributes to the aperture conductance, and Eq. (4-78) would be summed from $n = 2$ to ∞, and so on.

Another problem of practical interest is that of the waveguide junction of Fig. 4-18. Again we assume only the dominant mode propagates in each section. Take a wave incident on the junction from the smaller guide, and let the larger guide be matched. For an approximate solution, we assume E_y in the aperture to be that of the incident wave

$$E_y \big|_{z=0} \approx \begin{cases} \sin \dfrac{\pi x}{c} & x < c \\ 0 & x > c \end{cases} \quad (4\text{-}80)$$

From Eqs. (4-73), we determine the only nonzero mode amplitudes as

$$E_{m0} = \frac{2c \sin (m\pi c/a)}{\pi a [1 - (mc/a)^2]} \quad (4\text{-}81)$$

Thus, only the $n = 0$ term of the n summation remains in Eq. (4-72).

Again we can find an aperture admittance for the junction. From Eqs. (4-74) and (4-81), the complex power at $z = 0$ is

$$P = \frac{2bc^2}{\pi^2 a} \sum_{m=1}^{\infty} (Y_0)_{m0}^* \left[\frac{\sin (m\pi c/a)}{1 - (mc/a)^2} \right]^2$$

where, from Eqs. (4-36),

$$(Y_0)_{10} = \frac{k^2}{\omega\mu\beta} = \frac{\sqrt{1-(f_c/f)^2}}{\eta}$$

$$(Y_0)_{m0} = \frac{k^2-(m\pi/a)^2}{-j\omega\mu\alpha} = \frac{-j}{\eta}\sqrt{\left(\frac{m\lambda}{2a}\right)^2-1} \qquad m>1$$

The voltage across the center of the aperture is $V = b$. The aperture admittance referred to this voltage is therefore

$$Y_a = \frac{2c^2}{\pi^2 ab}\left\{\left[\frac{\sin(\pi c/a)}{1-(c/a)^2}\right]^2 (Y_0)_{10}\right.$$
$$\left. -\frac{j}{\eta}\sum_{m=2}^{\infty}\left[\frac{\sin(m\pi c/a)}{1-(mc/a)^2}\right]^2\sqrt{\left(\frac{m\lambda}{2a}\right)^2-1}\right\} \quad (4\text{-}82)$$

The imaginary part of this is the aperture susceptance

$$B_a = \frac{-2\lambda}{\eta\pi^2 b}\left(\frac{c}{a}\right)^2\sum_{m=2}^{\infty}\left[\frac{\sin(m\pi c/a)}{1-(mc/a)^2}\right]^2\sqrt{\left(\frac{m}{2}\right)^2-\left(\frac{a}{\lambda}\right)^2} \quad (4\text{-}83)$$

which is plotted in Fig. 4-19. The susceptance is inductive (negative); so the original junction is called an inductive waveguide junction. For single-mode propagation, we must have $a < \lambda$; so our explicit interpre-

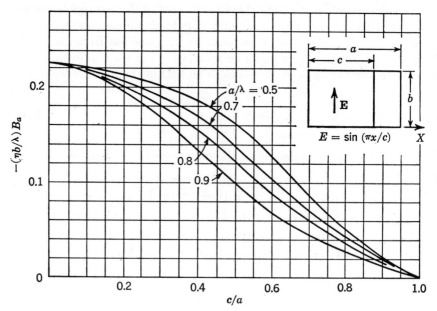

Fig. 4-19. Susceptance of an inductive aperture.

tation of the solution is restricted to this range. For wave propagation in the smaller guide, we must have $c > \lambda/2$ if it is air-filled. However, if the smaller guide is dielectric-filled, we can have wave propagation in it when $c < \lambda/2$. Moreover, the aperture susceptance is defined only in terms of E_y in the aperture and has significance independent of the manner in which this E_y is obtained.

4-10. Currents in Waveguides. The problems of the preceding section might be called "aperture excitation" of waveguides. We shall now consider "current excitation" of waveguides. This involves the determination of modal expansions in terms of current sheets over a guide cross section. The only difference between aperture excitation and current excitation is that the former assumes a knowledge of the tangential electric field and the latter assumes a knowledge of the discontinuity in the tangential magnetic field. The equivalence principle plus duality can be used to transform an aperture-type problem into a current-type problem, and vice versa.

To illustrate the solution, consider a rectangular waveguide with a sheet of x-directed electric currents over the $z = 0$ cross section. This is illustrated by Fig. 3-2, where $\mathbf{J}_s = \mathbf{u}_x f(x,y)$ is now arbitrary. We shall assume that only waves traveling outward from the current are present, that is, the guide is matched in both directions. At $z = 0$ we must have E_x, E_y, and H_x continuous. H_x must also be antisymmetric about $z = 0$; hence it must be identically zero, and it is convenient to use the TMx modes of Sec. 4-4. (Note that $\mathbf{J}$ and its images are x-directed; so it is to be expected that an x-directed $\mathbf{A}$ is sufficient for representing the field.) Superpositions of the TMx modes are

$$\psi^+ = \sum_{m=0}^{\infty} \sum_{n=1}^{\infty} B_{mn}^+ \cos \frac{m\pi x}{a} \sin \frac{n\pi y}{b} e^{-\gamma_{mn} z} \quad z > 0$$

$$\psi^- = \sum_{m=0}^{\infty} \sum_{n=1}^{\infty} B_{mn}^- \cos \frac{m\pi x}{a} \sin \frac{n\pi y}{b} e^{\gamma_{mn} z} \quad z < 0$$

(4-84)

where superscripts $+$ and $-$ refer to the regions $z > 0$ and $z < 0$, respectively. The field in terms of the ψ's is given by Eqs. (4-30). Continuity of E_x and E_y at $z = 0$ requires that

$$B_{mn}^+ = B_{mn}^- = B_{mn} \quad (4\text{-}85)$$

The remaining boundary condition is the discontinuity in H_y caused by J_x, which is

$$J_x = [H_y^- - H_y^+]_{z=0} = \sum_{m=0}^{\infty} \sum_{n=1}^{\infty} 2\gamma_{mn} B_{mn} \cos \frac{m\pi x}{a} \sin \frac{n\pi y}{b}$$

This is a Fourier cosine series in x and a Fourier sine series in y. It is evident that $2\gamma_{mn}B_{mn}$ are the Fourier coefficients of J_x, that is,

$$2\gamma_{mn}B_{mn} = J_{mn} = \frac{2\epsilon_m}{ab} \int_0^a dx \int_0^b dy\, J_x \cos \frac{m\pi x}{a} \sin \frac{n\pi y}{b} \qquad (4\text{-}86)$$

This completes the determination of the field. The solution for a y-directed current corresponds to a rotation of axes in the above solution. When both J_x and J_y exist, the solution is a superposition of the two cases $J_y = 0$ and $J_x = 0$. The solution for a magnetic current sheet in the waveguide is obtained in a dual manner. A z-directed electric current can be treated as a loop of magnetic current in the cross-sectional plane, according to Fig. 3-3. A z-directed magnetic current is the dual problem. Thus, we have the formal solution for all possible cases of currents in a rectangular waveguide.

It is also of interest to find the power supplied by the currents in a waveguide. This is most simply obtained from

$$P = -\iint_{z=0} \mathbf{E} \cdot \mathbf{J}_s^* \, ds = -\int_0^a dx \int_0^b dy\, J_x^* E_x \bigg|_{z=0}$$

We express J_x in its Fourier series and evaluate E_x by Eqs. (4-30) applied to the above solution. Because of the orthogonality relationships, the power reduces to

$$P = \sum_{m=0}^{\infty} \sum_{n=1}^{\infty} (Z_0)_{mn} |J_{mn}|^2 \frac{ab}{4\epsilon_m} \qquad (4\text{-}87)$$

where $(Z_0)_{mn}$ are the TMx wave impedances, given by Eqs. (4-35). This is a summation of the powers that each J_{mn} alone would produce in the guide. In a lossless guide, the power associated with each propagating mode is real, and that associated with a nonpropagating mode is imaginary.

As an example of the above theory, consider the coax to waveguide junction of Fig. 4-20. This is a waveguide "probe feed," the probe being the center conductor of the coax. If the probe is thin, the current on it will have approximately a sinusoidal distribution, as on the linear antenna. With the probe joined to the opposite waveguide wall, as shown in Fig. 4-20, the current maximum is at the joint $x = a$. We therefore assume a current on the probe

$$I(x) \approx \cos k(a - x) \qquad (4\text{-}88)$$

The current sheet approximating this probe is

$$J_x = I(x)\delta(y - c) \qquad (4\text{-}89)$$

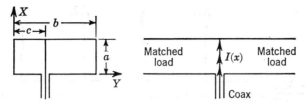

Fig. 4-20. A coax to waveguide junction.

where $\delta(y - c)$ is the impulse function, or delta function (see Appendix C). The Fourier coefficients for the current are then obtained from Eq. (4-86) as

$$J_{mn} = \frac{2\epsilon_m ka \sin ka \sin n\pi c/b}{b[(ka)^2 - (m\pi)^2]} \qquad (4\text{-}90)$$

This, coupled with our earlier formulas, determines the field.

In terms of this solution, let us consider the input impedance seen by the coaxial line. The power supplied by the stub is given by Eq. (4-87). The impedance seen by the coax is then

$$Z_i = \frac{P}{|I_i|^2} = R_i + jX_i$$

where, from Eq. (4-88), the input current is

$$I_i = \cos ka$$

Assume that the waveguide dimensions are such that only the TE_{01} mode propagates. Then only the $m = 0$, $n = 1$ term of Eq. (4-87) is real, and

$$\begin{aligned} R_i &= \frac{ab}{4} \left| \frac{J_{01}}{I_i} \right|^2 (Z_0)_{01} \\ &= \frac{a}{b} (Z_0)_{01} \left(\frac{\tan ka}{ka} \right)^2 \sin^2 \frac{\pi c}{b} \end{aligned} \qquad (4\text{-}91)$$

All other terms of the summation of Eq. (4-87) contribute to X_i. However, since we assumed a filamentary current, the series for X_i diverges. To obtain a finite X_i, we must consider a conductor of finite radius. For small a, the reactance will be capacitive. In the vicinity of $a = \lambda/4$, we have a resonance, above which the reactance is inductive. Note that Eq. (4-91) says that the input resistance is infinite at this resonance. This is incorrect for an actual junction, and the error lies in our assumed current. Equation (4-91) gives reliable input resistances only when we are somewhat removed from resonant points. [This is similar to our linear antenna solution (Sec. 2-10)]. Feeds in waveguides with arbitrary terminations are considered in Sec. 8-11.

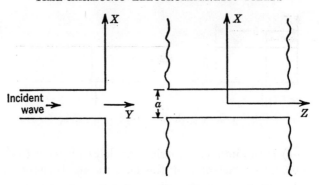

Fig. 4-21. A parallel-plate guide radiating into half-space.

4-11. Apertures in Ground Planes. We have already solved the problem of determining the field from apertures in ground planes, in Sec. 3-6. At this time, however, we shall take an alternative approach and obtain a different form of solution. By the uniqueness theorem, the two forms of solution must be equal. One form may be convenient for some calculations, and the other form for other calculations.

Let us demonstrate the theory for an aperture in the ground plane $y = 0$, illustrated by Fig. 4-21. We further restrict consideration to the case $E_z = 0$, there being only an E_x in the aperture. Taking a clue from our waveguide solution (Sec. 4-9), let us consider Fourier transforms (see Appendix C). The transform pair for E_x over the $y = 0$ plane is

$$E_x(x,0,z) = \frac{1}{4\pi^2} \int_{-\infty}^{\infty} dk_x \int_{-\infty}^{\infty} dk_z\, \bar{E}_x(k_x,k_z) e^{jk_x x} e^{jk_z z}$$
$$\bar{E}_x(k_x,k_z) = \int_{-\infty}^{\infty} dx \int_{-\infty}^{\infty} dz\, E_x(x,0,z) e^{-jk_x x} e^{-jk_z z} \qquad (4\text{-}92)$$

where a bar over a symbol denotes transform. The form of the transformation suggests that we choose as a wave function

$$\psi = \frac{1}{4\pi^2} \int_{-\infty}^{\infty} dk_x \int_{-\infty}^{\infty} dk_z\, f(k_x,k_z) e^{jk_x x} e^{jk_y y} e^{jk_z z} \qquad (4\text{-}93)$$

which is a superposition of the form of Eq. (4-9). For our present problem, we take Eq. (4-93) as representing a field TE to z, according to Eqs. (3-89). There is a one-to-one correspondence between a function and its transform; hence it is evident that the transform of ψ is

$$\bar{\psi} = f(k_x,k_z) e^{jk_y y} \qquad (4\text{-}94)$$

We also can rewrite Eqs. (3-89) in terms of transforms as

$$\bar{E}_x = -jk_y\bar{\psi} \qquad \bar{H}_x = \frac{-k_xk_z}{j\omega\mu}\bar{\psi}$$
$$\bar{E}_y = jk_x\bar{\psi} \qquad \bar{H}_y = \frac{-k_yk_z}{j\omega\mu}\bar{\psi} \qquad (4\text{-}95)$$
$$\bar{E}_z = 0 \qquad \bar{H}_z = \frac{k^2 - k_z^2}{j\omega\mu}\bar{\psi}$$

Specializing the above to the $y = 0$ plane, we have

$$\bar{E}_x\big|_{y=0} = -jk_y f(k_x, k_z)$$

A comparison of this with Eqs. (4-92) shows that

$$f(k_x, k_z) = \frac{-1}{jk_y}\bar{E}_x(k_x, k_z) \qquad (4\text{-}96)$$

where $\bar{E}_x$ is given by the second of Eqs. (4-92). This completes the solution. As a word of caution, $k_y = \pm\sqrt{k^2 - k_x^2 - k_z^2}$ is double-valued, and we must choose the correct root. For Eq. (4-94) to remain finite as $y \to \infty$, we must choose

$$k_y = \begin{cases} j\sqrt{k_x^2 + k_z^2 - k^2} & k < \sqrt{k_x^2 + k_z^2} \\ -\sqrt{k^2 - k_x^2 - k_z^2} & k > \sqrt{k_x^2 + k_z^2} \end{cases} \qquad (4\text{-}97)$$

The minus sign on the lower equality is necessary to remain on the same branch as designated by the upper equality.

The extension of this solution to problems in which both E_x and E_z exist over the $y = 0$ plane can be effected by adding the appropriate TE to x field to the above TE to z field. It can also be obtained as the sum of fields TE and TM to z, or to x, or to y. The case of H_x and H_y specified over the $y = 0$ plane is the dual problem and can be obtained by an interchange of symbols.

For simplicity, we shall choose our illustrative problems to be two-dimensional ones. Let Fig. 4-21 represent a parallel-plate waveguide opening onto a ground plane. If the incident wave is in the transmission-line mode (TEM to y), it is apparent from symmetry that H_z will be the only component of $\mathbf{H}$. Let us therefore take H_z as the scalar wave function and construct

$$H_z = \frac{1}{2\pi}\int_{-\infty}^{\infty} f(k_x)e^{jk_xx}e^{jk_yy}\,dk_x \qquad (4\text{-}98)$$

From this, it is evident that the transform of H_z is

$$\bar{H}_z = f(k_x)e^{jk_yy} \qquad (4\text{-}99)$$

From the field equations, we relate the transform of **E** to $\bar{H}_z$ as

$$\bar{E}_x = \frac{k_y}{\omega\epsilon}\bar{H}_z \qquad \bar{E}_y = -\frac{k_x}{\omega\epsilon}\bar{H}_z \tag{4-100}$$

Specializing $\bar{E}_x$ to $y = 0$, we have

$$\left.\bar{E}_x\right|_{y=0} = \frac{k_y}{\omega\epsilon}f(k_x) = \int_{-\infty}^{\infty} E_x(x,0)e^{-jk_x x}\,dx \tag{4-101}$$

from which $f(k_x)$ may be found. For an approximate solution to Fig. 4-21 for $y > 0$, we assume E_x in the aperture to be of the form of the incident mode, that is,

$$\left.E_x\right|_{y=0} \approx \begin{cases} 1 & |x| < \dfrac{a}{2} \\ 0 & |x| > \dfrac{a}{2} \end{cases} \tag{4-102}$$

Using this in Eq. (4-101), we find

$$\left.\bar{E}_x\right|_{y=0} = \frac{k_y}{\omega\epsilon}f(k_x) = \frac{2}{k_x}\sin\left(k_x\frac{a}{2}\right) \tag{4-103}$$

To complete the solution, we must also choose the root of k_y for proper behavior as $y \to \infty$. From Eq. (4-99), it is evident that this root is

$$k_y = \begin{cases} j\sqrt{k_x^2 - k^2} & k < |k_x| \\ -\sqrt{k^2 - k_x^2} & k > |k_x| \end{cases} \tag{4-104}$$

The fields are found from the transforms by inversion.

A parameter of interest to us in future work is the aperture admittance. To evaluate this, we shall make use of the integral form of Parseval's theorem (Appendix C), which is

$$\int_{-\infty}^{\infty} f(x)g^*(x)\,dx = \frac{1}{2\pi}\int_{-\infty}^{\infty} \bar{f}(k)\bar{g}^*(k)\,dk$$

We can express the power per unit width (z direction) transmitted by the aperture as

$$P = -\int_{-\infty}^{\infty} [E_x H_z^*]_{y=0}\,dx = -\frac{1}{2\pi}\int_{-\infty}^{\infty} [\bar{E}_x \bar{H}_z^*]_{y=0}\,dk_x$$

From Eqs. (4-100) and (4-102), this becomes

$$P = -\frac{\omega\epsilon}{2\pi}\int_{-\infty}^{\infty} \frac{1}{k_y^*}|\bar{E}_x|^2\,dk_x = -\frac{4}{\lambda\eta}\int_{-\infty}^{\infty} \frac{\sin^2(k_x a/2)}{k_y^* k_x^2}\,dk_x$$

PLANE WAVE FUNCTIONS   183

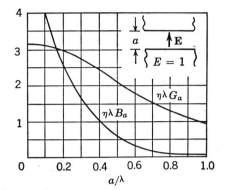

Fig. 4-22. Aperture admittance of a capacitive slot radiator.

We now define the aperture admittance referred to the aperture voltage $V = a$ as

$$Y_a = \frac{P^*}{|V|^2} = \frac{-4}{\lambda \eta a^2} \int_{-\infty}^{\infty} \frac{\sin^2 (k_x a/2)}{k_y k_x^2} \, dk_x$$

Note that, by Eq. (4-104), the above integrand is real for $|k_x| < k$ and imaginary for $|k_x| > k$. We can therefore separate Y_a into its real and imaginary parts as

$$G_a = \frac{4}{\lambda \eta a^2} \int_{-k}^{k} \frac{\sin^2 (k_x a/2)}{k_x^2 \sqrt{k^2 - k_x^2}} \, dk_x$$

$$B_a = \frac{4}{\lambda \eta a^2} \left(\int_{-\infty}^{-k} + \int_{k}^{\infty} \right) \frac{\sin^2 (k_x a/2)}{k_x^2 \sqrt{k_x^2 - k^2}} \, dk_x$$

The above integrals can be simplified to give

$$\begin{aligned} \lambda \eta G_a &= 2 \int_0^{ka/2} \frac{\sin^2 w \, dw}{w^2 \sqrt{(ka/2)^2 - w^2}} \\ \lambda \eta B_a &= 2 \int_{ka/2}^{\infty} \frac{\sin^2 w \, dw}{w^2 \sqrt{w^2 - (ka/2)^2}} \end{aligned} \quad (4\text{-}105)$$

For small ka, these are[1]

$$\left. \begin{aligned} \lambda \eta G_a &\approx \pi \left[1 - \frac{(ka)^2}{24} \right] \\ \lambda \eta B_a &\approx 3.135 - 2 \log ka \end{aligned} \right\} \quad \frac{a}{\lambda} < 0.1 \quad (4\text{-}106)$$

For intermediate ka, the aperture conductance and susceptance are plotted in Fig. 4-22. For large ka, we have

[1] The formula for B_a is a quasi-static result. The direct specialization of the second of Eqs. (4-105) to small ka gives a numerical factor of 4.232 instead of 3.135.

$$\left.\begin{aligned}\lambda\eta G_a &\approx \frac{\lambda}{a} \\ \lambda\eta B_a &\approx \left(\frac{\lambda}{\pi a}\right)^2 \left[1 - \frac{1}{2}\sqrt{\frac{\lambda}{a}} \cos\left(\frac{2a}{\lambda} + \frac{1}{4}\right)\pi\right]\end{aligned}\right\} \quad \frac{a}{\lambda} > 1 \quad (4\text{-}107)$$

The aperture is capacitive, since B_a is always positive.

Another problem of practical interest is that of Fig. 4-21 when the incident wave is in the dominant TE mode (TE to y). In this case, E_z will be the only component of $\mathbf{E}$, and we shall take E_z as our scalar wave function. Analogous to the preceding problem, we construct

$$E_z = \frac{1}{2\pi}\int_{-\infty}^{\infty} f(k_x) e^{jk_x x} e^{jk_y y} \, dk_x \tag{4-108}$$

In terms of Fourier transforms, this is

$$\bar{E}_z = f(k_x) e^{jk_y y} \tag{4-109}$$

From the field equations, we find the transform of $\mathbf{H}$ to be

$$\bar{H}_x = \frac{-k_y}{\omega\mu} \bar{E}_z \qquad \bar{H}_y = \frac{k_x}{\omega\mu} \bar{E}_z \tag{4-110}$$

The $f(k_x)$ is evaluated by specializing Eq. (4-109) to $y = 0$, which gives

$$\bar{E}_z \Big|_{y=0} = f(k_x) = \int_{-\infty}^{\infty} E_z(x,0) e^{-jk_x x} \, dx \tag{4-111}$$

For an approximate solution, we assume the E_z in the aperture of Fig. 4-21 to be that of the incident TE mode, that is,

$$E_z \Big|_{y=0} \approx \begin{cases} \cos\dfrac{\pi x}{a} & |x| < \dfrac{a}{2} \\ 0 & |x| > \dfrac{a}{2} \end{cases} \tag{4-112}$$

Substituting this into the preceding equation, we find

$$\bar{E}_z \Big|_{y=0} = f(k_x) = \frac{2\pi a \cos(k_x a/2)}{\pi^2 - (k_x a)^2} \tag{4-113}$$

The choice of the root for k_y is the same as in the preceding example, given by Eq. (4-104). This completes the formal solution.

Let us again calculate the aperture admittance. The power transmitted by the aperture is

$$P = \int_{-\infty}^{\infty} [E_z H_x^*]_{y=0} \, dx = \frac{1}{2\pi}\int_{-\infty}^{\infty} [\bar{E}_z \bar{H}_x^*]_{y=0} \, dk_x$$

where we have used Parseval's theorem. From Eqs. (4-110) and (4-113),

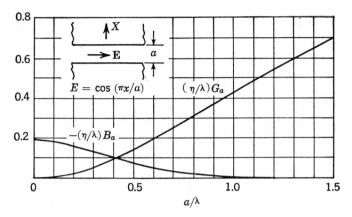

FIG. 4-23. Aperture admittance of an inductive slot radiator.

this becomes

$$P = \frac{-1}{2\pi\omega\mu} \int_{-\infty}^{\infty} k_y^* |\bar{E}_z|^2 \, dk_x = \frac{-2\pi a^2}{\omega\mu} \int_{-\infty}^{\infty} \frac{k_y^* \cos^2{(k_x a/2)}}{[\pi^2 - (k_x a)^2]^2} \, dk_x$$

We shall refer the aperture admittance to the voltage per unit length of the aperture, which is $V = 1$. This gives

$$Y_a = \frac{P^*}{|V|^2} = \frac{-2\pi a^2}{\omega\mu} \int_{-\infty}^{\infty} \frac{k_y \cos^2{(k_x a/2)}}{[\pi^2 - (k_x a)^2]^2} \, dk_x$$

The integrand is real for $|k_x| < k$ and imaginary for $|k_x| > k$. A separation of Y_a into real and imaginary parts is therefore accomplished in the same manner as in the preceding example. The result is

$$\frac{\eta}{\lambda} G_a = \frac{1}{2} \int_0^{ka/2} \frac{\sqrt{(ka/2)^2 - w^2} \cos^2 w}{[(\pi/2)^2 - w^2]^2} \, dw$$
$$\frac{\eta}{\lambda} B_a = \frac{-1}{2} \int_{ka/2}^{\infty} \frac{\sqrt{w^2 - (ka/2)^2} \cos^2 w}{[(\pi/2)^2 - w^2]^2} \, dw \qquad (4\text{-}114)$$

For small ka, we have

$$\left. \begin{array}{l} \dfrac{\eta}{\lambda} G_a \approx \dfrac{2}{\pi} \left(\dfrac{a}{\lambda} \right)^2 \\[6pt] \dfrac{\eta}{\lambda} B_a \approx -0.194 \end{array} \right\} \qquad \frac{a}{\lambda} < 0.1 \qquad (4\text{-}115)$$

For intermediate ka, the aperture conductance and susceptance are plotted in Fig. 4-23. For large ka,

$$\frac{\eta}{\lambda} G_a \approx \frac{a}{2\lambda} \qquad \frac{a}{\lambda} > 1.5 \qquad (4\text{-}116)$$

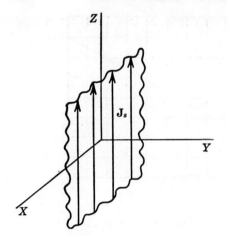

Fig. 4-24. A sheet of z-directed currents in the $y = 0$ plane.

and B_a is negligible. The aperture is inductive since B_a is always negative.

4-12. Plane Current Sheets. The field of plane sheets of current can, of course, be determined by the potential integral method of Sec. 2-9. We now reconsider the problem from the alternative approach of constructing transforms. The procedure is similar to that used in the preceding section for apertures. In fact, if the equivalence principle plus image theory is applied to the results of the preceding section, we have complete duality between apertures (magnetic current sheets) and electric current sheets. However, rather than taking this short cut, let us follow the more circuitous path of constructing the solution from basic concepts.

Suppose we have a sheet of z-directed electric currents over a portion of the $y = 0$ plane, as suggested by Fig. 4-24. The field can be expressed in terms of a wave function representing the z-component of magnetic vector potential. (This we know from the potential integral solution.) The problem is of the radiation type, requiring continuous distributions of eigenvalues. We anticipate the wave functions to be of the transform type, such as Eq. (4-93). From Eqs. (3-86), we have the transforms of the field components for the TM to z field, given by

$$\bar{H}_x = jk_y\bar{\psi} \qquad \bar{E}_x = \frac{-k_xk_z}{j\omega\epsilon}\bar{\psi}$$
$$\bar{H}_y = -jk_x\bar{\psi} \qquad \bar{E}_y = \frac{-k_yk_z}{j\omega\epsilon}\bar{\psi} \qquad (4\text{-}117)$$
$$\bar{H}_z = 0 \qquad \bar{E}_z = \frac{k^2 - k_z^2}{j\omega\epsilon}\bar{\psi}$$

These are dual to Eqs. (4-95). We construct the transform of ψ as

$$\begin{aligned}\psi^+ &= f^+(k_x,k_z)e^{jk_y^+ y} & y &> 0 \\ \psi^- &= f^-(k_x,k_z)e^{jk_y^- y} & y &< 0\end{aligned} \quad (4\text{-}118)$$

For the proper behavior of the fields at large $|y|$, we must choose k_y^+, as in Eq. (4-97), and k_y^- as the other root. That is,

$$k_y^+ = -k_y^- = \begin{cases} j\sqrt{k_x^2 + k_z^2 - k^2} & k < \sqrt{k_x^2 + k_z^2} \\ -\sqrt{k^2 - k_x^2 - k_z^2} & k > \sqrt{k_x^2 + k_z^2}\end{cases} \quad (4\text{-}119)$$

Our boundary conditions at the current sheet are continuity of E_x and E_y, and a discontinuity in H_x, according to Eq. (1-86). The boundary condition on E_x and E_y leads to $f^+ = f^-$, and the boundary condition on H_x then leads to

$$f^+(k_x,k_z) = f^-(k_x,k_z) = \frac{j}{2k_y^+}\bar{J}_z \quad (4\text{-}120)$$

where $\bar{J}_z$, the transform of J_z, is

$$\bar{J}_z(k_x,k_z) = \int_{-\infty}^{\infty}\int_{-\infty}^{\infty} J_z(x,z)e^{-jk_x x}e^{-jk_z z}\,dx\,dz \quad (4\text{-}121)$$

This completes the determination of the field transforms. The field is given by the inverse transformation.

Our two solutions (potential integral and transform) plus the uniqueness theorem can be used to establish mathematical identities. For example, consider the current element of Fig. 2-21. The potential integral solution is $\mathbf{A} = \mathbf{u}_z\psi$ where

$$\psi = \frac{Il e^{-jkr}}{4\pi r} \quad (4\text{-}122)$$
$$r = \sqrt{x^2 + y^2 + z^2}$$

For the transform solution,

$$J_z = Il\,\delta(x)\,\delta(z)$$
$$\bar{J}_z = \frac{1}{4\pi^2}\int_{-\infty}^{\infty}\int_{-\infty}^{\infty} J_z e^{-jk_x x}e^{-jk_z z}\,dx\,dz = \frac{Il}{4\pi^2}$$

Hence, for $y > 0$ we have $\mathbf{A} = \mathbf{u}_z\psi$ where

$$\psi = \frac{jIl}{8\pi^2}\int_{-\infty}^{\infty}\int_{-\infty}^{\infty}\frac{1}{k_y}e^{jk_x x}e^{jk_y y}e^{jk_z z}\,dk_x\,dk_z \quad (4\text{-}123)$$

where $k_y = k_y^+$ is given by Eq. (4-119). In this example, ψ as well as the field is unique. Hence, equating Eqs. (4-122) and (4-123), we have the identity

$$\frac{e^{-jkr}}{r} = \frac{1}{2\pi j}\int_{-\infty}^{\infty}\int_{-\infty}^{\infty}\frac{e^{-jy\sqrt{k^2-k_x^2-k_z^2}}}{\sqrt{k^2-k_x^2-k_z^2}}e^{jk_x x}e^{jk_z z}\,dk_x\,dk_z \quad (4\text{-}124)$$

This holds for all y, since k_y changes sign as y changes sign.

We have considered explicitly only sheets of z-directed current. The solution for x-directed current can be obtained by a rotation of coordinates. When the current sheet has both x and z components, the solution is a superposition of the x-directed case and the z-directed case. The solution for magnetic current sheets is dual to that for electric current sheets. Finally, if the sheet contains y-directed electric currents, we can convert to the equivalent x- and z-directed magnetic current sheet for a solution, and vice versa for y-directed magnetic currents.

A two-dimensional problem to which we shall have occasion to refer in the next chapter is that of a ribbon of axially directed current, uniformly distributed. This is shown in Fig. 4-25. The parameter of interest to us is the "impedance per unit length," defined by

$$Z = \frac{P}{|I|^2} \quad (4\text{-}125)$$

where P is the complex power per unit length and I is the total current. Rather than work through the details, let us apply duality to the aperture problem of Fig. 4-22. According to the concepts of Sec. 3-6, the field $y > 0$ is unchanged if the aperture is replaced by a magnetic current ribbon $K = 2V$. This ribbon radiates into whole space; so the power per unit length is twice that from the aperture. The admittance of the magnetic current ribbon is thus

$$Y_{\text{mag rib}} = \frac{P^*}{|K|^2} = \frac{2P^*_{\text{apert}}}{|2V|^2} = \tfrac{1}{2}Y_{\text{apert}}$$

where the aperture admittance

$$Y_{\text{apert}} = G_a + jB_a$$

is given by Eq. (4-105), which we can represent by

$$Y_{\text{apert}} = \frac{1}{\eta\lambda}f(ka)$$

By duality, we have the radiation impedance of the electric current ribbon given by

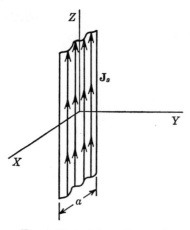

FIG. 4-25. A ribbon of current.

$$Z_{\text{elec rib}} = \frac{1}{2}\frac{\eta}{\lambda}f(ka) = \frac{\eta^2}{2}Y_{\text{apert}} \quad (4\text{-}126)$$

(Compare this with Prob. 3-7. The factor-of-two difference arises because the aperture of Fig. 4-22 radiates into half-space and the twin-slot line sees all-space.) For narrow ribbons, we have from Eqs. (4-106) and (4-126)

$$Z_{\text{elec rib}} \xrightarrow[ka \to 0]{} \frac{\eta}{2\lambda}[\pi + j(3.135 - 2 \log ka)] \tag{4-127}$$

This we shall compare to the corresponding Z for a cylinder of current in Sec. 5-6.

PROBLEMS

4-1. Show that Eq. (4-9) is a solution to the scalar Helmholtz equation.

4-2. For $k = \beta - j\alpha$, show that

$$\sin kx = \sin \beta x \cosh \alpha x - j \cos \beta x \sinh \alpha x$$
$$\cos kx = \cos \beta x \cosh \alpha x + j \sin \beta x \sinh \alpha x$$

4-3. Derive Eqs. (4-17).

4-4. Following the method used to establish Eq. (2-93), show that the attenuation constant due to conductor losses in a rectangular waveguide is given by Eq. (2-93) for all TE_{0n} modes and by

$$(\alpha_c)_{mn} = \frac{2\mathcal{R}}{\eta} \left[\frac{(a+b)(f_c/f)^2}{ab\sqrt{1-(f_c/f)^2}} + \sqrt{1-\left(\frac{f_c}{f}\right)^2} \frac{bm^2 + an^2}{b^2 m^2 + a^2 n^2} \right]$$

for TE_{mn} modes, m and n nonzero, and by

$$(\alpha_c)_{mn} = \frac{2\mathcal{R}}{\eta ab \sqrt{1-(f_c/f)^2}} \frac{m^2 b^3 + n^2 a^3}{m^2 b^2 + n^2 a^2}$$

for TM_{mn} modes.

4-5. An air-filled rectangular waveguide is needed for operation at 10,000 megacycles. It is desired to have single-mode operation over a 2:1 frequency range, with center frequency 10,000 megacycles. It is also desired to have maximum power-handling capacity under these conditions. Determine the waveguide dimensions and the attenuation constant of the propagating mode for copper walls.

4-6. For a parallel-plate waveguide formed by conductors covering the $y = 0$ and $y = b$ planes, show that

$$\psi_n^{TE} = \cos \frac{n\pi y}{b} e^{-jk_z z} \qquad n = 1, 2, 3, \ldots$$

are the mode functions generating the two-dimensional TE_n modes according to Eqs. (3-89), and

$$\psi_n^{TM} = \sin \frac{n\pi y}{b} e^{-jk_z z} \qquad n = 1, 2, 3, \ldots$$

are the mode functions generating the two-dimensional TM_n modes according to Eqs. (3-86). Show that the TEM mode is generated by

$$\psi_0^{TM} = y e^{-jkz}$$

4-7. Show that an alternative set of mode functions for the parallel-plate waveguide of Prob. 4-6 are

$$\psi_n^{TEx} = \cos\frac{n\pi y}{b} e^{-jk_z z} \qquad n = 0, 1, 2, \ldots$$

which generate the TEx_n modes according to Eqs. (4-32), and

$$\psi_n^{TMx} = \sin\frac{n\pi y}{b} e^{-jk_z z} \qquad n = 1, 2, 3, \ldots$$

which generate the TMx_n modes according to Eqs. (4-30). Note that $n = 0$ in the above TEx mode function gives the TEM mode.

4-8. Show that the TEx and TMx modes of Sec. 4-4 are linear combinations of the TE and TM modes of Sec. 4-3, that is,

$$\mathbf{E}_{mn}^{TEx} = A(\mathbf{E}_{mn}^{TE} + B\mathbf{E}_{mn}^{TM})$$
$$\mathbf{H}_{mn}^{TMx} = C(\mathbf{H}_{mn}^{TE} + D\mathbf{H}_{mn}^{TM})$$

Determine A, B, C, and D.

4-9. Show that the resonant frequencies of the two-dimensional (no z variation) resonator formed by conducting plates over the $x = 0$, $x = a$, $y = 0$, and $y = b$ planes are the cutoff frequencies of the rectangular waveguide.

4-10. Following the method used to establish Eq. (2-101), show that the Q due to conductor losses for the various modes in a rectangular cavity are

$$(Q_c)_{0np}^{TE} = \frac{\eta abck_r^3}{2\mathcal{R}(bck_r^2 + 2ack_y^2 + 2abk_z^2)}$$

$$(Q_c)_{m0p}^{TE} = \frac{\eta abck_r^3}{2\mathcal{R}(ack_r^2 + 2bck_x^2 + 2abk_z^2)}$$

$$(Q_c)_{mnp}^{TE} = \frac{\eta abck_{xy}^2 k_r^3}{4\mathcal{R}[bc(k_{xy}^4 + k_y^2 k_z^2) + ac(k_{xy}^4 + k_x^2 k_z^2) + abk_{xy}^2 k_z^2]}$$

$$(Q_c)_{mn0}^{TM} = \frac{\eta abck_r^3}{2\mathcal{R}(abk_r^2 + 2bck_x^2 + 2ack_y^2)}$$

$$(Q_c)_{mnp}^{TM} = \frac{\eta abck_{xy}^2 k_r}{4\mathcal{R}[b(a+c)k_x^2 + a(b+c)k_y^2]}$$

where

$$k_x = \frac{m\pi}{a} \qquad k_y = \frac{n\pi}{b} \qquad k_z = \frac{p\pi}{c}$$

$$k_{xy} = \sqrt{k_x^2 + k_y^2} \qquad k_r = \sqrt{k_x^2 + k_y^2 + k_z^2}$$

4-11. Calculate the first ten higher-order resonant frequencies for the rectangular cavity of Prob. 2-38.

4-12. Consider the two-dimensional parallel-plate waveguide formed by conductors over the $x = 0$ and $x = a$ planes, and dielectrics ϵ_1 for $0 < x < d$ and ϵ_2 for $d < x < a$. Show that for modes TM to x the characteristic equation is Eq. (4-45) with

$$k_{x1} = \sqrt{\omega^2 \epsilon_1 \mu_1 - k_z^2} \qquad k_{x2} = \sqrt{\omega^2 \epsilon_2 \mu_2 - k_z^2}$$

and for modes TE to x the characteristic equation is Eq. (4-47). Note that no mode TEM to z (the direction of propagation) is possible.

4-13. Show that the lowest-order TM to x mode of Prob. 4-12 reduces to the transmission-line mode either as $\epsilon_1 \to \epsilon_2$ and $\mu_1 \to \mu_2$ or as $d \to 0$. Show that, if $a \ll \lambda_2$,

$$k_z \approx \omega \sqrt{\frac{\epsilon_1 \epsilon_2 [\mu_1 d + \mu_2(a-d)]}{\epsilon_1(a-d) + \epsilon_2 d}}$$

for the dominant mode. Show that the static inductance and capacitance per unit width and length of the transmission line are

$$L = \mu_1 d + \mu_2(a-d) \qquad C = \frac{\epsilon_1 \epsilon_2}{\epsilon_1(a-d) + \epsilon_2 d}$$

The usual transmission-line formula $k_z = \omega \sqrt{LC}$ therefore applies if a is small. Also, the field is almost TEM.

4-14. Consider the dominant mode of the partially filled guide (Fig. 4-6) for $b > a$. When d is small, Eq. (4-45) can be approximated by Eq. (4-48) for the dominant mode. Denote the empty-guide propagation constant ($d = 0$) by

$$\beta_0 = \sqrt{k_2^2 - \left(\frac{\pi}{b}\right)^2}$$

and show, from the Taylor expansion of Eq. (4-48) about $d = 0$ and $k_z = \beta_0$, that for small d

$$k_z = \beta_0 + \frac{\epsilon_2}{\epsilon_1}\left(\frac{k_1^2 - k_2^2}{2\beta_0}\right)\frac{d}{a}$$

4-15. Consider the dominant mode of the partially filled guide (Fig. 4-6) for $a > b$. Denote the empty-guide propagation constant ($d = 0$) by

$$\beta_0 = \sqrt{k_2^2 - \left(\frac{\pi}{a}\right)^2}$$

and show, from the Taylor expansion of the reciprocal of Eq. (4-47) about $d = 0$ and $k_z = \beta_0$, that for small d

$$k_z = \beta_0 + \frac{\mu_1 - \mu_2}{\mu_2 \beta_0}\left(\frac{\pi}{a}\right)^2 \frac{d}{a} + \frac{\pi^2 \mu_1}{3\mu_2 \beta_0}(k_1^2 - k_2^2)\left(\frac{d}{a}\right)^3$$

4-16. Show that the resonant frequencies of a partially filled rectangular cavity (Fig. 4-6 with additional conductors covering the $z = 0$ and $z = c$ planes) are solutions to Eqs. (4-45) and (4-47) with

$$k_{x1}^2 + \left(\frac{n\pi}{b}\right)^2 + \left(\frac{p\pi}{c}\right)^2 = k_1^2$$

$$k_{x2}^2 + \left(\frac{n\pi}{b}\right)^2 + \left(\frac{p\pi}{c}\right)^2 = k_2^2$$

where $n = 0, 1, 2, \ldots$; $p = 0, 1, 2, \ldots$; $n = p = 0$ excepted.

4-17. For the partially filled cavity of Prob. 4-16, show that if $c > b > a$, the resonant frequency of the dominant mode for small d is given by

$$\omega_r = \omega_0 \left[1 - \frac{1}{2}\left(\frac{\mu_1}{\mu_2} - \frac{\epsilon_2}{\epsilon_1}\right)\frac{d}{a}\right]$$

where ω_0 is the resonant frequency of the empty cavity,

$$\omega_0 = \frac{1}{\sqrt{\epsilon_2\mu_2}}\sqrt{\left(\frac{\pi}{b}\right)^2 + \left(\frac{\pi}{c}\right)^2}$$

Hint: Use the results of Prob. 4-14.

4-18. For the partially filled cavity of Prob. 4-16, show that if $c > a > b$, the resonant frequency of the dominant mode for small d is given by

$$\omega = \omega_0\left[1 - \frac{\mu_1 - \mu_2}{\mu_2}\frac{c^2}{a^2 + c^2}\frac{d}{a} - \frac{\pi^2\mu_1}{3\mu_2}\left(\frac{\epsilon_1\mu_1}{\epsilon_2\mu_2} - 1\right)\left(\frac{d}{a}\right)^3\right]$$

where ω_0 is the resonant frequency of the empty cavity

$$\omega_0 = \frac{1}{\sqrt{\epsilon_2\mu_2}}\sqrt{\left(\frac{\pi}{a}\right)^2 + \left(\frac{\pi}{c}\right)^2}$$

Hint: Use the results of Prob. 4-15.

4-19. Consider a rectangular waveguide with a centered dielectric slab, as shown in the insert of Fig. 7-10. Show that the characteristic equation for determining the propagation constants of modes TE to x is

$$\frac{k_{x0}}{\mu_0}\cot\left(k_{x0}\frac{a-d}{2}\right) = \frac{k_{x1}}{\mu_1}\tan\left(k_{x1}\frac{d}{2}\right)$$

and for modes TM to x it is

$$\frac{k_{x0}}{\epsilon_0}\tan\left(k_{x0}\frac{a-d}{2}\right) = \frac{k_{x1}}{\epsilon_1}\cot\left(k_{x1}\frac{d}{2}\right)$$

where
$$k_{x0}^2 + \left(\frac{n\pi}{b}\right)^2 + k_z^2 = k_0^2 = \omega^2\epsilon_0\mu_0$$
$$k_{x1}^2 + \left(\frac{n\pi}{b}\right)^2 + k_z^2 = k_1^2 = \omega^2\epsilon_1\mu_1$$

The dominant mode is the lowest-order TE mode (smallest root for $n = 0$).

4-20. Derive Eq. (4-58).

4-21. A plane slab of polystyrene ($\epsilon_r = 2.56$) is $\frac{3}{4}$ centimeter thick. What slab-guide modes will propagate unattenuated at a frequency of 30,000 megacycles? Calculate the cutoff frequencies of these modes. Using Fig. 4-11, determine the propagation constants of the propagating TE modes at 30,000 megacycles. Determine the propagation constants of the propagating TM modes by numerical solution of Eq. (4-56) or (4-58). How can the cutoff frequencies of corresponding TE and TM modes be the same, yet the propagation constants be different?

4-22. By a Taylor expansion of Eq. (4-56) about $a = 0$, $v = 0$, show that the dominant TM mode of the slab guide (Fig. 4-10) is characterized by

$$v = \frac{\epsilon_0}{\epsilon_d}(k_d^2 - k_0^2)\frac{a}{2}$$

for small a. Similarly, show that the dominant TE mode is characterized by

$$v = \frac{\mu_0}{\mu_d}(k_d^2 - k_0^2)\frac{a}{2}$$

for small a. In each case, the propagation constant is given by

$$k_z = k_0 + \frac{v^2}{2k_0}$$

4-23. A plane conductor has been coated with shellac ($\epsilon_r = 3.0$) to a thickness of 0.005 inch. It is to be used in a 30,000-megacycle field. Will any tightly bound surface wave be possible? Calculate the attenuation constant in the direction perpendicular to the coated conductor.

4-24. For the corrugated conductor of Fig. 4-15, it is desired that the field be attenuated to 36.8 per cent of its surface value at one wavelength from the surface. Determine the minimum depth of slot needed.

4-25. Suppose that the slots of the corrugated conductor of Fig. 4-15 are filled with a dielectric characterized by ϵ_d, μ_d. Show that for this case

$$v = \frac{\epsilon_0}{\epsilon_d} k_d \tan k_d d$$

$$k_z = k_0 \sqrt{1 + \frac{\epsilon_0 \mu_d}{\epsilon_d \mu_0} \tan^2 k_d d}$$

where $k_d = \omega \sqrt{\epsilon_d \mu_d}$.

4-26. Use the TEx mode functions of Prob. 4-7 for the parallel-plate waveguide formed by conductors covering the $y = 0$ and $y = b$ planes. Show that a field having no E_z is given by Eqs. (4-32) with

$$\psi = \sum_{n=0}^{\infty} A_n \cos \frac{n\pi y}{b} e^{-\gamma_n z} \qquad z > 0$$

where

$$A_n = \frac{\epsilon_n}{b\gamma_n} \int_0^b E_y \bigg|_{z=0} \cos \frac{n\pi y}{b} \, dy$$

4-27. Consider the junction of two parallel-plate transmission lines of height c for $z < 0$ and height b for $z > 0$, with the bottom plate continuous. (The cross section is that of the second drawing of Fig. 4-16.) Using the formulation of Prob. 4-26, show that the aperture susceptance per unit width referred to the aperture voltage is

$$B_a \approx \frac{4}{\eta \lambda} \sum_{n=1}^{\infty} \frac{\sin^2 (n\pi c/b)}{(n\pi c/b)^2 \sqrt{n^2 - (2b/\lambda)^2}}$$

where a constant E_y has been assumed in the aperture. Compare this with Eq. (4-78).

4-28. The centered capacitive waveguide junction is shown in Fig. 4-26. Show that the aperture susceptance referred to the maximum aperture voltage is given by Eq. (4-78) with λ_g replaced by $2\lambda_g$. It is assumed that E_y in the aperture is that of the incident mode.

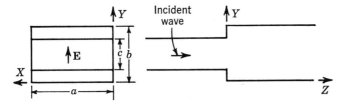

Fig. 4-26. A centered capacitive waveguide junction

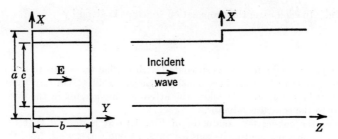

FIG. 4-27. A centered inductive waveguide junction.

4-29. Consider the centered inductive waveguide junction of Fig. 4-27. Assuming that E_y in the aperture is that of the incident mode, show that the aperture susceptance referred to the maximum aperture voltage is given by

$$B_a \approx \frac{-8\lambda}{\eta \pi^2 b} \left(\frac{c}{a}\right)^2 \sum_{3,5,7,\ldots}^{\infty} \left[\frac{\cos (m\pi c/2a)}{1 - (mc/a)^2}\right]^2 \sqrt{\left(\frac{m}{2}\right)^2 - \left(\frac{a}{\lambda}\right)^2}$$

4-30. In Eq. (4-83), note that as $c/a \to 0$ the summation becomes similar to an integration. Use the analogy $mc/a \sim x$ and $c/a \sim dx$ to show that

$$-\frac{b\eta}{\lambda} B_a \xrightarrow[c/a \to 0]{} \frac{1}{\pi^2} \int_0^\infty \left(\frac{\sin \pi x}{1 - x^2}\right)^2 x\, dx$$

Integrate by parts, and use the identity[1]

$$\int_0^\infty \frac{\sin 2\pi x}{x^2 - 1}\, dx = \int_0^{2\pi} \frac{\sin y}{y}\, dy = \mathrm{Si}(2\pi)$$

to show that

$$-\frac{b\eta}{\lambda} B_a \xrightarrow[c/a \to 0]{} \frac{\mathrm{Si}(2\pi)}{2\pi} = 0.226$$

4-31. Let there be a sheet of y-directed current J_y over the $z = 0$ plane of a parallel-plate waveguide formed by conductors over the $y = 0$ and $y = b$ planes. The guide is matched in both the $+z$ and $-z$ directions. Show that the field produced by the current sheet is

$$\sum_{n=0}^{\infty} A_n \cos \frac{n\pi y}{b} e^{-\gamma_n |z|} = \begin{cases} H_x & z > 0 \\ -H_x & z < 0 \end{cases}$$

where

$$A_n = \frac{\epsilon_n}{2b} \int_0^b J_y(y) \cos \frac{n\pi y}{b}\, dy$$

4-32. Let the current sheet of Prob. 4-31 be x-directed instead of y-directed. Show that field produced by this x-directed current sheet is

$$E_x = \sum_{n=1}^{\infty} B_n \sin \frac{n\pi y}{b} e^{-\gamma_n |z|}$$

where

$$B_n = \frac{j\omega\mu}{\gamma_n b} \int_0^b J_x(y) \sin \frac{n\pi y}{b}\, dy$$

---

[1] D. Bierens de Haan, "Nouvelles tables d'intégrales définies," p. 225, table 161, no. 3, Hafner Publishing Company, New York, 1939 (reprint).

4-33. Consider the coax to waveguide junction of Fig. 4-28a. Only the TE$_{01}$ mode propagates in the waveguide, which is matched in both directions. Assume that the current on the wire varies as $\cos{(kl)}$, where l is the distance from the end of the wire. Show that the input resistance seen by the coax is

$$R_i = \frac{a}{b}(Z_0)_{01}\left[\frac{\sin{(\pi c/b)}\sin{kd}}{ka\cos{k(c+d)}}\right]^2$$

where $(Z_0)_{01}$ is the TE$_{01}$ characteristic wave impedance.

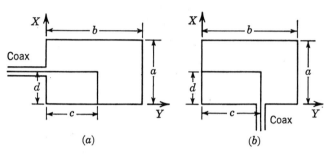

Fig. 4-28. Coax to waveguide junctions.

4-34. Suppose that the coax to waveguide junction of Prob. 4-33 is changed to that of Fig. 4-28b. Show that the input resistance seen by the coax is now

$$R_i = \frac{a}{b}(Z_0)_{01}\left\{\frac{\sin{(\pi c/b)}[\sin{k(c+d)} - \sin{kc}]}{ka\cos{k(c+d)}}\right\}^2$$

4-35. By expanding $(\sin{w}/w)^2$ in a Taylor series about $w = 0$, show that the first of Eqs. (4-105) becomes

$$\lambda\eta G_a = \pi\left[1 - \frac{1}{6}\left(\frac{ka}{2}\right)^2 + \frac{1}{60}\left(\frac{ka}{2}\right)^4 - \frac{1}{1008}\left(\frac{ka}{2}\right)^6 + \cdots\right]$$

4-36. Consider the second of Eqs. (4-105) as the contour integral

$$\lambda\eta B_a = \text{Re}\left[\int_{C_1}\frac{(1 - e^{j2w})dw}{w^2\sqrt{w^2 - (ka/2)^2}}\right]$$

where C_1 is shown in Fig. 4-29. Consider the closed contour $C_1 + C_2 + C_\infty + C_0$, and express $\lambda\eta B_a$ in terms of a contour integral over C_2 and C_0. Show that as $ka/2$ becomes large, this last contour integral reduces to the second of Eqs. (4-107).

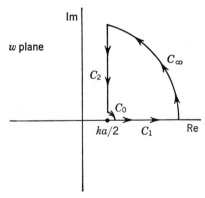

Fig. 4-29. Contours for Prob. 4-36.

4-37. By expanding $\cos^2 w/[(\pi/2)^2 - w^2]^2$ in a Taylor series about $w = 0$, show that the first of Eqs. (4-114) becomes

$$\frac{\eta}{\lambda} G_a = \frac{2}{\pi} \sum_{n=1}^{\infty} b_n \left(\frac{a}{\lambda}\right)^{2n}$$

where
$$b_1 = +1.0$$
$$b_2 = -0.467401$$
$$b_3 = +0.189108$$
$$b_4 = -0.055613$$
$$b_5 = +0.012182$$
$$b_6 = -0.002083$$

4-38. Specialize the second of Eqs. (4-114) to the case $a = 0$, integrate by parts, and use the identity (see Prob. 4-30)

$$\int_0^{\infty} \frac{\sin 2x \, dx}{(\pi/2)^2 - x^2} = \frac{2}{\pi} \int_0^{\pi} \frac{\sin y}{y} dy = \frac{2}{\pi} \text{Si}(\pi)$$

to show that
$$-\frac{\eta}{\lambda} B_a \xrightarrow[a \to 0]{} \frac{1}{2\pi} \text{Si}(\pi) - \frac{2}{\pi^2} = 0.194$$

4-39. Show that the first of Eqs. (4-114) reduces to the contour integral

$$\frac{\eta}{\lambda} G_a \xrightarrow[ka \to \infty]{} \frac{ka}{8} \text{Re} \left[\int_{C_1} \frac{(1 + e^{j2w}) \, dw}{[(\pi/2)^2 - w^2]^2} \right]$$

where C_1 is shown in Fig. 4-30. Consider the closed contour $C_1 + C_2 + C_\infty + C_0$, and express G_a in terms of a contour integral over C_2 and C_0. Evaluate this last contour integral, and show that

$$\frac{\eta}{\lambda} G_a \xrightarrow[ka \to \infty]{} \frac{ka}{4\pi}$$

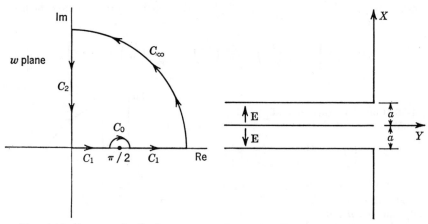

FIG. 4-30. Contours for Prob. 4-39.   FIG. 4-31. Two parallel-plate transmission lines radiating into half-space.

4-40. Two parallel-plate transmission lines opening onto a conducting plane are excited in opposite phase and equal magnitude, as shown in Fig. 4-31. Assume E_x in the aperture is a constant for each line, and show that the aperture susceptance referred to the aperture voltage of one line is

$$G_a = \frac{8}{\lambda \eta} \int_0^{ka} \frac{\sin^4 w \, dw}{w^2 \sqrt{(ka)^2 - w^2}}$$

$$B_a = \frac{8}{\lambda \eta} \int_{ka}^{\infty} \frac{\sin^4 w \, dw}{w^2 \sqrt{w^2 - (ka)^2}}$$

4-41. Construct the vector potential $\mathbf{A} = \mathbf{u}_z \psi$ for a sheet of z-directed currents over the $y = 0$ plane (Fig. 4-24) by (a) the potential integral method and (b) the transform method. Show by use of Green's second identity [Eq. (3-44)] that the two ψ's are equal. Specialize the potential integral solution to $r \to \infty$, and show that

$$\psi \xrightarrow[r \to \infty]{} \frac{e^{-jkr}}{4\pi r} \bar{J}_z(-k \cos \phi \sin \theta, -k \cos \theta)$$

where $\bar{J}_z(k_x, k_z)$ is given by Eq. (4-121).

4-42. Suppose that the current in Fig. 4-25 is x-directed rather than z-directed, and of magnitude

$$J_x = \cos \frac{\pi x}{a} \qquad |x| < \frac{a}{2}$$

Show that the impedance per unit length, defined by Eq. (4-125), where I is the current per unit length, is given by Eq. (4-126), where Y_{apert} is now the aperture admittance of Fig. 4-23.

CHAPTER 5

CYLINDRICAL WAVE FUNCTIONS

5-1. The Wave Functions. Problems having boundaries which coincide with cylindrical coordinate surfaces are usually solved in cylindrical coordinates.[1] We shall usually orient the cylindrical coordinate system as shown in Fig. 5-1. We first consider solutions to the scalar Helmholtz equation. Once we have these scalar wave functions, we can construct electromagnetic fields according to Eqs. (3-91).

The scalar Helmholtz equation in cylindrical coordinates is

$$\frac{1}{\rho}\frac{\partial}{\partial\rho}\left(\rho\frac{\partial\psi}{\partial\rho}\right) + \frac{1}{\rho^2}\frac{\partial^2\psi}{\partial\phi^2} + \frac{\partial^2\psi}{\partial z^2} + k^2\psi = 0 \tag{5-1}$$

which is Eq. (2-7) with the Laplacian expressed in cylindrical coordinates. Following the method of separation of variables, we seek to find solutions of the form

$$\psi = R(\rho)\Phi(\phi)Z(z) \tag{5-2}$$

Substitution of Eq. (5-2) into Eq. (5-1) and division by ψ yields

$$\frac{1}{\rho R}\frac{d}{d\rho}\left(\rho\frac{dR}{d\rho}\right) + \frac{1}{\rho^2\Phi}\frac{d^2\Phi}{d\phi^2} + \frac{1}{Z}\frac{d^2Z}{dz^2} + k^2 = 0$$

The third term is explicitly independent of ρ and ϕ. It must also be independent of z if the equation is to sum to zero for all ρ, ϕ, z. Hence,

$$\frac{1}{Z}\frac{d^2Z}{dz^2} = -k_z^2 \tag{5-3}$$

where k_z is a constant. Substitution of this into the preceding equation and multiplication by ρ^2 gives

$$\frac{\rho}{R}\frac{d}{d\rho}\left(\rho\frac{dR}{d\rho}\right) + \frac{1}{\Phi}\frac{d^2\Phi}{d\phi^2} + (k^2 - k_z^2)\rho^2 = 0$$

Now the second term is independent of ρ and z, and the other terms are

[1] The term "cylindrical" is often used in a more general sense to include cylinders of arbitrary cross section. We are at present using the term to mean "circularly cylindrical."

independent of ϕ. Hence,

$$\frac{1}{\Phi}\frac{d^2\Phi}{d\phi^2} = -n^2 \quad (5\text{-}4)$$

where n is a constant. The preceding equation then becomes

$$\frac{\rho}{R}\frac{d}{d\rho}\left(\rho\frac{dR}{d\rho}\right) - n^2 + (k^2 - k_z^2)\rho^2 = 0 \quad (5\text{-}5)$$

which is an equation in ρ only.

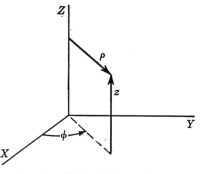

FIG. 5-1. Cylindrical coordinates.

The wave equation is now separated. To summarize, define k_ρ as

$$k_\rho^2 + k_z^2 = k^2 \quad (5\text{-}6)$$

and write the separated equations [Eqs. (5-3), (5-4), and (5-5)] as

$$\rho\frac{d}{d\rho}\left(\rho\frac{dR}{d\rho}\right) + [(k_\rho\rho)^2 - n^2]R = 0$$

$$\frac{d^2\Phi}{d\phi^2} + n^2\Phi = 0 \quad (5\text{-}7)$$

$$\frac{d^2Z}{dz^2} + k_z^2 Z = 0$$

The Φ and Z equations are harmonic equations, giving rise to harmonic functions. These we denote, in general, by $h(n\phi)$ and $h(k_z z)$. The R equation is *Bessel's equation* of order n, solutions of which we shall denote in general by $B_n(k_\rho\rho)$.[1] Commonly used solutions to Bessel's equation are

$$B_n(k_\rho\rho) \sim J_n(k_\rho\rho), \ N_n(k_\rho\rho), \ H_n^{(1)}(k_\rho\rho), \ H_n^{(2)}(k_\rho\rho) \quad (5\text{-}8)$$

where $J_n(k_\rho\rho)$ is the Bessel function of the first kind, $N_n(k_\rho\rho)$ is the Bessel function of the second kind, $H_n^{(1)}(k_\rho\rho)$ is the Hankel function of the first kind, and $H_n^{(2)}(k_\rho\rho)$ is the Hankel function of the second kind. These functions are considered in some detail in Appendix D, and we shall discuss them later in this section. Any two of the functions of Eq. (5-8) are linearly independent solutions; so $B_n(k_\rho\rho)$ is, in general, a linear combination of any two of them. According to Eq. (5-2), we can now form solutions to the Helmholtz equation as

$$\psi_{k_\rho, n, k_z} = B_n(k_\rho\rho)h(n\phi)h(k_z z) \quad (5\text{-}9)$$

[1] It is more usual to denote solutions to Bessel's equation by $Z_n(k_\rho\rho)$, but we wish to avoid confusion with our $Z(z)$ function and with impedances.

where k_ρ and k_z are interrelated by Eq. (5-6). We call these ψ *elementary wave functions*.

Linear combinations of the elementary wave functions are also solutions to the Helmholtz equation. We can sum over possible values (eigenvalues) of n and k_ρ, or of n and k_z (but not over k_ρ and k_z for they are interrelated). For example,

$$\psi = \sum_n \sum_{k_z} C_{n,k_z} \psi_{k_\rho,n,k_z}$$
$$= \sum_n \sum_{k_z} C_{n,k_z} B_n(k_\rho \rho) h(n\phi) h(k_z z) \qquad (5\text{-}10)$$

where the C_{n,k_z} are constants, is a solution to the Helmholtz equation. We can also integrate over the separation constants, although n is usually discrete (this is discussed below). We shall, however, have occasion to integrate over either k_ρ or k_z. Thus, possible solutions to the Helmholtz equation are

$$\psi = \sum_n \int_{k_z} f_n(k_z) B_n(k_\rho \rho) h(n\phi) h(k_z z) \, dk_z \qquad (5\text{-}11)$$

$$\psi = \sum_n \int_{k_\rho} g_n(k_\rho) B_n(k_\rho \rho) h(n\phi) h(k_z z) \, dk_\rho \qquad (5\text{-}12)$$

where the integrations are over any contour in the complex plane and $f_n(k_z)$ and $g_n(k_\rho)$ are functions to be determined from boundary conditions. We shall use Eq. (5-11) to construct Fourier integrals, as we did in Chap. 4. Equation (5-12) is used to construct Fourier-Bessel integrals.

We discussed the interpretation of the harmonic functions in Sec. 4-1, a summary being given in Table 4-1. The z coordinate of the cylindrical coordinate system is also one of the rectangular coordinates; so the same considerations that dictated the choice of $h(k_z z)$ in Chap. 4 apply at present. The ϕ coordinate is an angle coordinate and, as such, places restrictions on the choice of $h(n\phi)$ and n. For example, if we desire the field in a cylindrical region containing all ϕ from 0 to 2π, it is necessary that $\psi(\phi) = \psi(\phi + 2\pi)$ if ψ is to be single-valued. This means that $h(n\phi)$ must be periodic in ϕ, in which case n must be an integer. In most cases, we choose $\sin(n\phi)$ or $\cos(n\phi)$ or a linear combination of the two, although in some cases the exponentials $e^{jn\phi}$ and $e^{-jn\phi}$ are more descriptive, or easier to deal with analytically. Thus, the n summations of Eqs. (5-10) to (5-12) are usually Fourier series on ϕ.

Now, consider the various solutions to Bessel's equation. Graphs of the lower-order Bessel functions are given in Appendix D. We note that only the $J_n(k_\rho \rho)$ functions are nonsingular at $\rho = 0$. Hence, if a field is to be finite at $\rho = 0$, the $B_n(k_\rho \rho)$ must be $J_n(k_\rho \rho)$, and the elementary wave functions are of the form

$$\psi_{k_\rho,n,k_z} = J_n(k_\rho\rho)e^{jn\phi}e^{jk_z z} \qquad \rho = 0 \text{ included} \qquad (5\text{-}13)$$

We have written the harmonic functions in exponential form, which is still general since sines and cosines are linear combinations of them. Note from Eq. (5-6) that $k_\rho = \pm \sqrt{k^2 - k_z^2}$ is indeterminate with respect to sign. Our convention will be to choose the root whose real part is positive, that is, Re $(k_\rho) > 0$.[1] Now consider the asymptotic expressions for the various solutions to Bessel's equation [Eqs. (D-11) and (D-13)]. Note that $H_n^{(2)}(k_\rho\rho)$ are the only solutions which vanish for large ρ if k_ρ is complex. They represent outward-traveling waves if k_ρ is real. Therefore, if there are no sources at infinity, the $B_n(k_\rho\rho)$ must be $H_n^{(2)}(k_\rho\rho)$ if $\rho \to \infty$ is to be included. Hence, the elementary wave functions become

$$\psi_{k_\rho,n,k_z} = H_n^{(2)}(k_\rho\rho)e^{jn\phi}e^{jk_z z} \qquad \rho \to \infty \text{ included} \qquad (5\text{-}14)$$

Other choices of cylinder functions are convenient in certain cases, as we shall see when we apply them.

Insight into the behavior of solutions to Bessel's equation can be gained by noting their similarities to harmonic functions. It is evident from the asymptotic formulas of Appendix D that, except for an attenuation of $1/\sqrt{k\rho}$, the following qualitative analogies can be made:

$$\begin{aligned}
&J_n(k\rho) \text{ analogous to } \cos k\rho \\
&N_n(k\rho) \text{ analogous to } \sin k\rho \\
&H_n^{(1)}(k\rho) \text{ analogous to } e^{jk\rho} \\
&H_n^{(2)}(k\rho) \text{ analogous to } e^{-jk\rho}
\end{aligned} \qquad (5\text{-}15)$$

For example, J_n and N_n exhibit oscillatory behavior for real k, as do the sinusoidal functions. Hence, these solutions represent cylindrical standing waves. The $H_n^{(1)}$ and $H_n^{(2)}$ functions represent traveling waves for k real, as do the exponential functions. They therefore represent cylindrical traveling waves, $H_n^{(1)}$ representing inward-traveling waves and $H_n^{(2)}$ representing outward-traveling waves.[2] If k is complex, the traveling waves are attenuated or augmented in the direction of travel (in addition to the $1/\sqrt{k\rho}$ factor). When k is imaginary ($k = -j\alpha$), it is conventional to use the *modified Bessel functions* I_n and K_n, defined by

$$\begin{aligned}
I_n(\alpha\rho) &= j^n J_n(-j\alpha\rho) \\
K_n(\alpha\rho) &= \frac{\pi}{2}(-j)^{n+1} H_n^{(2)}(-j\alpha\rho)
\end{aligned} \qquad (5\text{-}16)$$

---

[1] If k_ρ is imaginary, choose the root according to the limit Im $(k) \to 0$.

[2] This direction of wave travel is a consequence of our choice of $e^{j\omega t}$ time variation. If we had initially chosen $e^{-j\omega t}$, then our interpretation of $H_n^{(1)}$ and $H_n^{(2)}$ would be reversed.

These are real when $\alpha\rho$ is real. From their asymptotic behavior, Eqs. (D-19), it is evident that we have the qualitative analogies

$$I_n(\alpha\rho) \text{ analogous to } e^{\alpha\rho}$$
$$K_n(\alpha\rho) \text{ analogous to } e^{-\alpha\rho} \quad (5\text{-}17)$$

From these it is apparent that the modified Bessel functions are used to represent evanescent-type fields. That the various analogies of Eqs. (5-15) and (5-17) exist is no coincidence. Both Bessel's equation and the harmonic equation are specializations of the wave equation. In the case of waves on water, a dropped stone would give rise to "Bessel function" waves, while the wind gives rise to "harmonic function" waves.

Table 5-1 summarizes the properties of solutions to Bessel's equation. Our understanding of the physical interpretation, given in the last column, will increase as we apply the various functions to specific problems. When $k = 0$, we have the degenerate Bessel functions

$$B_0(0\rho) \sim 1, \log \rho$$
$$B_n(0\rho) \sim \rho^n, \rho^{-n} \quad n \neq 0$$

Note that these are essentially the small-argument expressions for J_n and N_n.

To express an electromagnetic field in terms of the wave functions ψ, the method of Sec. 3-12 can be used. The unit z-coordinate vector is a constant vector; so we can obtain a field TM to z by letting $\mathbf{A} = \mathbf{u}_z\psi$ and expanding Eqs. (3-85) in cylindrical coordinates. The result is

$$E_\rho = \frac{1}{\hat{y}} \frac{\partial^2 \psi}{\partial \rho \, \partial z} \qquad H_\rho = \frac{1}{\rho} \frac{\partial \psi}{\partial \phi}$$
$$E_\phi = \frac{1}{\hat{y}\rho} \frac{\partial^2 \psi}{\partial \phi \, \partial z} \qquad H_\phi = -\frac{\partial \psi}{\partial \rho} \qquad (5\text{-}18)$$
$$E_z = \frac{1}{\hat{y}}\left(\frac{\partial^2}{\partial z^2} + k^2\right)\psi \qquad H_z = 0$$

which are sufficiently general to express any TM (no H_z) field existing in a homogeneous source-free region. Similarly, we can obtain a field TE to z by letting $\mathbf{F} = \mathbf{u}_z\psi$ and expanding Eqs. (3-88) in cylindrical coordinates. The result is

$$E_\rho = -\frac{1}{\rho} \frac{\partial \psi}{\partial \phi} \qquad H_\rho = \frac{1}{\hat{z}} \frac{\partial^2 \psi}{\partial \rho \, \partial z}$$
$$E_\phi = \frac{\partial \psi}{\partial \rho} \qquad H_\phi = \frac{1}{\hat{z}\rho} \frac{\partial^2 \psi}{\partial \phi \, \partial z} \qquad (5\text{-}19)$$
$$E_z = 0 \qquad H_z = \frac{1}{\hat{z}}\left(\frac{\partial^2}{\partial z^2} + k^2\right)\psi$$

TABLE 5-1. PROPERTIES OF SOLUTIONS TO BESSEL'S EQUATION ($\gamma = 1.781$)*†

$B_n(k\rho)$	Alternative representations	Small-argument formulas ($k\rho \to 0$)	Large-argument formulas ($|k\rho| \to \infty$)	Zeros	Infinities	Physical interpretation
$H_n^{(1)}(k\rho)$	$J_n(k\rho) + jN_n(k\rho)$	$1 - j\dfrac{2}{\pi}\log\left(\dfrac{2}{\gamma k\rho}\right) \quad n = 0$ $\dfrac{(k\rho)^n}{2^n n!} - j\dfrac{2^n(n-1)!}{\pi(k\rho)^n} \quad n > 0$	$\sqrt{\dfrac{-2j}{\pi k\rho}}\, j^{-n} e^{jk\rho}$	$k\rho \to j\infty$	$k\rho = 0$ $k\rho \to -j\infty$	k real—inward-traveling wave k imaginary—evanescent field k complex—attenuated traveling wave
$H_n^{(2)}(k\rho)$	$J_n(k\rho) - jN_n(k\rho)$	$1 + j\dfrac{2}{\pi}\log\left(\dfrac{2}{\gamma k\rho}\right) \quad n = 0$ $\dfrac{(k\rho)^n}{2^n n!} + j\dfrac{2^n(n-1)!}{\pi(k\rho)^n} \quad n > 0$	$\sqrt{\dfrac{2j}{\pi k\rho}}\, j^n e^{-jk\rho}$	$k\rho \to -j\infty$	$k\rho = 0$ $k\rho \to j\infty$	k real—outward-traveling wave k imaginary—evanescent field k complex—attenuated traveling wave
$J_n(k\rho)$	$\tfrac{1}{2}[H_n^{(1)}(k\rho) + H_n^{(2)}(k\rho)]$	$1 \quad n = 0$ $\dfrac{(k\rho)^n}{2^n n!} \quad n > 0$	$\sqrt{\dfrac{2}{\pi k\rho}}\cos\left(k\rho - \dfrac{n\pi}{2} - \dfrac{\pi}{4}\right)$	Infinite number along the real axis	$k\rho \to \pm j\infty$	k real—standing wave k imaginary—two evanescent fields k complex—localized standing wave
$N_n(k\rho)$	$\dfrac{1}{2j}[H_n^{(1)}(k\rho) - H_n^{(2)}(k\rho)]$	$-\dfrac{2}{\pi}\log\left(\dfrac{2}{\gamma k\rho}\right) \quad n = 0$ $-\dfrac{2^n(n-1)!}{\pi(k\rho)^n} \quad n > 0$	$\sqrt{\dfrac{2}{\pi k\rho}}\sin\left(k\rho - \dfrac{n\pi}{2} - \dfrac{\pi}{4}\right)$	Infinite number along the real axis	$k\rho = 0$ $k\rho \to \pm j\infty$	k real—standing wave k imaginary—two evanescent fields k complex—localized standing waves

* When $k = -j\alpha$, the functions $I_n(j k\rho) = I_n(\alpha\rho) = j^n J_n(-j\alpha\rho)$ and $K_n(j k\rho) = K_n(\alpha\rho) = \dfrac{\pi}{2}(-j)^{n+1} H_n^{(2)}(-j\alpha\rho)$ are used.

† When $k = 0$, the Bessel functions are 1 and $\log \rho$, $n = 0$, and ρ^n and ρ^{-n}, $n \neq 0$.

which are sufficiently general to express any TE (no E_z) field existing in a homogeneous source-free region. An arbitrary field (one having both an E_z and an H_z) can be expressed as a superposition of Eqs. (5-18) and (5-19).

5-2. The Circular Waveguide. The propagation of waves in a hollow conducting tube of circular cross section, called the *circular waveguide*, provides a good illustration of the use of cylindrical wave functions. Qualitatively, the phenomenon is similar to wave propagation in the rectangular waveguide, considered in Sec. 4-3. The coordinates to be used are shown in Fig. 5-2.

For modes TM to z, we may express the field in terms of an **A** having only a z component ψ. The field is finite at $\rho = 0$; so the wave functions must be of the form of Eqs. (5-13). It is conventional to express the ϕ variation by sinusoidal functions; hence

$$\psi = J_n(k_\rho\rho) \begin{Bmatrix} \sin n\phi \\ \cos n\phi \end{Bmatrix} e^{-jk_z z} \qquad (5\text{-}20)$$

is the desired form of the mode functions. Either $\sin n\phi$ or $\cos n\phi$ may be chosen; so we have a mode degeneracy except for the cases $n = 0$. The TM field is found from Eqs. (5-18) applied to the above ψ. In particular,

$$E_z = \frac{1}{\hat{y}} (k^2 - k_z^2)\psi$$

which must vanish at the conducting walls $\rho = a$. Hence, we must have

$$J_n(k_\rho a) = 0 \qquad (5\text{-}21)$$

from which eigenvalues for k_ρ may be determined. The functions $J_n(x)$ are shown in Fig. D-1. Note that for each n there are a denumerably infinite number of zeros. These are ordered and designated by x_{np}, the

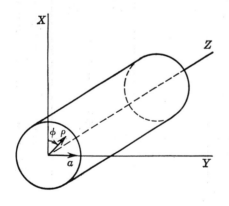

Fig. 5-2. The circular waveguide.

CYLINDRICAL WAVE FUNCTIONS

TABLE 5-2. ORDERED ZEROS x_{np} OF $J_n(x)$

n p	0	1	2	3	4	5
1	2.405	3.832	5.136	6.380	7.588	8.771
2	5.520	7.016	8.417	9.761	11.065	12.339
3	8.654	10.173	11.620	13.015	14.372	
4	11.792	13.324	14.796			

first subscript referring to the order of the Bessel function and the second to the order of the zero. The lower order x_{np} are tabulated in Table 5-2.

Equation (5-21) is now satisfied if we choose

$$k_\rho = \frac{x_{np}}{a} \tag{5-22}$$

Substituting this into Eq. (5-20), we have the TM_{np} mode functions

$$\psi_{np}^{TM} = J_n\left(\frac{x_{np}\rho}{a}\right) \begin{Bmatrix} \sin n\phi \\ \cos n\phi \end{Bmatrix} e^{-jk_z z} \tag{5-23}$$

where $n = 0, 1, 2, \ldots$, and $p = 1, 2, 3, \ldots$. The electromagnetic field is then determined from Eqs. (5-18) with the above ψ. The mode phase constant k_z is determined according to Eq. (5-6), that is,

$$\left(\frac{x_{np}}{a}\right)^2 + k_z^2 = k^2 \tag{5-24}$$

Subscripts np on the k_z are sometimes used to indicate explicitly that it depends on the mode number.

Modes TE to z are expressed in terms of an **F** having only a z component ψ. This wave function must be of the form of Eq. (5-20), with the field determined by Eqs. (5-19). The E_ϕ component is $\partial\psi/\partial\rho$, which must vanish at $\rho = a$; hence the condition

$$J_n'(k_\rho a) = 0 \tag{5-25}$$

must be satisfied. The J_n are oscillatory functions; hence, the J_n' also are oscillatory functions. (For example, $J_0' = -J_1$.). The $J_n'(x)$ have a denumerably infinite number of zeros, which we order as x'_{np}. (The prime is used to avoid confusion with the zeros of the Bessel function itself.) The lower-order zeros are tabulated in Table 5-3.

TABLE 5-3. ORDERED ZEROS x'_{np} OF $J_n'(x')$

n p	0	1	2	3	4	5
1	3.832	1.841	3.054	4.201	5.317	6.416
2	7.016	5.331	6.706	8.015	9.282	10.520
3	10.173	8.536	9.969	11.346	12.682	13.987
4	13.324	11.706	13.170			

We now satisfy Eq. (5-25) by choosing

$$k_\rho = \frac{x'_{np}}{a} \tag{5-26}$$

Using this in the wave function of Eq. (5-20), we have the TE$_{np}$ mode functions

$$\psi_{np}^{TE} = J_n\left(\frac{x'_{np}\rho}{a}\right) \begin{Bmatrix} \sin n\phi \\ \cos n\phi \end{Bmatrix} e^{-jk_z z} \tag{5-27}$$

where $n = 0, 1, 2, \ldots$, and $p = 1, 2, 3, \ldots$. The electromagnetic field is given by Eqs. (5-19) with the above ψ. The mode propagation constant is determined by Eq. (5-6), which with Eq. (5-26) becomes

$$\left(\frac{x'_{np}}{a}\right)^2 + k_z^2 = k^2 \tag{5-28}$$

This completes our determination of the mode spectrum for the circular waveguide.

The interpretation of the mode propagation constants is the same as for those of the rectangular guide and, in fact, is the same for all cylindrical guides of arbitrary cross section if the dielectric is homogeneous. (This we show in Sec. 8-1.) The cutoff wave number of a mode is that for which the mode propagation constant vanishes. Hence, from Eqs. (5-24) and (5-28), we have

$$(k_c)_{np}^{TM} = \frac{x_{np}}{a} \qquad (k_c)_{np}^{TE} = \frac{x'_{np}}{a} \tag{5-29}$$

If $k > k_c$, the mode propagates, and if $k < k_c$ the mode is cutoff. Letting $k_c = 2\pi f_c \sqrt{\epsilon\mu}$, we obtain the cutoff frequencies

$$(f_c)_{np}^{TM} = \frac{x_{np}}{2\pi a \sqrt{\epsilon\mu}} \qquad (f_c)_{np}^{TE} = \frac{x'_{np}}{2\pi a \sqrt{\epsilon\mu}} \tag{5-30}$$

Alternatively, setting $k_c = 2\pi/\lambda_c$, we obtain the cutoff wavelengths

$$(\lambda_c)_{np}^{TM} = \frac{2\pi a}{x_{np}} \qquad (\lambda_c)_{np}^{TE} = \frac{2\pi a}{x'_{np}} \tag{5-31}$$

Thus, the cutoff frequencies are proportional to the x_{np} for TM modes, and to the x'_{np} for the TE modes. Referring to Tables 5-2 and 5-3, we note that the zeros in ascending order of magnitude are $x'_{11}, x_{01}, x'_{21}, x_{11}$, and x'_{01}, etc. Hence, the modes in order of ascending cutoff frequencies are TE$_{11}$, TM$_{01}$, TE$_{21}$, TM$_{11}$, and TE$_{01}$ (a degeneracy), etc.

Circular waveguides are used in applications where rotational symmetry is needed. The dominant TE$_{11}$ "mode" is actually a pair of degenerate modes (sin ϕ and cos ϕ variation); hence there is no frequency

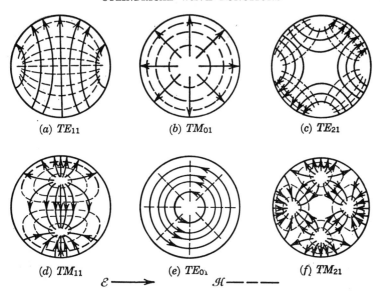

(a) TE_{11} (b) TM_{01} (c) TE_{21}

(d) TM_{11} (e) TE_{01} (f) TM_{21}

$\mathcal{E}$ ⟶ $\mathcal{H}$ -----

FIG. 5-3. Circular waveguide mode patterns.

range for single-mode propagation. (Recall that single-mode operation over a 2:1 frequency range is possible in the rectangular waveguide.) Note that, except for the degeneracies between TE_{0p} and TM_{1p} modes, TE and TM modes have different cutoff frequencies and hence different propagation constants. The modes of the circular waveguide have z-directed wave impedances of the same form as we found in the rectangular waveguide. For example, in a TE mode,

$$(Z_0)^{TE} = \frac{E_\rho}{H_\phi} = -\frac{E_\phi}{H_\rho} = \frac{\omega\mu}{k_z} \qquad (5\text{-}32)$$

which is the same as Eq. (4-27). The behavior of the Z_0's is therefore the same as in the rectangular waveguide, which is plotted in Fig. 4-3. Attenuation of waves in circular waveguides due to conduction losses in the walls is given in Prob. 5-9. Modal expansions in circular waveguides can be obtained by the general treatment of Sec. 8-2.

The mode patterns for some of the lower-order modes are shown in Fig. 5-3. These can be determined in the usual manner (find $\mathcal{E}$ and $\mathcal{H}$, and specialize to some instant of time). Field lines ending in the cross-sectional plane loop down the guide, in the same manner as they did in the rectangular waveguide.

Solutions for cylindrical waveguides of other cross sections also can be expressed in terms of elementary cylindrical wave functions. Representative cross sections are shown in Fig. 5-4. Note that all of these

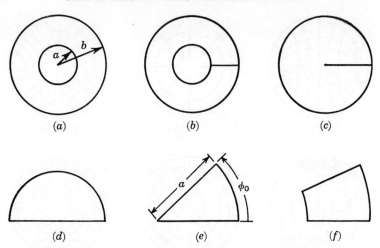

FIG. 5-4. Some waveguide cross sections for which the mode functions are elementary wave functions. (a) Coaxial; (b) coaxial with baffle; (c) circular with baffle; (d) semicircular; (e) wedge; (f) sectoral.

are formed by conductors covering complete ρ = constant and ϕ = constant coordinate surfaces. Wave functions for the guides of Fig. 5-4 are given in Probs. 5-5 to 5-7.

5-3. Radial Waveguides. In the circular waveguide we have plane waves, that is, the equiphase surfaces are parallel planes. Wave functions of the form

$$\psi = B_n(k_\rho\rho)h(k_z z)e^{\pm in\phi}$$

with $B_n(k_\rho\rho)$ and $h(k_z z)$ real, have equiphase surfaces which are intersecting planes (the ϕ = constant surfaces). Such waves travel in the circumferential direction, and we shall call them *circulating waves*. Examples are given in Prob. 5-10. Finally, we might have wave functions of the form

$$\psi = h(k_z z)h(n\phi) \begin{Bmatrix} H_n^{(1)}(k_\rho\rho) \\ H_n^{(2)}(k_\rho\rho) \end{Bmatrix}$$

with $h(k_z z)$ and $h(n\phi)$ real. These waves have cylindrical equiphase surfaces (ρ = constant), and travel in the radial direction. We shall call them *radial waves*.[1] In this section some simple waveguides capable of guiding radial waves will be considered.

Radial waves can be supported by parallel conducting plates. Depend-

---

[1] These are true cylindrical waves as defined in Sec. 2-11, but we are using the term "cylindrical wave function" to mean "a wave function in the cylindrical coordinate system," regardless of its equiphase surfaces.

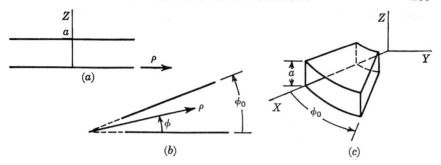

FIG. 5-5. Radial waveguides. (a) Parallel plate; (b) wedge; (c) horn.

ing upon the excitation, waves between the plates may be either plane or radial. When the waves are of the radial type, we call the guiding plates a *parallel-plate radial waveguide*. Figure 5-5a shows the coordinate system we shall use. The TM wave functions satisfying the boundary conditions $E_\rho = E_\phi = 0$ at $z = 0$ and $z = a$ are

$$\psi_{mn}^{TM} = \cos\left(\frac{m\pi}{a}z\right)\cos n\phi \begin{Bmatrix} H_n^{(1)}(k_\rho\rho) \\ H_n^{(2)}(k_\rho\rho) \end{Bmatrix} \quad (5\text{-}33)$$

where $m = 0, 1, 2, \ldots$, and $n = 0, 1, 2, \ldots$, and, by Eq. (5-6),

$$k_\rho = \sqrt{k^2 - \left(\frac{m\pi}{a}\right)^2} \quad (5\text{-}34)$$

The electromagnetic field is given by Eqs. (5-18) with the above ψ. The TE wave functions satisfying the boundary conditions are

$$\psi_{mn}^{TE} = \sin\left(\frac{m\pi}{a}z\right)\cos n\phi \begin{Bmatrix} H_n^{(1)}(k_\rho\rho) \\ H_n^{(2)}(k_\rho\rho) \end{Bmatrix} \quad (5\text{-}35)$$

where $m = 1, 2, 3, \ldots$, and $n = 0, 1, 2, \ldots$, and Eq. (5-34) still applies. The electromagnetic field for the TE modes is found from Eqs. (5-19) with the above ψ. In both the TM and TE cases, the $H_n^{(1)}(k_\rho\rho)$ represent inward-traveling waves (toward the z axis), and the $H_n^{(2)}(k_\rho\rho)$ represent outward-traveling waves. For a complete set of modes, those with $\sin n\phi$ variation must also be included.

Radial waves are characterized by a phase constant which is a function of radial distance. Following the general definition of Sec. 2-11, we have the phase constants for the above ψ's given by

$$\beta_\rho = \frac{\partial}{\partial\rho}\left[\tan^{-1}\frac{N_n(k_\rho\rho)}{J_n(k_\rho\rho)}\right]$$
$$= \frac{2}{\pi\rho}\frac{1}{J_n^2(k_\rho\rho) + N_n^2(k_\rho\rho)} \quad (5\text{-}36)$$

Using asymptotic formulas for the Bessel functions, we find that for real k_ρ

$$\beta_\rho \xrightarrow[k_\rho\rho \to \infty]{} k_\rho \qquad (5\text{-}37)$$

This is to be expected, because at large radii the waves should be similar to plane waves on the parallel-plate guide. Note that the phase constant of Eq. (5-36) is that of the mode function and not that for the field. Components of **E** and **H** transverse to ρ are not generally in phase. They become in phase at large radii.

Each mode of the radial waveguide is also characterized by a single radially directed wave impedance. Using Eqs. (5-33) and (5-18), we find for outward-traveling TM modes

$$Z_{+\rho}^{TM} = -\frac{E_z}{H_\phi} = \frac{k_\rho}{j\omega\epsilon} \frac{H_n^{(2)}(k_\rho\rho)}{H_n^{(2)\prime}(k_\rho\rho)} \qquad (5\text{-}38)$$

while for inward-traveling TM modes

$$Z_{-\rho}^{TM} = \frac{E_z}{H_\phi} = -\frac{k_\rho}{j\omega\epsilon} \frac{H_n^{(1)}(k_\rho\rho)}{H_n^{(1)\prime}(k_\rho\rho)} \qquad (5\text{-}39)$$

Note that for real k_ρ we have $Z_{-\rho}^{TM} = Z_{+\rho}^{TM*}$. Similarly, for TE modes we find

$$Z_{+\rho}^{TE} = \frac{E_\phi}{H_z} = \frac{j\omega\mu}{k_\rho} \frac{H_n^{(2)\prime}(k_\rho\rho)}{H_n^{(2)}(k_\rho\rho)}$$

$$Z_{-\rho}^{TE} = -\frac{E_\phi}{H_z} = \frac{-j\omega\mu}{k_\rho} \frac{H_n^{(1)\prime}(k_\rho\rho)}{H_n^{(1)}(k_\rho\rho)} \qquad (5\text{-}40)$$

where the first equation applies to outward-traveling waves and the second equation to inward-traveling waves. Note that the TE wave admittances are dual to the TM wave impedances.

It is seen from Eq. (5-34) that k_ρ is imaginary if $m\pi/a > k$. In this case, let $k_\rho = -j\alpha$, and

$$H_n^{(2)}(-j\alpha) = \frac{2}{\pi} j^{n+1} K_n(\alpha)$$

where K_n is the modified Bessel function (see Appendix D). The mode functions are now everywhere in phase, and there is no wave propagation. The radial wave impedances become imaginary, indicating no power flow. For example, from Eq. (5-38), if $k_\rho = -j\alpha$,

$$Z_{+\rho}^{TM} = \frac{-j\alpha}{j\omega\epsilon} \frac{H_n^{(2)}(-j\alpha\rho)}{H_n^{(2)\prime}(-j\alpha\rho)} = \frac{j\alpha}{\omega\epsilon} \frac{K_n(\alpha\rho)}{K_n'(\alpha\rho)} \qquad (5\text{-}41)$$

which are always capacitively reactive, since K_n is positive and K_n' is negative. Hence, whenever $a < \lambda/2$, the modes $m > 0$ are nonpropagating (evanescent). For small a, only the TM_{0n} modes propagate, for which Eq. (5-33) reduces to

$$\psi_{0n}{}^{\text{TM}} = \cos n\phi \begin{Bmatrix} H_n{}^{(1)}(k\rho) \\ H_n{}^{(2)}(k\rho) \end{Bmatrix} \quad (5\text{-}42)$$

From Eqs. (5-38) and (5-39) we have the wave impedances for these modes given by

$$Z_{+\rho}{}^{\text{TM}} = Z_{-\rho}{}^{\text{TM}*} = -j\eta \frac{H_n{}^{(2)}(k\rho)}{H_n{}^{(2)\prime}(k\rho)}$$

$$= \frac{\eta}{|H_n{}^{(2)\prime}(k\rho)|^2} \left\{ \frac{2}{\pi k\rho} - j[J_n(k\rho)J_n'(k\rho) + N_n(k\rho)N_n'(k\rho)] \right\} \quad (5\text{-}43)$$

A consideration of the behavior of the Bessel functions (Figs. D-1 and D-2) reveals that for arguments $k\rho < n$ the N_n functions and their derivatives become large in magnitude. Hence, when $2\pi\rho < n\lambda$, the wave impedances become predominantly reactive. Figure 5-6 illustrates this behavior by showing X/R, where $Z_{+\rho}{}^{\text{TM}} = R + jX$, for the first five TM$_{0n}$ modes. We shall call $k\rho = n$ the point of gradual cutoff, the wave impedances being predominantly resistive when $k\rho > n$ and predominantly reactive when $k\rho < n$. Note that these gradual cutoffs occur when the circumference of the radial waveguide is an integral number of wavelengths.

From the above discussion it is evident that the TM$_{00}$ mode is dominant, that is, propagates energy effectively at smaller radii than any other mode. For this mode we have

$$E_z^- = \frac{k^2}{j\omega\epsilon} H_0{}^{(1)}(k\rho) \qquad H_\phi^- = kH_1{}^{(1)}(k\rho) \quad (5\text{-}44)$$

representing inward-traveling waves, and

$$E_z^+ = \frac{k^2}{j\omega\epsilon} H_0{}^{(2)}(k\rho)$$
$$H_\phi^+ = kH_1{}^{(2)}(k\rho) \quad (5\text{-}45)$$

which represent outward-traveling waves. Note that there are no ρ components of **E** or **H**, the mode being TEM to ρ. It is called the *transmission-line mode* of the parallel-plate radial guide, because of its similarity with plane transmission-line modes. For example, at a given radius we can calculate a unique voltage between the plates and a net radially directed current on one of

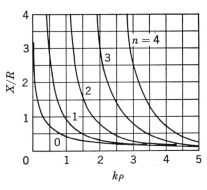

Fig. 5-6. Ratios of wave reactance to wave resistance for the TM$_{0n}$ radial modes on the parallel-plate waveguide.

the plates. Also, the radial transmission line can be analyzed by the classical transmission-line equations with L and C a function of ρ (Prob. 5-13).

Radial waves also can be supported by inclined conducting planes, called a *wedge radial waveguide*, as shown in Fig. 5-5b. We shall assume no z variation of the field, considering the problem as two-dimensional. TM wave functions satisfying the boundary condition $E_z = 0$ at $\phi = 0$ and $\phi = \phi_0$ are

$$\psi_p{}^{\text{TM}} = \sin\left(\frac{p\pi}{\phi_0}\phi\right) \begin{Bmatrix} H^{(1)}_{p\pi/\phi_0}(k\rho) \\ H^{(2)}_{p\pi/\phi_0}(k\rho) \end{Bmatrix} \tag{5-46}$$

where $p = 1, 2, 3, \ldots$, and the electromagnetic field is given by Eqs. (5-18). TE wave functions satisfying the boundary condition $E_\rho = 0$ at $\phi = 0$ and $\phi = \phi_0$ are

$$\psi_p{}^{\text{TE}} = \cos\left(\frac{p\pi}{\phi_0}\phi\right) \begin{Bmatrix} H^{(1)}_{p\pi/\phi_0}(k\rho) \\ H^{(2)}_{p\pi/\phi_0}(k\rho) \end{Bmatrix} \tag{5-47}$$

where $p = 0, 1, 2, \ldots$, and the electromagnetic field is given by Eqs. (5-19). The interpretation of the modes is essentially the same as that for the TM_{0n} parallel-plate modes, except that nonintegral orders of Hankel functions appear. This introduces no conceptual difficulties, but if numerical results are desired we would be hampered by a lack of tables for functions of arbitrary fractional order.

The radial wave impedances for the wedge-guide modes are of the same form as for the parallel-plate guide [Eqs. (5-38) to (5-40)]. We need only replace n by $p\pi/\phi_0$ and k_ρ by k. These wave impedances exhibit the same characteristic of gradual cutoff for fractional-order Hankel functions as they do for integral-order Hankel functions. Again the transitional point is that for which the argument and order are equal, that is, $p\pi/\phi_0 = k\rho$. The radii so determined correspond to those for which the arc subtending the wedge is an integral number of half-wavelengths long. This is as we should expect from our knowledge of plane waves between parallel plates (the limiting case $\phi_0 \to 0$).

The dominant mode is evidently the TE_0 mode, in which case, from Eqs. (5-47) and (5-19), we have

$$E_\phi{}^- = -k H_0{}^{(1)\prime}(k\rho) \qquad H_z{}^- = \frac{k^2}{j\omega\mu} H_0{}^{(1)}(k\rho) \tag{5-48}$$

for inward-traveling waves, and

$$E_\phi{}^+ = -k H_0{}^{(2)\prime}(k\rho) \qquad H_z{}^+ = \frac{k^2}{j\omega\mu} H_0{}^{(2)}(k\rho) \tag{5-49}$$

for outward-traveling waves. This is a transmission-line mode, characterized by no E_ρ or H_ρ and possessing a unique voltage and current at any given radii. This mode also can be analyzed by the classical transmission-line equations for nonuniform lines (L and C a function of ρ). Note that the field is dual to that of the parallel-plate line [Eqs. (5-44) and (5-45)].

Finally, simple radial waves can be supported by the horn-shaped guide of Fig. 5-5c, called a *sectoral horn waveguide*. The TM modes are specified by the wave functions

$$\psi_{mp}{}^{TM} = \cos\left(\frac{m\pi}{a}z\right)\sin\left(\frac{p\pi}{\phi_0}\phi\right)\begin{Bmatrix} H^{(1)}_{p\pi/\phi_0}(k_\rho\rho) \\ H^{(2)}_{p\pi/\phi_0}(k_\rho\rho) \end{Bmatrix} \quad (5\text{-}50)$$

where $m = 0, 1, 2, \ldots$, and $p = 1, 2, 3, \ldots$. The field is given by Eqs. (5-18), and

$$k_\rho = \sqrt{k^2 - \left(\frac{m\pi}{a}\right)^2} \quad (5\text{-}51)$$

The TE modes are specified by the mode functions

$$\psi_{mp}{}^{TE} = \sin\left(\frac{m\pi}{a}z\right)\cos\left(\frac{p\pi}{\phi_0}\phi\right)\begin{Bmatrix} H^{(1)}_{p\pi/\phi_0}(k_\rho\rho) \\ H^{(2)}_{p\pi/\phi_0}(k_\rho\rho) \end{Bmatrix} \quad (5\text{-}52)$$

where $m = 1, 2, 3, \ldots$, and $p = 0, 1, 2, \ldots$. The field is given by Eqs. (5-19), and k_ρ by Eq. (5-51). These modes are qualitatively similar to the hybrid modes of the rectangular waveguide (Sec. 4-4). There is, of course, no transmission-line mode, because of the single conducting boundary. Only the TM_{0p} modes propagate if $a < \lambda/2$; these plus the TM_{1p} and TE_{1p} modes propagate if $\lambda/2 < a < \lambda$; and so on. Each propagating mode has a radius of gradual cutoff, this being the radius at which the guide cross section is about the same size as a rectangular waveguide at cutoff. The TM_{01} mode is usually considered as the dominant mode. (If $a > \lambda/2$ one might argue that the TE_{10} mode is dominant at small radii.)

5-4. The Circular Cavity. If a section of circular waveguide is closed by conductors over two cross sections, we have a resonator known as the circular cavity. This is shown in Fig. 5-7. It is a simple matter to modify the circular waveguide mode functions to satisfy the addi-

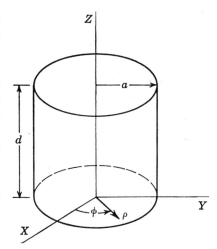

Fig. 5-7. The circular cavity.

tional boundary conditions of zero tangential **E** at $z = 0$ and $z = d$. The result is a set of modes TM to z, specified by

$$\psi_{npq}^{TM} = J_n\left(\frac{x_{np}\rho}{a}\right) \begin{Bmatrix} \sin n\phi \\ \cos n\phi \end{Bmatrix} \cos\left(\frac{q\pi}{d} z\right) \qquad (5\text{-}53)$$

where $n = 0, 1, 2, \ldots$; $p = 1, 2, 3, \ldots$; and $q = 0, 1, 2, \ldots$. The field is given by Eqs. (5-18). The set of modes TE to z is specified by

$$\psi_{npq}^{TE} = J_n\left(\frac{x'_{np}\rho}{a}\right) \begin{Bmatrix} \sin n\phi \\ \cos n\phi \end{Bmatrix} \sin\left(\frac{q\pi}{d} z\right) \qquad (5\text{-}54)$$

where $n = 0, 1, 2, \ldots$; $p = 1, 2, 3, \ldots$; $q = 1, 2, 3, \ldots$; and the field is given by Eqs. (5-19). The separation constant equation [Eq. (5-6)] becomes

$$\left(\frac{x_{np}}{a}\right)^2 + \left(\frac{q\pi}{d}\right)^2 = k^2$$

$$\left(\frac{x'_{np}}{a}\right)^2 + \left(\frac{q\pi}{d}\right)^2 = k^2$$

for the TM and TE modes, respectively. Setting $k = 2\pi f \sqrt{\epsilon\mu}$, we can solve for the resonant frequencies

$$(f_r)_{npq}^{TM} = \frac{1}{2\pi a \sqrt{\epsilon\mu}} \sqrt{x_{np}^2 + \left(\frac{q\pi a}{d}\right)^2}$$

$$(f_r)_{npq}^{TE} = \frac{1}{2\pi a \sqrt{\epsilon\mu}} \sqrt{x'^{2}_{np} + \left(\frac{q\pi a}{d}\right)^2} \qquad (5\text{-}55)$$

Each n except $n = 0$ denotes a pair of degenerate modes ($\cos n\phi$ or $\sin n\phi$ variation). The x_{np} and x'_{np} are given in Tables 5-2 and 5-3. The resonant frequencies for various ratios of d/a are tabulated in Table 5-4.

TABLE 5-4. $\dfrac{(f_r)_{npq}}{(f_r)_{dominant}}$ FOR THE CIRCULAR CAVITY OF RADIUS a AND LENGTH d

$\dfrac{d}{a}$	TM_{010}	TE_{111}	TM_{110}	TM_{011}	TE_{211}	TM_{111} TE_{011}	TE_{112}	TM_{210}	TM_{020}
0	1.0	∞	1.59	∞	∞	∞	∞	2.13	2.29
0.5	1.0	2.72	1.59	2.80	2.90	3.06	5.27	2.13	2.29
1.0	1.0	1.50	1.59	1.63	1.80	2.05	2.72	2.13	2.29
2.0	1.0	1.0	1.59	1.19	1.42	1.72	1.50	2.13	2.29
3.0	1.13	1.0	1.80	1.24	1.52	1.87	1.32	2.41	2.60
4.0	1.20	1.0	1.91	1.27	1.57	1.96	1.20	2.56	3.00
∞	1.31	1.0	2.08	1.31	1.66	2.08	1.0	2.78	3.00

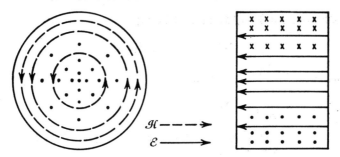

Fig. 5-8. Mode pattern for the TM_{010} mode (dominant when $d/a \leq 2$).

Note that for $d/a < 2$ the TM_{010} mode is dominant, while for $d/a \geq 2$ the TE_{111} mode is dominant. If $d/a < 1$, the second resonance is 1.59 times the first resonant frequency. Note that this is very similar to the square-base rectangular cavity of small height (the mode separation is 1.58 in that case).

The TM_{010} mode corresponds to the first resonance of a short-circuited radial transmission line. The field pattern of this mode, which is dominant for small d, is shown in Fig. 5-8. The TE_{111} mode corresponds to the first resonance of a short-circuited circular waveguide operating in the TE_{11} mode. Its mode pattern is thus that of a standing wave in a circular waveguide, similar to Fig. 5-3a. The case $d/a \to \infty$ corresponds to that of a two-dimensional circular resonator, for which the resonant frequencies are the cutoff frequencies of the circular waveguide. The last row of Table 5-4 therefore is also the cutoff frequency spectrum of the circular waveguide.

The Q's of the circular cavity are also of interest, especially the Q of the TM_{010} mode (dominant for small d). From Eqs. (5-53) and (5-18) we determine the field components of the mode as

$$E_z = \frac{k^2}{j\omega\epsilon} J_0\left(\frac{x_{01}\rho}{a}\right)$$

$$H_\phi = \frac{x_{01}}{a} J_1\left(\frac{x_{01}\rho}{a}\right)$$

Following the procedure of Sec. 2-8, we calculate the stored energy in the cavity as

$$\mathcal{W} = 2\overline{\mathcal{W}}_e = \epsilon \iiint |E|^2 \, d\tau$$

$$= \frac{k^4}{\omega^2\epsilon} 2\pi d \int_0^a \rho J_0^2\left(\frac{x_{01}\rho}{a}\right) d\rho$$

This is a known integral,[1] the result being

$$\mathcal{W} = \frac{\pi k^4 \, da^2}{\omega^2 \epsilon} J_1^{\,2}(x_{01}) \tag{5-56}$$

The power dissipated in the conducting walls is approximately

$$\bar{\mathcal{P}}_d = \mathcal{R} \oint |H|^2 \, ds$$
$$= \mathcal{R} \left(\frac{x_{01}}{a}\right)^2 2\pi \left[ad J_1^{\,2}(x_{01}) + 2 \int_0^a \rho J_1^{\,2}\left(\frac{x_{01}\rho}{a}\right) d\rho \right]$$

where $\mathcal{R}$ is the intrinsic wave resistance of the metal walls. The above integral is again known,[1] and we obtain

$$\bar{\mathcal{P}}_d = \mathcal{R} \left(\frac{x_{01}}{a}\right)^2 2\pi a(d + a) J_1^{\,2}(x_{01}) \tag{5-57}$$

The Q of the cavity is therefore

$$Q = \frac{\omega \mathcal{W}}{\bar{\mathcal{P}}_d} = \frac{k^4 \, da^3}{2\omega \epsilon \mathcal{R} x_{01}^{\,2}(d + a)}$$

Recalling that the condition for resonance is $ka = x_{01} = 2.405$, we can simplify this to

$$Q = \frac{1.202 \eta}{\mathcal{R}(1 + a/d)} \tag{5-58}$$

where η is the intrinsic impedance of the dielectric. This can be compared to the Q of a square-base rectangular cavity [Eq. (2-102)]. It is seen that, for the same height-to-diameter ratio, the circular cavity has an 8.3 per cent higher Q than the rectangular cavity. This is to be expected, since the volume-to-area ratio is higher for a circular cylinder than for a square cylinder. The Q's for the other modes of the circular cavity are given in Prob. 5-16.

5-5. Other Guided Waves. The geometries of some other cylindrical systems capable of supporting guided waves are shown in Figs. 5-9 and 5-10. We treated the analogous plane-wave systems in Chap. 4. The methods of solution for the systems of Figs. 5-9 and 5-10, as well as their qualitative behavior, are similar to those of Chap. 4.

For the partially filled radial waveguide of Fig. 5-9a, we can obtain fields TM to z which satisfy the conditions $E_\rho = E_\phi = 0$ at $z = 0$ and $z = a$ by choosing

$$\begin{aligned}\psi_1 &= C_1 \cos k_{z1} z \cos n\phi \, H_n^{(2)}(k_\rho \rho) \\ \psi_2 &= C_2 \cos [k_{z2}(a - z)] \cos n\phi \, H_n^{(2)}(k_\rho \rho)\end{aligned} \tag{5-59}$$

---

[1] E. Jahnke and F. Emde, "Tables of Functions," p. 146, Dover Publications, New York, 1945 (reprint).

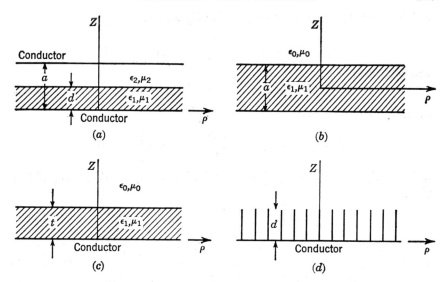

Fig. 5-9. Some radial waveguides. (a) Partially filled; (b) dielectric slab; (c) coated conductor; (d) corrugated conductor.

where $n = 0, 1, 2, \ldots$. The subscripts 1 and 2 refer to the regions $z < d$ and $z > d$, respectively. We have anticipated that the ρ and ϕ variations must be the same in both regions to satisfy boundary conditions at $z = d$. Equations (5-59) represent outward-traveling waves. Inward-traveling waves would be of the same form but with $H_n^{(2)}$ replaced by $H_n^{(1)}$. The k's in each region must, of course, satisfy the separation relationships

$$\begin{aligned}k_\rho^2 + k_{z1}^2 &= k_1^2 = \omega^2 \epsilon_1 \mu_1 \\ k_\rho^2 + k_{z2}^2 &= k_2^2 = \omega^2 \epsilon_2 \mu_2\end{aligned} \qquad (5\text{-}60)$$

The field vectors themselves are obtained from Eqs. (5-18), using the ψ's of Eqs. (5-59).

To evaluate the C's and k_ρ, we must satisfy the conditions that E_ρ, E_ϕ, H_ρ, and H_ϕ be continuous at $z = d$. For E_ρ we have

$$[E_{\rho 1} - E_{\rho 2}]_{z=d} = \frac{1}{j\omega}\left[\frac{\partial^2}{\partial \rho \, \partial z}\left(\frac{1}{\epsilon_1}\psi_1 - \frac{1}{\epsilon_2}\psi_2\right)\right]_{z=d} = 0$$

which reduces to

$$k_{z1}\frac{C_1}{\epsilon_1}\sin k_{z1}d = -k_{z2}\frac{C_2}{\epsilon_2}\sin k_{z2}(a - d) \qquad (5\text{-}61)$$

For E_ϕ we have

$$[E_{\phi 1} - E_{\phi 2}]_{z=d} = \frac{1}{j\omega d}\left[\frac{\partial^2}{\partial \phi \, \partial z}\left(\frac{1}{\epsilon_1}\psi_1 - \frac{1}{\epsilon_2}\psi_2\right)\right]_{z=d} = 0$$

which also reduces to Eq. (5-61). For H_ρ we have

$$[H_{\rho 1} - H_{\rho 2}]_{z=d} = \frac{1}{\rho}\left[\frac{\partial}{\partial \phi}(\psi_1 - \psi_2)\right]_{z=d} = 0$$

which reduces to

$$C_1 \cos k_{z1}d = C_2 \cos k_{z2}(a - d) \tag{5-62}$$

Finally, for H_ϕ we have

$$[H_{\phi 1} - H_{\phi 2}]_{z=d} = -\left[\frac{\partial}{\partial \rho}(\psi_1 - \psi_2)\right]_{z=d} = 0$$

which again reduces to Eq. (5-62). Division of Eq. (5-61) by Eq. (5-62) yields

$$\frac{k_{z1}}{\epsilon_1} \tan k_{z1}d = -\frac{k_{z2}}{\epsilon_2} \tan [k_{z2}(a - d)] \tag{5-63}$$

The k_{z1} and k_{z2} are functions of k_ρ according to Eq. (5-60); so Eq. (5-63) is a transcendental equation for determining possible k_ρ's. Once k_ρ is evaluated, the ratio C_1/C_2 may be obtained from either Eq. (5-61) or Eq. (5-62).

For fields TE to z we can satisfy the condition $E_\rho = E_\phi = 0$ at $z = a$ by choosing

$$\begin{aligned}\psi_1 &= C_1 \sin k_{z1}z \cos n\phi \, H_n^{(2)}(k_\rho \rho) \\ \psi_2 &= C_2 \sin k_{z2}(a - z) \cos n\phi \, H_n^{(2)}(k_\rho \rho)\end{aligned} \tag{5-64}$$

where $n = 0, 1, 2, \ldots$; and Eqs. (5-60) must again be satisfied. The field components are found from these ψ's by Eqs. (5-19). Matching tangential components of **E** and **H** at $z = d$ yields

$$\frac{k_{z1}}{\mu_1} \cot k_{z1}d = -\frac{k_{z2}}{\mu_2} \cot [k_{z2}(a - d)] \tag{5-65}$$

as the equation for determining k_ρ for TE modes. It is interesting to note that the characteristic equations for the partially filled radial waveguide [Eqs. (5-63) and (5-65)] are of the same form as those for the partially filled rectangular waveguide [Eqs. (4-45) and (4-47)]. This we could have anticipated, since at large ρ the Hankel functions reduce to plane waves, as shown by Eqs. (D-13).

The modes of the partially filled radial guide can be ordered in the same manner as were the modes of the partially filled rectangular waveguide. The dominant mode is the lowest-order TM mode (logically designated the TM_{00} mode). It reduces to the radial transmission-line mode in the empty guide and has no cutoff frequency. For $a \ll \lambda$ it can be analyzed by conventional transmission-line concepts.

It should be apparent from our treatment of the waveguide of Fig. 5-9a that the characteristic equations for the radial waveguides of Fig. 5-9b, c,

CYLINDRICAL WAVE FUNCTIONS

and d will be of the same form as those for the plane waveguides of Figs. 4-10, 4-13, and 4-15. We need only to replace the k_z's by k_ρ's. Hence, for the dielectric-slab radial waveguide of Fig. 5-9b, the characteristic equations are

$$\frac{\epsilon_d}{\epsilon_0} \frac{va}{2} = \begin{cases} \dfrac{ua}{2} \tan \dfrac{ua}{2} \\ -\dfrac{ua}{2} \cot \dfrac{ua}{2} \end{cases} \tag{5-66}$$

for modes TM to z, and

$$\frac{\mu_d}{\mu_0} \frac{va}{2} = \begin{cases} \dfrac{ua}{2} \tan \dfrac{ua}{2} \\ -\dfrac{ua}{2} \cot \dfrac{ua}{2} \end{cases} \tag{5-67}$$

for modes TE to z. The u and v are related to k_ρ by

$$\begin{aligned} u^2 + k_\rho^2 &= k_d^2 = \omega^2 \epsilon_d \mu_d \\ -v^2 + k_\rho^2 &= k_0^2 = \omega^2 \epsilon_0 \mu_0 \end{aligned} \tag{5-68}$$

Possible solutions to these equations can be obtained graphically by the method of Fig. 4-11. Just as in the plane-wave case, the lowest TE and TM modes have no cutoff frequencies. The cutoff frequencies of the modes in general are given by Eq. (4-63).

The modes of the coated-conductor radial waveguide of Fig. 5-9c are those of the slab waveguide having $E_\rho = E_\phi = 0$ over the mid-plane of the slab. The dominant mode is the lowest TM mode, which has no cutoff frequency. The cutoff frequencies of the modes in general are given by Eq. (4-64). Finally, for the corrugated-conductor radial line of Fig. 5-9d, the characteristic equation for the dominant mode is

$$k_\rho = k_0 \sqrt{1 + \tan^2 k_0 d} \tag{5-69}$$

This is analogous to Eq. (4-71) in the plane-wave case.

The circular waveguide systems of Fig. 5-10 are interesting, because, except for rotationally symmetric fields, the modes are neither TE nor TM to any cylindrical coordinate. The systems of Fig. 5-10a, b, and c have the common property that they are "two-dielectric" problems. We can consider them all at once, as follows. Let region 1 be the inner dielectric cylinder in each case and region 2 the outer one. We then choose electric and magnetic ψ's as

$$\begin{aligned} \psi^{m1} &= A B_n^{m1}(k_{\rho 1}\rho) \cos n\phi \, e^{-jk_z z} \\ \psi^{e1} &= B B_n^{e1}(k_{\rho 1}\rho) \sin n\phi \, e^{-jk_z z} \end{aligned} \tag{5-70}$$

in region 1, and

$$\begin{aligned} \psi^{m2} &= C B_n^{m2}(k_{\rho 2}\rho) \cos n\phi \, e^{-jk_z z} \\ \psi^{e2} &= D B_n^{e2}(k_{\rho 2}\rho) \sin n\phi \, e^{-jk_z z} \end{aligned} \tag{5-71}$$

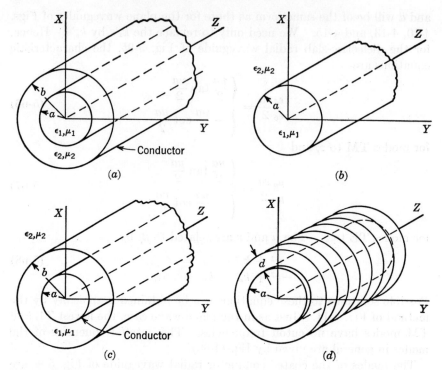

Fig. 5-10. Some circular waveguides. (a) Partially filled; (b) dielectric slab; (c) coated conductor; (d) corrugated conductor.

in region 2. The ψ^m determine partial fields according to Eqs. (5-18) and the ψ^e determine partial fields according to Eqs. (5-19). The total field is the sum of the two partial fields in each region. The $B_n(k_\rho \rho)$ denote appropriate solutions to Bessel's equation of order n, chosen so as to satisfy all boundary conditions except those at the interface $\rho = a$. In each region the ψ's must satisfy the separation relationships

$$k_{\rho 1}^2 + k_z^2 = k_1^2 = \omega^2 \epsilon_1 \mu_1 \\ k_{\rho 2}^2 + k_z^2 = k_2^2 = \omega^2 \epsilon_2 \mu_2 \qquad (5\text{-}72)$$

The requirements that H_z, E_z, H_ϕ, and E_ϕ be continuous at $\rho = a$ lead to

$$\epsilon_2 k_{\rho 1}^2 A B_n^{m1}(k_{\rho 1} a) = \epsilon_1 k_{\rho 2}^2 C B_n^{m2}(k_{\rho 2} a)$$
$$\mu_2 k_{\rho 1}^2 B B_n^{e1}(k_{\rho 1} a) = \mu_1 k_{\rho 2}^2 D B_n^{e2}(k_{\rho 2} a)$$
$$A k_{\rho 1} B_n^{m1\prime}(k_{\rho 1} a) + \frac{B k_z n}{\omega \mu_1 a} B_n^{e1}(k_{\rho 1} a) = C k_{\rho 2} B_n^{m2\prime}(k_{\rho 2} a) + \frac{D k_z n}{\omega \mu_2 a} B_n^{e2}(k_{\rho 2} a)$$
$$\frac{A K_z n}{\omega \epsilon_1 a} B_n^{m1}(k_{\rho 1} a) + B k_{\rho 1} B_n^{e1\prime}(k_{\rho 1} a) = \frac{C k_z n}{\omega \epsilon_2 a} B_n^{m2}(k_{\rho 2} a) + D k_{\rho 2} B_n^{e2\prime}(k_{\rho 2} a)$$

These equations have a nontrivial solution only if the determinant of the

CYLINDRICAL WAVE FUNCTIONS

coefficients of A, B, C, and D vanishes. Hence, defining,

$$F_1 = B_n{}^{m1}(k_{\rho 1}a) \qquad F_2 = B_n{}^{e1}(k_{\rho 1}a)$$
$$F_3 = B_n{}^{m2}(k_{\rho 2}a) \qquad F_4 = B_n{}^{e2}(k_{\rho 2}a) \tag{5-73}$$

The characteristic equation in determinantal form is

$$\begin{vmatrix} \epsilon_2 k_{\rho 1}{}^2 F_1 & 0 & \epsilon_1 k_{\rho 2}{}^2 F_3 & 0 \\ 0 & \mu_2 k_{\rho 1}{}^2 F_2 & 0 & \mu_1 k_{\rho 2}{}^2 F_4 \\ k_{\rho 1}F'_1 & \dfrac{k_z n}{\omega \mu_1 a} F_2 & k_{\rho 2}F'_3 & \dfrac{k_z n}{\omega \mu_2 a} F_4 \\ \dfrac{k_z n}{\omega \epsilon_1 a} F_1 & k_{\rho 1}F'_2 & \dfrac{k_z n}{\omega \epsilon_2 a} F_3 & k_{\rho 2}F'_4 \end{vmatrix} = 0 \tag{5-74}$$

When $n = 0$, the field separates into modes TE and TM to z, and the characteristic equation is much simpler. It is

$$k_{\rho 2}F_1 F'_3 - k_{\rho 1}F'_1 F_3 = 0 \tag{5-75}$$

for TM modes ($n = 0$), and

$$k_{\rho 2}F_2 F'_4 - k_{\rho 1}F'_2 F_4 = 0 \tag{5-76}$$

for TE modes ($n = 0$).

We must now pick the proper F functions for the various cases. For the partially filled circular waveguide (Fig. 5-10a), the field must be finite at $\rho = 0$; hence

$$F_1 = F_2 = J_n(k_{\rho 1}a) \tag{5-77}$$

To satisfy $E_z = 0$ at $\rho = b$, we choose

$$F_3 = J_n(k_{\rho 2}a)N_n(k_{\rho 2}b) - N_n(k_{\rho 2}a)J_n(k_{\rho 2}b) \tag{5-78}$$

Furthermore, to satisfy $E_\phi = 0$ at $\rho = b$, we choose

$$F_4 = J_n(k_{\rho 2}a)N'_n(k_{\rho 2}b) - N_n(k_{\rho 2}a)J'_n(k_{\rho 2}b) \tag{5-79}$$

The dominant mode is the lowest-order $n = 1$ mode, which reduces to the TE_{11} mode of the empty guide. A solution for the k_z of this dominant mode is plotted in Fig. 5-11 for the case $\epsilon_1 = 10\epsilon_0$, $\epsilon_2 = \epsilon_0$, $\mu_1 = \mu_2 = \mu_0$, $b = 0.4\lambda_0$.

For the dielectric-rod waveguide (Fig. 5-10b), the field must again be finite at $\rho = 0$; so Eqs. (5-77) still apply. However, external to the rod, the field must decay exponentially above the cutoff frequency and represent outward-traveling waves below the cutoff frequency. Hence, we choose

$$F_3 = F_4 = K_n(jk_{\rho 2}a) = \frac{\pi}{2}(-j)^{n+1}H_n{}^{(2)}(k_{\rho 2}a) \tag{5-80}$$

Once again, the dominant mode is the lowest $n = 1$ mode, and its cutoff

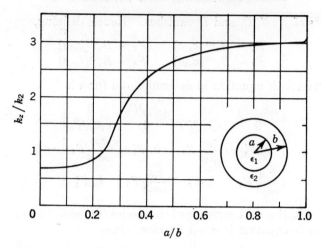

FIG. 5-11. Phase constant for the partially filled circular waveguide, $\epsilon_1 = 10\epsilon_2$, $b = 0.4\lambda_2$. (*After H. Seidel.*)

frequency is zero.[1] Some solutions for the k_z of the dominant mode are shown in Fig. 5-12 for the case $\epsilon_2 = \epsilon_0$ and $\mu_1 = \mu_2 = \mu_0$. Note that $k_0 < k_z < k_1$, which is the same relationship that applies to the dielectric-slab guide of Sec. 4-7.

For the coated conductor of Fig. 5-10c we must again have exponential decay of the field as $\rho \to \infty$; so Eqs. (5-80) still apply. However, to

[1] S. A. Schelkunoff, "Electromagnetic Waves," pp. 425–428, D. Van Nostrand Company, Inc., Princeton, N.J., 1943.

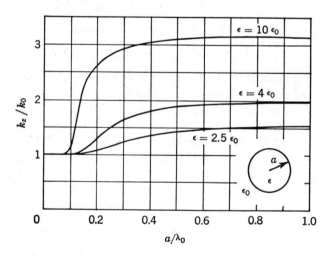

FIG. 5-12. Phase constant for the circular dielectric rod. (*After M. C. Gray.*)

satisfy the condition $E_z = 0$ at $\rho = b$, we should choose

$$F_1 = J_n(k_{\rho 1}a)N_n(k_{\rho 1}b) - N_n(k_{\rho 1}a)J_n(k_{\rho 1}b) \qquad (5\text{-}81)$$

and, to satisfy $E_\phi = 0$ at $\rho = b$,

$$F_2 = J_n(k_{\rho 1}a)N'_n(k_{\rho 1}b) - N_n(k_{\rho 1}a)J'_n(k_{\rho 1}b) \qquad (5\text{-}82)$$

For this guide the dominant mode is the lowest $n = 0$ TM mode, which has no cutoff frequency. (Compare it with the dominant mode of the plane coated conductor of Sec. 4-8.) Copper wire with an enamel coating can be used as an efficient waveguide for some applications.[1]

Finally, the corrugated wire of Fig. 5-10d can be analyzed in a manner similar to that used for the corrugated plane (Fig. 4-15). The field external to the corrugated wire will be essentially the dominant TM ($n = 0$) mode of the coated wire. The field in the corrugations will be essentially that of the shorted parallel-plate radial transmission line. The characteristic equation is obtained by matching wave impedances at the corrugated surface. As the radius of the corrugated cylinder becomes large, the solution approaches that for the corrugated plane.

5-6. Sources of Cylindrical Waves. In this section we shall consider two-dimensional sources of cylindrical waves, that is, sources independent of the z coordinate. The extension to three dimensions can be effected by a Fourier transformation with respect to z (see Sec. 5-11).

Suppose we have an infinitely long filament of constant a-c current along the z axis, as shown in Fig. 5-13a. From the theory of Sec. 2-9, we should expect the field to be TM to z, expressible in terms of an **A** having only a z component ψ. From symmetry, ψ should be independent

[1] G. Goubau, Surface-wave Transmission Lines, *Proc. IRE*, vol. 39, no. 6, pp. 619–624, June, 1951.

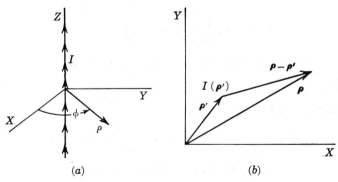

FIG. 5-13. An infinite filament of constant a-c current (a) along the z axis and (b) displaced parallel to the z axis.

of ϕ and z. To represent outward-traveling waves, we choose
$$A_z = \psi = CH_0^{(2)}(k\rho)$$
where C is a constant to be determined according to
$$\lim_{\rho \to 0} \oint H_\phi \rho \, d\phi = I$$
Evaluating $\mathbf{H} = \nabla \times \mathbf{A}$, we find
$$H_\phi = -\frac{\partial \psi}{\partial \rho} = -C\frac{\partial}{\partial \rho}[H_0^{(2)}(k\rho)] \xrightarrow[k\rho \to 0]{} \frac{j2C}{\pi\rho}$$
The preceding equation then yields
$$C = \frac{I}{4j}$$
Hence,
$$A_z = \psi = \frac{I}{4j} H_0^{(2)}(k\rho) \tag{5-83}$$
is the desired solution. The line current is the elemental two-dimensional source, just as the current element (Sec. 2-9) is the elemental three-dimensional source.

The electromagnetic field is obtained from Eqs. (5-18), using the ψ of Eq. (5-83). The result is
$$E_z = \frac{-k^2 I}{4\omega\epsilon} H_0^{(2)}(k\rho) \qquad H_\phi = \frac{-kI}{4j} H_0^{(2)\prime}(k\rho) \tag{5-84}$$
Thus, lines of electric intensity run parallel to the current, and lines of magnetic intensity encircle it. Equiphase surfaces are cylinders, but $\mathbf{E}$ and $\mathbf{H}$ are not in general in phase. However, at large distances we have
$$\left. \begin{aligned} E_z &= -\eta kI \sqrt{\frac{j}{8\pi k\rho}} e^{-jk\rho} \\ H_\phi &= kI \sqrt{\frac{j}{8\pi k\rho}} e^{-jk\rho} \end{aligned} \right\} \qquad \rho \gg \lambda \tag{5-85}$$
which is essentially an outward-traveling plane wave. The amplitude of the wave decreases as $\rho^{-1/2}$, in contrast to the r^{-1} variation in the three-dimensional case. The outward-directed complex power crossing a cylinder of unit length and radius ρ is
$$P_f = \oiint \mathbf{E} \times \mathbf{H}^* \cdot d\mathbf{s} = -\int_0^{2\pi} E_z H_\phi^* \rho \, d\phi$$
$$= \frac{\eta\pi\rho}{8j} |kI|^2 H_0^{(2)}(k\rho)[H_0^{(2)\prime}(k\rho)]^* \tag{5-86}$$
The real part of this is the time-average power flow $\bar{\mathcal{P}}_f$, which, by virtue

of the Wronskian [Eq. (D-17)], reduces to

$$\bar{\mathcal{P}}_f = \operatorname{Re}(P_f) = \frac{\eta k}{4}|I|^2 \qquad (5\text{-}87)$$

Hence, the time-average power is independent of the distance from the source, as we should expect. It could be more simply obtained from Eqs. (5-85).

If the current filament is not along the z axis but parallel to it, we can extend Eq. (5-83) by replacing ρ by the distance from the current to the field point. In radius vector notation, we specify the field point by

$$\boldsymbol{\varrho} = \mathbf{u}_x x + \mathbf{u}_y y$$

and the source point (current filament) by

$$\boldsymbol{\varrho}' = \mathbf{u}_x x' + \mathbf{u}_y y'$$

as shown in Fig. 5-13b. The distance from the source point to the field point is then

$$|\boldsymbol{\varrho} - \boldsymbol{\varrho}'| = \sqrt{(x - x')^2 + (y - y')^2}$$
$$= \sqrt{\rho^2 + \rho'^2 - 2\rho\rho' \cos(\phi - \phi')}$$

We emphasize that A_z is evaluated at $\boldsymbol{\varrho}$ by writing $A_z(\boldsymbol{\varrho})$ and that I is located at $\boldsymbol{\varrho}'$ by writing $I(\boldsymbol{\varrho}')$. We can now generalize Eq. (5-83) to read

$$A_z(\boldsymbol{\varrho}) = \frac{I(\boldsymbol{\varrho}')}{4j} H_0^{(2)}(k|\boldsymbol{\varrho} - \boldsymbol{\varrho}'|) \qquad (5\text{-}88)$$

This is our free-space Green's function for two-dimensional fields.

The solution for two or more filaments of z-directed current can be represented by a summation of the A_z's from each current element. Suppose we have two filaments of equal magnitude but opposite phase, as represented by Fig. 5-14a. As the separation $s \to 0$ and the magnitude $I \to \infty$ such that Is remains constant, we have a two-dimensional dipole

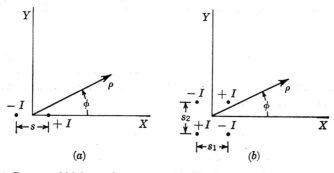

FIG. 5-14. Sources of higher-order waves. (a) Dipole source; (b) quadrupole source.

source. Note that A_z at a point (x,y) due to a current filament at $(x',0)$ is the same as A_z at $(x - x', y)$ due to a current filament at $(0,0)$. Hence, for Fig. 5-14a, the vector potential is

$$A_z = A_z{}^1\left(x - \frac{s}{2}, y\right) - A_z{}^1\left(x + \frac{s}{2}, y\right)$$

where $A_z{}^1$ is that due to a single current filament at the origin [Eq. (5-83)]. In the limit $s \to 0$ the above equation becomes

$$A_z \xrightarrow[s \to 0]{} -s\frac{\partial A_z{}^1}{\partial x} = -\frac{Is}{4j}\frac{\partial}{\partial x}[H_0{}^{(2)}(k\rho)]$$

The differentiation yields

$$A_z = \frac{kIs}{4j} H_1{}^{(2)}(k\rho) \cos \phi \tag{5-89}$$

Thus, the vector potential of a dipole line source is a cylindrical wave function of order $n = 1$.

For the quadrupole source of Fig. 5-14b we have, by reasoning similar to that above,

$$A_z \xrightarrow[\substack{s_1 \to 0 \\ s_2 \to 0}]{} s_1 s_2 \frac{\partial^2 A_z{}^1}{\partial x \, \partial y} = -s_2 \frac{\partial A_z{}^{(2)}}{\partial y}$$

where $A_z{}^{(2)}$ is the vector potential of the dipole source, given by Eq. (5-89). Hence,

$$A_z = \frac{-kIs_1s_2}{4j} \frac{\partial}{\partial y}[H_1{}^{(2)}(k\rho) \cos \phi]$$

which reduces to
$$A_z = \frac{k^2 I s_1 s_2}{8j} H_2{}^{(2)}(k\rho) \sin 2\phi \tag{5-90}$$

Thus, the vector potential of a quadrupole line source is a wave function of order $n = 2$.

This procedure can be continued to obtain sources for the higher-order wave functions. It can be shown (Prob. 5-29) that, when A_z is a wave function of order n, a possible source consists of $2n$ current filaments equispaced on an infinitesimal cylinder. We shall call such a source a *multipole source of order n*. The dual analysis applies to the case of magnetic current filaments. It is merely necessary to replace I by K and **A** by **F** in the various vector-potential formulas of this section. For example, from Eq. (5-88), the electric vector potential at ϱ due to a magnetic current filament at ϱ' is

$$F_z(\varrho) = \frac{K(\varrho')}{4j} H_0{}^{(2)}(k|\varrho - \varrho'|) \tag{5-91}$$

Using both electric and magnetic multipoles, we can generate an arbitrary source-free field in homogeneous space ($\rho > 0$).

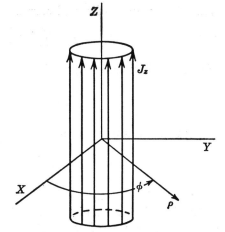

Fig. 5-15. A cylinder of uniform current.

The field due to a cylinder of currents can be obtained quite simply by treating the problem as a boundary-value problem. We shall consider here only a cylinder of uniform z-directed surface current. (The general case is considered in Prob. 5-30.) The geometry of the problem is illustrated by Fig. 5-15. Because of the rotational symmetry, we choose

$$\psi = \begin{cases} A_z^- = C_1 J_0(k\rho) & \rho < a \\ A_z^+ = C_2 H_0^{(2)}(k\rho) & \rho > a \end{cases}$$

The boundary conditions to be satisfied are

$$E_z^+ = E_z^- \qquad H_\phi^+ - H_\phi^- = J_z$$

where J_z is the density of the z-directed current sheet. Using Eqs. (5-18) with the above ψ, and satisfying the boundary conditions, we obtain

$$E_z = \begin{cases} -\dfrac{\pi}{2}\eta k a J_z H_0^{(2)}(ka) J_0(k\rho) & \rho < a \\ -\dfrac{\pi}{2}\eta k a J_z J_0(ka) H_0^{(2)}(k\rho) & \rho > a \end{cases} \qquad (5\text{-}92)$$

as the only component of **E**. Let us calculate an impedance per unit length for this source, as we did for the ribbon of current in Sec. 4-12. By definition,

$$Z = \frac{P}{|I|^2}$$

where P is the complex power per unit length

$$P = -\int_0^{2\pi} E_z J_z^* a \, d\phi = -2\pi a J_z^* E_z \Big|_{\rho=a}$$

and I is the total z-directed current

$$I = \int_0^{2\pi} J_z a \, d\phi = 2\pi a J_z$$

Hence, the impedance per unit length is

$$Z = \frac{\eta k}{4} J_0(ka) H_0^{(2)}(ka) \qquad (5\text{-}93)$$

Using small-argument formulas for J_0 and $H_0^{(2)}$, we obtain

$$Z \xrightarrow[ka \to 0]{} \frac{\eta}{2\lambda}\left(\pi - j2 \log \frac{\gamma ka}{2}\right) \qquad (5\text{-}94)$$

where $\gamma = 1.781$. Compare this with the Z of a ribbon of current [Eq. (4-127)]. The resistances (real parts) are identical. The reactance of a cylinder of current of small diameter d is approximately equal to the reactance of a ribbon of current of width $w = 2d$. More generally, it can be shown[1] by a quasi-static approximation that the impedance per unit length of a small elliptic cylinder of minor axis a and major axis b is the same as that of a circular cylinder of diameter

$$d = \tfrac{1}{2}(a + b)$$

A ribbon is the special case $a = 0$ and $b = w$.

5-7. Two-dimensional Radiation. We can construct the solution for an arbitrary two-dimensional distribution of currents by dividing the source into elemental filaments of current and summing the fields from all elements. For example, if we have a J_z, independent of z, each element $J_z \, ds'$ produces a vector potential

$$dA_z = \frac{J_z \, ds'}{4j} H_0^{(2)}(k|\varrho - \varrho'|)$$

where ds' is an element of area perpendicular to z. Summing over the entire source, we have

$$A_z = \frac{1}{4j} \iint J_z(\varrho') H_0^{(2)}(k|\varrho - \varrho'|) \, ds'$$

where the integration extends over a cross section of the source. Since the equations for A_x due to J_x and for A_y due to J_y are of the same form as those for A_z due to J_z, the above equation also applies for z replaced by x or y. Combining components, we have the vector equation

$$\mathbf{A}(\varrho) = \frac{1}{4j} \iint \mathbf{J}(\varrho') H_0^{(2)}(k|\varrho - \varrho'|) \, ds' \qquad (5\text{-}95)$$

[1] R. W. P. King, "The Theory of Linear Antennas," pp. 16–20, Harvard University Press, Cambridge, Mass., 1956.

representing the solution for an arbitrary two-dimensional distribution of electric currents. The cases of surface currents and current filaments are included by implication. The electromagnetic field is obtained, as usual, from $\mathbf{H} = \nabla \times \mathbf{A}$. The electric vector potential due to two-dimensional magnetic currents $\mathbf{M}$ is given by the formula dual to Eq. (5-95), or

$$\mathbf{F}(\varrho) = \frac{1}{4j} \iint \mathbf{M}(\varrho') H_0^{(2)}(k|\varrho - \varrho'|) \, ds' \qquad (5\text{-}96)$$

The electromagnetic field in this case is given by $\mathbf{E} = -\nabla \times \mathbf{F}$.

When the field point is distant from the source, our formulas simplify to a form similar to those for three-dimensional radiation (Sec. 3-13). For $k|\varrho - \varrho'|$ large, the Hankel function can be represented by the asymptotic formula

$$H_0^{(2)}(k|\varrho - \varrho'|) \rightarrow \sqrt{\frac{2j}{\pi k |\varrho - \varrho'|}} \, e^{-jk|\varrho - \varrho'|}$$

Furthermore, when $\rho \gg \rho'$, as shown in Fig. 5-16, we have

$$|\varrho - \varrho'| \rightarrow \rho - \rho' \cos(\phi - \phi') \qquad (5\text{-}97)$$

The second term must be retained in the phase factor, $\exp(-jk|\varrho - \varrho'|)$, but not in the magnitude factor, $|\varrho - \varrho'|^{-\frac{1}{2}}$. Hence, the vector potentials of Eqs. (5-95) and (5-96) reduce to

$$\begin{aligned}\mathbf{A} &= \frac{e^{-jk\rho}}{\sqrt{8j\pi k \rho}} \iint \mathbf{J}(\varrho') e^{jk\rho' \cos(\phi - \phi')} \, ds' \\ \mathbf{F} &= \frac{e^{-jk\rho}}{\sqrt{8j\pi k \rho}} \iint \mathbf{M}(\varrho') e^{jk\rho' \cos(\phi - \phi')} \, ds'\end{aligned} \qquad (5\text{-}98)$$

provided $\rho \gg \rho'_{\max}$. These are the radiation-zone formulas corresponding to Eqs. (3-95) in the three-dimensional case.

Fig. 5-16. Geometry for determining the radiation field.

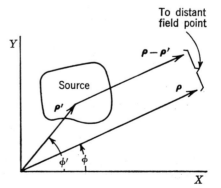

We now have the ρ variation explicitly shown in Eqs. (5-98), and simplified formulas for the radiation field can be obtained. As evidenced by Eq. (5-85), the distant field of a single current filament is essentially an outward-traveling plane wave; so the superposition of fields from all current elements should also be of this type. Hence, in the radiation zone,

$$E_\phi = \eta H_z \qquad E_z = -\eta H_\phi \qquad (5\text{-}99)$$

which can be verified by direct expansion of Eqs. (3-4), using Eqs. (5-98). To obtain the field components, let us again divide the field into that due to **J**, given by $\mathbf{H}' = \nabla \times \mathbf{A}$, and that due to **M**, given by $\mathbf{E}'' = -\nabla \times \mathbf{F}$. Retaining only the dominant terms ($\rho^{-1/2}$ variation), we obtain

$$H'_\phi = jkA_z \qquad E''_\phi = -jkF_z$$
$$H'_z = -jkA_\phi \qquad E''_z = jkF_\phi$$

in the radiation zone. The corresponding E'_ϕ, E'_z, H''_ϕ, and H''_z can be determined from Eqs. (5-99). The total field is simply the sum of the primed and double-primed components, or

$$E_\phi = -j\omega\mu A_\phi - jkF_z$$
$$E_z = -j\omega\mu A_z + jkF_\phi \qquad (5\text{-}100)$$

in the radiation zone, with **H** given by Eqs. (5-99). These formulas correspond to Eqs. (3-97) in the three-dimensional case. Note that, except for the contrasting $\rho^{-1/2}$ and r^{-1} dependences, the radiation fields are of similar mathematical forms in two and three dimensions.

5-8. Wave Transformations. It is often convenient to express the elementary wave functions of one coordinate system in terms of those of another coordinate system.[1] We refer to expressions of this type as *wave transformations*. Some representative wave transformations are derived in this section. Others will be derived as they are needed.

Suppose we have the plane wave e^{-jx}, which we wish to express in terms of cylindrical waves. (The conventional coordinate orientation of Fig. 5-1 is assumed.) This wave is finite at the origin and periodic in 2π on ϕ. Hence, it must be expressible as

$$e^{-jx} = e^{-j\rho\cos\phi} = \sum_{n=-\infty}^{\infty} a_n J_n(\rho) e^{jn\phi}$$

where the a_n are constants. To evaluate the a_n, multiply each side by $e^{-jm\phi}$ and integrate from 0 to 2π on ϕ. This gives

$$\int_0^{2\pi} e^{-j\rho\cos\phi} e^{-jm\phi} \, d\phi = 2\pi a_m J_m(\rho)$$

---

[1] Two coordinate systems are considered to be distinct if their origins or orientations are different, even though they may be geometrically the same.

The left-hand side is actually a well-known integral, but we need not recognize this. The mth derivative of the left-hand side with respect to ρ evaluated at $\rho = 0$ is

$$j^{-m} \int_0^{2\pi} \cos^m \phi \, e^{-jm\phi} \, d\phi = \frac{2\pi j^{-m}}{2^m}$$

The mth derivative of the right-hand side evaluated at $\rho = 0$ is $2\pi a_m/2^m$. Hence,

$$a_m = j^{-m}$$

and we have shown that

$$e^{-jx} = e^{-j\rho \cos \phi} = \sum_{n=-\infty}^{\infty} j^{-n} J_n(\rho) e^{jn\phi} \quad (5\text{-}101)$$

and also that

$$J_n(\rho) = \frac{j^n}{2\pi} \int_0^{2\pi} e^{-j\rho \cos \phi} e^{-jn\phi} \, d\phi \quad (5\text{-}102)$$

Equation (5-101) is the wave transformation expressing the plane wave e^{-jx} in terms of cylindrical wave functions. It is closely related to the so-called "generating function" of Bessel functions.[1]

Another wave transformation of interest is that which corresponds to a translation of cylindrical coordinate origin. Consider the wave function

$$\psi = H_0^{(2)}(|\varrho - \varrho'|) = H_0^{(2)}[\sqrt{\rho^2 + \rho'^2 - 2\rho\rho' \cos(\phi - \phi')}]$$

where ρ and ρ' are as defined in Fig. 5-13b. We can think of ψ as the field of a line source at ρ' in terms of a cylindrical wave function having its origin at the source. We shall reexpress ψ in terms of wave functions referred to $\rho = 0$. In the region $\rho < \rho'$, permissible wave functions are $J_n(\rho) e^{jn\phi}$, n an integer, for ψ is finite at $\rho = 0$ and periodic in 2π on ϕ. In the region $\rho > \rho'$, permissible wave functions are $H_n^{(2)}(\rho) e^{jn\phi}$, n an integer, for ψ must represent outward-traveling waves. Also, ψ must be symmetric in primed and unprimed coordinates (reciprocity). Hence, ψ is of the form

$$\psi = \begin{cases} \sum_{n=-\infty}^{\infty} b_n H_n^{(2)}(\rho') J_n(\rho) e^{jn(\phi-\phi')} & \rho < \rho' \\ \sum_{n=-\infty}^{\infty} b_n J_n(\rho') H_n^{(2)}(\rho) e^{jn(\phi-\phi')} & \rho > \rho' \end{cases}$$

where the b_n are constants. To evaluate the b_n, let $\rho' \to \infty$ and $\phi' = 0$, and use the asymptotic formulas for the Hankel functions. Our original

[1] R. V. Churchill, "Fourier Series and Boundary Value Problems," p. 147, McGraw-Hill Book Company, Inc., New York, 1941.

expression for ψ then becomes

$$\psi = H_0^{(2)}(|\varrho - \varrho'|) \xrightarrow[\substack{\rho' \to \infty \\ \phi' = 0}]{} \sqrt{\frac{2j}{\pi \rho'}} \, e^{-j\rho'} e^{j\rho \cos \phi}$$

and our constructed expression for ψ becomes

$$\psi \xrightarrow[\substack{\rho' \to \infty \\ \phi' = 0}]{} \sqrt{\frac{2j}{\pi \rho'}} \, e^{-j\rho'} \sum_{n=-\infty}^{\infty} b_n j^n J_n(\rho) e^{jn\phi}$$

These are now representations of a plane wave, and, from Eq. (5-101), it follows that $b_n = 1$. Thus,

$$H_0^{(2)}(|\varrho - \varrho'|) = \begin{cases} \sum_{n=-\infty}^{\infty} H_n^{(2)}(\rho') J_n(\rho) e^{jn(\phi - \phi')} & \rho < \rho' \\ \sum_{n=-\infty}^{\infty} J_n(\rho') H_n^{(2)}(\rho) e^{jn(\phi - \phi')} & \rho > \rho' \end{cases} \quad (5\text{-}103)$$

This equation is known as the *addition theorem* for Hankel functions. It is also valid for superscripts (2) replaced by superscripts (1), since $H_n^{(1)} = H_n^{(2)*}$. Adding the addition theorem for $H_0^{(2)}$ to that for $H_0^{(1)}$, we obtain

$$J_0(|\varrho - \varrho'|) = \sum_{n=-\infty}^{\infty} J_n(\rho') J_n(\rho) e^{jn(\phi - \phi')} \quad (5\text{-}104)$$

which is the addition theorem for Bessel functions of the first kind. An addition theorem for Bessel functions of the second kind is obtained by subtracting that for $H_0^{(2)}$ from that for $H_0^{(1)}$.

5-9. Scattering by Cylinders. A source radiating in the presence of a conducting cylinder is one of the simplest "wave-scatter" problems for which an exact solution can be obtained. We shall at present consider only two-dimensional cases. Extension to three-dimensional cases can be effected by the method of Sec. 5-12.

Let us first consider a plane wave incident upon a conducting cylinder, as represented by Fig. 5-17. Take the incident wave to be z-polarized, that is,

$$E_z^i = E_0 e^{-jkx} = E_0 e^{-jk\rho \cos \phi} \quad (5\text{-}105)$$

Using the wave transformation of Eq. (5-101), we can express the incident field as

$$E_z^i = E_0 \sum_{n=-\infty}^{\infty} j^{-n} J_n(k\rho) e^{jn\phi}$$

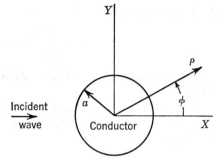

Fig. 5-17. A plane wave incident upon a conducting cylinder.

The total field with the conducting cylinder present is the sum of the incident and scattered fields, that is,

$$E_z = E_z{}^i + E_z{}^s$$

To represent outward-traveling waves, the scattered field must be of the form

$$E_z{}^s = E_0 \sum_{n=-\infty}^{\infty} j^{-n} a_n H_n^{(2)}(k\rho) e^{jn\phi} \qquad (5\text{-}106)$$

hence the total field is

$$E_z = E_0 \sum_{n=-\infty}^{\infty} j^{-n} [J_n(k\rho) + a_n H_n^{(2)}(k\rho)] e^{jn\phi} \qquad (5\text{-}107)$$

At the cylinder the boundary condition $E_z = 0$ at $\rho = a$ must be met. It is evident from the above equation that this condition is met if

$$a_n = \frac{-J_n(ka)}{H_n^{(2)}(ka)} \qquad (5\text{-}108)$$

which completes the solution.

The surface current on the cylinder may be obtained from

$$J_z = H_\phi \bigg|_{\rho=a} = \frac{1}{j\omega\mu} \frac{\partial E_z}{\partial \rho} \bigg|_{\rho=a}$$

Using Eqs. (5-107) and (5-108), and simplifying the result by Eq. (D-17), we obtain

$$J_z = \frac{-2E_0}{\omega\mu a} \sum_{n=-\infty}^{\infty} \frac{j^{-n} e^{jn\phi}}{H_n^{(2)}(ka)} \qquad (5\text{-}109)$$

In a thin wire the $n = 0$ term becomes dominant, and we have essentially a filament of current. Using the small-argument formula for $H_0^{(2)}$, we find the total current as

$$I = \int_0^{2\pi} J_z a \, d\phi = \frac{2\pi E_0}{j\omega\mu \log ka} \tag{5-110}$$

Hence, the current in a thin wire is 90° out of phase with the incident field.

The pattern of the scattered field is also of interest. At large distances from the cylinder we can use the asymptotic formulas for $H_n^{(2)}$, and Eq. (5-106) becomes

$$E_z^s \xrightarrow[k\rho \to \infty]{} E_0 \sqrt{\frac{2j}{\pi k\rho}} e^{-jk\rho} \sum_{n=-\infty}^{\infty} a_n e^{jn\phi}$$

where the a_n are given by Eq. (5-108). The magnitude of the ratio of the scattered field to the incident field is therefore

$$\frac{|E_z^s|}{|E_z^i|} = \sqrt{\frac{2}{\pi k\rho}} \left| \sum_{n=-\infty}^{\infty} \frac{J_n(ka)}{H_n^{(2)}(ka)} e^{jn\phi} \right| \tag{5-111}$$

This is the scattered-field pattern. For small ka, the $n = 0$ term becomes dominant and

$$\frac{|E_z^s|}{|E_z^i|} \xrightarrow[ka \to 0]{} \frac{\pi}{2 \log ka} \sqrt{\frac{2}{\pi k\rho}} \tag{5-112}$$

The scattered-field pattern for a thin wire is therefore a circle, which is to be expected, since the wire is essentially a filament of current.

When the incident field is polarized transversely to z, it can be expressed as

$$H_z^i = H_0 e^{-jkx} = H_0 \sum_{n=-\infty}^{\infty} j^{-n} J_n(k\rho) e^{jn\phi} \tag{5-113}$$

Again, the total field is considered as the sum of the incident and reflected fields, that is,

$$H_z = H_z^i + H_z^s$$

To represent outward-traveling waves, the scattered field is of the form

$$H_z^s = H_0 \sum_{n=-\infty}^{\infty} j^{-n} b_n H_n^{(2)}(k\rho) e^{jn\phi}$$

and the total field is given by

$$H_z = H_0 \sum_{n=-\infty}^{\infty} j^{-n} [J_n(k\rho) + b_n H_n^{(2)}(k\rho)] e^{jn\phi} \tag{5-114}$$

CYLINDRICAL WAVE FUNCTIONS

This time our boundary condition is $E_\phi = 0$ at $\rho = a$. From the field equations

$$E_\phi = \frac{1}{j\omega\epsilon}(\nabla \times \mathbf{u}_z H_z)_\phi$$

$$= \frac{jk}{\omega\epsilon} H_0 \sum_{n=0}^{\infty} j^{-n}[J_n'(k\rho) + b_n H_n^{(2)\prime}(k\rho)]e^{jn\phi}$$

and the boundary condition is met if

$$b_n = \frac{-J_n'(ka)}{H_n^{(2)\prime}(ka)} \tag{5-115}$$

An incident wave of arbitrary polarization can be treated as a superposition of Eqs. (5-105) and (5-113).

When the incident wave is polarized transversely to z, the surface current on the cylinder is

$$J_\phi = H_z\bigg|_{\rho=a} = \frac{j2H_0}{\pi ka} \sum_{n=-\infty}^{\infty} \frac{j^{-n}e^{jn\phi}}{H_n^{(2)\prime}(ka)} \tag{5-116}$$

For small ka, the $n = 0$ term becomes dominant. However, the $n = \pm 1$ terms radiate more efficiently and cannot be neglected, as we shall now show. At large distances from the cylinder, the scattered field becomes

$$H_z^s \xrightarrow[k\rho \to \infty]{} H_0 \sqrt{\frac{2j}{\pi k\rho}} e^{-jk\rho} \sum_{n=-\infty}^{\infty} b_n e^{jn\phi}$$

with b_n given by Eq. (5-115). The magnitude of the ratio of the scattered to incident field is thus

$$\frac{|H_z^s|}{|H_z^i|} = \sqrt{\frac{2}{\pi k\rho}} \left|\sum_{n=-\infty}^{\infty} \frac{J_n'(ka)}{H_n^{(2)\prime}(ka)} e^{jn\phi}\right| \tag{5-117}$$

For small ka we find

$$\frac{J_n'(ka)}{H_n^{(2)\prime}(ka)} = \begin{cases} -\dfrac{j\pi(ka)^2}{4} & n = 0 \\ \dfrac{j\pi(ka)^2}{4} & |n| = 1 \\ \dfrac{j\pi(ka/2)^{2|n|}}{|n|!(|n|-1)!} & |n| > 1 \end{cases}$$

Hence, for thin wires the scattered-field pattern is

$$\frac{|H_z^s|}{|H_z^i|} \xrightarrow[ka \to 0]{} \frac{\pi(ka)^2}{4} \sqrt{\frac{2}{\pi k\rho}} |1 - 2\cos\phi| \tag{5-118}$$

The $n = 0$ term of Eq. (5-116) is equivalent to a z-directed magnetic

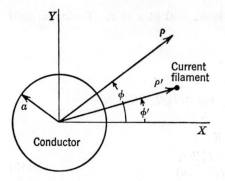

Fig. 5-18. A current filament parallel to a conducting cylinder.

current filament, while the $n = \pm 1$ terms are equivalent to a y-directed electric dipole.

A more general problem is that of a current filament parallel to a conducting cylinder, as shown in Fig. 5-18. (Plane-wave incidence is the special case $\rho' \to \infty$.) When the filament is an electric current I, the incident field is

$$E_z{}^i = \frac{-k^2 I}{4\omega\epsilon} H_0^{(2)}(k|\varrho - \varrho'|) \tag{5-119}$$

For $\rho < \rho'$ we have, by the addition theorem [Eq. (5-103)],

$$E_z{}^i = \frac{-k^2 I}{4\omega\epsilon} \sum_{n=-\infty}^{\infty} H_n^{(2)}(k\rho') J_n(k\rho) e^{jn(\phi-\phi')}$$

To this we must add a scattered field of the same form, but with the J_n replaced by $H_n^{(2)}$, namely,

$$E_z{}^s = \frac{-k^2 I}{4\omega\epsilon} \sum_{n=-\infty}^{\infty} c_n H_n^{(2)}(k\rho') H_n^{(2)}(k\rho) e^{jn(\phi-\phi')} \tag{5-120}$$

From the preceding two equations it is evident that

$$c_n = -\frac{J_n(ka)}{H_n^{(2)}(ka)} \tag{5-121}$$

satisfies the boundary condition $E_z = E_z{}^i + E_z{}^s = 0$. Thus, our final solution is

$$E_z = \begin{cases} \dfrac{-k^2 I}{4\omega\epsilon} \displaystyle\sum_{n=-\infty}^{\infty} H_n^{(2)}(k\rho')[J_n(k\rho) + c_n H_n^{(2)}(k\rho)] e^{jn(\phi-\phi')} & \rho < \rho' \\ \dfrac{-k^2 I}{4\omega\epsilon} \displaystyle\sum_{n=-\infty}^{\infty} H_n^{(2)}(k\rho)[J_n(k\rho') + c_n H_n^{(2)}(k\rho')] e^{jn(\phi-\phi')} & \rho > \rho' \end{cases}$$

$$\tag{5-122}$$

Note that our answer is symmetrical in ρ, ϕ and ρ', ϕ' (reciprocity). Note also that the "reflection coefficients" of Eq. (5-121) are equal to those of Eq. (5-108) and are, in general, independent of the incident field.

Specializing the second of Eqs. (5-122) to the far zone, we have

$$E_z \xrightarrow[k\rho \to \infty]{} f(\rho) \sum_{n=-\infty}^{\infty} j^n \left[J_n(k\rho') - \frac{J_n(ka)}{H_n^{(2)}(ka)} H_n^{(2)}(k\rho') \right] e^{jn(\phi-\phi')}$$

The magnitude of this is the radiation field pattern. Figure 5-19 shows the radiation pattern of a current filament 0.25λ away from a conducting cylinder of radius 3.75λ. The radiation pattern of a current filament 0.25λ in front of a plane reflector is shown for comparison. The patterns of Fig. 5-19 are also valid for current elements of finite length as long as the reflector is of infinite extent.

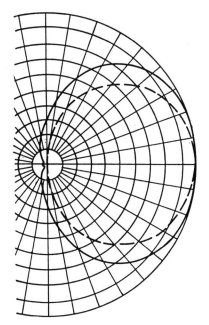

FIG. 5-19. Radiation pattern for a current filament 0.25λ away from a cylindrical reflector of radius 3.75λ (plane reflector case shown dashed).

If the line source of Fig. 5-18 is a magnetic current filament K, we have

$$H_z{}^i = \frac{-k^2 K}{4\omega\mu} H_0^{(2)}(k|\varrho - \varrho'|)$$

instead of Eq. (5-119). The problem is dual to the electric current case, except that the reflection coefficients at the conducting cylinder must be those of Eq. (5-115) instead of those of Eq. (5-121). Therefore, the final solution will be dual to Eq. (5-122), or

$$H_z = \begin{cases} \dfrac{-k^2 K}{4\omega\mu} \displaystyle\sum_{n=-\infty}^{\infty} H_n^{(2)}(k\rho')[J_n(k\rho) + b_n H_n^{(2)}(k\rho)]e^{jn(\phi-\phi')} & \rho < \rho' \\[2ex] \dfrac{-k^2 K}{4\omega\mu} \displaystyle\sum_{n=-\infty}^{\infty} H_n^{(2)}(k\rho)[J_n(k\rho') + b_n H_n^{(2)}(k\rho')]e^{jn(\phi-\phi')} & \rho > \rho' \end{cases}$$

(5-123)

where the b_n are given by Eq. (5-115). According to the equivalence principle, the field of a narrow slot in a conducting cylinder is the same as the field of a magnetic current on the surface of a conducting cylinder. Specializing the second of Eqs. (5-123) to the case $\rho' = a$, $\phi' = 0$, $\rho \to \infty$, we have

$$H_z = f(\rho) \sum_{n=-\infty}^{\infty} \frac{j^n e^{jn\phi}}{H_n^{(2)'}(ka)}$$

The magnitude of this is the radiation pattern of a "slitted cylinder." Figure 5-20 shows a slitted-cylinder pattern for the case $a = 2\lambda$. The pattern for a slit in an infinite ground plane is shown for comparison. The patterns of Fig. 5-20 are also valid for slits of finite length as long as the conductor is of infinite extent.

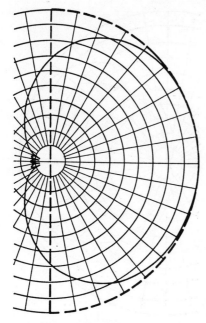

FIG. 5-20. Radiation pattern for a slitted cylinder of radius 2λ (slit in a ground plane shown dashed).

5-10. Scattering by Wedges. A source radiating in the presence of a conducting wedge is also a relatively simple problem.[1] We again restrict consideration to the two-dimensional case at this time. We shall solve for the field of current filaments in the vicinity of wedges and obtain solutions for plane-wave illumination and aperture radiation as special cases. A wedge of vanishingly small angle is the classical conducting half-plane problem.

Consider first the case of a filament of electric current at ρ', ϕ' adjacent to a conducting wedge defined by $\phi = \alpha$ and $\phi = 2\pi - \alpha$ (wedge angle = 2α). This is shown in Fig. 5-21. The incident field is given by Eq. (5-119) and has only a z component of **E**. The total field also will have only a z component of **E**, since this is sufficient to satisfy the boundary conditions. We construct

$$E_z = \begin{cases} \sum_v a_v H_v^{(2)}(k\rho') J_v(k\rho) \sin v(\phi' - \alpha) \sin v(\phi - \alpha) & \rho < \rho' \\ \sum_v a_v J_v(k\rho') H_v^{(2)}(k\rho) \sin v(\phi' - \alpha) \sin v(\phi - \alpha) & \rho > \rho' \end{cases}$$

(5-124)

---

[1] Problems involving conductors over *complete* coordinate surfaces are usually easy to solve. In this case the wedge covers two ϕ = constant coordinate surfaces.

which satisfies reciprocity and insures continuity of E_z at $\rho = \rho'$. To satisfy the boundary conditions $E_z = 0$ over $\phi = \alpha$ and $\phi = 2\pi - \alpha$, we choose

$$v = \frac{m\pi}{2(\pi - \alpha)} \qquad m = 1, 2, 3, \ldots \qquad (5\text{-}125)$$

The a_v are determined by the nature of the source.

To evaluate the a_v, we view the current element as an impulse of current on the surface $\rho = \rho'$. The boundary condition to be satisfied at a current sheet is

$$J_z = H_\phi(\rho'+) - H_\phi(\rho'-)$$

Using the field equations and Eq. (5-124), we find

$$H_\phi = \begin{cases} \dfrac{k}{j\omega\mu} \displaystyle\sum_v a_v H_v^{(2)}(k\rho') J_v'(k\rho) \sin v(\phi' - \alpha) \sin v(\phi - \alpha) & \rho < \rho' \\ \dfrac{k}{j\omega\mu} \displaystyle\sum_v a_v J_v(k\rho') H_v^{(2)\prime}(k\rho) \sin v(\phi' - \alpha) \sin v(\phi - \alpha) & \rho > \rho' \end{cases}$$

Thus, using the Wronskian [Eq. (D-17)], we have the surface current given by

$$J_z = \frac{-2}{\omega\mu\pi\rho'} \sum_v a_v \sin v(\phi' - \alpha) \sin v(\phi - \alpha)$$

This is simply a Fourier series for the current on $\rho = \rho'$. The Fourier sine series for an impulsive current of strength I at $\phi = \phi'$ on $\rho = \rho'$ is

$$J_z = \frac{I}{(\pi - \alpha)\rho'} \sum_v \sin v(\phi' - \alpha) \sin v(\phi - \alpha)$$

By comparison of the preceding two equations it is evident that

$$a_v = \frac{-\omega\mu\pi I}{2(\pi - \alpha)} \qquad (5\text{-}126)$$

This completes the solution.

To obtain the radiation pattern of a current I near a wedge, use the asymptotic formula for $H_v^{(2)}(k\rho)$ in the second of Eqs. (5-124). This, with Eq. (5-126), gives

$$E_z \xrightarrow[k\rho \to \infty]{} f(\rho) \sum_v j^v J_v(k\rho')$$
$$\sin v(\phi' - \alpha) \sin v(\phi - \alpha)$$

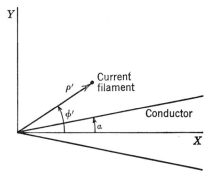

Fig. 5-21. A current filament adjacent to a conducting wedge.

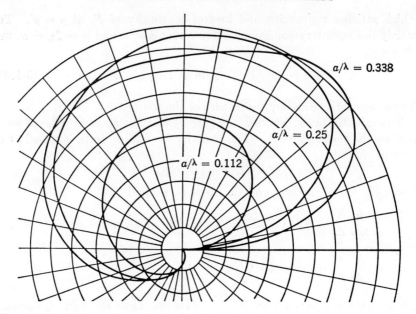

FIG. 5-22. Radiation patterns for an electric current filament adjacent to a conducting half plane, $\rho' = a$, $\phi' = \pi/4$. (*After J. R. Wait.*)

where v is given by Eq. (5-125). Figure 5-22 shows some radiation patterns for the special case $\alpha = 0$ (the conducting half plane).

Another special case of interest is that of plane-wave illumination. This is obtained by letting the source recede to infinity. In this case, the incident field becomes

$$E_z{}^i = \frac{-k^2 I}{4\omega\epsilon} H_0{}^{(2)}(k|\boldsymbol{\varrho} - \boldsymbol{\varrho}'|)$$

$$\xrightarrow[k\rho' \to \infty]{} \frac{-\omega\mu I}{4} \sqrt{\frac{2j}{\pi k \rho'}} \, e^{-jk\rho'} e^{jk\rho \cos(\phi - \phi')}$$

This is recognized as the plane-wave field

$$E_z{}^i = E_0 e^{jk\rho \cos(\phi - \phi')}$$

where
$$E_0 = \frac{-\omega\mu I}{4} \sqrt{\frac{2j}{\pi k \rho'}} \, e^{-jk\rho'} \quad (5\text{-}127)$$

The total field in the vicinity of the wedge is obtained by specializing the first of Eqs. (5-124) to large ρ'. This gives

$$E_z \xrightarrow[k\rho' \to \infty]{} \sqrt{\frac{2j}{\pi k \rho'}} \, e^{-jk\rho'} \sum_v a_v j^v J_v(k\rho) \sin v(\phi' - \alpha) \sin v(\phi - \alpha)$$

Finally, substituting for a_v from Eq. (5-126) and for I from Eq. (5-127), we obtain

$$E_z = \frac{2\pi E_0}{\pi - \alpha} \sum_v j^v J_v(k\rho) \sin v(\phi' - \alpha) \sin v(\phi - \alpha) \qquad (5\text{-}128)$$

where v is given by Eq. (5-125). This is the solution for a plane z-polarized wave incident at the angle ϕ' on a wedge of angle 2α. For $\alpha = 0$ we have

$$E_z = 2E_0 \sum_{n=1}^{\infty} j^{n/2} J_{n/2}(k\rho) \sin \frac{n\phi'}{2} \sin \frac{n\phi}{2} \qquad (5\text{-}129)$$

which is the solution for a plane wave incident on a conducting half plane.

The "almost dual" problem (dual except for boundary conditions) is that of a magnetic-current filament K at ρ', ϕ' in Fig. 5-21. We construct a solution

$$H_z = \begin{cases} \sum_v b_v H_v^{(2)}(k\rho') J_v(k\rho) \cos v(\phi' - \alpha) \cos v(\phi - \alpha) & \rho < \rho' \\ \sum_v b_v J_v(k\rho') H_v^{(2)}(k\rho) \cos v(\phi' - \alpha) \cos v(\phi - \alpha) & \rho > \rho' \end{cases}$$

(5-130)

which is similar to Eq. (5-124) except for the sines replaced by cosines. The boundary conditions $E_\rho = 0$ at $\phi = \alpha$ and $\phi = 2\pi - \alpha$ can now be satisfied by choosing

$$v = \frac{m\pi}{2(\pi - \alpha)} \qquad m = 0, 1, 2, \ldots \qquad (5\text{-}131)$$

The coefficients b_v are determined by the nature of the source, in a manner analogous to that used to obtain Eq. (5-126). The result is

$$b_v = \begin{cases} \dfrac{\omega \epsilon \pi K}{4(\pi - \alpha)} & v = 0 \\ \dfrac{\omega \epsilon \pi K}{2(\pi - \alpha)} & v > 0 \end{cases} \qquad (5\text{-}132)$$

which completes the solution.

The radiation pattern of a magnetic current K near a wedge is obtained from the second of Eqs. (5-130) by using the asymptotic expression for $H_v^{(2)}(k\rho)$. The result is

$$H_z \xrightarrow[k\rho \to \infty]{} f(\rho) \sum_v \epsilon_v j^v J_v(k\rho') \cos v(\phi' - \alpha) \cos v(\phi - \alpha)$$

where Neumann's number ϵ_v is 1 for $v = 0$ and 2 for $v > 0$. Figure 5-23 shows some radiation patterns for the special case $\alpha = 0$. When $\phi' = \alpha$ we have the solution for a radiating slit in a conducting wedge.

Finally, for plane-wave incidence we can specialize the first of Eqs. (5-130) to the case $\rho' \to \infty$. The procedure is analogous to that used to establish Eq. (5-128), and the result is

$$H_z = \frac{\pi H_0}{\pi - \alpha} \sum_v \epsilon_v j^v J_v(k\rho) \cos v(\phi' - \alpha) \cos v(\phi - \alpha) \quad (5\text{-}133)$$

This is the field due to a plane wave polarized orthogonally to z incident at an angle ϕ' on a wedge of angle 2α. The case $\alpha = 0$ gives

$$H_z = H_0 \sum_{n=0}^{\infty} \epsilon_n j^{n/2} J_{n/2}(k\rho) \cos \frac{n\phi'}{2} \cos \frac{n\phi}{2} \quad (5\text{-}134)$$

which is the solution for a plane wave incident on a conducting half plane.

5-11. Three-dimensional Radiation. A three-dimensional problem having cylindrical boundaries can be reduced to a two-dimensional problem by applying a Fourier transformation with respect to z (the cylinder

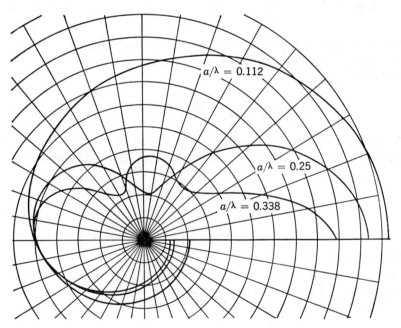

Fig. 5-23. Radiation patterns for a magnetic current filament adjacent to a conducting half plane, $\rho' = a$, $\phi' = \pi/4$. (*After J. R. Wait.*)

CYLINDRICAL WAVE FUNCTIONS

axis).[1] For example, if $\psi(x,y,z)$ is a solution to the three-dimensional wave equation

$$\left(\frac{\partial^2}{\partial x^2} + \frac{\partial^2}{\partial y^2} + \frac{\partial^2}{\partial z^2} + k^2\right)\psi = 0$$

then

$$\bar\psi(x,y,w) = \int_{-\infty}^{\infty} \psi(x,y,z)e^{-jwz}\,dz$$

will be a solution to the two-dimensional wave equation

$$\left(\frac{\partial^2}{\partial x^2} + \frac{\partial^2}{\partial y^2} + \kappa^2\right)\bar\psi = 0$$

where $\kappa^2 = k^2 - w^2$. Once the two-dimensional problem for $\bar\psi$ is solved, the three-dimensional solution is obtained from the inversion

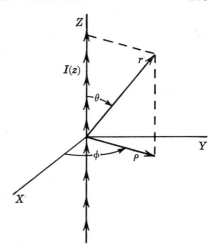

FIG. 5-24. A filament of current along the z axis.

$$\psi(x,y,z) = \frac{1}{2\pi}\int_{-\infty}^{\infty} \bar\psi(x,y,w)e^{jwz}\,dw$$

This is usually a difficult operation. Fortunately, in the radiation zone the inversion becomes quite simple. We shall now obtain this far-zone inversion formula.

Consider the problem of a filament of z-directed current along the z axis, as illustrated by Fig. 5-24. The only restriction placed on the current $I(z)$ is that it be Fourier-transformable. In the usual way, we construct a solution

$$\mathbf{H} = \nabla \times \mathbf{A} \qquad \mathbf{A} = \mathbf{u}_z\psi \qquad (5\text{-}135)$$

where ψ is a wave function independent of ϕ and representing outward-traveling waves at large ρ. Anticipating the need for Fourier transforms, we construct

$$\psi = \frac{1}{2\pi}\int_{-\infty}^{\infty} f(w)H_0^{(2)}(\rho\sqrt{k^2 - w^2})e^{jwz}\,dw$$

which is of the general form of Eq. (5-11). The Fourier transform of ψ is evidently

$$\bar\psi = f(w)H_0^{(2)}(\rho\sqrt{k^2 - w^2})$$

The $f(w)$ is determined by the nature of the source, according to

$$\int_0^{2\pi} \bar H_\phi \rho\,d\phi \xrightarrow[\rho \to 0]{} \bar I(w)$$

[1] This applies to cylinders of arbitrary cross section as well as to circular cylinders.

where $\bar{H}_\phi$ and $\bar{I}$ are the transforms of H_ϕ and I. From the small-argument formula for $H_0^{(2)}$, we have

$$\bar{H}_\phi = -\frac{\partial \bar{\psi}}{\partial \rho} \xrightarrow[\rho \to 0]{} \frac{2j}{\pi \rho} f(w)$$

and the preceding equation yields

$$f(w) = \frac{\bar{I}(w)}{4j}$$

Hence, the "transform solution" to the problem of Fig. 5-24 is

$$\psi = \frac{1}{8\pi j} \int_{-\infty}^{\infty} \bar{I}(w) H_0^{(2)}(\rho \sqrt{k^2 - w^2}) e^{jwz}\, dw \qquad (5\text{-}136)$$

where

$$\bar{I}(w) = \int_{-\infty}^{\infty} I(z') e^{-jwz'}\, dz' \qquad (5\text{-}137)$$

The field is obtained from ψ according to Eqs. (5-135). Compare the equations of this paragraph to those of the second paragraph of Sec. 5-6. The transformed equations in the three-dimensional problem are of the same form as the equations in the two-dimensional problem.

Another solution to the problem of Fig. 5-24 is the "potential integral solution" of Sec. 2-9. This is

$$\psi = \int_{-\infty}^{\infty} I(z') \frac{e^{-jk\sqrt{\rho^2 + (z-z')^2}}}{4\pi \sqrt{\rho^2 + (z-z')^2}}\, dz' \qquad (5\text{-}138)$$

with the field given by Eqs. (5-135). It can be shown that the ψ is unique in this problem. Hence, Eqs. (5-136) and (5-138) are equal, giving us a mathematical identity. For example, if $I(z)$ is a short current element of moment Il, then $\bar{I}(w) = Il$ and Eq. (5-136) becomes

$$\psi = \frac{Il}{8\pi j} \int_{-\infty}^{\infty} H_0^{(2)}(\rho \sqrt{k^2 - w^2}) e^{jwz}\, dw$$

and Eq. (5-138) becomes
$$\psi = \frac{Il e^{-jkr}}{4\pi r}$$

Equating these two ψ's we have the identity

$$\frac{e^{-jkr}}{r} = \frac{1}{2j} \int_{-\infty}^{\infty} H_0^{(2)}(\rho \sqrt{k^2 - w^2}) e^{jwz}\, dw \qquad (5\text{-}139)$$

Many other identities can be established in a similar fashion.

It is convenient to have two forms for ψ because some operations are easier to perform on one form than on the other. For example, it is simple to specialize Eq. (5-138) to the radiation zone, and we did so in Sec. 2-10. In particular, the specialization is given by Eq. (2-122), which can be written as

$$\psi \xrightarrow[r\to\infty]{} \frac{e^{-jkr}}{4\pi r} \bar{I}(-k\cos\theta) \tag{5-140}$$

where $\bar{I}(w)$ is given by Eq. (5-137). By Eq. (3-97) we have

$$E_\theta \xrightarrow[r\to\infty]{} -j\omega\mu A_\theta = j\omega\mu \sin\theta\, \psi$$

or

$$E_\theta \xrightarrow[r\to\infty]{} j\omega\mu \frac{e^{-jkr}}{4\pi r} \sin\theta\, \bar{I}(-k\cos\theta) \tag{5-141}$$

Hence, the radiation field is simply related to the transform of the source evaluated at $w = -k\cos\theta$. More important, the specialization of Eq. (5-140) must also be the corresponding specialization of Eq. (5-136). We therefore have the identity

$$\int_{-\infty}^{\infty} \bar{I}(w) H_0^{(2)}(\rho\sqrt{k^2-w^2})\, e^{jwz}\, dw \xrightarrow[r\to\infty]{} 2j \frac{e^{-jkr}}{r} \bar{I}(-k\cos\theta) \tag{5-142}$$

which holds for any function $\bar{I}(w)$. Equation (5-142) can also be established by contour integration, using the method of steepest descent.[1]

Finally, we shall need a formula similar to Eq. (5-142) valid for Hankel functions of arbitrary order. The desired generalization can be effected by considering the asymptotic expression

$$H_n^{(2)}(x) \xrightarrow[x\to\infty]{} \sqrt{\frac{2j}{\pi x}}\, j^n e^{-jx}$$

from which it is evident that

$$H_n^{(2)}(x) \xrightarrow[x\to\infty]{} j^n H_0^{(2)}(x)$$

As long as $\theta \neq 0$ or π, we have $\rho \to \infty$ as $r \to \infty$, since $\rho = r\sin\theta$. Also, if k is complex (some dissipation assumed), then $\sqrt{k^2 - w^2}$ is never zero on the path of integration. We are then justified in using the asymptotic formula for Hankel functions and can replace the $H_0^{(2)}$ of Eq. (5-142) by $j^{-n} H_n^{(2)}$. The result is

$$\int_{-\infty}^{\infty} \bar{I}(w) H_n^{(2)}(\rho\sqrt{k^2-w^2}) e^{jwz}\, dw \xrightarrow[r\to\infty]{} 2\frac{e^{-jkr}}{r} j^{n+1} \bar{I}(-k\cos\theta) \tag{5-143}$$

We shall have use for this formula in the radiation problems that follow.

5-12. Apertures in Cylinders.[2] Consider a conducting cylinder of infinite length in which one or more apertures exist. The geometry is

---

[1] A. Erdelyi, "Asymptotic Expansions," pp. 26–27, Dover Publications, New York, 1956.

[2] Silver and Saunders, The External Field Produced by a Slot in an Infinite Circular Cylinder, *J. Appl. Phy.*, vol. 21, no. 5, pp. 153–158, February, 1950.

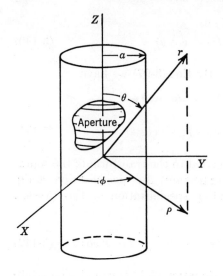

Fig. 5-25. An aperture in a conducting cylinder.

shown in Fig. 5-25. We seek a solution for the field external to the cylinder in terms of the tangential components of **E** over the apertures.

Anticipating that we shall use transforms of the fields, let us define the "cylindrical transforms" of the tangential components of **E** on the cylinder as

$$\bar{E}_z(n,w) = \frac{1}{2\pi} \int_0^{2\pi} d\phi \int_{-\infty}^{\infty} dz\, E_z(a,\phi,z) e^{-jn\phi} e^{-jwz}$$
$$\bar{E}_\phi(n,w) = \frac{1}{2\pi} \int_0^{2\pi} d\phi \int_{-\infty}^{\infty} dz\, E_\phi(a,\phi,z) e^{-jn\phi} e^{-jwz}$$
(5-144)

The inverse transformation is

$$E_z(a,\phi,z) = \frac{1}{2\pi} \sum_{n=-\infty}^{\infty} e^{jn\phi} \int_{-\infty}^{\infty} \bar{E}_z(n,w) e^{jwz}\, dw$$
$$E_\phi(a,\phi,z) = \frac{1}{2\pi} \sum_{n=-\infty}^{\infty} e^{jn\phi} \int_{-\infty}^{\infty} \bar{E}_\phi(n,w) e^{jwz}\, dw$$
(5-145)

Note that these are Fourier series on ϕ and Fourier integrals on z. The field external to the cylinder can be expressed as the sum of a TE component and TM component. According to the concepts of Sec. 3-12, the field is given by

$$\mathbf{E} = -\nabla \times \mathbf{F} - j\omega\mu\mathbf{A} + \frac{1}{j\omega\epsilon}\nabla\nabla \cdot \mathbf{A}$$
$$\mathbf{H} = \nabla \times \mathbf{A} - j\omega\epsilon\mathbf{F} + \frac{1}{j\omega\mu}\nabla\nabla \cdot \mathbf{F}$$
(5-146)

where

$$\mathbf{A} = \mathbf{u}_z A_z \qquad \mathbf{F} = \mathbf{u}_z F_z$$
(5-147)

We now construct the wave functions A_z and F_z as

$$A_z = \frac{1}{2\pi} \sum_{n=-\infty}^{\infty} e^{jn\phi} \int_{-\infty}^{\infty} f_n(w) H_n^{(2)}(\rho \sqrt{k^2 - w^2}) e^{jwz} \, dw$$

$$F_z = \frac{1}{2\pi} \sum_{n=-\infty}^{\infty} e^{jn\phi} \int_{-\infty}^{\infty} g_n(w) H_n^{(2)}(\rho \sqrt{k^2 - w^2}) e^{jwz} \, dw$$

(5-148)

which are of the form of Eq. (5-11). We choose the Bessel functions as $H_n^{(2)}$ to represent outward-traveling waves. We choose the ϕ and z functions such that the field will be of the same form as Eqs. (5-145).

To determine the $f_n(w)$ and $g_n(w)$ in Eqs. (5-148), let us calculate E_z and E_ϕ according to Eqs. (5-146). The result is

$$E_z(\rho,\phi,z) = \frac{1}{2\pi j\omega\epsilon} \sum_{n=-\infty}^{\infty} e^{jn\phi} \int_{-\infty}^{\infty} (k^2 - w^2) f_n(w) H_n^{(2)}(\rho \sqrt{k^2 - w^2}) e^{jwz} \, dw$$

$$E_\phi(\rho,\phi,z) = \frac{1}{2\pi} \sum_{n=-\infty}^{\infty} e^{jn\phi} \int_{-\infty}^{\infty} \left[-\frac{nw}{j\omega\epsilon} f_n(w) H_n^{(2)}(\rho \sqrt{k^2 - w^2}) \right.$$
$$\left. + g_n(w) \sqrt{k^2 - w^2} \, H_n^{(2)\prime}(\rho \sqrt{k^2 - w^2}) \right] e^{jwz} \, dw$$

Since these equations specialized to $\rho = a$ must equal Eqs. (5-145), we have

$$f_n(w) = \frac{j\omega\epsilon \bar{E}_z(n,w)}{(k^2 - w^2) H_n^{(2)}(a \sqrt{k^2 - w^2})}$$

$$g_n(w) = \frac{1}{\sqrt{k^2 - w^2} \, H_n^{(2)\prime}(a \sqrt{k^2 - w^2})} \left[\bar{E}_\phi(n,w) + \frac{nw}{a(k^2 - w^2)} \bar{E}_z(n,w) \right]$$

(5-149)

This completes the solution.

The inversions of Eqs. (5-148) are difficult except for the far zone, in which case we can use Eq. (5-143). Hence, we have

$$A_z \xrightarrow[r \to \infty]{} \frac{e^{-jkr}}{\pi r} \sum_{n=-\infty}^{\infty} e^{jn\phi} j^{n+1} f_n(-k \cos \theta)$$

$$F_z \xrightarrow[r \to \infty]{} \frac{e^{-jkr}}{\pi r} \sum_{n=-\infty}^{\infty} e^{jn\phi} j^{n+1} g_n(-k \cos \theta)$$

(5-150)

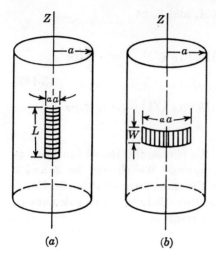

Fig. 5-26. A conducting cylinder and (a) an axial slot, (b) a circumferential slot.

Finally, in the radiation zone Eqs. (3-97) apply; hence

$$E_\theta \xrightarrow[r\to\infty]{} j\omega\mu \frac{e^{-jkr}}{\pi r} \sin\theta \sum_{n=-\infty}^{\infty} e^{jn\phi} j^{n+1} f_n(-k\cos\theta)$$

$$E_\phi \xrightarrow[r\to\infty]{} -jk \frac{e^{-jkr}}{\pi r} \sin\theta \sum_{n=-\infty}^{\infty} e^{jn\phi} j^{n+1} g_n(-k\cos\theta)$$

(5-151)

Thus, the radiation pattern of apertures in cylinders is relatively easy to calculate. The only difficulty is that the number of significant terms in the summation becomes very large for cylinders of large diameter.

To illustrate the theory, let us consider the thin rectangular slot in the two orientations shown in Fig. 5-26. For the axial slot we shall assume in the aperture

$$E_\phi = \frac{V}{\alpha a} \cos\frac{\pi z}{L} \qquad \begin{cases} -\frac{L}{2} < z < \frac{L}{2} \\ -\frac{\alpha}{2} < \phi < \frac{\alpha}{2} \end{cases}$$

(5-152)

and $E_z = 0$. (This approximates the case of excitation by a rectangular waveguide.) For a very narrow slot ($\alpha \to 0$) the transforms of Eq. (5-144) become

$$\bar{E}_\phi(n,w) = \frac{VL}{a} \frac{\cos(wL/2)}{\pi^2 - (Lw)^2}$$

and $\bar{E}_z(n,w) = 0$. From Eqs. (5-149) we then have $f_n(w) = 0$ and

$$g_n(w) = \frac{VL \cos{(wL/2)}}{[\pi^2 - (Lw)^2]a\sqrt{k^2 - w^2}\,H_n^{(2)\prime}(a\sqrt{k^2 - w^2})}$$

Finally, by Eqs. (5-151) we have the radiation field given by $E_\theta = 0$ and

$$E_\phi = \frac{VLe^{-jkr}}{\pi^3 ar}\left[\frac{\cos\left(\dfrac{kL}{2}\cos\theta\right)}{1 - \left(\dfrac{kL}{\pi}\cos\theta\right)^2}\right] \sum_{n=-\infty}^{\infty} \frac{j^n e^{jn\phi}}{H_n^{(2)\prime}(ka\sin\theta)} \quad (5\text{-}153)$$

which can be further simplified to a cosine series in ϕ. The radiation pattern in the plane $\theta = 90°$ is identical to that of the slitted cylinder; so for $a = 2\lambda$ the pattern is given in Fig. 5-20. The "vertical" pattern in the $\phi = 0$ plane is almost indistinguishable from the radiation pattern of the same slot in an infinite ground plane.[1]

For the circumferential slot of Fig. 5-26b, we assume in the aperture

$$E_z = \frac{V}{W}\cos\frac{\pi\phi}{\alpha} \quad \begin{cases} -\dfrac{W}{2} < z < \dfrac{W}{2} \\ -\dfrac{\alpha}{2} < \phi < \dfrac{\alpha}{2} \end{cases} \quad (5\text{-}154)$$

and $E_\phi = 0$. (Again this approximates excitation by a rectangular waveguide.) For a narrow slot ($W \to 0$) the transforms of Eq. (5-144) become

$$\bar{E}_z(n,w) = \frac{V\alpha \cos{(n\alpha/2)}}{\pi^2 - (n\alpha)^2}$$

and $\bar{E}_\phi(n,w) = 0$. Then from Eqs. (5-149) and (5-151) we can calculate the radiation field as

$$E_\theta = \frac{kV\alpha e^{-jkr}}{j\pi r \sin\theta} \sum_{n=-\infty}^{\infty} \frac{j^n \cos{(n\alpha/2)}\,e^{jn\phi}}{[\pi^2 - (n\alpha)^2]H_n^{(2)}(ka\sin\theta)}$$

$$E_\phi = -\frac{V\alpha e^{-jkr}\cot\theta}{\pi rka \sin\theta} \sum_{n=-\infty}^{\infty} \frac{nj^n \cos{(n\alpha/2)}\,e^{jn\phi}}{[\pi^2 - (n\alpha)^2]H_n^{(2)\prime}(ka\sin\theta)} \quad (5\text{-}155)$$

In the principal planes $\theta = \pi/2$ and $\phi = 0$, the field is entirely θ-polarized. However, in other directions, the cross-polarized component E_ϕ may be appreciable. The radiation patterns for circumferential slots in reasonably large cylinders are very close to the radiation patterns for the same

[1] L. L. Bailin, The Radiation Field Produced by a Slot in a Large Circular Cylinder, *IRE Trans.*, vol. AP-3, no. 3, pp. 128–137, July, 1955.

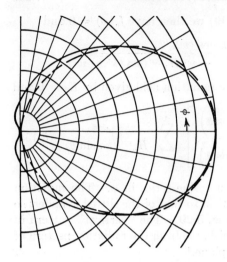

Fig. 5-27. Radiation pattern for a circumferential slot of length 0.65λ in a conducting cylinder of diameter 3λ (same slot in a ground plane shown dashed).

slot in an infinite ground plane. To illustrate this, Fig. 5-27 shows the radiation pattern in the plane $\theta = \pi/2$ for a circumferential slot 0.65λ long in a cylinder 3λ in diameter. The radiation pattern for the same slot in an infinite ground plane is shown dashed.

5-13. Apertures in Wedges. The problem of diffraction by a conductor is reciprocal to the problem of radiation by apertures in the conductor. By this, we mean that a solution to one of these problems is readily converted to a solution to the other by using the reciprocity theorem. We shall illustrate the procedure for the case of conducting wedges.

Figure 5-28 shows the reciprocal problems of (a) a current element and a conducting wedge and (b) an aperture in a conducting wedge. To keep the theory simple, we shall consider only the case of a distant current element and the radiation field of the aperture. For the z-directed electric-current element of Fig. 5-28a the field will be TM to z, expressible in terms of an $\mathbf{A} = \mathbf{u}_z \psi$. The incident field is

$$\psi^i = Il \frac{e^{-jk|\mathbf{r}-\mathbf{r}'|}}{4\pi|\mathbf{r}-\mathbf{r}'|}$$

which, when $r \gg r'$, reduces to

$$\psi^i = Il \frac{e^{-jkr}}{4\pi r} e^{jkz' \cos\theta} e^{jk\rho' \sin\theta \cos(\phi-\phi')} \qquad (5\text{-}156)$$

This is simply a plane wave incident upon the wedge. The ψ in this three-dimensional problem is subject to the same boundary condition ($\psi = 0$) on the wedge as is E_z in the two-dimensional problem of Sec.

5-10. Hence the solution must be of the same form as Eq. (5-128), that is,

$$\psi = \frac{2\pi\psi_0}{\pi - \alpha} \sum_v j^v J_v(k\rho' \sin\theta) \sin v(\phi' - \alpha) \sin v(\phi - \alpha) \quad (5\text{-}157)$$

where
$$\psi_0 = Il\frac{e^{-jkr}}{4\pi r} e^{jkz'\cos\theta}$$
$$v = \frac{m\pi}{2(\pi - \alpha)} \quad m = 1, 2, 3, \ldots \quad (5\text{-}158)$$

In terms of ψ, the field is given by Eqs. (5-18). This completes the solution to Fig. 5-28a.

To obtain the solution to Fig. 5-28b, we apply reciprocity [Eq. (3-35)] to the region bounded by the conducting wedge. Because of the boundary conditions on $\mathbf{E}$ at the conductor, Eq. (3-35) reduces to

$$-\iint_{\text{apert}} E_z{}^b H_\rho{}^a \, ds = Il E_z{}^b \quad (5\text{-}159)$$

where the superscripts a and b refer to the fields of Figs. 5-28a and b, respectively. From Eqs. (5-18) and (5-157) we calculate

$$H_\rho{}^a = \frac{2\pi\psi_0}{\rho'(\pi - \alpha)} \sum_v v j^v J_v(k\rho' \sin\theta) \cos v(\phi' - \alpha) \sin v(\phi - \alpha)$$

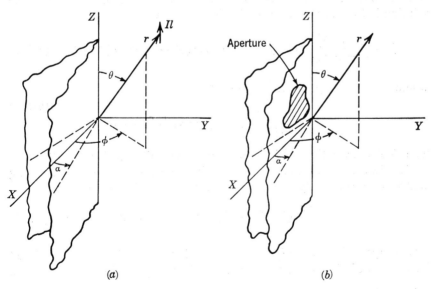

Fig. 5-28. The reciprocal problems of (a) a current element and a conducting wedge and (b) an aperture in a conducting wedge.

Specializing this to the surface $\phi' = \alpha$, we can reduce Eq. (5-159) to

$$IlE_z = -\int_{-\infty}^{\infty} dz' \int_0^{\infty} d\rho' \frac{2\pi\psi_0 E_z^b}{\rho'(\pi - \alpha)} \sum_v v j^v J_v(k\rho' \sin\theta) \sin v(\phi - \alpha)$$

Finally, ψ_0 is given by Eq. (5-158), and in the radiation zone

$$E_\theta = -\frac{E_z}{\sin\theta}$$

Hence, the θ component of **E** in the radiation zone is given by

$$E_\theta = \frac{e^{-jkr}}{2r(\pi - \alpha)\sin\theta} \sum_v v j^v \sin v(\phi - \alpha) f_v(k\cos\theta, k\sin\theta) \quad (5\text{-}160)$$

where
$$f_v(w,u) = \int_{-\infty}^{\infty} e^{jwz} dz \int_0^{\infty} J_v(u\rho) \, d\rho \, \frac{1}{\rho} E_z(\rho,\alpha,z) \quad (5\text{-}161)$$

Note that $f_v(w,u)$ is of the form of a Fourier transform on z and a Fourier-Bessel (or Hankel) transform on ρ.[1]

In a similar manner, the E_ϕ component of the radiation field can be obtained by applying reciprocity to Fig. 5-28a with Il replaced by Kl. This z-directed magnetic-current element gives rise to a field TE to z, expressible according to $\mathbf{F} = \mathbf{u}_z \psi$. The incident field is then specified by Eq. (5-156) with I replaced by K. Again the three-dimensional problem is essentially the same as the two-dimensional problem of Sec. 5-10. The solution is then of the form of Eq. (5-133), that is,

$$\psi = \frac{\pi\psi_0}{\pi - \alpha} \sum_v \epsilon_v j^v J_v(k\rho' \sin\theta) \cos v(\phi' - \alpha) \cos v(\phi - \alpha) \quad (5\text{-}162)$$

where
$$\psi_0 = Kl \frac{e^{-jkr}}{4\pi r} e^{jkz' \cos\theta}$$
$$v = \frac{m\pi}{2(\pi - \alpha)} \qquad m = 0, 1, 2, \ldots \quad (5\text{-}163)$$

The electromagnetic field is found from ψ according to Eqs. (5-19).

To relate this solution to the field from an aperture in a conducting wedge, we again apply reciprocity [Eq. (3-35)]. This reduces to

$$\iint_{\text{apert}} (E_z^b H_\rho^a - E_\rho^b H_z^a) \, ds = K l H_z^b \quad (5\text{-}164)$$

where superscripts a and b refer to the fields of Fig. 5-28a with Il replaced by Kl, and of Fig. 5-28b, respectively. From Eqs. (5-19) and (5-162) we

---

[1] I. N. Sneddon, "Fourier Transforms," p. 6, McGraw-Hill Book Company, Inc., New York, 1951.

calculate

$$H_\rho{}^a = \frac{\pi k^2 \sin\theta \cos\theta}{\omega\mu(\pi-\alpha)} \psi_0 \sum_v \epsilon_v j^v J_v'(k\rho' \sin\theta) \cos v(\phi'-\alpha) \cos v(\phi-\alpha)$$

$$H_z{}^a = \frac{\pi k^2 \sin^2\theta}{j\omega\mu(\pi-\alpha)} \psi_0 \sum_v \epsilon_v j^v J_v(k\rho' \sin\theta) \cos v(\phi'-\alpha) \cos v(\phi-\alpha)$$

Finally, we evaluate Eq. (5-164) and use the radiation-zone relationship

$$E_\phi = -\eta H_\theta = \frac{\eta H_z}{\sin\theta}$$

The result is

$$E_\phi = \frac{ke^{-jkr}}{4r(\pi-\alpha)} \sum_v \epsilon_v j^v \cos v(\phi-\alpha)[\cos\theta\, g_v(k\cos\theta, k\sin\theta) \quad (5\text{-}165)$$
$$+ j\sin\theta\, h_v(k\cos\theta, k\sin\theta)]$$

where
$$g_v(w,u) = \int_{-\infty}^{\infty} e^{jwz}\, dz \int_0^{\infty} J_v'(u\rho)\, d\rho\, E_z(\rho,\alpha,z)$$
$$h_v(w,u) = \int_{-\infty}^{\infty} e^{jwz}\, dz \int_0^{\infty} J_v(u\rho)\, d\rho\, E_\rho(\rho,\alpha,z)$$
(5-166)

We now have a complete solution for the radiation field from apertures in conducting wedges.

As an example, let us calculate the radiation from a narrow axial slot of length L, as shown in Fig. 5-29. We shall assume that in the slot

$$E_\rho = V\delta(\rho-a)\cos\frac{\pi z}{L} \quad (5\text{-}167)$$

is the only tangential component of **E**. The f, g, and h functions [Eqs. (5-161) and (5-166)] are then found to be

$$f_v = 0 \qquad g_v = 0$$
$$h_v = \frac{2\pi V L \cos(wL/2)}{\pi^2 - (Lw)^2} J_v(ua)$$

From Eq. (5-160) we see that $E_\theta = 0$, and from Eq. (5-165) we have

$$E_\phi = f(r)\sin\theta\, \frac{\cos[k(L/2)\cos\theta]}{\pi^2 - (kL\cos\theta)^2}$$

$$\sum_v \epsilon_v j^v \cos v(\phi-\alpha) J_v(ka\sin\theta) \quad (5\text{-}168)$$

where $v = \frac{1}{2}, 1, \frac{3}{2}, \ldots$. Plots of

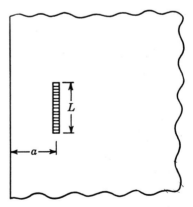

Fig. 5-29. A narrow axial slot in a conducting half plane.

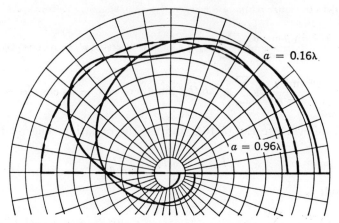

FIG. 5-30. Radiation patterns for axial slots in a conducting half plane (the slot in an infinite ground plane is shown dashed).

the radiation pattern in the plane $\theta = 90°$ are shown in Fig. 5-30 for the case $\alpha = 0$ (half plane). The cases $a = 0.16\lambda$ and $a = 0.96\lambda$ are shown, with the infinite ground-plane pattern shown dashed for comparison.

PROBLEMS

5-1. Show that Eq. (5-12) is a solution to the scalar Helmholtz equation.

5-2. Show that $\psi = (\log \rho)e^{-jkz}$ is a solution to the scalar Helmholtz equation. Determine the TM field generated by this ψ according to Eqs. (5-18). Sketch the $\mathcal{E}$ and $\mathcal{H}$ lines in a $z = $ constant plane. What physical system supports this wave? Repeat for the TE case.

5-3. For two-dimensional fields (no z variation) show that an arbitrary field in a source-free homogeneous region can be expressed in terms of two scalar wave functions, ψ_1 and ψ_2, according to Eqs. (3-79) where

$$\mathbf{A} = \mathbf{u}_\rho \rho \psi_1 \qquad \mathbf{F} = \mathbf{u}_\rho \rho \psi_2$$

Note that this corresponds to choosing

$$\Phi^a = -\frac{\rho}{\hat{y}}\frac{\partial}{\partial \rho}\left(\frac{A_\rho}{\rho}\right) \qquad \Phi^f = -\frac{\rho}{\hat{z}}\frac{\partial}{\partial \rho}\left(\frac{F_\rho}{\rho}\right)$$

instead of Eqs. (3-80).

5-4. A circular waveguide has a dominant mode cutoff frequency of 9000 megacycles. What is its inside diameter if it is air-filled? Determine the cutoff frequencies for the next ten lowest-order modes. Repeat for the case $\epsilon_r = 4$.

5-5. All the waveguides whose cross sections are shown in Fig. 5-4 are characterized by wave functions of the form

$$\psi = B_n(k_\rho \rho) h(n\phi) e^{\pm jk_z z}$$

where TM modes are determined by Eqs. (5-18) and TE modes by Eqs. (5-19). The phase constant is given by

$$k_z = \sqrt{k^2 - k_\rho^2}$$

CYLINDRICAL WAVE FUNCTIONS

Let a denote the inner radius and b the outer radius of the coaxial waveguide of Fig. 5-4a. Show that for TM modes

$$B_n(k_\rho \rho) = N_n(k_\rho a)J_n(k_\rho \rho) - J_n(k_\rho a)N_n(k_\rho \rho)$$
$$h(n\phi) = \sin n\phi \quad \text{or} \quad \cos n\phi$$

where $n = 0, 1, 2, \ldots$, and k_ρ is a root of

$$J_n(k_\rho a)N_n(k_\rho b) - J_n(k_\rho b)N_n(k_\rho a) = 0$$

Show that for TE modes

$$B_n(k_\rho \rho) = N_n'(k_\rho a)J_n(k_\rho \rho) - J_n'(k_\rho a)N_n(k_\rho \rho)$$
$$h(n\phi) = \sin n\phi \quad \text{or} \quad \cos n\phi$$

where $n = 0, 1, 2, \ldots$, and k_ρ is a root of

$$J_n'(k_\rho a)N_n'(k_\rho b) - N_n'(k_\rho a)J_n'(k_\rho b) = 0$$

5-6. Show that the modes of the coaxial waveguide with a baffle (Fig. 5-4b) are characterized by the same $B_n(k_\rho \rho)$ functions as the coaxial guide (Prob. 5-5), but for TM modes

$$h(n\phi) = \sin n\phi \qquad n = \tfrac{1}{2}, 1, \tfrac{3}{2}, 2, \ldots$$

and for TE modes

$$h(n\phi) = \cos n\phi \qquad n = 0, \tfrac{1}{2}, 1, \tfrac{3}{2}, \ldots$$

where the baffle is at $\phi = 0$. The dominant mode is the lowest TE mode with $n = \tfrac{1}{2}$.

5-7. Show that the wedge waveguide of Fig. 5-4e supports TM modes specified by

$$\psi^{\text{TM}} = J_n(k_\rho \rho) \sin n\phi \, e^{\pm jk_z z}$$

where
$$n = \frac{\pi}{\phi_0}, \frac{2\pi}{\phi_0}, \frac{3\pi}{\phi_0}, \ldots$$

and $k_\rho a$ is a zero of $J_n(k_\rho a)$. Show that it supports TE modes specified by

$$\psi^{\text{TE}} = J_n(k_\rho \rho) \cos n\phi \, e^{\pm jk_z z}$$

where
$$n = 0, \frac{\pi}{\phi_0}, \frac{2\pi}{\phi_0}, \ldots$$

and $k_\rho a$ is a zero of $J_n'(k_\rho a)$. The guides of Figs. 5-4c and d are the special cases $\phi_0 = 2\pi$ and π, respectively.

5-8. Show that the cutoff wavelength for the dominant mode of the circular waveguide with baffle (Fig. 5-4c) is

$$\lambda_c = \frac{2\pi a}{1.16}$$

5-9. Using the perturbational method of Sec. 2-7, show that the attenuation constants due to conductor losses in a circular waveguide are given by

$$\alpha_c = \frac{\mathcal{R}}{\eta a \sqrt{1 - (f_c/f)^2}}$$

for all TM modes, and by

$$\alpha_c = \frac{\mathcal{R}}{\eta a \sqrt{1 - (f_c/f)^2}} \left[\frac{n^2}{(x_{np}')^2 - n^2} + \left(\frac{f_c}{f}\right)^2 \right]$$

for all TE modes. Note that for the "circular electric" modes ($n = 0$) the attenuation decreases without limit as $f \to \infty$.

5-10. Consider the two-dimensional "circulating waveguide" formed of concentric conducting cylinders $\rho = a$ and $\rho = b$. Show that the wave function

$$\psi = [AJ_n(k\rho) + BN_n(k\rho)]e^{-jn\phi}$$

specifies circulating modes TM to z according to Eqs. (5-18) if n is a root of

$$-\frac{B}{A} = \frac{J_n(ka)}{N_n(ka)} = \frac{J_n(kb)}{N_n(kb)}$$

Show that the above wave function specifies modes TE to z according to Eqs. (5-19) if n is a root of

$$-\frac{B}{A} = \frac{J_n'(ka)}{N_n'(ka)} = \frac{J_n'(kb)}{N_n'(kb)}$$

5-11. For the TM radial wave specified by Eq. (5-33), show that the radial phase constant of E_z is given by Eq. (5-36), while the radial phase constant of H_ϕ is

$$\beta_\rho = \frac{2}{\pi\rho}\left[1 - \left(\frac{n}{k_\rho\rho}\right)^2\right]\frac{1}{[J_n'(k_\rho\rho)]^2 + [N_n'(k_\rho\rho)]^2}$$

Show that Eq. (5-37) is also valid for this phase constant.

5-12. Consider the TM radial wave impedances of Eqs. (5-38) and (5-39). Show that for large radii

$$Z_{+\rho}^{TM} = Z_{-\rho}^{TM} \xrightarrow[k_\rho\rho \to \infty]{} \eta$$

and that for small radii

$$Z_{+\rho}^{TM} = Z_{-\rho}^{TM\,*} \xrightarrow[k_\rho\rho \to 0]{} \begin{cases} \dfrac{\eta}{2}k_\rho\rho\left(\pi + j\log\dfrac{2}{\gamma k_\rho\rho}\right) & n = 0 \\ \eta k_\rho\rho\left[\left(\dfrac{2\pi}{n!}\right)^2\left(\dfrac{k_\rho\rho}{2}\right)^{2n} + \dfrac{j}{n}\right] & n > 0 \end{cases}$$

where $\gamma = 1.781$.

5-13. Consider the radial parallel-plate waveguide of Fig. 5-5a. For the transmission-line mode [Eqs. (5-45)], one can define a voltage and current as

$$V(\rho) = -aE_z \qquad I(\rho) = 2\pi\rho H_\phi$$

Show that V and I satisfy the transmission-line equations

$$\frac{dV}{d\rho} = -j\omega LI \qquad \frac{dI}{d\rho} = -j\omega CV$$

where L and C are the "static" parameters

$$L = \frac{\mu a}{2\pi\rho} \qquad C = \frac{2\pi\epsilon\rho}{a}$$

Why should we expect circuit concepts to apply for this mode?

5-14. Consider the wedge guide of Fig. 5-5b. For the dominant mode [Eq. (5-49)], one can define a voltage and current as

$$V(\rho) = E_\phi\rho\phi_0 \qquad I(\rho) = H_z a$$

Show that V and I satisfy the transmission-line equation (Prob. 5-13) with

$$L = \frac{\mu\rho\phi_0}{a} \qquad C = \frac{\epsilon a}{\rho\phi_0}$$

5-15. Show that the resonant frequencies of the two-dimensional cylindrical cavity (no z variation, conductor over $\rho = a$) are equal to the cutoff frequencies of the circular waveguide.

5-16. Following the perturbational method used to derive Eq. (5-58), show that the Q due to conductor losses for the various modes in the circular cavity of Fig. 5-7 are

$$(Q_c)_{np0}^{\text{TM}} = \frac{\eta x_{np}}{2\Re(1+a/d)}$$

$$(Q_c)_{npq}^{\text{TM}} = \frac{\eta\sqrt{x_{np}^2 + \left(\dfrac{q\pi a}{d}\right)^2}}{2\Re(1+2a/d)}$$

$$(Q_c)_{npq}^{\text{TE}} = \frac{\eta[x_{np}'^2 + (q\pi a/d)^2]^{3/2}(x_{np}'^2 - n^2)}{2\Re\left[\left(\dfrac{nq\pi a}{d}\right)^2 + x_{np}'^4 + \dfrac{2a}{d}\left(\dfrac{q\pi a}{d}\right)^2(x_{np}'^2 - n^2)\right]}$$

5-17. The circular cavity of Fig. 5-7 has dimensions $a = d = 3$ centimeters. Determine the first ten resonant frequencies and the Q of the dominant mode if the walls are copper.

5-18. Consider the dominant mode of the partially filled radial waveguide of Fig. 5-9a. Show that for small a and large ρ the phase constant is

$$\beta \approx k_2 \sqrt{\frac{1+(\mu_1/\mu_2 - 1)d/a}{1+(\epsilon_2/\epsilon_1 - 1)d/a}}$$

Compare this to the uniform transmission-line formula [Eq. (2-66)], using the static approximations

$$C = \frac{2\pi\epsilon_1\epsilon_2\rho}{\epsilon_1(a-d)+\epsilon_2 d} \qquad L = \frac{\mu_1 d + \mu_2(a-d)}{2\pi\rho}$$

5-19. Consider the dielectric-slab radial guide of Fig. 5-9b. Let $\epsilon_1 = 4\epsilon_0$ and $\mu_1 = \mu_0$ and $a = \lambda_0$. Which modes can propagate unattenuated in the slab? Repeat the problem for the coated-conductor guide of Fig. 5-9c with $t = a/2$.

5-20. For the partially filled circular waveguide (Fig. 5-10a), show that the characteristic equation [Eq. (5-74)] for the $n = 1$ modes reduces to

$$[AN_1(k_{\rho 2}b) + BJ_1(k_{\rho 2}b)][AN_1'(k_{\rho 2}b) + BJ_1'(k_{\rho 2}b)] = 0$$

where
$$A = k_{\rho 1}J_1'(k_{\rho 1}a)J_1(k_{\rho 2}a) - k_{\rho 2}J_1'(k_{\rho 2}a)J_1(k_{\rho 1}a)$$
$$B = k_{\rho 2}N_1'(k_{\rho 2}a)J_1(k_{\rho 1}a) - k_{\rho 1}J_1'(k_{\rho 1}a)N_1(k_{\rho 2}a)$$

5-21. Consider the dominant $(n = 1)$ mode of the dielectric-rod waveguide of Fig. 5-10b. Show that for small a the characteristic equation becomes

$$ua = \sqrt{\frac{(\mu_1+\mu_2)(\epsilon_1+\epsilon_2)}{2\mu_2\epsilon_2 K_0(va)}}$$

where
$$u^2 = k_1^2 - k_z^2 \qquad v^2 = k_z^2 - k_2^2$$

Note that there is no cutoff frequency.

5-22. The field external to a dielectric-rod waveguide varies as $K_1(v\rho)$. Using the results of Prob. 5-21, show that for a small ($a \ll \lambda_1$), nonmagnetic ($\mu_1 = \mu_2$) rod

$$\log \frac{2}{\gamma v a} \approx \frac{1}{(k_2 a)^2} \frac{\epsilon_1 + \epsilon_2}{\epsilon_1 - \epsilon_2}$$

where $\gamma = 1.781$. Take $\epsilon_1 = 9\epsilon_2$ and $a = 0.1\lambda_1$, and calculate the distance from the axis for which the field is 10 per cent of its value at the surface of the rod.

5-23. Consider the circular cavity with concentric dielectric rod, as shown in Fig. 5-31a. Show that the dominant resonant frequency is the smallest root of

$$\frac{1}{\eta} \frac{J_0'(kc)}{J_0(kc)} = \frac{1}{\eta_0} \left[\frac{N_0(k_0 a) J_0'(k_0 c) - J_0(k_0 a) N_0'(k_0 c)}{N_0(k_0 a) J_0(k_0 c) - J_0(k_0 a) N_0(k_0 c)} \right]$$

For small c/a, show that resonant frequency ω_r is related to the empty-cavity resonance

$$\omega_0 = \frac{x_{01}}{a \sqrt{\epsilon_0 \mu_0}} \qquad x_{01} = 2.405$$

according to
$$\frac{\omega_r - \omega_0}{\omega_0} = \frac{\pi}{4} x_{01} \frac{N_0(x_{01})}{J_0'(x_{01})} (\epsilon_r - 1) \left(\frac{c}{a}\right)^2$$
$$= -1.86(\epsilon_r - 1) \left(\frac{c}{a}\right)^2$$

where $\epsilon_r = \epsilon/\epsilon_0$.

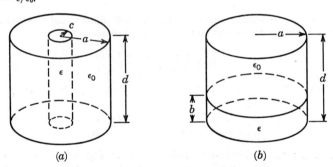

Fig. 5-31. Partially filled cavities.

5-24. Consider the circular cavity with a dielectric slab, as shown in Fig. 5-31b. Show that the characteristic equation for the resonant frequency of the dominant mode is

$$\frac{k_{z0}}{\epsilon_0} \tan k_{z0}(d - b) = -\frac{k_z}{\epsilon} \tan k_z b$$

where
$$k_{z0}{}^2 = k_0{}^2 - \left(\frac{x_{01}}{a}\right)^2 \qquad k_z{}^2 = k^2 - \left(\frac{x_{01}}{a}\right)^2$$

Show that when both d and b are small

$$\omega_r \approx \omega_0 \sqrt{\frac{1 - (1 - 1/\epsilon_r) b/d}{1 + (\mu_r - 1) b/d}}$$

where ω_0 is the empty-cavity resonant frequency, given in Prob. 5-23, and $\epsilon_r = \epsilon/\epsilon_0$ and $\mu_r = \mu/\mu_0$.

CYLINDRICAL WAVE FUNCTIONS

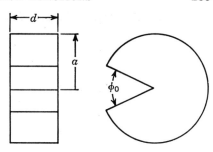

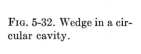

Fig. 5-32. Wedge in a circular cavity.

5-25. Consider the circular cavity with a conducting wedge, as shown in Fig. 5-32. Show that, for d small, the resonant frequency of the dominant mode is given by

$$f_r = \frac{w}{2\pi a \sqrt{\epsilon\mu}}$$

where w is the first root of $J_v(w) = 0$ and $v = \pi/(2\pi - \phi_0)$. Some representative values of w are

v	0.5	0.6	0.7	0.8	0.9	1.0
w	3.14	3.28	3.42	3.56	3.70	3.83

5-26. Figure 5-33a shows a linear density of x-directed current elements along the z axis. Show that the field is given by $\mathbf{H} = \nabla \times \mathbf{A}$ where

$$A_x = \frac{J_x l}{4j} H_0^{(2)}(k\rho)$$

Show that the field is identical to that produced by the magnetic dipole formed of z-directed magnetic currents $+K$ at $y = -s/2$ and $-K$ at $y = s/2$ in the limit $s \to 0$.

5-27. Show that the field of the magnetic-dipole source of Fig. 5-33b in the limit $s \to 0$ is given by $\mathbf{E} = -\nabla \times \mathbf{u}_z\psi$ where

$$\psi = \frac{kKs}{4j} H_1^{(2)}(k\rho) \sin\phi$$

5-28. Consider the quadrupole source of Fig. 5-33c in the limit $s_1 \to 0$ and $s_2 \to 0$. Show that the field is given by $\mathbf{H} = \nabla \times \mathbf{u}_z\psi$ where

$$\psi = \frac{k^2 I s_1 s_2}{8j} [-H_0^{(2)}(k\rho) + H_2^{(2)}(k\rho) \cos 2\phi]$$

5-29. Figure 5-33d represents a source of $2n$ current filaments, equal in amplitude but alternating in sign, on a cylinder of radius $\rho = a$. Show that, in the limit $a \to 0$,

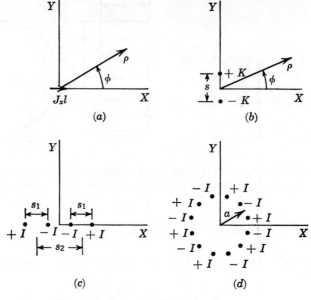

Fig. 5-33. Some two-dimensional sources.

the field is given by $\mathbf{H} = \nabla \times \mathbf{u}_z \psi$ where

$$\psi = \frac{\pi I}{2j(n-1)!} \left(\frac{ka}{2}\right)^n H_n^{(2)}(k\rho) \sin n\phi$$

5-30. Let the cylinder of current in Fig. 5-15 be an arbitrary function of ϕ, but still independent of z. Show that the field is given by $\mathbf{H} = \nabla \times \mathbf{u}_z \psi$ with

$$\psi = \begin{cases} \dfrac{\pi a}{2j} \displaystyle\sum_{n=-\infty}^{\infty} A_n J_n(ka) H_n^{(2)}(k\rho) e^{jn\phi} & \rho > a \\ \dfrac{\pi a}{2j} \displaystyle\sum_{n=-\infty}^{\infty} A_n H_n^{(2)}(ka) J_n(k\rho) e^{jn\phi} & \rho < a \end{cases}$$

where

$$A_n = \frac{1}{2\pi} \int_0^{2\pi} J_z e^{-jn\phi} \, d\phi$$

A cylinder of z-directed magnetic currents is dual to this problem.

5-31. Show that the radiation field from a ribbon of uniform z-directed current (Fig. 4-25) is given by

$$E_z = \frac{-j\omega\mu a e^{-jk\rho}}{\sqrt{8\pi jk\rho}} J_z \frac{\sin\left(\dfrac{ka}{2} \cos\phi\right)}{(ka/2) \cos\phi}$$

and $H_\phi = -E_z/\eta$.

5-32. Consider the slot antenna of Fig. 4-21, and make the assumption that tangential **E** in the slot is $\mathbf{u}_x E_0$, a constant. Show that the radiation field is

$$H_z = \frac{-j\omega\epsilon a e^{-jk\rho}}{\sqrt{2\pi jk\rho}} E_0 \frac{\sin\left(\frac{ka}{2}\cos\phi\right)}{(ka/2)\cos\phi}$$

and $E_\phi = \eta H_z$.

5-33. Derive the following wave transformations:

$$\cos(\rho \sin\phi) = \sum_{n=0}^{\infty} \epsilon_n J_{2n}(\rho) \cos 2n\phi$$

$$\sin(\rho \sin\phi) = 2\sum_{n=0}^{\infty} J_{2n+1}(\rho) \sin(2n+1)\phi$$

5-34. Let the cylinder of Fig. 5-17 be dielectric with parameters ϵ_d, μ_d. For a TM incident plane wave [Eq. (5-105)], show that the scattered field is given by Eq. (5-106) with

$$a_n = \frac{-J_n(ka)}{H_n^{(2)}(ka)} \left[\frac{\epsilon_d J_n'(k_d a)/\epsilon k_d a J_n(k_d a) - J_n'(ka)/ka J_n(ka)}{\epsilon_d J_n'(k_d a)/\epsilon k_d a J_n(k_d a) - H_n^{(2)'}(ka)/ka H_n^{(2)}(ka)} \right]$$

and that the field internal to the cylinder is given by

$$E_z = E_0 \sum_{n=-\infty}^{\infty} j^n c_n J_n(k_d \rho) e^{jn\phi}$$

with

$$c_n = \frac{1}{J_n(k_d a)} [J_n(ka) + a_n H_n^{(2)}(ka)]$$

Note that this solution reduces to the solution for the conducting cylinder when $\epsilon_d \to \infty$.

5-35. Repeat Prob. 5-34 for the opposite polarization, that is, when the incident field is given by Eq. (5-113). Note that this problem is completely dual to Prob. 5-34; so the solution is obtainable by using the interchange of symbols of Table 3-2. Note that the solution reduces to the solution for a conducting cylinder as $\mu_d \to 0$.

5-36. Show that the solution of Prob. 5-34 in the nonmagnetic case reduces to

$$E_z^s \xrightarrow[ka \to 0]{} \frac{-j\pi E_0}{4} (ka)^2 (\epsilon_r - 1) H_0^{(2)}(k\rho)$$

where $\epsilon_r = \epsilon_d/\epsilon_0$. Repeat for the opposite polarization, using the result of Prob. 5-35.

5-37. Consider a conducting half plane covering the $\phi = 0$ surface and a z-polarized plane wave of magnitude E_0 incident at an angle ϕ'. The solution is given by Eq. (5-129). Show that the current on the half plane is

$$J_z = \frac{2E_0}{j\omega\mu\rho} \sum_{n=1}^{\infty} nj^{n/2} J_{n/2}(k\rho) \sin\frac{n\phi'}{2}$$

Show that near the edge of the half plane

$$J_z \xrightarrow[k\rho \to 0]{} \frac{2}{\eta} E_0 \sqrt{\frac{2}{j\pi k\rho}} \sin \frac{\phi'}{2}$$

and

$$E_z \xrightarrow[k\rho \to 0]{} 2E_0 \sqrt{\frac{2jk\rho}{\pi}} \sin \frac{\phi'}{2} \sin \frac{\phi}{2}$$

Hence, E_z vanishes as $\sqrt{k\rho}$, and J_z becomes infinite as $1/\sqrt{k\rho}$. This is a general characteristic of knife edges.

5-38. Consider the half plane of Prob. 5-37 with the incident plane wave polarized transverse to z. The solution is given by Eq. (5-134). Show that the current on the half plane is

$$J_\rho = 2H_0 \sum_{n=0}^{\infty} \epsilon_n j^{n/2} J_{n/2}(k\rho) \cos \frac{n\phi'}{2}$$

Show that near the knife edge

$$J_\rho \xrightarrow[k\rho \to 0]{} 2H_0$$

$$E_\rho \xrightarrow[k\rho \to 0]{} -\eta H_0 \sqrt{\frac{2}{j\pi k\rho}} \cos \frac{\phi'}{2} \sin \frac{\phi}{2}$$

where ϕ' is the angle of incidence and ϕ the angle to the field point. Note that J_ρ is finite at $\rho = 0$, while E_ρ becomes infinite as $1/\sqrt{k\rho}$. This is also a general characteristic of knife edges.

5-39. Figure 5-34a shows a conducting cylinder with an axially pointing magnetic dipole Kl on its surface at $\phi = 0$, $z = 0$. Show that the radiation field is given by

$$E_\phi = -\frac{Kle^{-jkr}}{2\pi^2 ar} \sum_{n=0}^{\infty} \frac{\epsilon_n j^n \cos n\phi}{H_n^{(2)\prime}(ka \sin \theta)}$$

where ϵ_n is Neumann's number.

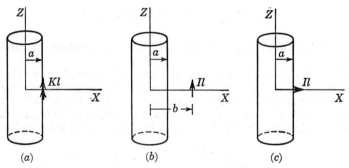

Fig. 5-34. Conducting cylinder with (a) axial magnetic dipole on its surface, (b) axial electric dipole a distance b from the axis, and (c) radial electric dipole on its surface.

5-40. Consider the axially pointing electric dipole a distance b from the axis of a conducting cylinder of radius a, as shown in Fig. 5-34b. Show that the radiation field is given by

$$E_\theta = f(r) \sin \theta \sum_{n=-\infty}^{\infty} \frac{J_n(\alpha)N_n(\beta) - N_n(\alpha)J_n(\beta)}{H_n^{(2)}(\alpha)} j^n e^{jn\phi}$$

where $\alpha = ka \sin \theta$ and $\beta = kb \sin \theta$.

CYLINDRICAL WAVE FUNCTIONS 263

5-41. Consider the radially pointing electric dipole on a conducting cylinder of radius a, as shown in Fig. 5-34c. Show that in the $z = 0$ plane (in which Il lies) the radiation field is given by

$$E_\phi = f(\rho) \sum_{n=1}^{\infty} \frac{nj^n \sin n\phi}{H_n^{(2)\prime}(ka)}$$

The field in other directions has both θ and ϕ components.

5-42. Figure 5-35a shows a conducting half plane with a magnetic dipole parallel to the edge, a distance a from it, and on the side $\phi = 0$. Show that the radiation field is

$$E_\phi = \frac{jkKl}{4\pi r} e^{-jkr} \sin \theta \sum_{n=0}^{\infty} \epsilon_n j^{n/2} J_{n/2}(ka \sin \theta) \cos \frac{n\phi}{2}$$

where ϵ_n is Neumann's number.

5-43. Suppose that the magnetic dipole of Fig. 5-35a points in the x direction instead of the z direction. Show that the radiation field is then given by

$$E_\theta = \frac{Kle^{-jkr}}{4\pi ar \sin \theta} \sum_{n=1}^{\infty} nj^{n/2} J_{n/2}(ka \sin \theta) \sin \frac{n\phi}{2}$$

$$E_\phi = \frac{jkKl}{4\pi r} e^{-jkr} \sin \theta \sum_{n=0}^{\infty} \epsilon_n j^{n/2} J'_{n/2}(ka \sin \theta) \cos \frac{n\phi}{2}$$

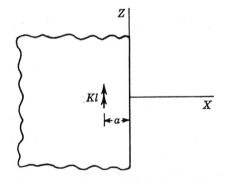
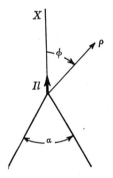

FIG. 5-35. A conducting half plane with a magnetic dipole on the side $\phi = 0$ a distance a from the edge.

FIG. 5-36. Electric current element on the edge of a conducting wedge.

5-44. Consider the x-directed electric dipole on the edge of a conducting wedge, as shown in Fig. 5-36. Show that in the plane of the element the radiation field is given by

$$E_\phi = f(\rho) \sin \frac{\pi \phi}{2\pi - \alpha}$$

For a half plane, the pattern is a cardioid with a null in the $\phi = 0$ direction.

CHAPTER 6

SPHERICAL WAVE FUNCTIONS

6-1. The Wave Functions. The spherical coordinate system is the simplest one for which a coordinate surface (r = constant) is of finite extent. The usual definition of spherical coordinates is shown in Fig. 6-1. Once again we must determine solutions to the scalar Helmholtz equation, from which we may construct electromagnetic fields.

In spherical coordinates the Helmholtz equation is

$$\frac{1}{r^2}\frac{\partial}{\partial r}\left(r^2 \frac{\partial \psi}{\partial r}\right) + \frac{1}{r^2 \sin \theta}\frac{\partial}{\partial \theta}\left(\sin \theta \frac{\partial \psi}{\partial \theta}\right) + \frac{1}{r^2 \sin^2 \theta}\frac{\partial^2 \psi}{\partial \phi^2} + k^2 \psi = 0 \quad (6\text{-}1)$$

Again we use the method of separation of variables and let

$$\psi = R(r)H(\theta)\Phi(\phi) \quad (6\text{-}2)$$

Substituting this into Eq. (6-1), dividing by ψ, and multiplying by $r^2 \sin^2 \theta$, we obtain

$$\frac{\sin^2 \theta}{R}\frac{d}{dr}\left(r^2 \frac{dR}{dr}\right) + \frac{\sin \theta}{H}\frac{d}{d\theta}\left(\sin \theta \frac{dH}{d\theta}\right) + \frac{1}{\Phi}\frac{d^2\Phi}{d\phi^2} + k^2 r^2 \sin^2 \theta = 0$$

The ϕ dependence is now separated out, and we let

$$\frac{1}{\Phi}\frac{d^2\Phi}{d\phi^2} = -m^2 \quad (6\text{-}3)$$

where m is a constant. Substitution of this into the preceding equation and division by $\sin^2 \theta$ yields

$$\frac{1}{R}\frac{d}{dr}\left(r^2 \frac{dR}{dr}\right) + \frac{1}{H \sin \theta}\frac{d}{d\theta}\left(\sin \theta \frac{dH}{d\theta}\right) - \frac{m^2}{\sin^2 \theta} + k^2 r^2 = 0$$

This separates the r and θ dependence. An apparently strange choice of separation constant n is made according to

$$\frac{1}{H \sin \theta}\frac{d}{d\theta}\left(\sin \theta \frac{dH}{d\theta}\right) - \frac{m^2}{\sin^2 \theta} = -n(n+1) \quad (6\text{-}4)$$

because the properties of the H functions depend upon whether or not n

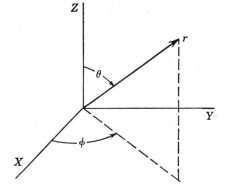

FIG. 6-1. The spherical coordinate system.

is an integer. With this choice the preceding equation becomes

$$\frac{1}{R}\frac{d}{dr}\left(r^2 \frac{dR}{dr}\right) - n(n+1) + k^2 r^2 = 0 \quad (6\text{-}5)$$

which completes the separation procedure.

Collecting the above results, we have the trio of separated equations

$$\frac{d}{dr}\left(r^2 \frac{dR}{dr}\right) + [(kr)^2 - n(n+1)]R = 0$$

$$\frac{1}{\sin\theta}\frac{d}{d\theta}\left(\sin\theta \frac{dH}{d\theta}\right) + \left[n(n+1) - \frac{m^2}{\sin^2\theta}\right]H = 0 \quad (6\text{-}6)$$

$$\frac{d^2\Phi}{d\phi^2} + m^2\Phi = 0$$

Note that there is now no interrelationship between separation constants. The Φ equation is the familiar harmonic equation, giving rise to solutions $h(m\phi)$. The R equation is closely related to Bessel's equation. Its solutions are called *spherical Bessel functions*, denoted $b_n(kr)$, which are related to ordinary Bessel functions by

$$b_n(kr) = \sqrt{\frac{\pi}{2kr}}\, B_{n+\frac{1}{2}}(kr) \quad (6\text{-}7)$$

(see Appendix D). The θ equation is related to Legendre's equation, and its solutions are called *associated Legendre functions*. We shall denote solutions in general by $L_n{}^m(\cos\theta)$. Commonly used solutions are

$$L_n{}^m(\cos\theta) \sim P_n{}^m(\cos\theta),\ Q_n{}^m(\cos\theta) \quad (6\text{-}8)$$

where $P_n{}^m(\cos\theta)$ are the associated Legendre functions of the first kind and $Q_n{}^m(\cos\theta)$ are the associated Legendre functions of the second kind. These are considered in some detail in Appendix E. We can now form product solutions to the Helmholtz equation as

$$\psi_{m,n} = b_n(kr)L_n{}^m(\cos\theta)h(m\phi) \qquad (6\text{-}9)$$

These are the elementary wave functions for the spherical coordinate system.

Again we can construct more general solutions to the Helmholtz equation by forming linear combinations of the elementary wave functions. The most general form that we shall have occasion to use is a summation over possible values of m and n

$$\begin{aligned}\psi &= \sum_m \sum_n C_{m,n} \psi_{m,n} \\ &= \sum_m \sum_n C_{m,n} b_n(kr) L_n{}^m(\cos\theta) h(m\phi)\end{aligned} \qquad (6\text{-}10)$$

where the $C_{m,n}$ are constants. Integrations over m and n are also solutions to the Helmholtz equation, but such forms are not needed for our purposes.

The harmonic functions $h(m\phi)$ have already been considered in Sec. 4-1. If a single-valued ψ in the range 0 to 2π on ϕ is desired, we must choose $h(m\phi)$ to be a linear combination of $\sin(m\phi)$ and $\cos(m\phi)$, or of $e^{jm\phi}$ and $e^{-jm\phi}$, with m an integer. A study of solutions to the associated Legendre equation shows that all solutions have singularities at $\theta = 0$ or $\theta = \pi$ except the $P_n{}^m(\cos\theta)$ with n an integer. Thus, if ψ is to be finite in the range 0 to π on θ, then n must also be an integer and $L_n{}^m(\cos\theta)$ must be $P_n{}^m(\cos\theta)$. The spherical Bessel functions behave qualitatively in the same manner as do the corresponding cylindrical Bessel functions. Thus, for k real, $j_n(kr)$ and $n_n(kr)$ represent standing waves, $h_n{}^{(1)}(kr)$ represents an inward-traveling wave, and $h_n{}^{(2)}(kr)$ represents an outward-traveling wave. Incidentally, it turns out that the spherical Bessel functions are simpler in form than the cylindrical Bessel functions. For example, the zero-order functions are

$$\begin{aligned}j_0(kr) &= \frac{\sin kr}{kr} & h_0{}^{(1)}(kr) &= \frac{e^{jkr}}{jkr} \\ n_0(kr) &= -\frac{\cos kr}{kr} & h_0{}^{(2)}(kr) &= -\frac{e^{-jkr}}{jkr}\end{aligned} \qquad (6\text{-}11)$$

The higher-order functions are polynomials in $1/kr$ times $\sin(kr)$ and $\cos(kr)$, which can be readily obtained from the recurrence formula. The only spherical Bessel functions finite at $r = 0$ are the $j_n(kr)$. Thus, to represent a finite field inside a sphere, the elementary wave functions are

$$\psi_{m,n} = j_n(kr)P_n{}^m(\cos\theta)e^{jm\phi} \qquad r = 0 \text{ included} \qquad (6\text{-}12)$$

with m and n integers. To represent a finite field outside of a sphere, we must choose outward-traveling waves (proper behavior at infinity). Hence,

$$\psi_{m,n} = h_n^{(2)}(kr) P_n^m(\cos\theta) e^{jm\phi} \qquad r \to \infty \text{ included} \qquad (6\text{-}13)$$

with m and n integers, are the desired elementary wave functions.

To represent electromagnetic fields in terms of the wave functions ψ, we can use the method of Sec. 3-12. This involves letting ψ be a *rectangular* component of **A** or **F**. The z component is most simply related to spherical components; hence the logical choice is

$$\mathbf{A} = \mathbf{u}_z \psi = \mathbf{u}_r \psi \cos\theta - \mathbf{u}_\theta \psi \sin\theta \qquad (6\text{-}14)$$

which generates a field TM to z. Explicit expressions for the field components in terms of ψ are given in Prob. 6-1. The dual choice is

$$\mathbf{F} = \mathbf{u}_z \psi = \mathbf{u}_r \psi \cos\theta - \mathbf{u}_\theta \psi \sin\theta \qquad (6\text{-}15)$$

which generates a field TE to z. Explicit expressions for the field components are given in Prob. 6-1. An arbitrary electromagnetic field in terms of spherical wave functions can be constructed as a superposition of its TM and TE parts.

An alternative, and somewhat simpler, representation of an arbitrary electromagnetic field is also possible in spherical coordinates. Suppose we attempt to construct the field as a superposition of two parts, one TM to r and the other TE to r. For this we choose $\mathbf{A} = \mathbf{u}_r A_r$ and $\mathbf{F} = \mathbf{u}_r F_r$, with the field being given by Eq. (3-79). The A_r and F_r are *not* solutions to the scalar Helmholtz equation, because $\nabla^2 A_r \neq (\nabla^2 \mathbf{A})_r$. To determine the equations that A_r and F_r must satisfy, we return to the general equations for vector potentials [Eqs. (3-78)]. For the magnetic vector potential we let $\mathbf{A} = \mathbf{u}_r A_r$ and expand the first of Eqs. (3-78). The θ and ϕ components of the resulting equation are, respectively,

$$\frac{\partial^2 A_r}{\partial r \, \partial \theta} = -\hat{y} \frac{\partial \Phi^a}{\partial \theta} \qquad \frac{\partial^2 A_r}{\partial r \, \partial \phi} = -\hat{y} \frac{\partial \Phi^a}{\partial \phi}$$

where Φ^a is an arbitrary scalar. Note that the above two equations are satisfied identically if we choose

$$-\hat{y}\Phi^a = \frac{\partial A_r}{\partial r} \qquad (6\text{-}16)$$

Substituting this into the r-component equation obtained from the expansion of Eq. (3-78), we have

$$\frac{\partial^2 A_r}{\partial r^2} + \frac{1}{r^2 \sin\theta} \frac{\partial}{\partial \theta}\left(\sin\theta \frac{\partial A_r}{\partial \theta}\right) + \frac{1}{r^2 \sin^2\theta} \frac{\partial^2 A_r}{\partial \phi^2} + k^2 A_r = 0 \qquad (6\text{-}17)$$

It readily can be shown that this equation is

$$(\nabla^2 + k^2)\frac{A_r}{r} = 0 \tag{6-18}$$

so A_r/r is a solution to the scalar Helmholtz equation. A dual development applies to the electric vector potential. To be explicit, if we take $\mathbf{F} = \mathbf{u}_r F_r$, substitute into the second of Eqs. (3-78), and choose

$$-\hat{z}\Phi^b = \frac{\partial F_r}{\partial r} \tag{6-19}$$

we find that
$$(\nabla^2 + k^2)\frac{F_r}{r} = 0 \tag{6-20}$$

is the equation for F_r. Thus, electromagnetic fields can be constructed by choosing

$$\mathbf{A} = \mathbf{r}\psi^a \qquad \mathbf{F} = \mathbf{r}\psi^f \tag{6-21}$$

where $\mathbf{r} = \mathbf{u}_r r$ is the radius vector from the origin and the ψ's are solutions to the Helmholtz equation. The field is found from the above vector potentials by Eq. (3-79), which is explicitly

$$\begin{aligned}\mathbf{E} &= -\nabla \times \mathbf{r}\psi^f + \frac{1}{\hat{y}}\nabla \times \nabla \times \mathbf{r}\psi^a \\ \mathbf{H} &= \nabla \times \mathbf{r}\psi^a + \frac{1}{\hat{z}}\nabla \times \nabla \times \mathbf{r}\psi^f\end{aligned} \tag{6-22}$$

These we shall find sufficiently general to express any a-c field in a source-free homogeneous region of space.

The ψ's of Eqs. (6-22) are always multiplied by r, and, because of this, it is convenient to introduce another type of spherical Bessel function, defined as

$$\hat{B}_n(kr) = krb_n(kr) = \sqrt{\frac{\pi kr}{2}} B_{n+\frac{1}{2}}(kr) \tag{6-23}$$

These are the spherical Bessel functions used by Schelkunoff.[1] Their qualitative behavior is the same as the corresponding cylindrical Bessel function. The differential equation that they satisfy is

$$\left[\frac{d^2}{dr^2} + k^2 - \frac{n(n+1)}{r^2}\right]\hat{B}_n = 0 \tag{6-24}$$

which can be obtained by substituting for b_n in terms of $\hat{B}_n$ in the first of Eqs. (6-6). General forms for the A_r and F_r in terms of the spherical

---

[1] S. A. Schelkunoff, "Electromagnetic Waves," pp. 51–52, D. Van Nostrand Company, Inc., Princeton, N.J., 1943.

Bessel functions of Eq. (6-23) are

$$\sum_{m,n} C_{m,n} \hat{B}_n(kr) L_n{}^m(\cos\theta) h(m\phi) \qquad (6\text{-}25)$$

where the $C_{m,n}$ are constants. The considerations involved in choosing specific forms for $\hat{B}_n(kr)$, $L_n{}^m(\cos\theta)$, and $h(m\phi)$ are the same as those used in Eqs. (6-12) and (6-13).

For future reference, let us tabulate explicit formulas for finding the field components in terms of A_r and F_r. Letting $\mathbf{A} = \mathbf{u}_r A_r$ and $\mathbf{F} = \mathbf{u}_r F_r$, and expanding Eqs. (3-79), we obtain

$$\begin{aligned}
E_r &= \frac{1}{\hat{y}}\left(\frac{\partial^2}{\partial r^2} + k^2\right) A_r \\
E_\theta &= \frac{-1}{r\sin\theta}\frac{\partial F_r}{\partial \phi} + \frac{1}{\hat{y}r}\frac{\partial^2 A_r}{\partial r\,\partial\theta} \\
E_\phi &= \frac{1}{r}\frac{\partial F_r}{\partial \theta} + \frac{1}{\hat{y}r\sin\theta}\frac{\partial^2 A_r}{\partial r\,\partial\phi} \\
H_r &= \frac{1}{\hat{z}}\left(\frac{\partial^2}{\partial r^2} + k^2\right) F_r \\
H_\theta &= \frac{1}{r\sin\theta}\frac{\partial A_r}{\partial \phi} + \frac{1}{\hat{z}r}\frac{\partial^2 F_r}{\partial r\,\partial\theta} \\
H_\phi &= -\frac{1}{r}\frac{\partial A_r}{\partial \theta} + \frac{1}{\hat{z}r\sin\theta}\frac{\partial^2 F_r}{\partial r\,\partial\phi}
\end{aligned} \qquad (6\text{-}26)$$

When $F_r = 0$, that is, when only A_r exists, we have a field TM to r. Similarly, when $A_r = 0$, the above equations represent a field TE to r.

6-2. The Spherical Cavity. Figure 6-2 shows the spherical cavity, formed of a conducting sphere of radius a enclosing a homogeneous dielectric ϵ, μ. We shall find it possible to satisfy the boundary conditions (tangential components of $\mathbf{E}$ vanish at $r = a$) using single wave functions. For modes TE to r we choose

$$F_r = \hat{J}_n(kr) P_n{}^m(\cos\theta) \begin{Bmatrix} \cos m\phi \\ \sin m\phi \end{Bmatrix} \qquad (6\text{-}27)$$

where m and n are integers. The $\hat{J}_n$ is chosen because the field must be finite at $r = 0$; the $P_n{}^m$ is chosen because the field must be finite at $\theta = 0$ and π. The field components are then found from Eq. (6-26) with $A_r = 0$ and F_r as given above. Note that $E_\theta = E_\phi = 0$ at $r = a$ if

$$\hat{J}_n(ka) = 0 \qquad (6\text{-}28)$$

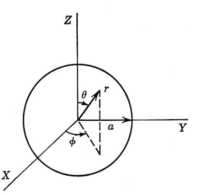

Fig. 6-2. The spherical cavity.

TABLE 6-1. ORDERED ZEROS u_{np} OF $\hat{J}_n(u)$

p \ n	1	2	3	4	5	6	7	8
1	4.493	5.763	6.988	8.183	9.356	10.513	11.657	12.791
2	7.725	9.095	10.417	11.705	12.967	14.207	15.431	16.641
3	10.904	12.323	13.698	15.040	16.355	17.648	18.923	20.182
4	14.066	15.515	16.924	18.301	19.653	20.983	22.295	
5	17.221	18.689	20.122	21.525	22.905			
6	20.371	21.854						

Hence ka must be a zero of the spherical Bessel function. The denumerably infinite set of zeros of $\hat{J}_n(u)$ are ordered as u_{np}. Table 6-1 gives the lower-order zeros.

We now satisfy the boundary conditions by choosing

$$k = \frac{u_{np}}{a} \qquad (6\text{-}29)$$

which is the condition for resonance. Hence, the TE to r mode functions are

$$(F_r)_{mnp} = \hat{J}_n\left(u_{np}\frac{r}{a}\right) P_n^m(\cos\theta) \begin{Bmatrix} \cos m\phi \\ \sin m\phi \end{Bmatrix} \qquad (6\text{-}30)$$

where $m = 0, 1, 2, \ldots$; $n = 1, 2, 3, \ldots$; and $p = 1, 2, 3, \ldots$. The field is given by Eqs. (6-26) with $A_r = 0$.

If an A_r is chosen of the form of Eq. (6-27), we generate a field TM to r. The boundary conditions $E_\theta = E_\phi = 0$ at $r = a$ are then satisfied if

$$\hat{J}_n'(ka) = 0 \qquad (6\text{-}31)$$

so ka must be a zero of the derivative of the spherical Bessel function for TM modes. The denumerably infinite set of zeros of $\hat{J}_n'(u')$ are ordered as u'_{np}, and the lower-order ones are given in Table 6-2.

TABLE 6-2. ORDERED ZEROS u'_{np} OF $\hat{J}_n'(u')$

p \ n	1	2	3	4	5	6	7	8
1	2.744	3.870	4.973	6.062	7.140	8.211	9.275	10.335
2	6.117	7.443	8.722	9.968	11.189	12.391	13.579	14.753
3	9.317	10.713	12.064	13.380	14.670	15.939	17.190	18.425
4	12.486	13.921	15.314	16.674	18.009	19.321	20.615	21.894
5	15.644	17.103	18.524	19.915	21.281	22.626		
6	18.796	20.272	21.714	23.128				
7	21.946							

Our boundary conditions are now satisfied by choosing

$$k = \frac{u'_{np}}{a} \tag{6-32}$$

which is the condition for resonance. The TM to r mode functions are therefore

$$(A_r)_{mnp} = \hat{J}_n\left(u'_{np}\frac{r}{a}\right) P_n{}^m(\cos\theta) \begin{Bmatrix} \cos m\phi \\ \sin m\phi \end{Bmatrix} \tag{6-33}$$

where $m = 0, 1, 2, \ldots$; $n = 1, 2, 3, \ldots$; and $p = 1, 2, 3, \ldots$. The field is given by Eqs. (6-26) with $F_r = 0$.

The resonant frequencies of the TE and TM modes are found from Eqs. (6-29) and (6-32), respectively. Letting $k = 2\pi f_r \sqrt{\epsilon\mu}$, we have

$$\begin{aligned} (f_r)_{mnp}^{\text{TE}} &= \frac{u_{np}}{2\pi a \sqrt{\epsilon\mu}} \\ (f_r)_{mnp}^{\text{TM}} &= \frac{u'_{np}}{2\pi a \sqrt{\epsilon\mu}} \end{aligned} \tag{6-34}$$

Note that there are numerous degeneracies (same resonant frequencies) among the modes, since f_r is independent of m. For example, the three lowest-order TE modes are defined by

$$(F_r)_{0,1,1} = \hat{J}_1\left(4.493\frac{r}{a}\right) \cos\theta$$

$$(F_r)_{1,1,1}^{\text{even}} = \hat{J}_1\left(4.493\frac{r}{a}\right) \sin\theta \cos\phi$$

$$(F_r)_{1,1,1}^{\text{odd}} = \hat{J}_1\left(4.493\frac{r}{a}\right) \sin\theta \sin\phi$$

where superscripts "even" and "odd" have been added to denote the choice $\cos m\phi$ and $\sin m\phi$, respectively. These three modes have the same mode patterns except that they are rotated 90° in space from each other. The next higher TE resonance has a fivefold degeneracy, the modes being ordered (0,2,1), (1,2,1) even, (1,2,1) odd, (2,2,1) even, and (2,2,1) odd. In this case there are two characteristic mode patterns. For each integer increase in n, the degeneracy increases by two, since $P_n{}^m(\cos\theta)$ exists only for $m \leq n$. The situation for TM modes is analogous.

We see by Eqs. (6-34) that the resonant frequencies are proportional to the u_{np} and u'_{np}. Hence, from Tables 6-1 and 6-2 it is evident that the modes in order of ascending resonant frequencies are $\text{TM}_{m,1,1}$, $\text{TM}_{m,2,1}$, $\text{TE}_{m,1,1}$, $\text{TM}_{m,3,1}$, $\text{TE}_{m,2,1}$, and so on. The lowest-order modes are there-

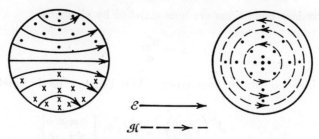

Fig. 6-3. Mode pattern for the dominant modes of the spherical cavity.

fore the three $TM_{m,1,1}$ modes. Except for a rotation in space, these three modes have the same mode pattern, which is sketched in Fig. 6-3.

The Q of the lowest-order modes is also of interest. For this calculation, consider the $TM_{0,1,1}$ mode. The magnetic field is given by

$$H_\phi = \frac{1}{r} \hat{J}_1 \left(2.744 \frac{r}{a}\right) \sin \theta$$

Following the procedure of Sec. 2-8, we calculate the stored energy as

$$\mathcal{W} = 2\overline{\mathcal{W}}_m = \mu \iiint |H|^2 \, d\tau$$
$$= \mu \int_0^{2\pi} d\phi \int_0^\pi d\theta \int_0^a dr \, H_\phi{}^2 \, r^2 \sin \theta$$

The θ and ϕ integrations are easily performed, giving

$$\mathcal{W} = \frac{8\pi}{3} \mu \int_0^a \hat{J}_1{}^2 \left(2.744 \frac{r}{a}\right) dr$$

This last integral is evaluated as[1]

$$\int_0^a \hat{J}_1{}^2(kr) \, dr = \frac{a}{2} [\hat{J}_1{}^2(ka) - \hat{J}_0(ka)\hat{J}_2(ka)]$$

which, for $ka = 2.744$, is numerically equal to $1.14/k$. Thus, the stored energy is

$$\mathcal{W} = \frac{8\pi\mu}{3k} (1.14) \qquad (6\text{-}35)$$

The power dissipated in the conducting walls is approximately

$$\bar{\mathcal{P}}_d = \mathcal{R} \oiint |H|^2 \, ds = \mathcal{R} \frac{8\pi}{3} \hat{J}_1{}^2(2.744) \qquad (6\text{-}36)$$

Hence, the Q of the cavity is

$$Q = \frac{\omega \mathcal{W}}{\bar{\mathcal{P}}_d} = \frac{\omega\mu(1.14)}{k\mathcal{R}\hat{J}_1{}^2(2.744)} = 1.01 \frac{\eta}{\mathcal{R}} \qquad (6\text{-}37)$$

[1] E. Jahnke and F. Emde, "Tables of Functions," p. 146, Dover Publications, New York, 1945 (reprint).

Comparing this with Eqs. (5-58) and (2-102), we see that the spherical cavity has a Q that is 25 per cent higher than the Q of a circular cavity of height equal to its diameter and 35 per cent higher than the Q of a cubic cavity. The Q's of higher-order modes are given in Prob. 6-4.

6-3. Orthogonality Relationships. In many ways the Legendre polynomials are qualitatively similar to sinusoidal functions. For example, the $P_n(\cos\theta)$, sometimes called *zonal harmonics*, form a complete orthogonal set in the interval 0 to π on θ. An arbitrary function can therefore be expanded in a series of Legendre polynomials in this interval, similar to the Fourier series in sinusoidal functions. The functions $P_n{}^m(\cos\theta)\cos m\phi$ and $P_n{}^m(\cos\theta)\sin m\phi$, sometimes called *tesseral harmonics*, form a complete orthogonal set on the surface of a sphere. Hence, an arbitrary function defined over the surface of a sphere can be expanded in a series of tesseral harmonics. We shall, in this section, derive the necessary orthogonality relationships.

For our proof it is convenient to use Green's theorem [Eq. (3-44)], which is

$$\oint\left(\psi_1\frac{\partial\psi_2}{\partial n} - \psi_2\frac{\partial\psi_1}{\partial n}\right)ds = \iiint(\psi_1\nabla^2\psi_2 - \psi_2\nabla^2\psi_1)\,d\tau \quad (6\text{-}38)$$

The right-hand side vanishes if ψ_1 and ψ_2 are well behaved solutions to the same Helmholtz equation. Assuming this to be the case and applying Eq. (6-38) to a sphere of radius r, we have

$$r^2\int_0^{2\pi}d\phi\int_0^\pi d\theta\sin\theta\left(\psi_1\frac{\partial\psi_2}{\partial r} - \psi_2\frac{\partial\psi_1}{\partial r}\right) = 0 \quad (6\text{-}39)$$

In particular, choose

$$\psi_1 = j_n(kr)P_n(\cos\theta) \qquad \psi_2 = j_q(kr)P_q(\cos\theta)$$

which are solutions to the Helmholtz equation. Equation (6-39) then becomes

$$2\pi kr^2(j_n j_q' - j_q j_n')\int_0^\pi P_n P_q \sin\theta\,d\theta = 0$$

This must be valid for all r; so, if $n \neq q$, the integral itself must vanish. Hence,

$$\int_0^\pi P_n(\cos\theta)P_q(\cos\theta)\sin\theta\,d\theta = 0 \qquad n \neq q \quad (6\text{-}40)$$

When $n = q$, we have

$$\int_0^\pi [P_n(\cos\theta)]^2 \sin\theta\,d\theta = \frac{2}{2n+1} \quad (6\text{-}41)$$

which can be obtained by using Eq. (E-10) and integrating by parts.

To obtain a Legendre polynomial representation of a function $f(\theta)$ in 0 to π on θ, we assume

$$f(\theta) = \sum_{n=0}^{\infty} a_n P_n(\cos \theta) \qquad 0 \leq \theta \leq \pi \qquad (6\text{-}42)$$

Multiply each side by $P_q(\cos \theta) \sin \theta$ and integrate from 0 to π on θ.

$$\int_0^\pi f(\theta) P_q(\cos \theta) \sin \theta \, d\theta = \sum_{n=0}^{\infty} a_n \int_0^\pi P_n(\cos \theta) P_q(\cos \theta) \sin \theta \, d\theta$$

Each integral on the right vanishes by Eq. (6-40), except the one $n = p$, which is given by Eq. (6-41). The result is

$$a_n = \frac{2n+1}{2} \int_0^\pi f(\theta) P_n(\cos \theta) \sin \theta \, d\theta \qquad (6\text{-}43)$$

Equation (6-42) with the coefficients determined by Eq. (6-43) is called a *Fourier-Legendre series*. It converges in the same sense as the usual Fourier series.

For a more general result, define the tesseral harmonics as

$$\begin{aligned} T_{mn}^e(\theta,\phi) &= P_n^m(\cos \theta) \cos m\phi \\ T_{mn}^o(\theta,\phi) &= P_n^m(\cos \theta) \sin m\phi \end{aligned} \qquad (6\text{-}44)$$

and assume two solutions to the Helmholtz equation as

$$\psi_1 = j_n(kr) T_{mn}^i(\theta,\phi) \qquad \psi_2 = j_q(kr) T_{pq}^j(\theta,\phi)$$

These are well behaved within a sphere of radius r; hence Eq. (6-39) applies and reduces to

$$kr^2(j_n j_q' - j_q j_n') \int_0^{2\pi} d\phi \int_0^\pi d\theta \, T_{mn}^i T_{pq}^j \sin \theta \, d\theta = 0$$

The term outside the integral vanishes for arbitrary r only when $n = q$; hence

$$\int_0^{2\pi} d\phi \int_0^\pi d\theta \, T_{mn}^i(\theta,\phi) T_{pq}^j(\theta,\phi) \sin \theta = 0 \qquad n \neq q$$

For the ϕ integration, we have the known orthogonality relationships

$$\int_0^{2\pi} \sin m\phi \sin p\phi \, d\phi = 0$$

$$\int_0^{2\pi} \sin m\phi \sin p\phi \, d\phi = \int_0^{2\pi} \cos m\phi \cos p\phi \, d\phi = \begin{cases} 0 & m \neq p \\ \pi & m = p \neq 0 \end{cases}$$

$$(6\text{-}45)$$

SPHERICAL WAVE FUNCTIONS

Hence, the final orthogonality can be expressed as

$$\int_0^{2\pi} d\phi \int_0^{\pi} d\theta \, T_{mn}^e(\theta,\phi) T_{pq}^o(\theta,\phi) \sin \theta = 0 \qquad (6\text{-}46)$$
$$\int_0^{2\pi} d\phi \int_0^{\pi} d\theta \, T_{mn}^i(\theta,\phi) T_{pq}^i(\theta,\phi) \sin \theta = 0 \qquad m, n \neq p, q$$

where $i = e$ or o. When $m, n = p, q$, we have

$$\int_0^{2\pi} d\phi \int_0^{\pi} d\theta \, [T_{mn}^i(\theta,\phi)]^2 \sin \theta = \begin{cases} \dfrac{4\pi}{2n+1} & m = 0, i = e \\ \dfrac{2\pi}{2n+1} \dfrac{(n+m)!}{(n-m)!} & m \neq 0 \end{cases} \qquad (6\text{-}47)$$

which can be obtained by using Eq. (E-16) for P_n^m and integrating on θ by parts.

A two-dimensional Fourier-Legendre series can now be obtained for a function $f(\theta,\phi)$ on a spherical surface. For this we assume

$$f(\theta,\phi) = \sum_{n=0}^{\infty} \sum_{m=0}^{\infty} (a_{mn} T_{mn}^e + b_{mn} T_{mn}^o)$$
$$= \sum_{n=0}^{\infty} \sum_{m=0}^{\infty} (a_{mn} \cos m\phi + b_{mn} \sin m\phi) P_n^m(\cos \theta) \qquad (6\text{-}48)$$

multiply each side by $T_{p,q}^i \sin \theta$, and integrate over 0 to 2π on ϕ and 0 to π on θ. All terms except those having $m, n = p, q$ vanish by Eqs. (6-46), and by Eqs. (6-47)

$$a_{0n} = \frac{2n+1}{4\pi} \int_0^{2\pi} d\phi \int_0^{\pi} d\theta \, f(\theta,\phi) P_n(\cos \theta)$$
$$a_{mn} = \frac{2n+1}{2\pi} \frac{(n-m)!}{(n+m)!} \int_0^{2\pi} d\phi \int_0^{\pi} d\theta \, f(\theta,\phi) T_{mn}^e(\theta,\phi) \sin \theta \qquad (6\text{-}49)$$
$$b_{mn} = \frac{2n+1}{2\pi} \frac{(n-m)!}{(n+m)!} \int_0^{2\pi} d\phi \int_0^{\pi} d\theta \, f(\theta,\phi) T_{mn}^o(\theta,\phi) \sin \theta$$

The series Eq. (6-48) with coefficients Eqs. (6-49) converges in the same sense as the usual Fourier series.

Still another orthogonality relationship is of interest when dealing with vector fields. To establish the desired relationship, we start from the Lorentz reciprocity theorem [Eq. (3-34)], which is

$$\oint (\mathbf{E}^a \times \mathbf{H}^b - \mathbf{E}^b \times \mathbf{H}^a) \cdot d\mathbf{s} = 0 \qquad (6\text{-}50)$$

valid when no sources are within the surface of integration.[1] For the

---
[1] We could just as well use the vector Green's theorem, Eq. (3-46).

a and b fields, choose those obtained from Eqs. (6-26) with $F_r = 0$ and

$$A_r{}^a = \hat{J}_n(kr)T_{mn}{}^i(\theta,\phi) \qquad A_r{}^b = \hat{J}_q(kr)T_{pq}{}^j(\theta,\phi)$$

respectively. Applying Eq. (6-50) to a sphere of radius r, we obtain

$$\frac{1}{\hat{y}}(\hat{J}'_n\hat{J}_q - \hat{J}'_q\hat{J}_n)\int_0^{2\pi}d\phi\int_0^{\pi}d\theta\left(\sin\theta\frac{\partial T_{mn}{}^i}{\partial\theta}\frac{\partial T_{pq}{}^j}{\partial\theta} + \frac{1}{\sin\theta}\frac{\partial T_{mn}{}^i}{\partial\phi}\frac{\partial T_{pq}{}^j}{\partial\phi}\right) = 0$$

For arbitrary r and $n \neq q$ this equation can be satisfied only if the integral vanishes. Also, by the orthogonality relationships of Eqs. (6-45) the integral vanishes if $m \neq p$ and $i \neq j$. Thus,

$$\int_0^{2\pi}d\phi\int_0^{\pi}d\theta\left(\sin\theta\frac{\partial T_{mn}{}^i}{\partial\theta}\frac{\partial T_{pq}{}^j}{\partial\theta} + \frac{1}{\sin\theta}\frac{\partial T_{mn}{}^i}{\partial\phi}\frac{\partial T_{pq}{}^j}{\partial\phi}\right) = 0$$
$$m, n, i \neq p, q, j \quad (6\text{-}51)$$

When $m, n, i = p, q, j$, we have

$$\int_0^{2\pi}d\phi\int_0^{\pi}d\theta\left[\sin\theta\left(\frac{\partial T_{mn}{}^i}{\partial\theta}\right)^2 + \frac{1}{\sin\theta}\left(\frac{\partial T_{mn}{}^i}{\partial\phi}\right)^2\right]$$
$$= \begin{cases} \dfrac{4\pi n(n+1)}{2n+1} & m = 0, i = e \\ \dfrac{2\pi n(n+1)}{2n+1}\dfrac{(n+m)!}{(n-m)!} & m \neq 0 \end{cases} \quad (6\text{-}52)$$

which can be obtained by integrating once by parts and using Eq. (6-47).

6-4. Space as a Waveguide. We have seen that in a complete spherical-shell region ($0 \leq \theta \leq \pi$, $0 \leq \phi \leq 2\pi$) only spherical wave functions of integral m and n give a finite field. The fields specified by these wave functions can be thought of as the "modes of free space." When viewed in this manner, the space is often called a *spherical waveguide*, even though there is no material guiding the waves.

The spherical coordinate system is defined in Fig. 6-1. There exists a set of modes TM to r, generated by

$$(A_r)_{mn}{}^i = T_{mn}{}^i(\theta,\phi)\begin{Bmatrix}\hat{H}_n^{(1)}(kr) \\ \hat{H}_n^{(2)}(kr)\end{Bmatrix} \quad (6\text{-}53)$$

where $n = 1, 2, 3, \ldots$; $m = 0, 1, 2, \ldots, n$; and $i = e$ or o. The T functions are defined by Eqs. (6-44), and the field is given by

$$\mathbf{H}_{mn}^{\text{TM}i} = \nabla \times \mathbf{u}_r(A_r)_{mn}{}^i \qquad \mathbf{E}_{mn}^{\text{TM}i} = \frac{1}{j\omega\epsilon}\nabla \times \mathbf{H}_{mn}^{\text{TM}i} \quad (6\text{-}54)$$

Inward-traveling waves are represented by the $\hat{H}_n^{(1)}$ and outward-traveling waves by the $\hat{H}_n^{(2)}$. In the dual sense there exists a set of

SPHERICAL WAVE FUNCTIONS 277

modes TE to r, generated by

$$(F_r)_{mn}{}^i = T_{mn}{}^i(\theta,\phi) \begin{Bmatrix} \hat{H}_n{}^{(1)}(kr) \\ \hat{H}_n{}^{(2)}(kr) \end{Bmatrix} \tag{6-55}$$

where $n = 1, 2, 3, \ldots$; $m = 0, 1, 2, \ldots, n$; and $i = e$ or o. The field is given by

$$\mathbf{E}_{mn}^{\text{TE}i} = -\nabla \times \mathbf{u}_r(F_r)_{mn}{}^i \qquad \mathbf{H}_{mn}^{\text{TE}i} = -\frac{1}{j\omega\mu}\nabla \times \mathbf{E}_{mn}^{\text{TE}i} \tag{6-56}$$

The set of TM plus TE modes is complete, that is, a summation of them can be used to represent an arbitrary field in a source-free region. Mode patterns for the TM_{01} and TE_{02} modes are sketched in Fig. 6-4. The TM and TE modes are dual to each other; so an interchange of **E** by **H** and **H** by $-\mathbf{E}$ in Fig. 6-4 gives the TE_{01} and TM_{02} mode patterns.

The spherical modes are qualitatively similar to the radial modes of Sec. 5-3. There is no well-defined cutoff wavelength but rather a "cutoff radius." To illustrate, consider the radially directed wave impedances for the TM modes

$$\begin{aligned} Z_{+r}{}^{\text{TM}} &= \frac{E_\theta{}^+}{H_\phi{}^+} = -\frac{E_\phi{}^+}{H_\theta{}^+} = j\eta\,\frac{\hat{H}_n{}^{(2)\prime}(kr)}{\hat{H}_n{}^{(2)}(kr)} \\ Z_{-r}{}^{\text{TM}} &= -\frac{E_\theta{}^-}{H_\phi{}^-} = \frac{E_\phi{}^-}{H_\theta{}^-} = -j\eta\,\frac{\hat{H}_n{}^{(1)\prime}(kr)}{\hat{H}_n{}^{(1)}(kr)} \end{aligned} \tag{6-57}$$

where the superscripts $+$ and $-$ denote outward- and inward-traveling waves, respectively. Note that, for real η and k, $Z_{-r}{}^{\text{TM}} = (Z_{+r}{}^{\text{TM}})^*$. For

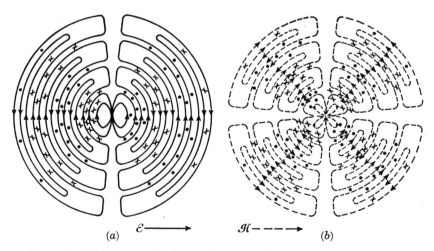

(a) $\mathcal{E} \longrightarrow$ $\mathcal{H} \text{-----} \rightarrow$ (b)

FIG. 6-4. Mode patterns for the (a) TM_{01} and (b) TE_{02} modes of free space.

the TE modes the radially directed wave impedances are

$$Z_{+r}^{TE} = \frac{E_\theta^+}{H_\phi^+} = -\frac{E_\phi^+}{H_\theta^+} = -j\eta \frac{\hat{H}_n^{(2)}(kr)}{\hat{H}_n^{(2)\prime}(kr)}$$

$$Z_{-r}^{TE} = -\frac{E_\theta^-}{H_\phi^-} = \frac{E_\phi^-}{H_\theta^-} = j\eta \frac{\hat{H}_n^{(1)}(kr)}{\hat{H}_n^{(1)\prime}(kr)}$$
(6-58)

The behavior of these wave impedances is qualitatively similar to the behavior of the two-dimensional wave impedances, illustrated by Fig. 5-6. In other words, the wave impedances of Eqs. (6-57) and (6-58) are predominantly reactive when $kr < n$, and predominantly resistive when $kr > n$. The value $kr = n$ is the point of gradual cutoff. Note that this cutoff is independent of the mode number m.

The frequency derivative of the various wave impedances is of interest for determining the bandwidth of various devices (see Sec. 6-13). A novel way of representing this frequency derivative, which also illustrates the above cutoff phenomenon, was devised by Professor Chu.[1] He took the wave impedances and, using the recurrence formulas for spherical Bessel functions, obtained a partial fraction expansion. For example, for the TM impedance of outward-traveling waves

$$Z_{+r}^{TM} = \eta \left\{ \frac{n}{jkr} + \cfrac{1}{\cfrac{2n-1}{jkr} + \cfrac{1}{\cfrac{2n-3}{jkr} + \cfrac{\cdot}{\cdot \cdot + \cfrac{1}{\cfrac{3}{jkr} + \cfrac{1}{\cfrac{1}{jkr}+1}}}}} \right\}$$
(6-59)

This can be interpreted as a ladder network of series capacitances and shunt inductances, as shown in Fig. 6-5a. The equivalent circuit for the TE_{mn} modes is shown in Fig. 6-5b. Those of us familiar with filter theory will recognize the equivalent circuits as high-pass filters. The dissipation in the resistive element at the end of the network represents the transmitted power in the field problem. It is therefore apparent that, for fixed r, the higher the mode number n the less easily power is transmitted by a spherical waveguide mode.

[1] L. J. Chu, Physical Limitations of Omnidirectional Antennas, *J. Appl. Phy.*, vol. 19, pp. 1163–1175, December, 1948.

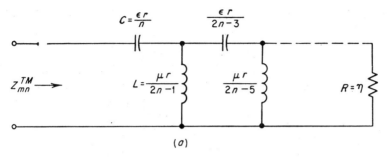

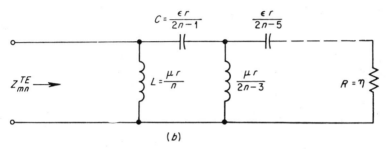

FIG. 6-5. Equivalent circuits for the (a) TM_{mn} and (b) TE_{mn} modes of free space.

A quality factor Q_n for modes of order n can now be defined as

$$Q_n = \begin{cases} \dfrac{2\omega \bar{\mathcal{W}}_e}{\bar{\mathcal{P}}} & \bar{\mathcal{W}}_e > \bar{\mathcal{W}}_m \\ \dfrac{2\omega \bar{\mathcal{W}}_m}{\bar{\mathcal{P}}} & \bar{\mathcal{W}}_m > \bar{\mathcal{W}}_e \end{cases} \qquad (6\text{-}60)$$

where $\bar{\mathcal{W}}_e$ and $\bar{\mathcal{W}}_m$ are the average electric and magnetic energies stored in the C's and L's, and $\bar{\mathcal{P}}$ is the power dissipated in the resistance. In TM waves $\bar{\mathcal{W}}_e > \bar{\mathcal{W}}_m$, while in TE waves $\bar{\mathcal{W}}_m > \bar{\mathcal{W}}_e$. However, the two cases are dual to each other; so the Q's of TM waves are equal to the Q's of the corresponding TE waves. An approximate calculation of the Q's for $Q > 1$ is shown in Fig. 6-6. Note that for $kr > n$ the wave impedances are low Q and for $kr < n$ they are high Q. This again illustrates the cutoff phenomenon that occurs at $kr = n$.

6-5. Other Radial Waveguides. A number of structures capable of supporting radially traveling waves can be obtained by covering $\theta = $ constant and $\phi = $ constant surfaces with conductors. Such "radial waveguides" are shown in Fig. 6-7.

We can have waves outside or inside a single conducting cone, as shown in Fig. 6-7a and b. These two cases are actually a single problem with two different values of θ_1. The fields must be periodic in 2π on ϕ and

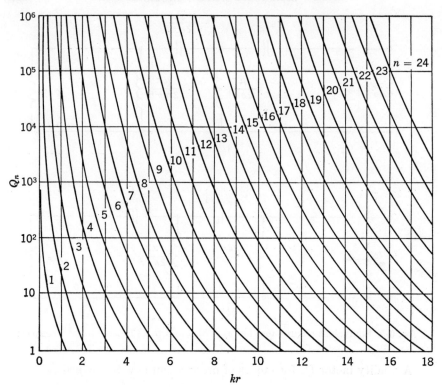

Fig. 6-6. Quality factors Q_n for the TM_{mn} and TE_{mn} modes of free space.

finite at $\theta = 0$. Hence, we choose the TM to r mode functions

$$(A_r)_{mv} = P_v^m(\cos\theta) \begin{Bmatrix} \cos m\phi \\ \sin m\phi \end{Bmatrix} \hat{H}_v^{(1)}(kr) \quad (6\text{-}61)$$

where $m = 0, 1, 2, \ldots$. To satisfy the boundary condition $E_r = E_\phi = 0$ at $\theta = \theta_1$, the parameter v must be a solution to

$$P_v^m(\cos\theta_1) = 0 \quad (6\text{-}62)$$

Also, we choose the TE to r mode functions

$$(F_r)_{mv} = P_v^m(\cos\theta) \begin{Bmatrix} \cos m\phi \\ \sin m\phi \end{Bmatrix} \hat{H}_v^{(1)}(kr) \quad (6\text{-}63)$$

where $m = 0, 1, 2, \ldots$. To satisfy the boundary condition $E_\phi = 0$ at $\theta = \theta_1$, the parameter v must be a solution to

$$\left[\frac{d}{d\theta} P_v^m(\cos\theta)\right]_{\theta=\theta_1} = 0 \quad (6\text{-}64)$$

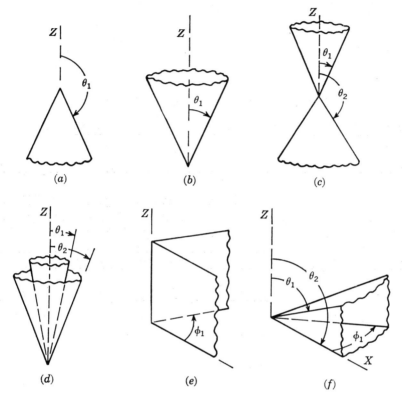

Fig. 6-7. Some spherically radial waveguides. (a) Conical (waves external); (b) conical (waves internal); (c) biconical; (d) coaxial; (e) wedge; (f) horn.

Because of a scarcity of tables for the eigenvalues v, it is difficult to obtain numerical values. The field components are, of course, obtained from the A_r and F_r by Eqs. (6-26).

The biconical and coaxial guides of Fig. 6-7c and d are again a single mathematical problem. Now both $\theta = 0$ and $\theta = \pi$ are excluded from the region of field; so two Legendre solutions, $P_v^m(\cos\theta)$ and $Q_v^m(\cos\theta)$, or $P_v^m(\cos\theta)$ and $P_v^m(-\cos\theta)$, are needed. Choosing the latter two solutions, we find modes TM to r defined by

$$(A_r)_{mv} = [P_v^m(\cos\theta) \, P_v^m(-\cos\theta_1) - P_v^m(-\cos\theta) \, P_v^m(\cos\theta_1)]$$
$$\begin{Bmatrix}\cos m\phi \\ \sin m\phi\end{Bmatrix} \hat{H}_v^{(1)}{}^{(2)}(kr) \quad (6\text{-}65)$$

where $m = 0, 1, 2, \ldots$, and the v are determined by the roots of

$$P_v^m(\cos\theta_2)P_v^m(-\cos\theta_1) - P_v^m(-\cos\theta_2)P_v^m(\cos\theta_1) = 0 \quad (6\text{-}66)$$

Similarly, for the modes TE to r we have

$$(F_r)_{mv} = \left[P_v^m(\cos\theta) \frac{dP_v^m(-\cos\theta_1)}{d\theta_1} - P_v^m(-\cos\theta) \frac{dP_v^m(\cos\theta_1)}{d\theta_1} \right] \begin{Bmatrix} \cos m\phi \\ \sin m\phi \end{Bmatrix} \hat{H}_v^{(1)}{}_{(2)}(kr) \quad (6\text{-}67)$$

where $m = 0, 1, 2, \ldots$, and the v are determined by the roots of

$$\frac{dP_v^m(\cos\theta_2)}{d\theta_2} \frac{dP_v^m(-\cos\theta_1)}{d\theta_1} - \frac{dP_v^m(-\cos\theta_2)}{d\theta_2} \frac{dP_v^m(\cos\theta_1)}{d\theta_1} = 0 \quad (6\text{-}68)$$

Again the field components are found from the A_r and F_r of Eqs. (6-65) and (6-67) according to Eqs. (6-26).

The dominant mode of the biconical and coaxial guides is a TEM, or transmission-line, mode. The eigenvalues $m = 0$, $v = 0$ satisfy both Eqs. (6-66) and (6-68), but the A_r and F_r of Eqs. (6-65) and (6-67) vanish. We could redefine Eq. (6-65) such that the limit $v \to 0$ exists, but instead let us separately define the TEM mode as a TM$_{00}$ mode defined by

$$(A_r)_{00} = Q_0(\cos\theta) \hat{H}_0^{(1)}{}_{(2)}(kr) = \log \cot \frac{\theta}{2} (\mp j) e^{\pm jkr} \quad (6\text{-}69)$$

The field components of this mode, determined from Eqs. (6-26), are

$$\begin{aligned} E_\theta^{\mp} &= \frac{jk}{\omega\epsilon r \sin\theta} e^{\pm jkr} \\ H_\phi^{\mp} &= \mp \frac{j}{r \sin\theta} e^{\pm jkr} \end{aligned} \quad (6\text{-}70)$$

where the upper signs refer to inward-traveling waves and the lower signs to outward-traveling waves. The wave impedance in the direction of travel is

$$\left. \begin{aligned} Z_r^+ &= \frac{E_\theta^+}{H_\phi^+} \\ Z_{-r}^- &= \frac{-E_\theta^-}{H_\phi^-} \end{aligned} \right\} = \eta \quad (6\text{-}71)$$

which is the same as for TEM waves on ordinary transmission lines. The characteristic impedance defined in terms of voltage and current is of greater interest. At a given r, the voltage is defined as

$$V = \int_{\theta_1}^{\pi-\theta_2} E_\theta r\, d\theta = j\eta \log \frac{\cot(\theta_1/2)}{\cot(\theta_2/2)} e^{\pm jkr} \quad (6\text{-}72)$$

and the current as

$$I = \int_0^{2\pi} H_\phi r \sin\theta\, d\phi = \mp 2\pi j e^{\pm jkr} \quad (6\text{-}73)$$

At small r these are the usual circuit quantities. The characteristic impedance is

$$Z_0 = \frac{V^+}{I^+} = -\frac{V^-}{I^-} = \frac{\eta}{2\pi} \log \frac{\cot(\theta_1/2)}{\cot(\theta_2/2)} \qquad (6\text{-}74)$$

Note that the various equations are the same as for the usual uniform transmission lines. For this reason the biconical and coaxial radial lines are called uniform radial transmission lines.

Spherical waves on the wedge waveguide of Fig. 6-7e exist for all θ but only for restricted ϕ. Hence, the wave functions will contain only the $P_n^w(\cos\theta)$ with n an integer and w determined by the boundary conditions. We then find TM modes defined by

$$(A_r)_{nw} = P_n^w(\cos\theta)\sin w\phi \, \hat{H}_n^{(2)}(kr) \qquad (6\text{-}75)$$

where $n = 1, 2, 3, \ldots$, and

$$w = \frac{p\pi}{\phi_1} \qquad (6\text{-}76)$$

with $p = 1, 2, 3, \ldots$. The TE modes are defined by

$$(F_r)_{nw} = P_n^w(\cos\theta)\cos w\phi \, \hat{H}_n^{(2)}(kr) \qquad (6\text{-}77)$$

where $n = 1, 2, 3, \ldots$, and w is given by Eq. (6-76) with $p = 0, 1, 2, \ldots$. There is no TEM spherical mode, the TEM mode being a cylindrical wave defined by Eqs. (5-48) and (5-49).

Finally, the spherical-horn waveguide of Fig. 6-7f will require Legendre functions $L_v^w(\cos\theta)$ of nonintegral v and w. The TM modes can be defined by Eqs. (6-65) and (6-66) with m changed to w and only the $\sin w\phi$ functions allowed. The values of w are those of Eq. (6-76). Similarly, the TE modes can be defined by Eqs. (6-67) and (6-68) with m changed to w and only the $\cos w\phi$ functions allowed. Again, w is given by Eq. (6-76). There will, of course, be no TEM mode.

6-6. Other Resonators. Resonators having modes expressible in terms of single spherical wave functions can be obtained by closing each of the radial waveguides of Fig. 6-7 by one or two conducting spheres. Some examples are shown in Fig. 6-8. The fields in each case can be expressed in terms of mode functions which are the same as for the radial waveguides of the preceding section, except that the traveling-wave functions $\hat{H}_n^{(1)}(kr)$ and $\hat{H}_n^{(2)}(kr)$ are replaced by standing-wave functions $\hat{J}_n(kr)$ and $\hat{N}_n(kr)$. Numerical calculations are hampered by a scarcity of tables of eigenvalues.

Let us calculate the Q's for the dominant modes of the first three cavities of Fig. 6-8. For the hemispherical cavity of Fig. 6-8a, the dominant mode is the dominant TM to r mode of the complete spherical cavity,

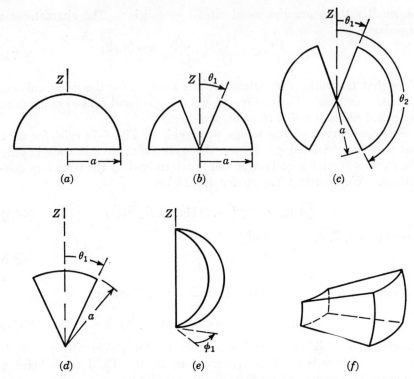

Fig. 6-8. Some cavities having modes expressible in terms of single spherical wave functions. (a) Hemispherical; (b) hemisphere with cone; (c) biconical; (d) conical; (e) wedge; (f) segment.

considered in Sec. 6-2. The magnetic field is

$$H_\phi = \frac{1}{r} \hat{J}_1\left(2.744 \frac{r}{a}\right) \sin \theta$$

and the stored energy is one-half that for the complete spherical cavity [Eq. (6-35)]; hence

$$\mathcal{W} = \frac{4\pi\mu}{3k} \quad (1.14)$$

The power dissipated in the hemispherical part of the walls is one-half that dissipated in the walls of the complete spherical cavity; hence

$$(\bar{\mathcal{P}}_d)_{\text{hemisphere}} = \mathcal{R} \frac{4\pi}{3} \quad (1.13)$$

The power dissipated in the plane wall is

$$(\bar{\mathcal{P}}_d)_{\text{plane}} = \mathcal{R} 2\pi \int_0^a H_\phi^2 \bigg|_{\theta=\pi/2} r\, dr = \mathcal{R} 2\pi (0.571)$$

Thus, the Q of the resonator is

$$Q = \frac{\omega \mathcal{W}}{\mathcal{P}_d} = 0.573 \frac{\eta}{\mathcal{R}} \qquad (6\text{-}78)$$

If we compare this with the Q of a rectangular cavity [Eq. (2-102)] and with the Q of a circular cavity [Eq. (5-58)] we see that, for the same height-to-diameter ratios, the hemispherical cavity Q is only 3.2 per cent higher than the rectangular cavity Q, and 4.5 per cent lower than the circular cavity Q. The hemispherical cavity Q is 54 per cent less than the spherical cavity Q, but we have removed the mode degeneracy. From Tables 6-1 and 6-2 we find that the second resonant frequency is 1.41 times the lowest resonant frequency for the hemispherical cavity, compared to approximately 1.58 for the rectangular and circular cavities.

The cavities of Fig. 6-8b and c are theoretically important because they have circuit terminals available. In other words, a voltage and current calculated at the cone tips have the usual circuit theory interpretation. The dominant mode

$$H_\phi = \frac{A \cos k(a-r)}{r \sin \theta} \qquad E_\theta = j\eta A \frac{\sin k(a-r)}{r \sin \theta}$$

will be excited if the cavity is fed across the cone tips. The voltage seen by the source is

$$V_{in} = \lim_{r \to 0} \int_{\theta_1}^{\theta_2} E_\theta r \, d\theta = 2\pi j A Z_0 \sin ka$$

where Z_0 is the characteristic impedance [Eq. (6-74)]. The current at the source is

$$I_{in} = \lim_{r \to 0} \int_0^{2\pi} H_\phi r \, d\phi = 2\pi A \cos ka$$

Hence, the input impedance seen by the source is

$$Z_{in} = \frac{V_{in}}{I_{in}} = jZ_0 \tan ka \qquad (6\text{-}79)$$

which is the usual formula for the input impedance of a short-circuited uniform transmission line. (We saw in the preceding section that the TEM mode of the biconical guide is a uniform transmission-line mode.) The resonances occur when $ka = n\pi/2$, or

$$\omega_r = \frac{n\pi}{2a \sqrt{\epsilon\mu}} \qquad (6\text{-}80)$$

where $n = 1, 2, 3, \ldots$. In the loss-free case, the input impedance is infinite for n odd (antiresonance) and zero for n even. When small losses are present, the input impedance is large for n odd and small for n even.

Let us consider the lowest resonance ($n = 1$) in more detail. The input conductance at resonance can be determined from the power losses as

$$G_{in} = \frac{\bar{\mathcal{P}}_d}{|V_{in}|^2} = \frac{\omega \mathcal{W}}{Q|V_{in}|^2}$$

The energy stored $\mathcal{W}$ is simply calculated as

$$\mathcal{W} = \mu \iiint |H|^2 \, d\tau = \frac{\pi^3}{\omega} |A|^2 Z_0$$

Thus
$$G_{in} = \frac{\pi^3 Z_0}{Q(2\pi Z_0)^2} = \frac{\pi}{4 Z_0 Q} \qquad (6\text{-}81)$$

where Z_0 is given by Eq. (6-74) and Q can be calculated in the usual manner as[1]

$$Q = \frac{\pi \eta}{4 \mathcal{R}} \left\{ 1 + 0.824 \frac{\csc \theta_1 + \csc \theta_2}{\log \left[\cot (\theta_1/2) \tan (\theta_2/2) \right]} \right\}^{-1} \qquad (6\text{-}82)$$

This Q is maximum when $\theta_1 = \pi - \theta_2 = 33.5°$, in which case

$$Q = 0.350 \frac{\eta}{\mathcal{R}}$$

Note that this is smaller than the Q's of other cavities that we have considered because of the introduction of the biconical feed system. In the special case $\theta_2 = 90°$, we have the cone-fed hemispherical cavity of Fig. 6-8b, for which

$$Q = \frac{\pi \eta}{4 \mathcal{R}} \left[1 + 0.824 \frac{1 + \csc \theta_1}{\log \cot (\theta_1/2)} \right]^{-1} \qquad (6\text{-}83)$$

This Q is maximum when $\theta_1 = 24.1°$, in which case

$$Q = 0.276 \frac{\eta}{\mathcal{R}}$$

This is a lower Q than that for the hemispherical cavity without the cone [Eq. (6-78)], because of the feed system. The input conductance [Eq. (6-81)] is not minimum when Q is maximum, because Z_0 is also a function of θ_1 and θ_2. For the biconical resonator (Fig. 6-8c), the input conductance is minimum when the cone angles are $\theta_1 = \pi - \theta_2 = 9.2°$. For the cone-fed hemispherical cavity (Fig. 6-8b), the minimum conductance is obtained when $\theta_1 = 7.5°$.

6-7. Sources of Spherical Waves. The sources of the lowest-order spherical waves are current elements, treated in Sec. 2-9. For exam-

---

[1] S. A. Schelkunoff, "Electromagnetic Waves," pp. 288–290, D. Van Nostrand Company, Inc., Princeton, N.J., 1943.

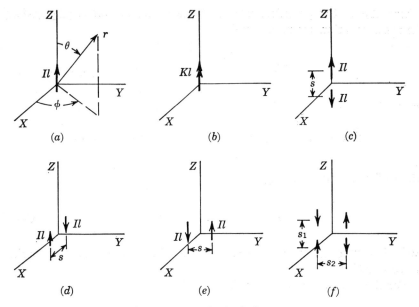

Fig. 6-9. Some sources of spherical waves.

ple, the electric-current element of Fig. 6-9a radiates a field given by $\mathbf{H} = \nabla \times \mathbf{A}$ with

$$A_z = \frac{Il}{4\pi r} e^{-jkr} = \frac{kIl}{4\pi j} h_0^{(2)}(kr) \qquad (6\text{-}84)$$

where $h_0^{(2)}$ is the spherical Hankel function of Eq. (6-11). Alternatively, the field can be represented by a radially directed $\mathbf{A}$ given by

$$\begin{aligned} A_r &= \frac{jkIl}{4\pi}\left(1 + \frac{1}{jkr}\right) e^{-jkr} \cos\theta \\ &= \frac{jkIl}{4\pi} \hat{H}_1^{(2)}(kr) P_1(\cos\theta) \end{aligned} \qquad (6\text{-}85)$$

The field of the current element is discussed in detail in Sec. 2-9. The dual source is the magnetic-current element of Fig. 6-9b. The field of this source is given by $\mathbf{E} = -\nabla \times \mathbf{F}$ where F_z or F_r is the same as A_z or A_r with I replaced by K.

The fields of the dipole and higher-multipole sources, represented by Fig. 6-9c to f, can be obtained by the same method as used in Sec. 5-6. For example, for the dipole source of Fig. 6-9c,

$$A_z = A_z^1\left(x, y, z - \frac{s}{2}\right) - A_z^1\left(x, y, z + \frac{s}{2}\right)$$

where $A_z{}^1$ is the potential from a single current element [Eq. (6-84)]. As the separation s is made small,

$$A_z \xrightarrow[s \to 0]{} -s \frac{\partial A_z{}^1}{\partial z} = \frac{jkIls}{4\pi} \frac{\partial}{\partial z} h_0{}^{(2)}(kr)$$

where $r = \sqrt{x^2 + y^2 + z^2}$. Also,

$$\frac{\partial}{\partial z} h_0{}^{(2)}(kr) = k h_0{}^{(2)\prime}(kr) \cos\theta = -k h_1{}^{(2)}(kr) P_1(\cos\theta)$$

Hence for the dipole of Fig. 6-9c

$$A_z = \frac{k^2 Ils}{4\pi j} h_1{}^{(2)}(kr) P_1(\cos\theta) \tag{6-86}$$

and $\mathbf{H} = \nabla \times \mathbf{A}$. Thus, the vector potential is a first-order spherical wave function.

For the dipole source of Fig. 6-9d, we have

$$A_z \xrightarrow[s \to 0]{} -s \frac{\partial A_z{}^1}{\partial x} = \frac{jkIls}{4\pi} \frac{\partial}{\partial x} h_0{}^{(2)}(kr)$$

$$= \frac{jk^2 Ils}{4\pi} h_0{}^{(2)\prime}(kr) \sin\theta \cos\phi$$

which can be written as

$$A_z = \frac{k^2 Ils}{4\pi j} h_1{}^{(2)}(kr) P_1^1(\cos\theta) \cos\phi \tag{6-87}$$

This is a first-order wave function of $n = 1$, $m = 1$. Similarly, for the dipole source of Fig. 6-9e, we find

$$A_z = \frac{k^2 Ils}{4\pi j} h_1{}^{(2)}(kr) P_1^1(\cos\theta) \sin\phi \tag{6-88}$$

Thus, all wave functions of order one can be interpreted as the A_z of dipole sources.

This procedure can be extended to higher-multipole sources in a straightforward manner. For example, for the quadrupole source of Fig. 6-9f, we have

$$A_z = s_1 s_2 \frac{\partial^2 A_z{}^1}{\partial y\, \partial z} = -s_2 \frac{\partial A_z{}^{(2)}}{\partial y}$$

where $A_z{}^{(2)}$ is for the dipole of Fig. 6-9c, given by Eq. (6-86). We also have

$$\frac{\partial}{\partial y}[h_1^{(2)}(kr)P_1(\cos\theta)] = \frac{y}{r}\frac{\partial}{\partial r}\left[h_1^{(2)}(kr)\frac{z}{r}\right]$$

$$= -\frac{kyz}{r^2}h_2^{(2)}(kr) = -kh_2^{(2)}(kr)\sin\theta\cos\theta\sin\phi$$

$$= \frac{k}{3}h_2^{(2)}(kr)P_2^1(\cos\theta)\sin\phi$$

Hence the vector potential of the quadrupole of Fig. 6-9f is

$$A_z = \frac{jk^3Ils_1s_2}{12\pi}h_2^{(2)}(kr)P_2^1(\cos\theta)\sin\phi \qquad (6\text{-}89)$$

In this manner we can identify each wave function of order n with the A_z of a multipole source of $2n$ z-directed current elements.

6-8. Wave Transformations. Now that we have wave functions in three basic coordinate geometries available, the number of possible wave transformations becomes very large. We shall here establish only a few representative transformations involving spherical wave functions. A convenient method of obtaining the desired results is that of Sec. 5-8.

Let us first consider the plane wave e^{jz} and express it in terms of spherical wave functions. This wave is finite at the origin and independent of ϕ; hence an expansion of the form

$$e^{jz} = e^{jr\cos\theta} = \sum_{n=0}^{\infty} a_n j_n(r) P_n(\cos\theta)$$

must be possible (see Fig. 6-1 for the coordinate orientation). To evaluate the a_n, multiply each side by $P_q(\cos\theta)\sin\theta$ and integrate from 0 to π on θ. Because of orthogonality [Eq. (6-40)], all terms except $q = n$ vanish, and by Eq. (6-41) we have

$$\int_0^\pi e^{jr\cos\theta} P_n(\cos\theta)\sin\theta\, d\theta = \frac{2a_n}{2n+1} j_n(r)$$

The nth derivative of the left-hand side with respect to r evaluated at $r = 0$ is

$$j^n \int_0^\pi \cos^n\theta\, P_n(\cos\theta)\sin\theta\, d\theta = \frac{j^n\, 2^{n+1}(n!)^2}{(2n+1)!}$$

The nth derivative of the right-hand side evaluated at $r = 0$ is

$$\frac{2^{n+1}(n!)^2}{(2n+1)(2n+1)!} a_n$$

Hence, equating the preceding two expressions, we obtain

$$a_n = j^n(2n+1)$$

which, substituted back into our starting equation, gives

$$e^{jz} = e^{jr\cos\theta} = \sum_{n=0}^{\infty} j^n(2n+1)j_n(r)P_n(\cos\theta) \qquad (6\text{-}90)$$

Note that we have also established the identity

$$j_n(r) = \frac{j^{-n}}{2}\int_0^\pi e^{jr\cos\theta} P_n(\cos\theta)\sin\theta\,d\theta \qquad (6\text{-}91)$$

Equation (6-90) is the desired transformation expressing a plane wave in terms of spherical wave functions.

Transformations from cylindrical waves to spherical waves can be obtained in a similar fashion. For example, consider the cylindrical wave $J_0(\rho)$, which is finite at $r = 0$, independent of ϕ, and symmetrical about $\theta = \pi/2$. Hence, there exists an expansion

$$J_0(\rho) = J_0(r\sin\theta) = \sum_{n=0}^{\infty} b_n j_{2n}(r)P_{2n}(\cos\theta)$$

As before, we multiply each side by $P_q(\cos\theta)\sin\theta$ and integrate from 0 to π on θ. The result is

$$\int_0^\pi J_0(r\sin\theta)P_{2n}(\cos\theta)\sin\theta\,d\theta = \frac{2b_n}{4n+1}j_{2n}(r)$$

To determine the b_n, we differentiate each side $2n$ times with respect to r and set $r = 0$. This gives

$$b_n = \frac{(-1)^n(4n+1)(2n-1)!}{2^{2n-1}n!(n-1)!}$$

Hence the desired wave transformation is

$$J_0(\rho) = J_0(r\sin\theta) = \sum_{n=0}^{\infty} \frac{(-1)^n(4n+1)(2n-1)!}{2^{2n-1}n!(n-1)!} j_{2n}(r)P_{2n}(\cos\theta)$$

$$(6\text{-}92)$$

Note also that the two equations preceding Eq. (6-92) establish an integral formula for $j_{2n}(r)$.

Now let us consider wave transformations corresponding to changes from one spherical coordinate system to another. To illustrate, consider the field of a point source at $\mathbf{r}'$

$$h_0^{(2)}(|\mathbf{r}-\mathbf{r}'|) = -\frac{e^{-j|\mathbf{r}-\mathbf{r}'|}}{j|\mathbf{r}-\mathbf{r}'|}$$

SPHERICAL WAVE FUNCTIONS

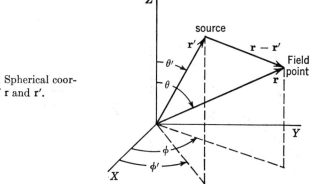

Fig. 6-10. Spherical coordinates of **r** and **r**′.

where **r** and **r**′ are defined in Fig. 6-10. We desire to express this field in terms of wave functions referred to $r = 0$. The field has rotational symmetry about the **r**′ axis; so let us express the wave functions in terms of the angle ξ where

$$\cos \xi = \cos \theta \cos \theta' + \sin \theta \sin \theta' \cos (\phi - \phi') \qquad (6\text{-}93)$$

Allowable wave functions in the region $r < r'$ are $j_n(r)P_n(\cos \xi)$, and allowable wave functions $r > r'$ are $h_n^{(2)}(r)P_n(\cos \xi)$. Furthermore, the field is symmetric in **r** and **r**′; hence we construct

$$h_0^{(2)}(|\mathbf{r} - \mathbf{r}'|) = \begin{cases} \displaystyle\sum_{n=0}^{\infty} c_n h_n^{(2)}(r') j_n(r) P_n(\cos \xi) & r < r' \\ \displaystyle\sum_{n=0}^{\infty} c_n j_n(r') h_n^{(2)}(r) P_n(\cos \xi) & r > r' \end{cases}$$

where the c_n are constants. If we let the source recede to infinity, the field in the vicinity of the origin is a plane wave. Using the asymptotic formula

$$h_n^{(2)}(z) = \frac{j^{n+1}}{z} e^{-jz}$$

we have for the left-hand side of the preceding equation

$$h_0^{(2)}(|\mathbf{r} - \mathbf{r}'|) \xrightarrow[\substack{r' \to \infty \\ \theta' \to 0}]{} \frac{je^{-jr'}}{r'} e^{jr \cos \theta}$$

and for the right-hand side

$$\xrightarrow[\substack{r' \to \infty \\ \theta' \to 0}]{} \frac{je^{-jr'}}{r'} \sum_{n=0}^{\infty} c_n j^n(r) P_n(\cos \theta)$$

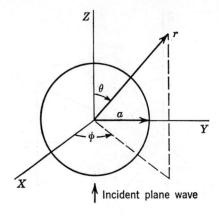

Fig. 6-11. A plane wave incident on a conducting sphere.

A comparison of these two expressions with Eq. (6-90) shows that $c_n = 2n + 1$; hence

$$h_0^{(2)}(|\mathbf{r} - \mathbf{r}'|) = \begin{cases} \sum_{n=0}^{\infty} (2n + 1)h_n^{(2)}(r')j_n(r)P_n(\cos \xi) & r < r' \\ \sum_{n=0}^{\infty} (2n + 1)j_n(r')h_n^{(2)}(r)P_n(\cos \xi) & r > r' \end{cases} \quad (6\text{-}94)$$

This is the *addition theorem* for spherical Hankel functions. Since $h_n^{(1)} = h_n^{(2)*}$, Eq. (6-94) is also valid for superscripts (2) replaced by (1). The real part of Eq. (6-94) is an addition theorem for $j_0(|\mathbf{r} - \mathbf{r}'|)$, and the imaginary part is an addition theorem for $n_0(|\mathbf{r} - \mathbf{r}'|)$.

Finally, one can express the zonal harmonics $P_n(\cos \xi)$ in terms of the tesseral harmonics $P_n^m(\cos \theta)h(m\phi)$. In other words, a wave function referred to the $\xi = 0$ axis of Fig. 6-10 can be expressed in terms of wave functions referred to the $\theta = 0$ axis. The identity is

$$P_n(\cos \xi) = \sum_{m=1}^{n} \epsilon_m \frac{(n-m)!}{(n+m)!} P_n^m(\cos \theta) P_n^m(\cos \theta') \cos m(\phi - \phi') \quad (6\text{-}95)$$

where ϵ_m is Neumann's number (1 for $m = 0$ and 2 for $m > 0$). The proof of Eq. (6-95), plus some other wave transformations that we have not treated explicitly, can be found in Stratton's book.[1] Equation (6-95) is an addition theorem for Legendre polynomials.

6-9. Scattering by Spheres. Figure 6-11 represents a conducting sphere illuminated by an incident plane wave. Take the incident wave

[1] J. A. Stratton, "Electromagnetic Theory," pp. 406–414, McGraw-Hill Book Company, Inc., New York, 1941.

to be x-polarized and z-traveling, that is,

$$E_x{}^i = E_0 e^{-jkz} = E_0 e^{-jkr\cos\theta}$$
$$H_y{}^i = \frac{E_0}{\eta} e^{-jkz} = \frac{E_0}{\eta} e^{-jkr\cos\theta} \tag{6-96}$$

For convenience in applying boundary conditions, we express this incident field as the sum of components TM and TE to r, that is, in terms of an F_r and an A_r. From Eqs. (6-26) we see that A_r can be obtained from E_r, and F_r from H_r. The r component of $\mathbf{E}^i$ is

$$E_r{}^i = \cos\phi \sin\theta\, E_x{}^i = E_0 \frac{\cos\phi}{jkr} \frac{\partial}{\partial\theta}(e^{-jkr\cos\theta})$$

Using Eq. (6-90), we can write this as

$$E_r{}^i = E_0 \frac{\cos\phi}{jkr} \sum_{n=0}^{\infty} j^{-n}(2n+1) j_n(kr) \frac{\partial}{\partial\theta} P_n(\cos\theta)$$

Finally, using Eq. (6-23) and the relationship $\partial P_n/\partial\theta = P_n{}^1$, we obtain[1]

$$E_r{}^i = -\frac{jE_0 \cos\phi}{(kr)^2} \sum_{n=1}^{\infty} j^{-n}(2n+1) \hat{J}_n(kr) P_n{}^1(\cos\theta)$$

Noting the form of $E_r{}^i$, we construct the magnetic vector potential as

$$A_r{}^i = \frac{E_0}{\omega\mu} \cos\phi \sum_{n=1}^{\infty} a_n \hat{J}_n(kr) P_n{}^1(\cos\theta) \tag{6-97}$$

and evaluate $E_r{}^i$ by Eqs. (6-26). Simplifying the result by Eq. (6-24), we obtain

$$E_r{}^i = -\frac{jE_0 \cos\phi}{(kr)^2} \sum_{n=1}^{\infty} a_n n(n+1) \hat{J}_n(kr) P_n{}^1(\cos\theta)$$

Comparing this expression with the preceding formula for $E_r{}^i$, we see that

$$a_n = \frac{j^{-n}(2n+1)}{n(n+1)} \tag{6-98}$$

A similar procedure using $H_r{}^i$ and $F_r{}^i$ gives

$$F_r{}^i = \frac{E_0}{k} \sin\phi \sum_{n=1}^{\infty} a_n \hat{J}_n(kr) P_n{}^1(\cos\theta) \tag{6-99}$$

where the a_n are again given by Eq. (6-98).

[1] Note that the $n = 0$ term of the summation drops out because $P_0{}^1 = 0$.

Now that the incident field is expressed in terms of radially TE and TM modes, the rest of the solution parallels the cylinder problem (Sec. 5-9). The scattered field will be generated by an A_r and F_r of the same form as the incident field with $\hat{J}_n$ replaced by $\hat{H}_n^{(2)}$. Hence, we construct scattered potentials as

$$A_r^s = \frac{E_0}{\omega\mu} \cos\phi \sum_{n=1}^{\infty} b_n \hat{H}_n^{(2)}(kr) P_n^1(\cos\theta)$$

$$F_r^s = \frac{E_0}{k} \sin\phi \sum_{n=1}^{\infty} c_n \hat{H}_n^{(2)}(kr) P_n^1(\cos\theta)$$

(6-100)

The total field is, of course, the sum of the incident and scattered fields. Therefore **E** and **H** are given by Eqs. (6-26) where

$$A_r = \frac{E_0}{\omega\mu} \cos\phi \sum_{n=1}^{\infty} [a_n \hat{J}_n(kr) + b_n \hat{H}_n^{(2)}(kr)] P_n^1(\cos\theta)$$

$$F_r = \frac{E_0}{k} \sin\phi \sum_{n=1}^{\infty} [a_n \hat{J}_n(kr) + c_n \hat{H}_n^{(2)}(kr)] P_n^1(\cos\theta)$$

(6-101)

The boundary conditions are $E_\theta = E_\phi = 0$ at $r = a$, which require that

$$b_n = -a_n \frac{\hat{J}_n'(ka)}{\hat{H}_n^{(2)'}(ka)}$$

$$c_n = -a_n \frac{\hat{J}_n(ka)}{\hat{H}_n^{(2)}(ka)}$$

(6-102)

This completes the solution. Note that the problem can be viewed as a short-circuited radial transmission line (Sec. 6-4) with many modes superimposed.

The surface current on the sphere can be found according to $\mathbf{J}_s = \mathbf{u}_r \times \mathbf{H}$ at $r = a$. The result is

$$J_\theta = \frac{j}{\eta} E_0 \frac{\cos\phi}{ka} \sum_{n=1}^{\infty} a_n \left[\frac{\sin\theta\, P_n^{1'}(\cos\theta)}{\hat{H}_n^{(2)'}(ka)} + \frac{jP_n^1(\cos\theta)}{\sin\theta\, \hat{H}_n^{(2)}(ka)} \right]$$

$$J_\phi = \frac{j}{\eta} E_0 \frac{\sin\phi}{ka} \sum_{n=1}^{\infty} a_n \left[\frac{P_n^1(\cos\theta)}{\sin\theta\, \hat{H}_n^{(2)'}(ka)} - \frac{\sin\theta\, P_n^{1'}(\cos\theta)}{j\hat{H}_n^{(2)}(ka)} \right]$$

(6-103)

where the a_n are given by Eq. (6-98). The distant scattered field can be found from the general expressions by using the asymptotic formula

$$\hat{H}_n^{(2)}(kr) \xrightarrow[kr\to\infty]{} j^{n+1} e^{-jkr}$$

and retaining only the terms varying as $1/r$. The result is

$$E_\theta{}^s = \frac{jE_0}{kr} e^{-jkr} \cos\phi \sum_{n=1}^{\infty} j^n \left[b_n \sin\theta\, P_n{}^{1\prime}(\cos\theta) - c_n \frac{P_n{}^1(\cos\theta)}{\sin\theta} \right] \quad (6\text{-}104)$$

$$E_\phi{}^s = \frac{jE_0}{kr} e^{-jkr} \sin\phi \sum_{n=1}^{\infty} j^n \left[b_n \frac{P_n{}^1(\cos\theta)}{\sin\theta} - c_n \sin\theta\, P_n{}^{1\prime}(\cos\theta) \right]$$

where the b_n and c_n are given by Eqs. (6-102). Of particular interest is the back-scattered field

$$E_x{}^s = E_\theta{}^s \bigg|_{\substack{\theta=\pi\\\phi=\pi}} = E_\phi{}^s \bigg|_{\substack{\theta=\pi\\\phi=-\pi/2}}$$

From this we can calculate the echo area according to Eq. (3-30), which is

$$A_e = \lim_{r\to\infty} \left(4\pi r^2 \frac{|E_x{}^s|^2}{|E_0|^2} \right)$$

Making use of the relationships

$$\frac{P_n{}^1(\cos\theta)}{\sin\theta} \xrightarrow[\theta\to\pi]{} \frac{(-1)^n}{2} n(n+1)$$

$$\sin\theta\, P_n{}^{1\prime}(\cos\theta) \xrightarrow[\theta\to\pi]{} \frac{(-1)^n}{2} n(n+1)$$

and the Wronskian of the spherical Bessel functions, we find

$$A_e = \frac{\lambda^2}{4\pi} \left| \sum_{n=1}^{\infty} \frac{(-1)^n (2n+1)}{\hat{H}_n^{(2)}(ka)\hat{H}_n^{(2)\prime}(ka)} \right|^2 \quad (6\text{-}105)$$

A plot of A_e/λ^2 is shown in Fig. 6-12. For small ka, the $n = 1$ term of Eq. (6-105) becomes dominant and

$$A_e \xrightarrow[ka\to 0]{} \frac{9\lambda^2}{4\pi} (ka)^6 \quad (6\text{-}106)$$

which is a good approximation when $a/\lambda < 0.1$. Equation (6-106) is known as the *Rayleigh scattering law*. It states that the echo area of small spheres varies as λ^{-4} and was

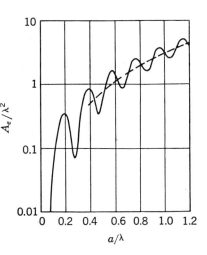

FIG. 6-12. Echo area of a conducting sphere of radius a (optical approximation shown dashed).

first used to explain the blueness of the sky. For large spheres

$$A_e \xrightarrow[ka \to \infty]{} \pi a^2 \qquad (6\text{-}107)$$

which is the physical optics solution. The region between the Rayleigh and optical approximations is called the resonance region and is characterized by oscillations of the echo area.

Let us now look at the field scattered by the small conducting sphere. Using small-argument formulas for the spherical Bessel functions, we find from Eq. (6-102) and (6-98) that

$$b_n \xrightarrow[ka \to 0]{} -\frac{n+1}{n} c_n \xrightarrow[ka \to 0]{} \left[\frac{2^n(n-1)!}{(2n)!}\right]^2 \frac{(ka)^{2n+1}}{j^{n+1}} \qquad (6\text{-}108)$$

so the $n = 1$ terms of Eqs. (6-104) become dominant for small ka. Hence, at large distances from small spheres,

$$E_\theta^s \xrightarrow[ka \to 0]{} E_0 \frac{e^{-jkr}}{kr} (ka)^3 \cos \phi \, (\cos \theta - \tfrac{1}{2})$$
$$E_\phi^s \xrightarrow[ka \to 0]{} E_0 \frac{e^{-jkr}}{kr} (ka)^3 \sin \phi \, (\tfrac{1}{2} \cos \theta - 1) \qquad (6\text{-}109)$$

A comparison of this result with the radiation field of dipoles shows that the scattered field is the field of an x-directed electric dipole

$$Il = E_0 \frac{4\pi j}{\eta k^2} (ka)^3 \qquad (6\text{-}110)$$

plus the field of a y-directed magnetic dipole

$$Kl = E_0 \frac{2\pi}{jk^2} (ka)^3 \qquad (6\text{-}111)$$

The ratio of the magnetic to electric dipole moments is $|Kl/Il| = \eta/2$. Figure 6-13 illustrates the origin of these two dipole moments. A surface

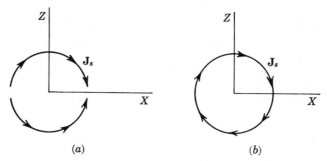

Fig. 6-13. Components of surface current giving rise to the dipole moments of a conducting sphere. (a) Electric moment; (b) magnetic moment.

current in the same direction on each side of the sphere gives rise to the electric moment, while a circulating current gives rise to the magnetic moment. In general, the scattered field of any small body can be expressed in terms of an electric dipole and a magnetic dipole. For a conducting body, the magnetic moment may vanish, but the electric moment must always exist.

Now consider the case of a dielectric sphere, that is, let the region $r < a$ of Fig. 6-11 be characterized by ϵ_d, μ_d, and the region $r > a$ by ϵ_0, μ_0. In addition to the field external to the sphere, specified by potentials of the form of Eqs. (6-101), there will be a field internal to the sphere, specified by

$$A_r^- = \frac{E_0}{\omega\mu_0} \cos \phi \sum_{n=1}^{\infty} d_n \hat{J}_n(k_d r) P_n^1(\cos \theta)$$

$$F_r^- = \frac{E_0}{k_0} \sin \phi \sum_{n=1}^{\infty} e_n \hat{J}_n(k_d r) P_n^1(\cos \theta)$$

(6-112)

The superscripts $-$ denote the region $r < a$, and superscripts $+$ denote the region $r > a$. Boundary conditions to be met at $r = a$ are

$$E_\theta^+ = E_\theta^- \qquad H_\theta^+ = H_\theta^-$$
$$E_\phi^+ = E_\phi^- \qquad H_\phi^+ = H_\phi^-$$

that is, tangential components of **E** and **H** must be continuous. Determining the field components by Eqs. (6-26), using Eqs. (6-101) for $r > a$ and Eqs. (6-112) for $r < a$, and imposing the above boundary conditions, we find

$$b_n = \frac{-\sqrt{\epsilon_d\mu_0}\, \hat{J}'_n(k_0 a)\hat{J}_n(k_d a) + \sqrt{\epsilon_0\mu_d}\, \hat{J}_n(k_0 a)\hat{J}'_n(k_d a)}{\sqrt{\epsilon_d\mu_0}\, \hat{H}_n^{(2)\prime}(k_0 a)\hat{J}_n(k_d a) - \sqrt{\epsilon_0\mu_d}\, \hat{H}_n^{(2)}(k_0 a)\hat{J}'_n(k_d a)} a_n$$

$$c_n = \frac{-\sqrt{\epsilon_d\mu_0}\, \hat{J}_n(k_0 a)\hat{J}'_n(k_d a) + \sqrt{\epsilon_0\mu_d}\, \hat{J}'_n(k_0 a)\hat{J}_n(k_d a)}{\sqrt{\epsilon_d\mu_0}\, \hat{H}_n^{(2)}(k_0 a)\hat{J}'_n(k_d a) - \sqrt{\epsilon_0\mu_d}\, \hat{H}_n^{(2)\prime}(k_0 a)\hat{J}_n(k_d a)} a_n$$

$$d_n = \frac{-j\sqrt{\epsilon_d\mu_0}}{\sqrt{\epsilon_d\mu_0}\, \hat{H}_n^{(2)\prime}(k_0 a)\hat{J}_n(k_d a) - \sqrt{\epsilon_0\mu_d}\, \hat{H}_n^{(2)}(k_0 a)\hat{J}'_n(k_d a)} a_n$$

$$e_n = \frac{j\sqrt{\epsilon_0\mu_d}}{\sqrt{\epsilon_d\mu_0}\, \hat{H}_n^{(2)}(k_0 a)\hat{J}'_n(k_d a) - \sqrt{\epsilon_0\mu_d}\, \hat{H}_n^{(2)\prime}(k_0 a)\hat{J}_n(k_d a)} a_n$$

(6-113)

where a_n is given by Eq. (6-98). The conducting sphere can be obtained as the specialization $\mu_d \to 0$, $\epsilon_d \to \infty$, such that k_d remains finite. Note that, in contrast to static-field problems, $\epsilon_d \to \infty$ is *not* sufficient to specialize to a conductor.

In the special case of a small dielectric sphere, the $n = 1$ coefficients are dominant and reduce to

$$b_1 \xrightarrow[k_0a\to 0]{} -(k_0a)^3 \frac{\epsilon_r - 1}{\epsilon_r + 2}$$

$$c_1 \xrightarrow[k_0a\to 0]{} -(k_0a)^3 \frac{\mu_r - 1}{\mu_r + 2}$$

$$d_1 \xrightarrow[k_0a\to 0]{} \frac{9}{2j\mu_r(2 + \epsilon_r)} \qquad (6\text{-}114)$$

$$e_1 \xrightarrow[k_0a\to 0]{} \frac{9}{2j\epsilon_r(2 + \mu_r)}$$

where $\epsilon_r = \epsilon_d/\epsilon_0$ and $\mu_r = \mu_d/\mu_0$. A calculation of the scattered field reveals that it is the field of the two dipoles

$$Il = \mathbf{u}_x E_0 \frac{4\pi j}{\eta k^2} (ka)^3 \frac{\epsilon_r - 1}{\epsilon_r + 2}$$

$$Kl = \mathbf{u}_y E_0 \frac{4\pi j}{k^2} (ka)^3 \frac{\mu_r - 1}{\mu_r + 2} \qquad (6\text{-}115)$$

Note that the magnetic dipole vanishes if the dielectric is nonmagnetic, that is, if $\mu_r = 1$. Similarly, a magnetic material with $\epsilon_r = 1$ would scatter no electric dipole field. The field internal to the sphere is uniform in both **E** and **H** for the small sphere. In fact, the specialization represented by Eqs. (6-114) is the "quasi-static" solution. It can be obtained by taking the d-c electric and magnetic polarizations and assuming that they vibrate in phase quadrature with the incident field.

6-10. Dipole and Conducting Sphere. Figure 6-14a shows a radially directed electric dipole near a conducting sphere. Figure 6-14b shows a problem reciprocal to that of Fig. 6-14a in the following sense. The component of $\mathbf{E}^a$ in the direction of Il^b equals the component of $\mathbf{E}^b$ in

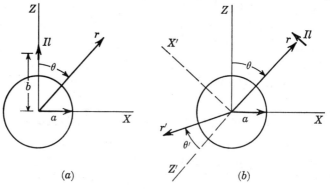

FIG. 6-14. The conducting sphere and a radially directed dipole. (a) Original problem; (b) reciprocal problem.

SPHERICAL WAVE FUNCTIONS

the direction of Il^a. (Superscripts refer to Fig. 6-14a and b.) If the Il of Fig. 6-14b recedes to infinity, we have the plane-wave scatter problem treated in the preceding section. Hence, the radiation field of Fig. 6-14a can be simply obtained from the results of Sec. 6-9.

In particular, in the vicinity of the conducting sphere we have

$$(E_{z'}{}^i)^b \xrightarrow[r \to \infty]{} \frac{-j\omega\mu Il}{4\pi r} e^{-jkr} e^{-jkr' \cos\theta'}$$

which is a plane wave. Letting

$$E_0 = \frac{-j\omega\mu Il}{4\pi r} e^{-jkr} \qquad (6\text{-}116)$$

we have the wave of Eq. (6-96). Hence, the field of Fig. 6-14b is specified by Eqs. (6-101) with coordinates primed. To relate this solution to that of Fig. 6-14a, we need the r' component of $\mathbf{E}$, which is

$$E_{r'}{}^b = \frac{1}{j\omega\epsilon}\left(\frac{\partial^2}{\partial r'^2} + k^2\right) A_{r'}{}^b$$

$$= \frac{E_0}{jk^2} \cos\phi' \sum_{n=1}^{\infty} n(n+1)[a_n \hat{J}_n(kb) + b_n \hat{H}_n{}^{(2)}(kb)] P_n{}^1(\cos\theta')$$

Finally, by reciprocity, $E_{r'}{}^b$ evaluated at $r' = b$, $\theta' = \pi - \theta$, $\phi' = 0$ equals $-E_\theta{}^a$ at r, θ, ϕ. Hence,

$$E_\theta{}^a = \frac{jE_0}{k^2} \sum_{n=1}^{\infty} n(n+1)[a_n \hat{J}_n(kb) + b_n \hat{H}_n{}^{(2)}(kb)](-1)^n P_n{}^1(\cos\theta) \qquad (6\text{-}117)$$

where a_n, b_n, and E_0 are given by Eqs. (6-98), (6-102), and (6-116), respectively. In the special case $b = a$, that is, when the current element is on the surface of the sphere, Eq. (6-117) reduces to

$$E_\theta = \frac{\eta Il}{4\pi jkr} e^{-jkr} \sum_{n=1}^{\infty} \frac{j^n(2n+1)}{\hat{H}_n{}^{(2)'}(ka)} P_n{}^1(\cos\theta) \qquad (6\text{-}118)$$

This is the radiation field of a radially directed electric dipole on the surface of a conducting sphere. Figure 6-15 shows the radiation patterns for spheres of radii $a = \lambda/4$ and $a = 2\lambda$. The pattern for the very small sphere is the usual dipole pattern. For a very large sphere it approaches the pattern of a dipole on a ground plane but always with some diffraction around the sphere. The radiation field for dipoles of other orientations, and also for magnetic dipoles, can be obtained in a similar manner. The field in the entire region $r > b$ can be determined from the radiation

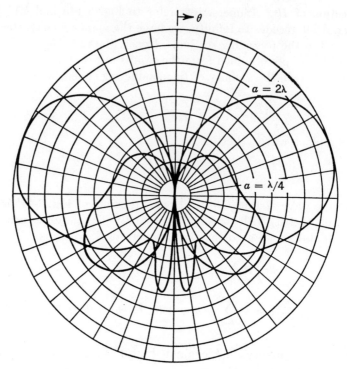

FIG. 6-15. Radiation patterns for the radially directed dipole on a conducting sphere of radius a.

field as follows. From symmetry considerations (Fig. 6-14a) we conclude that $\mathbf{H} = \mathbf{u}_\phi H_\phi$, and therefore the field can be expressed in terms of an $\mathbf{A} = \mathbf{u}_r A_r$. Also, A_r must be independent of ϕ and represent outward traveling waves; hence

$$A_r = \sum_{n=1}^{\infty} a_n \hat{H}_n^{(2)}(kr) P_n(\cos\theta) \qquad r > b \qquad (6\text{-}119)$$

From this we can calculate E_θ by Eqs. (6-26), obtaining

$$E_\theta = \frac{\eta e^{-jkr}}{jr} \sum_{n=1}^{\infty} a_n j^n \hat{H}_n^{(2)\prime}(kr) P_n^1(\cos\theta)$$

$$\xrightarrow[r\to\infty]{} \frac{\eta e^{-jkr}}{jr} \sum_{n=1}^{\infty} a_n j^n P_n^1(\cos\theta) \qquad (6\text{-}120)$$

The a_n are then evaluated by equating this expression to the radiation field previously determined. For example, in the special case $b = a$ we equate Eq. (6-120) to Eq. (6-118) and obtain

$$a_n = \frac{Il(2n+1)}{4\pi k \hat{H}_n^{(2)\prime}(ka)} \quad (6\text{-}121)$$

The field everywhere can now be obtained from Eqs. (6-26), (6-119), and (6-121).

6-11. Apertures in Spheres. In Sec. 4-9 we saw how to express the field in a matched rectangular waveguide in terms of the field over a cross section of the guide. In Sec. 6-4 we saw that space could be viewed

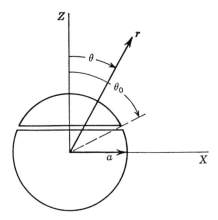

FIG. 6-16. Slotted conducting sphere.

as a spherical waveguide. A given sphere $r = a$ is a cross section of the spherical guide. If $r > a$ contains only free space, then the guide is matched, that is, there are no incoming waves. By writing the general expansion for outward-traveling waves and specializing to $r = a$, we obtain the field $r > a$. When apertures exists in a conducting sphere of radius $r = a$, the tangential components of **E** are zero except in the apertures. Our formulas for the field $r > a$ then reduce to ones involving only the tangential components of **E** over the apertures.

A general treatment of the problem is messy; so let us restrict consideration to the rotationally symmetric TM case, that is, one having only an H_ϕ. The slotted conducting sphere of Fig. 6-16 gives rise to such a field if there exists only an E_θ independent of ϕ in the slot. The field is expressible in terms of an A_r of the form

$$A_r = \sum_{n=1}^{\infty} a_n \hat{H}_n^{(2)}(kr) P_n(\cos \theta) \quad (6\text{-}122)$$

From Eqs. (6-26) we calculate

$$E_\theta = \frac{k}{j\omega\epsilon r} \sum_{n=1}^{\infty} a_n \hat{H}_n^{(2)\prime}(kr) \frac{\partial}{\partial \theta} P_n(\cos \theta) \quad (6\text{-}123)$$

Noting $\partial P_n/\partial \theta = P_n^1$, we multiply each side of the above equation by $P_m^1(\cos \theta) \sin \theta$ and integrate from 0 to π on θ. By the orthogonality relationship [Eqs. (6-46) and (6-47)], we obtain

$$\int_0^\pi E_\theta P_n^1(\cos \theta) \sin \theta \, d\theta = \frac{\eta}{jr} a_n \hat{H}_n^{(2)\prime}(kr) \frac{2\pi n(n+1)}{2n+1}$$

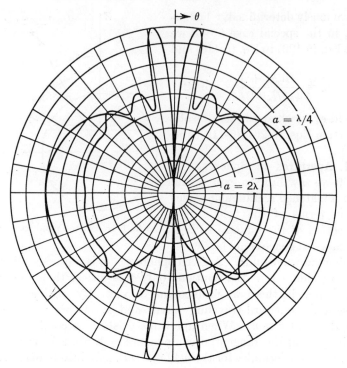

Fig. 6-17. Radiation patterns for the slotted sphere, $\theta_0 = \pi/2$.

Specializing this to $r = a$, we have the coefficients a_n determined as

$$a_n = \frac{ja(2n+1)}{\eta 2\pi n(n+1)\hat{H}_n^{(2)\prime}(ka)} \int_0^\pi E_\theta \bigg|_{r=a} P_n^1(\cos\theta) \sin\theta \, d\theta \quad (6\text{-}124)$$

The field simplifies to some extent in the radiation zone. Using the asymptotic forms for $\hat{H}_n^{(2)}$ in Eq. (6-123), we obtain

$$E_\theta \xrightarrow[kr\to\infty]{} \frac{\eta}{r} e^{-jkr} \sum_{n=1}^\infty a_n j^n P_n^1(\cos\theta) \quad (6\text{-}125)$$

This result could also be obtained from the plane-wave scatter result of Sec. 6-9, using reciprocity.

For the slotted sphere of Fig. 6-16, let us assume a small slot width, so that E_θ is essentially an impulse function at $r = a$. Hence, we assume

$$E_\theta \bigg|_{r=a} = \frac{V}{a} \delta(\theta - \theta_0) \quad (6\text{-}126)$$

where V is the voltage across the slot. Then Eq. (6-124) reduces to

$$a_n = \frac{jV(2n+1)P_n{}^1(\cos\theta_0)\sin\theta_0}{\eta 2\pi n(n+1)\hat{H}_n{}^{(2)'}(ka)}$$

and the radiation field [Eq. (6-125)] becomes

$$E_\theta = \frac{jVe^{-jkr}}{2\pi r}\sin\theta_0 \sum_{n=1}^{\infty} \frac{j^n(2n+1)P_n{}^1(\cos\theta_0)}{n(n+1)\hat{H}_n{}^{(2)'}(ka)} P_n{}^1(\cos\theta) \quad (6\text{-}127)$$

Figure 6-17 shows radiation patterns for the case $\theta_0 = \pi/2$, that is, when the conductor is divided into hemispheres. Patterns for spheres of radii $\lambda/4$ and 2λ are shown. Very small spheres produce a dipole pattern, while very large spheres produce an almost omnidirectional pattern with severe interference phenomena in the $\theta = 0$ and $\theta = \pi$ directions. In the limit $\theta_0 \to 0$ we obtain the patterns of Fig. 6-15, which is to be expected in view of the equivalence of a small magnetic current loop and an electric current element.

The general problem of finding the field in terms of arbitrary tangential components of **E** over a sphere is treated in the literature.[1]

6-12. Fields External to Cones. The general treatment of the problem of sources external to a conducting cone is also messy but can be found in the literature.[1] We shall here restrict consideration to the rotationally symmetric case of "ring-source" excitation of a conducting cone. The geometry of the problem is shown in Fig. 6-18. The special case of a magnetic current ring on the conical surface gives the field of a slotted cone. The limit as the magnetic current ring approaches the cone tip gives the field of an axially directed electric current element on the tip.

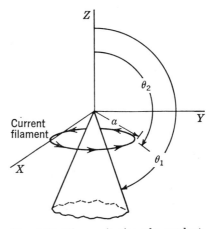

Fig. 6-18. Ring excitation of a conducting cone.

Consider first the case of an electric current ring. From symmetry considerations, it is evident that **E** will have only a ϕ component; so the field is TE to r. The modes of the "conical waveguide" are considered in Sec. 6-5, Eqs. (6-61) to (6-64). In the region $r < a$ we have standing waves, while in the region $r > a$ we have outward-traveling waves.

[1] L. Bailin and S. Silver, Exterior Electromagnetic Boundary Value Problems for Spheres and Cones, *IRE Trans.*, vol. AP-4, no. 1, pp. 5–15, January, 1956.

Hence, we construct

$$F_r = \begin{cases} \sum_v a_v P_v(\cos\theta)\hat{H}_v^{(2)}(kr) & r > a \\ \sum_v b_v P_v(\cos\theta)\hat{J}_v(kr) & r < a \end{cases} \quad (6\text{-}128)$$

where the v are ordered solutions to

$$\left[\frac{d}{d\theta}P_v(\cos\theta)\right]_{\theta=\theta_1} = 0 \quad (6\text{-}129)$$

Continuity of E_ϕ at $r = a$ requires that

$$a_v \hat{H}_v^{(2)}(ka) = b_v \hat{J}_v(ka) \quad (6\text{-}130)$$

Finally, H_θ at $r = a$ must be discontinuous by an amount equal to the surface-current density (in our case it is an impulse function). Thus,

$$J_\phi = \frac{k}{j\omega\mu a}\sum_v \frac{\partial}{\partial\theta}P_v(\cos\theta)[a_v\hat{H}_v^{(2)'}(ka) - b_v\hat{J}_v'(ka)]$$

which, using Eq. (6-130) and the Wronskian of the spherical Bessel functions, becomes

$$J_\phi = \frac{1}{\eta a}\sum_v \frac{\partial}{\partial\theta}P_v(\cos\theta)\frac{a_v}{\hat{J}_v(ka)} \quad (6\text{-}131)$$

By the methods of Sec. 6-3 the following orthogonality relationship can be derived:

$$\int_0^{\theta_1}\left(\frac{\partial}{\partial\theta}P_v\right)\left(\frac{\partial}{\partial\theta}P_w\right)\sin\theta\,d\theta = \begin{cases} 0 & w \neq v \\ N_v & w = v \end{cases} \quad (6\text{-}132)$$

where

$$N_v = -\frac{v(v+1)}{2v+1}\left[\sin\theta\,P_v\frac{\partial^2 P_v}{\partial\theta\,\partial v}\right]_{\theta=\theta_1} \quad (6\text{-}133)$$

Hence, multiplying each side of Eq. (6-131) by $P_w(\cos\theta)\sin\theta$ and integrating from 0 to θ_1 on θ, we obtain

$$a_v = \frac{\eta a}{N_v}\hat{J}_v(ka)\int_0^{\theta_1} J_\phi \frac{\partial}{\partial\theta}[P_v(\cos\theta)]\sin\theta\,d\theta \quad (6\text{-}134)$$

This completes the solution for an arbitrary ϕ-directed current sheet at $r = a$. For the current filament,

$$J_\phi = \frac{I}{a}\delta(\theta - \theta_2) \quad (6\text{-}135)$$

and Eq. (6-134) reduces to

$$a_v = \frac{\eta I}{N_v}\hat{J}_v(ka)\sin\theta_2\frac{\partial}{\partial\theta_2}P_v(\cos\theta_2) \quad (6\text{-}136)$$

Numerical calculations are difficult because of the problem of obtaining the eigenvalues v and the eigenfunctions P_v.

When the ring source of Fig. 6-18 is a magnetic current, the problem is dual to the electric-current case, except for boundary conditions. Hence, we construct

$$A_r = \begin{cases} \sum_u c_u P_u(\cos\theta)\hat{H}_u^{(2)}(kr) & r > a \\ \sum_u d_u P_u(\cos\theta)\hat{J}_u(kr) & r < a \end{cases} \quad (6\text{-}137)$$

where the u are ordered solutions to

$$P_u(\cos\theta_1) = 0 \quad (6\text{-}138)$$

in contrast to the v which were solutions to Eq. (6-129). Continuity of H_ϕ at $r = a$ requires that

$$c_u \hat{H}_u^{(2)}(ka) = d_u \hat{J}_u(ka) \quad (6\text{-}139)$$

At $r = a$ we have E_θ discontinuous by an amount equal to the surface-current density. Thus, analogous to Eq. (6-131), we have

$$M_\phi = -\frac{\eta}{a} \sum_u \frac{\partial}{\partial \theta} P_u(\cos\theta) \frac{c_u}{\hat{J}_u(ka)} \quad (6\text{-}140)$$

The orthogonality relationship for the eigenvalues defined by Eq. (6-138) is

$$\int_0^{\theta_1} \left(\frac{\partial}{\partial \theta} P_u\right)\left(\frac{\partial}{\partial \theta} P_w\right) \sin\theta\, d\theta = \begin{cases} 0 & w \neq u \\ M_u & w = u \end{cases} \quad (6\text{-}141)$$

where

$$M_u = \frac{u(u+1)}{2u+1} \left[\sin\theta \frac{\partial P_u}{\partial \theta} \frac{\partial P_u}{\partial u}\right]_{\theta=\theta_1} \quad (6\text{-}142)$$

Multiplying each side of Eq. (6-140) by $P_w(\cos\theta)\sin\theta$ and integrating from 0 to θ_1 on θ, we obtain

$$c_u = \frac{-a}{\eta M_u} \hat{J}_u(ka) \int_0^{\theta_1} M_\phi \frac{\partial}{\partial \theta}[P_v(\cos\theta)] \sin\theta\, d\theta \quad (6\text{-}143)$$

This completes the solution for an arbitrary ϕ-directed magnetic current sheet at $r = a$. For the magnetic current filament,

$$M_\phi = \frac{K}{a} \delta(\theta - \theta_2) \quad (6\text{-}144)$$

and Eq. (6-143) reduces to

$$c_u = \frac{-K}{\eta M_u} \hat{J}_u(ka) \sin\theta_2 \frac{\partial}{\partial \theta_2} P_u(\cos\theta_2) \quad (6\text{-}145)$$

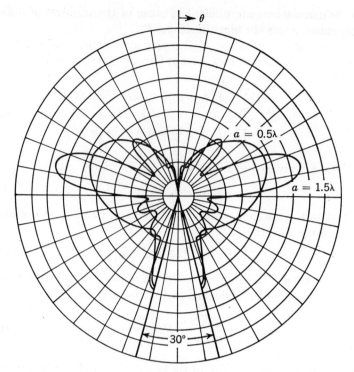

Fig. 6-19. Radiation patterns for the slotted conducting cone. (*After Bailin and Silver.*)

Again a calculation of the eigenvalues u and the eigenfunctions P_u is difficult.

If we now let $\theta_2 = \theta_1$ and set $K = V$ in the magnetic current solution, we have the case of a cone slotted at $r = a$ with a voltage V across the slot. For $r > a$ Eq. (6-137) becomes

$$A_r = \frac{V}{\eta} \sin^2 \theta_1 \sum_u \frac{1}{M_u} P'_u(\cos\theta_1) P_u(\cos\theta) \hat{J}_u(ka) \hat{H}_u^{(2)}(kr)$$

Using the asymptotic form for $\hat{H}_n^{(2)}$ and evaluating E_θ by Eq. (6-26), we find for the radiation field

$$E_\theta = \frac{V}{jr} e^{-jkr} \sum_u \frac{j^u(2u+1)[\partial P_u(\cos\theta)/\partial\theta]}{u(u+1)[\partial P_u(\cos\theta_1)/\partial u]} \hat{J}_u(ka) \quad (6\text{-}146)$$

Some radiation patterns for slotted cones with cone angle 30° are shown in Fig. 6-19. A discussion of the problem of plane-wave scattering by a cone is given by Mentzer.[1]

[1] J. R. Mentzer, "Scattering and Diffraction of Radio Waves," pp. 81–93, Pergamon Press, Inc., New York, 1955.

6-13. Maximum Antenna Gain.

The general form of the field in a spherical space external to all sources is Eqs. (6-26) with

$$A_r = \sum_{m,n} a_{mn}\hat{H}_n^{(2)}(kr)P_n^m(\cos\theta)\cos(m\phi + \alpha_{mn})$$

$$F_r = \sum_{m,n} b_{mn}\hat{H}_n^{(2)}(kr)P_n^m(\cos\theta)\cos(m\phi + \beta_{mn})$$

(6-147)

Given an arbitrary field at $r = r_1$, the field can be projected backward toward the origin as far as desired. At some sphere $r = a$ we can determine sources by the equivalence principle (Sec. 3-5), which will support this field. Hence, it appears that sources on an arbitrarily small sphere can support any desired radiation field.

The gain of an antenna is defined by Eq. (2-130) in general. We shall here consider the largest gain

$$g = \frac{4\pi r^2(S_r)_{\max}}{\bar{\mathcal{O}}_f}$$

(6-148)

where $(S_r)_{\max}$ is the maximum power density in the radiation zone and $\bar{\mathcal{O}}_f$ is the power radiated. By the discussion of the preceding paragraph, it appears that arbitrarily high gain can be obtained, regardless of antenna size. In practice, however, the gain of a directive antenna is found to be related to its size. A uniformly illuminated aperture[1] type of antenna is found to give the highest practical gain. This apparent discrepancy between theory and practice can be resolved if the concepts of cutoff and Q of spherical waves are properly applied.

Let us orient our spherical coordinate system so that maximum radiation is in the $\theta = 0$ direction. The radially directed power flux in this direction is then

$$(S_r)_{\max} = E_x H_y^* - E_y H_x^*$$

(6-149)

From Eqs. (6-147) and (6-26) we find

$$E_x = \frac{e^{-jkr}}{2jr} \sum_n n(n+1)j^n(\eta\, a_{1n}\cos\alpha_{1n} - b_{1n}\sin\beta_{1n})$$

$$E_y = \frac{e^{-jkr}}{2jr} \sum_n n(n+1)j^n(-\eta\, a_{1n}\sin\alpha_{1n} - b_{1n}\cos\beta_{1n})$$

$$H_x = \frac{e^{-jkr}}{2jr} \sum_n n(n+1)j^n \left(a_{1n}\sin\alpha_{1n} - \frac{1}{\eta}b_{1n}\cos\beta_{1n}\right)$$

$$H_y = \frac{e^{-jkr}}{2jr} \sum_n n(n+1)j^n \left(a_{1n}\cos\alpha_{1n} + \frac{1}{\eta}b_{1n}\sin\beta_{1n}\right)$$

(6-150)

---

[1] The term "uniformly illuminated aperture" is used to describe antennas for which the source (primary or secondary) is constant in amplitude and phase over a given area on a plane, and zero elsewhere.

in the $\theta = 0$ direction of the radiation zone. The total radiated power is found by integrating the Poynting vector over a large sphere. The result is

$$\bar{\mathcal{P}}_f = 4\pi \sum_{m,n} \frac{n(n+1)(n+m)!}{\epsilon_m(2n+1)(n-m)!} \left(\eta |a_{mn}|^2 + \frac{1}{\eta}|b_{mn}|^2 \right) \quad (6\text{-}151)$$

where $\epsilon_m = 1$ for $m = 0$ and $\epsilon_m = 2$ for $m > 0$. We used the orthogonality relationships of Eqs. (6-51) in the derivation of Eq. (6-151).

Equations (6-148) to (6-151) give a general formula for gain in terms of spherical waves. We shall now consider under what conditions g is a maximum. Note that the numerator of Eq. (6-148) involves only the a_{1n} and b_{1n} coefficients. Hence, the denominator can be decreased without changing the numerator, by setting

$$a_{mn} = b_{mn} = 0 \qquad m \neq 1 \quad (6\text{-}152)$$

Also, both numerator and denominator of g are independent of α_{1n} and β_{1n}; so they may be chosen for convenience without loss of generality. In particular, let $\alpha_{1n} = \pi$ and $\beta_{1n} = \pi/2$, and the gain formula reduces to

$$g = \frac{\left| \sum_n (A_n + B_n) \right|^2}{2 \sum_n \frac{1}{2n+1}(|A_n|^2 + |B_n|^2)} \quad (6\text{-}153)$$

where $\qquad A_n = j^n \eta n(n+1) a_{1n} \qquad B_n = j^n n(n+1) b_{1n} \quad (6\text{-}154)$

The denominator of Eq. (6-153) is independent of the phases of A_n and B_n; so we can maximize the numerator by choosing A_n and B_n real. Furthermore, g is symmetric in A_n and B_n; hence the maximum exists when

$$A_n = B_n = \text{real} \quad (6\text{-}155)$$

The maximum gain therefore will be found among those specified by

$$g = \frac{\left(\sum_n A_n \right)^2}{\sum_n A_n^2 \left(\frac{1}{2n+1} \right)} \quad (6\text{-}156)$$

where A_n is real. As long as n is unrestricted, this g is unbounded, as we anticipated earlier.

If the field, specified by Eqs. (6-147), contains only wave functions of order $n \leq N$, then an upper limit to g exists. Setting $\partial g/\partial A_i = 0$ for all A_i, we find

$$g_{\max} = \sum_{n=1}^{N} (2n + 1) = N^2 + 2N \qquad (6\text{-}157)$$

and also
$$A_n = \frac{2n + 1}{3} A_1 \qquad (6\text{-}158)$$

Equation (6-157) represents the highest possible gain using spherical waveguide modes of order $n \leq N$. A similar limitation to the near-zone gain also exists.[1]

To relate gain to antenna size, we define the radius a of an antenna as the radius of the smallest sphere that can contain the antenna. We saw in Sec. 6-4 that spherical modes of order n were rapidly cut off when $ka < n$. Hence, it is reasonable to assume that modes of order $n > ka$ are not normally present to any significant extent in the field of an antenna of radius a. We define the *normal gain* of an antenna of radius a as

$$g_{\text{normal}} = (ka)^2 + 2ka \qquad (6\text{-}159)$$

which is obtained by setting $N = ka$ in Eq. (6-157). Hence, the normal gain is maximum gain obtainable when only uncutoff modes are present. It is interesting to note that, for large ka, a circular, uniformly illuminated aperture of radius a has the same gain as the above-defined normal gain.[2] It is therefore not surprising that the uniformly illuminated aperture gives the highest antenna gain in practice.

The normal gain is not an absolute upper limit to the gain of an antenna. Antennas having higher gain are a distinct possibility and will be called *supergain antennas*. We shall use the Q concept of Sec. 6-4 to show that (1) supergain antennas must necessarily be narrow-band devices, and (2) supergain techniques yield only a small increase in gain over normal gain for large antennas. Other characteristics which we shall not demonstrate here are (3) supergain antennas have high field intensities at the antenna structure and (4) they tend to have excessive power loss in the antenna structure.

The Q of a loss-free antenna is defined as

$$Q = \begin{cases} \dfrac{2\omega \overline{\mathcal{W}}_e}{\overline{\mathcal{P}}_f} & \overline{\mathcal{W}}_e > \overline{\mathcal{W}}_m \\ \dfrac{2\omega \overline{\mathcal{W}}_m}{\overline{\mathcal{P}}_f} & \overline{\mathcal{W}}_m > \overline{\mathcal{W}}_e \end{cases} \qquad (6\text{-}160)$$

[1] R. F. Harrington, Effect of Antenna Size on Gain, Bandwidth, and Efficiency, *J. Research NBS*, vol. 64D, no. 1, pp. 1–12, January, 1960.

[2] S. Ramo and J. R. Whinnery, "Fields and Waves in Modern Radio," 2d ed., p. 533, John Wiley & Sons, Inc., New York, 1953.

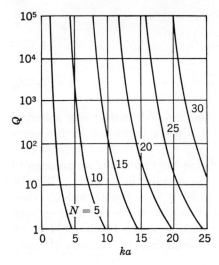

Fig. 6-20. Quality factors for ideal loss-free antennas adjusted for maximum gain using modes of order $n \leq N$.

where $\overline{\mathcal{W}}_e$ and $\overline{\mathcal{W}}_m$ are the time-average electric and magnetic energies and $\overline{\mathcal{P}}_f$ is the power radiated. We shall define an ideal loss-free antenna of radius a as one having no energy storage $r < a$. The Q of this ideal antenna must be less than or equal to the Q of any other loss-free antenna of radius a having the same field $r > a$, since fields $r < a$ can only add to energy storage. If the Q of an antenna is large, it can be interpreted as the reciprocal of the fractional bandwidth of the input impedance. If the Q is small, the antenna has broadband potentialities.

Antennas adjusted for maximum gain according to Eq. (5-158) have equal excitation of TM and TE modes. The Q_n of spherical modes, defined by Eq. (6-60) and plotted in Fig. 6-6, involve $\overline{\mathcal{W}}_e$ for TM modes and $\overline{\mathcal{W}}_m$ for TE modes. We need Q's defined in terms of the same energy for all modes, and it is convenient to deal with Q's for equal TM and TE modes. The Q for equal TM_n and TE_n modes is

$$Q_n^{\text{TM+TE}} \approx \tfrac{1}{2} Q_n \qquad ka < N \qquad (6\text{-}161)$$

because the $\overline{\mathcal{W}}_e$ is essentially that of the TM_n mode alone and the $\overline{\mathcal{P}}_f$ is twice that of the TM_n mode alone. When $Q_n < 1$, we take it as unity.

Because of the orthogonality of energy and power in the spherical modes, the total energy and power in any field is the sum of the modal energies and powers. Hence, the Q of our ideal loss-free antenna is

$$Q = \frac{\sum P_n Q_n^{\text{TM+TE}}}{\sum P_n} = \frac{\sum A_n^2 \left(\dfrac{1}{2n+1}\right) Q_n}{2 \sum A_n^2 \left(\dfrac{1}{2n+1}\right)}$$

where P_n is the transmitted power in the TM_n and TE_n modes. Using Eq. (6-158), this becomes

$$Q = \frac{\sum_{n=1}^{N}(2n+1)Q_n(ka)}{2N^2 + 4N} \tag{6-162}$$

where the Q_n are given in Fig. 6-6. Curves of antenna Q for several N are shown in Fig. 6-20. Note that the Q rises sharply for $ka < N$, showing that supergain antennas must necessarily be high Q, or frequency sensitive.

The Q of Fig. 6-20 is a lower bound to the Q of any loss-free antenna of radius a. By picking a Q, we can calculate an upper bound to the gain of an antenna of radius a. Figure 6-21 shows the ratio of this upper bound to the normal gain. Note that for large ka the increase in gain over normal gain possible by supergain techniques is small. For small ka supergain can give considerable improvement over normal gain. In fact, as $ka \to 0$ the supergain condition is unavoidable. All very small antennas are supergain antennas by our definition. The problems of narrow bandwidth and high losses associated with small antennas are well-known in practical antenna work.

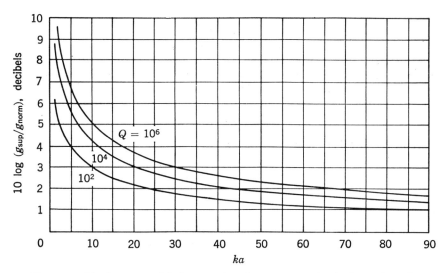

FIG. 6-21. Maximum possible increase in gain over normal gain for a given Q.

PROBLEMS

6-1. Use Eqs. (3-85) and the wave potential of Eq. (6-14) to show that a general expression for fields TM to z is

$$E_r = -j\omega\mu\psi \cos\theta + \frac{1}{j\omega\epsilon}\frac{\partial}{\partial r}\left[\frac{\cos\theta}{r^2}\frac{\partial}{\partial r}(r^2\psi) - \frac{1}{r\sin\theta}\frac{\partial}{\partial\theta}(\psi\sin^2\theta)\right]$$

$$E_\theta = j\omega\mu\psi \sin\theta + \frac{1}{j\omega\epsilon r}\frac{\partial}{\partial\theta}\left[\frac{\cos\theta}{r^2}\frac{\partial}{\partial r}(r^2\psi) - \frac{1}{r\sin\theta}\frac{\partial}{\partial\theta}(\psi\sin^2\theta)\right]$$

$$E_\phi = \frac{1}{j\omega\epsilon r\sin\theta}\frac{\partial}{\partial\phi}\left[\frac{\cos\theta}{r^2}\frac{\partial}{\partial r}(r^2\psi) - \frac{1}{r\sin\theta}\frac{\partial}{\partial\theta}(\psi\sin^2\theta)\right]$$

$$H_r = \frac{1}{r}\frac{\partial\psi}{\partial\phi}$$

$$H_\theta = \frac{\cot\theta}{r}\frac{\partial\psi}{\partial\phi}$$

$$H_\phi = \frac{-1}{r}\left[\sin\theta\frac{\partial}{\partial r}(r\psi) + \frac{\partial}{\partial\theta}(\psi\cos\theta)\right]$$

where ψ is a solution to the scalar Helmholtz equation.

6-2. Verify that Eqs. (6-17) and (6-18) are identical.

6-3. Consider an air-filled spherical resonator of radius 5 centimeters bounded by copper walls. Determine the first ten resonant frequencies and the Q of the dominant mode.

6-4. For the spherical cavity of Fig. 6-2, show that the Q due to conductor losses is, for TM modes,

$$(Q_c)_{mnp}^{TM} = \frac{\eta}{2\mathfrak{R}}\left[u'_{np} - \frac{n(n+1)}{u'_{np}}\right]$$

where the u'_{np} are given in Table 6-2, and, for TE modes,

$$(Q_c)_{mnp}^{TE} = \frac{\eta u_{np}}{2\mathfrak{R}}$$

where the u_{np} are given in Table 6-1.

6-5. Consider the cavity lying between concentric conducting spheres $r = a$ and $r = b$, with $b > a$. Show that the characteristic equation for modes TM to r is

$$\frac{\hat{J}'_n(kb)}{\hat{J}'_n(ka)} = \frac{\hat{N}'_n(kb)}{\hat{N}'_n(ka)}$$

and for modes TE to r it is

$$\frac{\hat{J}_n(kb)}{\hat{J}_n(ka)} = \frac{\hat{N}_n(kb)}{\hat{N}_n(ka)}$$

6-6. In the concentric-sphere cavity of Prob. 6-5 let $a \ll b$, and show that the resonant frequency ω is related to the empty cavity resonant frequency ω_0 by

$$\frac{\omega - \omega_0}{\omega_0} \approx \tfrac{2}{3}(2.744)^2 \frac{\hat{N}'_1(2.744)}{\hat{J}''_1(2.744)}\left(\frac{a}{b}\right)^3$$

where $\omega_0 = 2.744/b\sqrt{\epsilon\mu}$. [*Hint*: Express the characteristic equation in the form $f(k,a) = 0$, and expand in a Taylor series about $k_0 = \omega_0\sqrt{\epsilon\mu}$.]

6-7. Consider the partially filled spherical cavity formed by a conductor covering $r = b$ and containing a dielectric ϵ_1, μ_1 for $r < a$ and a dielectric ϵ_2, μ_2 for $a < r < b$. Show that the characteristic equation for the dominant mode is

$$\frac{\hat{N}'_1(k_2b)\hat{J}'_1(k_2a) - \hat{J}'_1(k_2b)\hat{N}'_1(k_2a)}{\hat{N}'_1(k_2b)\hat{J}_1(k_2a) - \hat{J}'_1(k_2b)\hat{N}_1(k_2a)} = \frac{\eta_1}{\eta_2}\frac{\hat{J}'_1(k_1a)}{\hat{J}_1(k_1a)}$$

where $k_1 = \omega\sqrt{\epsilon_1\mu_1}$ and $k_2 = \omega\sqrt{\epsilon_2\mu_2}$.

6-8. In the partially filled spherical cavity of Prob. 6-7, let $a \ll b$ and $\epsilon_2 = \epsilon_0$ and $\mu_2 = \mu_0$. By expanding the characteristic equation in a Taylor series about the empty-cavity resonant frequency ω_0, show that the resonant frequency ω is given by

$$\frac{\omega - \omega_0}{\omega_0} \approx \tfrac{2}{3}(2.744)^2 \frac{\hat{N}'_1(2.744)}{\hat{J}''_1(2.744)} \frac{\epsilon_r - 1}{\epsilon_r + 2} \left(\frac{a}{b}\right)^3$$

where $\epsilon_r = \epsilon_1/\epsilon_0$ and $\omega_0 = 2.744/b\sqrt{\epsilon_0\mu_0}$. Compare this with the answer to Prob. 6-6.

6-9. Consider the function

$$f(\theta,\phi) = \begin{cases} 1 & 0 < \theta < \dfrac{\pi}{2} \\ 0 & \dfrac{\pi}{2} < \theta < \pi \end{cases}$$

and determine the coefficients a_{mn} and b_{mn} for the two-dimensional Fourier-Legendre series of the form of Eq. (6-48).

6-10. Let **A** and **B** be two vectors and θ be the angle between them. Define $\mathbf{C} = \mathbf{A} - \mathbf{B}$ and show that, for $B > A$,

$$\frac{1}{C} = \frac{1}{\sqrt{A^2 + B^2 - 2AB\cos\theta}} = \frac{1}{B}\sum_{n=0}^{\infty}\left(\frac{A}{B}\right)^n P_n(\cos\theta)$$

6-11. Consider the characteristic impedances of the spherical modes of space [Eqs. (6-57)]. Show that

$$Z_{+r}^{TM} = Z_{-r}^{TM*} \begin{cases} \xrightarrow[kr\to\infty]{} \eta \\ \xrightarrow[kr\to 0]{} -j\eta\dfrac{n}{kr} \end{cases}$$

and $Z^{TE} = \eta^2/Z^{TM}$. Show also that the change from primarily resistive to primarily reactive wave impedances occurs at $kr \approx n$.

6-12. Show that the field of an electric current element Il is the dominant TM spherical mode of space, and the field of a magnetic-current element Kl is the dominant TE mode.

6-13. Using the usual perturbational method, show that the attenuation constant due to conductor losses for the TEM mode of the biconical or coaxial radial guide (Fig. 6-7c and d) is given by

$$\alpha = \frac{\mathcal{R}}{2\eta r}\frac{\csc\theta_1 + \csc\theta_2}{\log\dfrac{\cot\theta_1/2}{\cot\theta_2/2}}$$

6-14. Show that the dominant spherical TE mode of the wedge guide (Fig. 6-7e) is the free-space field of a z-directed magnetic-current element.

6-15. Use the qualitative behavior of the spherical Hankel functions to justify the statement that the spherical-horn guide of Fig. 6-7f has a "cutoff radius" approximately equal to that radius for which the cross section is the same as a rectangular guide at cutoff.

6-16. Consider a hemispherical cavity (Fig. 6-8a) constructed of copper with $a = 10$ centimeters, and air-filled. Determine the first ten resonant frequencies and the Q of the dominant mode.

6-17. Consider the second resonance $[n = 2$ in Eq. (6-80)] of the biconical cavity of Fig. 6-8c. Calculate the Q of the mode and the input resistance seen at the cone tips.

6-18. Consider the conical cavity of Fig. 6-8d. Show that modes TM to r are given by $\mathbf{H} = \nabla \times \mathbf{u}_r A_r$ where

$$(A_r)_{mvp} = P_v^m(\cos\theta) \cos m\phi \, \hat{J}_v\left(w'_{vp}\frac{r}{a}\right)$$

where w'_{vp} is the pth zero of $\hat{J}'_v(w)$ and v is a solution to Eq. (6-62). Similarly, show that modes TE to r are given by $\mathbf{E} = -\nabla \times \mathbf{u}_r F_r$ where

$$(F_r)_{mvp} = P_v^m(\cos\theta) \cos m\phi \, \hat{J}_v\left(w_{vp}\frac{r}{a}\right)$$

where w_{vp} is the pth zero of $\hat{J}_v(w)$ and v is a solution to Eq. (6-64). For a complete set of modes the $\sin m\phi$ variation must also be included.

6-19. Let the current elements of Fig. 6-9c be replaced by magnetic-current elements Kl. Show that, in the limit $s \to 0$, the field is given by $\mathbf{E} = -\nabla \times \mathbf{u}_z F_z$ where

$$F_z = \frac{k^2 K l s}{4\pi j} h_1^{(2)}(kr) P_1(\cos\theta)$$

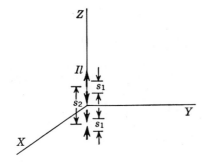

6-20. Consider the quadrupole source of Fig. 6-22 where each element is an electric current Il. Show that, in the limit $s_1 \to 0$ and $s_2 \to 0$, the field is given by $\mathbf{H} = \nabla \times \mathbf{u}_z A_z$ where

$$A_z = \frac{k^3 I l s_1 s_2}{24\pi j} [h_2^{(2)}(kr) P_2(\cos\theta) - \tfrac{1}{2} h_0^{(2)}(kr)]$$

Fig. 6-22. A quadrupole source.

6-21. Derive the following wave transformation:

$$\frac{e^{-j|\mathbf{r}-\mathbf{r}'|}}{|\mathbf{r}-\mathbf{r}'|} = \frac{1}{jrr'} \sum_{n=0}^{\infty} (2n+1) \hat{J}_n(r') \hat{H}_n^{(2)}(r) P_n(\cos\xi) \qquad r > r'$$

where ξ is the angle between $\mathbf{r}$ and $\mathbf{r}'$.

6-22. Derive the following wave transformation:

$$J_n(\rho) = \sum_{m=0}^{\infty} A_m \hat{j}_{2m+n}(r) P_{2m+n}^n(\cos\theta)$$

where
$$A_m = \frac{(-1)^{m+n}(4m+2n+1)(2m)!}{2^{2m+n}(m+n)!m!}$$

6-23. Derive the following formula:

$$\int_{-1}^{1} h_0^{(2)}(|\mathbf{r} - \mathbf{r}'|) \, d(\cos \xi) = \begin{cases} 2j_0(r')h_0^{(2)}(r) & r > r' \\ 2j_0(r)h_0^{(2)}(r') & r' > r \end{cases}$$

where ξ is the angle between $\mathbf{r}$ and $\mathbf{r}'$.

6-24. Consider the scattering of a plane-polarized wave by a small conducting sphere (Fig. 6-11). Show that the distant scattered field is plane polarized in the direction $\theta = 60°$.

6-25. Consider an x-polarized, z traveling plane wave incident on a conducting sphere encased in a concentric dielectric coating, as shown in Fig. 6-23. Show that the field is given by Eqs. (6-26), where for $r > b$ the A_r and F_r are given by Eqs. (6-101), and for $a < r < b$

$$A_r = \frac{E_0}{\omega\mu_0} \cos\phi \sum_{n=1}^{\infty} d_n[\hat{N}_n'(ka)\hat{J}_n(kr)$$
$$- \hat{J}_n'(ka)\hat{N}_n(kr)]P_n^1(\cos\theta)$$

$$F_r = \frac{E_0}{k} \sin\phi \sum_{n=1}^{\infty} e_n[\hat{N}_n(ka)\hat{J}_n(kr)$$
$$- \hat{J}_n(ka)\hat{N}_n(kr)]P_n^1(\cos\theta)$$

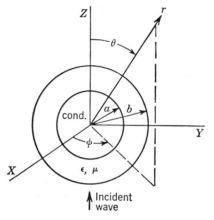

Fig. 6-23. A plane wave incident on a coated conducting sphere.

Impose boundary conditions on the tangential components of **E** at **H** at $r = b$, and obtain expressions for b_n, c_n, d_n, and e_n in terms of a_n, given by Eq. (6-98).

6-26. Consider a radially directed magnetic dipole Kl adjacent to a conducting sphere (Fig. 6-14 with Il replaced by Kl). Show that the radiation field is given by $E_\phi = -\eta H_\theta$ and

$$H_\theta = \frac{Kl}{4\pi\eta kr} e^{-jkr} \sum_{n=1}^{\infty} n(n+1)[a_n\hat{J}_n(kb) + c_n\hat{H}_n^{(2)}(kb)](-1)^n P_n^1(\cos\theta)$$

where a_n is given by Eq. (6-98) and c_n by Eq. (6-102).

6-27. Consider a radially directed electric dipole adjacent to a dielectric sphere (Fig. 6-14 with the sphere now dielectric). Show that the radiation field is then given by Eq. (6-117) if b_n is given by Eq. (6-113) instead of Eq. (6-102).

6-28. Consider a loop of uniform current I of radius a, as shown in Fig. 2-26. Show that the radiation field is given by

$$E_\phi = \frac{\eta I}{jr} e^{-jkr} \sum_{n=1}^{\infty} \frac{2n+1}{2n(n+1)} j^n A_n P_n^1(0) P_n^1(\cos\theta)$$

where
$$A_n = \hat{J}_n(ka)$$

and $\eta H_\theta = -E_\phi$.

6-29. Figure 6-24 shows a conducting sphere of radius R concentric with a loop of uniform current I of radius a. Show that the radiation field is of the same form as given in Prob. 6-28 except that

$$A_n^{-1} = \hat{H}_n^{(2)\prime}(ka) - \frac{\hat{J}_n(kR)\hat{N}_n'(ka) - \hat{N}_n(kR)\hat{J}_n'(ka)}{\hat{J}_n(kR)\hat{N}_n(ka) - \hat{N}_n(kR)\hat{J}_n(ka)} \hat{H}_n^{(2)}(ka)$$

Show that this reduces to the answer for Prob. 6-28 as $R \to 0$.

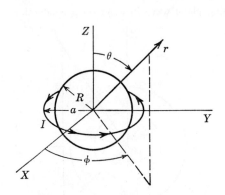 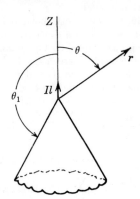

Fig. 6-24. A conducting sphere with a concentric ring of electric current.

Fig. 6-25. Current element at the tip of a conducting cone.

6-30. Figure 6-25 shows a current element Il at the tip of a conducting cone. Show that the radiation field is given by

$$E_\theta = f(r) \sin \theta \, P_u'(\cos \theta)$$

where u is the first root of $P_u(\cos \theta_1) = 0$. Some approximate eigenvalues are

$\pi - \theta_1$	1°	10°	24°	37°	49°	60°	69°	77°	84°	90°
u	0.1	0.2	0.3	0.4	0.5	0.6	0.7	0.8	0.9	1.0

6-31. By considering the equivalent circuit of Fig. 6-5 and the definition of Eq. (6-60) for Q, show that the Q of the $n = 1$ spherical mode is

$$Q_1 = \frac{1}{kr} + \frac{1}{(kr)^3}$$

If equal TE and TM waves are present, the total Q is approximately one-half this value. A small antenna (say $ka < 1$) will have minimum Q if only the $n = 1$ modes are present in its field. Hence, the minimum possible Q for a small loss-free antenna is

$$Q_{min} = \frac{1}{2}\left[\frac{1}{ka} + \frac{1}{(ka)^3}\right]$$

where a is the radius of the smallest sphere that can contain the antenna.

CHAPTER 7

PERTURBATIONAL AND VARIATIONAL TECHNIQUES

7-1. Introduction. The differential equation approach of the preceding three chapters leads to an exact solution of the mathematical problem. However, many problems cannot be treated by this method. We saw in Sec. 3-11 that electromagnetic field problems can be expressed in integral equation form. This form is particularly useful for (1) obtaining approximate solutions and (2) for general expositions of theory. In this chapter, we shall consider two techniques useful for integral equations arising in electromagnetic theory.

Perturbational Methods. The word "perturb" means to disturb or to change slightly. The perturbational methods are useful for calculating changes in some quantity due to small changes in the problem. Usually two problems are involved: the "unperturbed" problem, for which the solution is known, and the "perturbed" problem, which is slightly different from the unperturbed one. We have already used perturbational methods for calculating resonator quality factors and waveguide attenuation constants. Further uses are given in Secs. 7-2 to 7-4.

Variational Methods. The variational methods are useful for determining characteristic quantities, such as resonant frequencies, impedances, and so on. In contrast to the perturbational procedure, the variational procedure gives an approximation to the desired quantity itself, rather than to changes in the quantity. The variational procedure differs from other approximation methods in that the formula is "stationary" about the correct solution. This means that the formula is relatively insensitive to variations in an assumed field about the correct field. If the desired quantity is real, the variational formula may be an upper or lower bound to the quantity. Furthermore, if an assumed field is expressed as a series of functions with undetermined coefficients, then the coefficients can be adjusted by the Ritz procedure (Sec. 7-6). In fact, if a complete set of functions is used for the assumed field, the exact solution can sometimes be obtained, at least in principle.

7-2. Perturbations of Cavity Walls. Figure 7-1a represents a resonant cavity formed by a conductor covering S and enclosing the loss-free region τ. Figure 7-1b represents a deformation of the original cavity

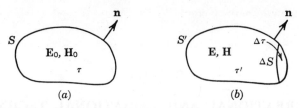

Fig. 7-1. Perturbation of cavity walls. (a) Original cavity; (b) perturbed cavity.

such that the conductor covers $S' = S - \Delta S$ and encloses $\tau' = \tau - \Delta\tau$. We wish to determine the change in the resonant frequency due to the change of the cavity wall.

Let $\mathbf{E}_0$, $\mathbf{H}_0$, ω_0 represent the field and resonant frequency of the original cavity, and let $\mathbf{E}$, $\mathbf{H}$, ω represent the corresponding quantities of the perturbed cavity. In both cases the field equations must be satisfied, that is,

$$\begin{array}{ll} -\nabla \times \mathbf{E}_0 = j\omega_0\mu\mathbf{H}_0 & -\nabla \times \mathbf{E} = j\omega\mu\mathbf{H} \\ \nabla \times \mathbf{H}_0 = j\omega_0\epsilon\mathbf{E}_0 & \nabla \times \mathbf{H} = j\omega\epsilon\mathbf{E} \end{array} \quad (7\text{-}1)$$

We scalarly multiply the last equation by $\mathbf{E}_0^*$ and the conjugate of the first equation by $\mathbf{H}$. The resulting two equations are

$$\mathbf{E}_0^* \cdot \nabla \times \mathbf{H} = j\omega\epsilon\mathbf{E} \cdot \mathbf{E}_0^*$$
$$-\mathbf{H} \cdot \nabla \times \mathbf{E}_0^* = -j\omega_0\mu\mathbf{H}_0^* \cdot \mathbf{H}$$

Adding these and applying the identity

$$\nabla \cdot (\mathbf{A} \times \mathbf{B}) = \mathbf{B} \cdot \nabla \times \mathbf{A} - \mathbf{A} \cdot \nabla \times \mathbf{B}$$

we have

$$\nabla \cdot (\mathbf{H} \times \mathbf{E}_0^*) = j\omega\epsilon\mathbf{E} \cdot \mathbf{E}_0^* - j\omega_0\mu\mathbf{H}_0^* \cdot \mathbf{H}$$

By analogous operations on the second and third of Eqs. (7-1), we obtain

$$\nabla \cdot (\mathbf{H}_0^* \times \mathbf{E}) = j\omega\mu\mathbf{H} \cdot \mathbf{H}_0^* - j\omega_0\epsilon\mathbf{E}_0^* \cdot \mathbf{E}$$

These last two equations are now added, and the sum integrated throughout the volume of the perturbed cavity. The divergence theorem is applied to the left-hand terms, one of which vanishes, because $\mathbf{n} \times \mathbf{E} = 0$ on S'. The resulting equation is

$$\oint_{S'} \mathbf{H} \times \mathbf{E}_0^* \cdot d\mathbf{s} = j(\omega - \omega_0) \iiint (\epsilon\mathbf{E} \cdot \mathbf{E}_0^* + \mu\mathbf{H} \cdot \mathbf{H}_0^*)\, d\tau \quad (7\text{-}2)$$

Finally, since $\mathbf{n} \times \mathbf{E}_0 = 0$ on S, we have

$$\oint_S \mathbf{H} \times \mathbf{E}_0^* \cdot d\mathbf{s} = 0$$

and the left-hand side of Eq. (7-2) can be written as

$$\oiint_{S'} \mathbf{H} \times \mathbf{E}_0^* \cdot d\mathbf{s} = \oiint_{S'-S} \mathbf{H} \times \mathbf{E}_0^* \cdot d\mathbf{s} = -\oiint_{\Delta S} \mathbf{H} \times \mathbf{E}_0^* \cdot d\mathbf{s}$$

The last term is taken as negative, to conform to the convention that $d\mathbf{s}$ points outward. We can now rewrite Eq. (7-2) as

$$\omega - \omega_0 = \frac{j \oiint_{\Delta S} \mathbf{H} \times \mathbf{E}_0^* \cdot d\mathbf{s}}{\iiint_{\tau'} (\epsilon \mathbf{E} \cdot \mathbf{E}_0^* + \mu \mathbf{H} \cdot \mathbf{H}_0^*) \, d\tau} \quad (7\text{-}3)$$

This is an exact formula for the change in resonant frequency due to an inward perturbation of the cavity walls. Note that our development assumes that ϵ and μ are real, that is, we have assumed no losses. Problem 7-1 gives the general formulation in the lossy case.

The crudest approximation to be made in Eq. (7-3) is that of replacing $\mathbf{E}$, $\mathbf{H}$ by the unperturbed field $\mathbf{E}_0$, $\mathbf{H}_0$. For small perturbations this is certainly reasonable in the denominator and should be valid in the numerator if the deformation is shallow and smooth. With this approximation the integral in the numerator of Eq. (7-3) becomes

$$\oiint_{\Delta S} \mathbf{H} \times \mathbf{E}_0^* \cdot d\mathbf{s} \approx \oiint_{\Delta S} (\mathbf{H}_0 \times \mathbf{E}_0^*) \cdot d\mathbf{s}$$
$$= j\omega_0 \iiint_{\Delta \tau} (\epsilon |E_0|^2 - \mu |H_0|^2) \, d\tau$$

The last equality follows from the conservation of complex power [Eq. (1-62)]. Substituting this into Eq. (7-3), and also substituting $\mathbf{E}_0$, $\mathbf{H}_0$ for $\mathbf{E}$, $\mathbf{H}$ in the denominator, we have

$$\frac{\omega - \omega_0}{\omega_0} \approx \frac{\iiint_{\Delta \tau} (\mu |H_0|^2 - \epsilon |E_0|^2) \, d\tau}{\iiint_{\tau} (\mu |H_0|^2 + \epsilon |E_0|^2) \, d\tau} \quad (7\text{-}4)$$

Note that the terms in the numerator are proportional to the electric and magnetic energies "removed" by the perturbation, while the denominator is proportional to the total energy stored. Hence, Eq. (7-4) can be written as

$$\frac{\omega - \omega_0}{\omega_0} \approx \frac{\Delta \bar{\mathcal{W}}_m - \Delta \bar{\mathcal{W}}_e}{\mathcal{W}} \quad (7\text{-}5)$$

where $\Delta \bar{\mathcal{W}}_m$ and $\Delta \bar{\mathcal{W}}_e$ are time-average electric and magnetic energies originally contained in $\Delta \tau$ and $\mathcal{W}$ is the total energy stored in the original cavity. Finally, if $\Delta\tau$ is of small extent, we can approximate the $\Delta\mathcal{W}$'s by $\Delta\tau$ times the energy densities at the position of $\Delta\tau$. Furthermore, $\mathcal{W}$ can be written as τ times a *space-average* energy density $\hat{w}$. Thus, Eq. (7-5) can be written as

$$\frac{\omega - \omega_0}{\omega_0} \approx \frac{(\bar{w}_m - \bar{w}_e)\Delta\tau}{\hat{w}\,\tau} = C\frac{\Delta\tau}{\tau} \qquad (7\text{-}6)$$

where C depends only on the cavity geometry and the position of the perturbation.

It is evident from the preceding equations that an *inward* perturbation will *raise* the resonant frequency if it is made at a point of large H (high $\bar{w}_m$), and will *lower* the resonant frequency if it is made at a point of large E (high $\bar{w}_e$). The opposite behavior results from an outward perturbation. It is also evident that the greatest changes in resonant frequency will occur when the perturbation is at a position of maximum E and zero H, or vice versa.

Numerical calculations using Eqs. (7-4) to (7-6) are easy for the cavities treated previously, because we calculated $\mathcal{W}$ when we determined the Q's. For the dominant mode of the rectangular cavity of Fig. 2-19, $\mathcal{W}$ is given by Eq. (2-98), or

$$\mathcal{W} = \frac{\epsilon}{4}|E_0|^2\tau$$

For $\Delta\tau$ located at the mid-point of the base (maximum E) we use Eqs. (2-96) to find $\Delta\bar{\mathcal{W}}_m = 0$, and

$$\Delta\bar{\mathcal{W}}_e = \frac{\epsilon}{2}|E_0|^2\Delta\tau$$

Hence, from Eq. (7-5) we find

$$\frac{\omega - \omega_0}{\omega_0} \approx -2\frac{\Delta\tau}{\tau} \qquad (7\text{-}7)$$

If the perturbation occurs at the mid-point of the longer side wall (maximum H), we have $\Delta\bar{\mathcal{W}}_e = 0$ and

$$\Delta\bar{\mathcal{W}}_m = \frac{\epsilon|E_0|^2}{2(1 + c^2/b^2)}\Delta\tau$$

Hence, from Eq. (7-5) we find

$$\frac{\omega - \omega_0}{\omega_0} \approx \frac{2}{1 + (c/b)^2}\frac{\Delta\tau}{\tau} \qquad (7\text{-}8)$$

Note that for a square-base cavity ($b = c$) the change in resonant frequency due to $\Delta\tau$ at maximum H is only one-half as great (and in the opposite direction) as that due to $\Delta\tau$ at maximum E.

TABLE 7-1. THE PARAMETER C OF EQ. (7-6) FOR DEFORMATIONS (a) AT MAXIMUM E AND (b) AT MAXIMUM H OF THE DOMINANT MODE

Cavity	Geometry	C
Rectangular ($a \leq b \leq c$)		(a) -2 (b) $\dfrac{2}{1 + (c/b)^2}$
Short cylinder ($d < 2a$)		(a) -1.85 (b) 0.5
Long cylinder ($d \geq 2a$)		(a) -0.843 (b) $\dfrac{2.86}{1 + (1.71a/d)^2}$
Spherical		(a) -0.361 (b) 0.680
Hemispherical		(a) -2.02 (b) 0.680

Table 7-1 gives the value of C in Eqs. (7-6) for cavities of several geometries for $\Delta\tau$ located at (a) maximum E and (b) maximum H. These values have been obtained using the crude approximations of replacing E, H by E_0, H_0 in Eq. (7-3). They are therefore valid only for smooth, shallow deformations. In general, the frequency shift depends on the shape of the deformation as well as on the shape of the cavity. The formulas for deformations of the form of small spheres or small cylinders can be obtained from the results of the next section by letting $\epsilon \to \infty$ and $\mu \to 0$.

7-3. Cavity-material Perturbations. Let us now investigate the change in the resonant frequency of a cavity due to a perturbation of the material within the cavity. Figure 7-2a represents the original cavity containing matter ϵ, μ. Figure 7-2b represents the same cavity but with the matter changed to $\epsilon + \Delta\epsilon$, $\mu + \Delta\mu$.

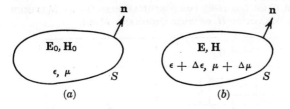

Fig. 7-2. Perturbation of matter in a cavity. (a) Original cavity; (b) perturbed cavity.

Let $\mathbf{E}_0, \mathbf{H}_0, \omega_0$ represent the field and resonant frequency of the original cavity, and let $\mathbf{E}, \mathbf{H}, \omega$ represent the corresponding quantities of the perturbed cavity. Within S the field equations apply, that is,

$$\begin{array}{ll} -\nabla \times \mathbf{E}_0 = j\omega_0 \mu \mathbf{H}_0 & -\nabla \times \mathbf{E} = j\omega(\mu + \Delta\mu)\mathbf{H} \\ \nabla \times \mathbf{H}_0 = j\omega_0 \epsilon \mathbf{E}_0 & \nabla \times \mathbf{H} = j\omega(\epsilon + \Delta\epsilon)\mathbf{E} \end{array} \quad (7\text{-}9)$$

As in the preceding section, we scalarly multiply the last equation by $\mathbf{E}_0^*$ and the conjugate of the first equation by $\mathbf{H}$, and add the resulting two equations. This gives

$$\nabla \cdot (\mathbf{H} \times \mathbf{E}_0^*) = j\omega(\epsilon + \Delta\epsilon)\mathbf{E} \cdot \mathbf{E}_0^* - j\omega_0 \mu \mathbf{H}_0^* \cdot \mathbf{H}$$

Analogous operation on the second and third of Eqs. (7-9) gives

$$\nabla \cdot (\mathbf{H}_0^* \times \mathbf{E}) = j\omega(\mu + \Delta\mu)\mathbf{H} \cdot \mathbf{H}_0^* - j\omega_0 \epsilon \mathbf{E}_0^* \cdot \mathbf{E}$$

The sum of the preceding two equations is integrated throughout the cavity, and the divergence theorem is applied to the left-hand terms. The left-hand terms then vanish, because both $\mathbf{n} \times \mathbf{E} = 0$ on S and $\mathbf{n} \times \mathbf{E}_0 = 0$ on S. The result is

$$0 = \iiint \{[\omega(\epsilon + \Delta\epsilon) - \omega_0 \epsilon]\mathbf{E} \cdot \mathbf{E}_0^* + [\omega(\mu + \Delta\mu) - \omega_0 \mu]\mathbf{H} \cdot \mathbf{H}_0^*\} \, d\tau$$

Finally, this can be rearranged as

$$\frac{\omega - \omega_0}{\omega} = -\frac{\iiint (\Delta\epsilon \mathbf{E} \cdot \mathbf{E}_0^* + \Delta\mu \mathbf{H} \cdot \mathbf{H}_0^*) \, d\tau}{\iiint (\epsilon \mathbf{E} \cdot \mathbf{E}_0^* + \mu \mathbf{H} \cdot \mathbf{H}_0^*) \, d\tau} \quad (7\text{-}10)$$

This is an exact formula for the change in resonant frequency, due to a change in ϵ and/or μ within a cavity. Once again our development has assumed the loss-free case, that is, ϵ and μ are real. The general formulation when losses are present is given in Prob. 7-5.

In the limit, as $\Delta\epsilon \to 0$ and $\Delta\mu \to 0$, we can approximate $\mathbf{E}, \mathbf{H}, \omega$ by $\mathbf{E}_0, \mathbf{H}_0, \omega_0$ and obtain

$$\frac{\omega - \omega_0}{\omega_0} \approx -\frac{\iiint (\Delta\epsilon |E_0|^2 + \Delta\mu |H_0|^2) \, d\tau}{\iiint (\epsilon |E_0|^2 + \mu |H_0|^2) \, d\tau} \quad (7\text{-}11)$$

This states that any small *increase* in ϵ and/or μ can only *decrease* the resonant frequency. Any large change in ϵ and/or μ can be considered as a succession of many small changes. Hence, *any increase in ϵ and/or μ within a cavity can only decrease the resonant frequency.*

We can recognize the various terms of Eq. (7-11) as energy expressions and rewrite it as

$$\frac{\omega - \omega_0}{\omega_0} \approx -\frac{1}{\mathcal{W}} \iiint \left(\frac{\Delta\epsilon}{\epsilon} \bar{w}_e + \frac{\Delta\mu}{\mu} \bar{w}_m \right) d\tau \qquad (7\text{-}12)$$

where $\mathcal{W}$ is the total energy contained in the original cavity. Now if the change in ϵ and μ occupies only a small region $\Delta\tau$, we can further approximate Eq. (7-12) by

$$\frac{\omega - \omega_0}{\omega} \approx -\frac{1}{\hat{w}} \left(\frac{\Delta\epsilon}{\epsilon} \bar{w}_e + \frac{\Delta\mu}{\mu} \bar{w}_m \right) \frac{\Delta\tau}{\tau}$$
$$= -\left(C_1 \frac{\Delta\epsilon}{\epsilon} + C_2 \frac{\Delta\mu}{\mu} \right) \frac{\Delta\tau}{\tau} \qquad (7\text{-}13)$$

where $\hat{w}$ is the space average of $\mathcal{W}$. The parameters C_1 and C_2 depend only on the cavity geometry and the position of $\Delta\tau$. Note that a small change in ϵ at a point of zero E or a small change in μ at a point of zero H does not change the resonant frequency. If we compare Eq. (7-13) with Eq. (7-6), it is evident that $C = C_2 - C_1$. For the cases considered in Table 7-1, $\Delta\tau$ is either at a point of zero H, in which case $C_2 = 0$, or at a point of zero E, in which case $C_1 = 0$. To be explicit, for a material perturbation at (*a*) of Table 7-1 we have $C_1 = -C$ and $C_2 = 0$, while for a material perturbation at (*b*) of Table 7-1 we have $C_1 = 0$ and $C_2 = C$.

The preceding approximations require that $\Delta\epsilon$, $\Delta\mu$, and $\Delta\tau$ all be small. We shall now consider a procedure for removing these restrictions on $\Delta\epsilon$ and $\Delta\mu$. This introduces the further complication that the change in frequency depends on the shape of $\Delta\tau$, as well as on its location. The modification is accomplished by using a quasi-static approximation to the field internal to $\Delta\tau$. This assumes that the field internal to $\Delta\tau$ is related to the field external to $\Delta\tau$ in the same manner as for static fields. The procedure is justifiable, because, in a region small compared to wavelength, the Helmholtz equation can be approximated by Laplace's equation.

There are four types of samples for which this quasi-static modification to the perturbational solution is very simply accomplished. These are shown in Fig. 7-3 for the dielectric case. For the magnetic case, it is merely necessary to replace **E** by **H** and ϵ by μ. For the thin slab with **E** normal to it (Fig. 7-3a), we must have continuity of the normal com-

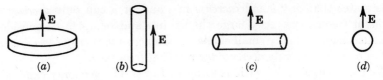

FIG. 7-3. Some small dielectric objects for which the quasi-static solutions are simple.

ponent of **D**, so that

$$E_{\text{int}} = \frac{1}{\epsilon_r} E_{\text{ext}} \tag{7-14}$$

This approximation is valid regardless of the cross-sectional shape of the cylinder. For the long thin cylinder with **E** tangential to it (Fig. 7-3b), we must have continuity of the tangential component of **E**, so that

$$E_{\text{int}} = E_{\text{ext}} \tag{7-15}$$

Again this approximation is independent of the cross-sectional shape of the cylinder. For **E** normal to a long thin *circular* cylinder (Fig. 7-3c), we can use the static solution,[1] which is

$$E_{\text{int}} = \frac{2}{1 + \epsilon_r} E_{\text{ext}} \tag{7-16}$$

Finally, for **E** normal to a small sphere (Fig. 7-3d), we can use the static solution,[2] which is

$$E_{\text{int}} = \frac{3}{2 + \epsilon_r} E_{\text{ext}} \tag{7-17}$$

The static solution for a dielectric ellipsoid in a uniform field is also known but is not very simple in form.[2]

To use the above quasi-static approximations, we approximate **E** (and **H** in the magnetic case) in the *numerator* of Eq. (7-10) by E_{int} of the preceding equations. In the denominator we can still use the approximations $\mathbf{E} = \mathbf{E}_0$ and $\mathbf{H} = \mathbf{H}_0$, because the contribution from $\Delta\tau$ is small compared to that from the rest of τ. Hence, our quasi-static correction to the perturbational formula is

$$\frac{\omega - \omega_0}{\omega_0} \approx -\frac{\iiint \Delta\epsilon \mathbf{E}_{\text{int}} \cdot \mathbf{E}_0^* \, d\tau}{2 \iiint \epsilon |E_0|^2 \, d\tau} \tag{7-18}$$

[1] W. R. Smyth, "Static and Dynamic Electricity," pp. 67–68, McGraw-Hill Book Company, Inc., New York, 1950.

[2] J. A. Stratton, "Electromagnetic Theory," pp. 205–213, McGraw-Hill Book Company, Inc., New York, 1941.

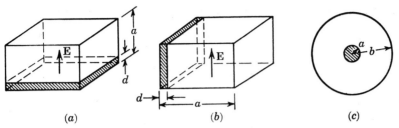

Fig. 7-4. Cavities used to illustrate the perturbational formulas.

for the case $\Delta\mu = 0$. (The denominator has been simplified by equating $\bar{\mathcal{W}}_m$ to $\bar{\mathcal{W}}_e$.) The corresponding formula for the frequency shift due to a magnetic material would be of same form, but with **E** replaced by **H** and ϵ by μ throughout.

Equation (7-18) is, of course, most valuable for problems for which the exact solution is not known. However, so that we may gain confidence in the results as well as practice in the procedure, let us apply Eq. (7-18) to problems for which we have the exact solution. These are illustrated in Fig. 7-4. For a dielectric slab on the base of a rectangular cavity (Fig. 7-4a), we have E_{int} given by Eq. (7-14). The field and energy expressions for the unperturbed cavity are given in Sec. 2-8. Application of Eq. (7-18) then yields

$$\frac{\omega - \omega_0}{\omega_0} \approx -\frac{1}{2}\frac{\epsilon_r - 1}{\epsilon_r}\frac{d}{a} \qquad (7\text{-}19)$$

where d is the slab thickness and a is the cavity height. A comparison of this with the result of Prob. 4-17 for $\mu_1 = \mu_2 = \mu_0$ and $\epsilon_2 = \epsilon_0$ shows that our answer is identical to the first term of the expansion for ω in powers of d/a. In fact, if $\Delta\mu$ is also nonzero and we treat it to the same degree of approximation (match tangential H), we again get the correct first term of the expansion. To illustrate the improvement obtained by using the quasi-static field, we can compare Eq. (7-19) to the result obtained from Eq. (7-11), which is

$$\frac{\omega - \omega_0}{\omega_0} \approx -\frac{1}{2}(\epsilon_r - 1)\frac{d}{a}$$

It is apparent that the above formula is accurate only for $\epsilon_r \approx 1$, that is, when $\Delta\epsilon$ is small.

A nonmagnetic dielectric slab at a side wall of the rectangular cavity (Fig. 7-4b) has but little effect on the resonant frequency, because **E** is zero at the wall. In this case **E** is tangential to the air-dielectric interface; so Eq. (7-15) should apply. Note that Eqs. (7-18) and (7-11) give identical approximations in this case. In particular, we obtain

$$\frac{\omega - \omega_0}{\omega_0} \approx -\frac{(\epsilon_r - 1)}{a} \int_0^d \sin^2 \frac{\pi x}{a} \, dx$$

$$\approx -\frac{\pi^2}{3}(\epsilon_r - 1)\left(\frac{d}{a}\right)^3 \quad (7\text{-}20)$$

A comparison of this with the answer to Prob. 4-18 shows that we again have the correct first term of the expansion when $\Delta\mu = 0$.

As a final example, consider the spherical cavity with a concentric dielectric sphere (Fig. 7-4c). The field of the unperturbed cavity is defined by

$$H_\phi = \frac{1}{r} \hat{J}_1\left(2.744 \frac{r}{b}\right) \sin \theta$$

and the stored energy is given by Eq. (6-35). Applying Eq. (7-18), using the quasi-static Eq. (7-17), we obtain

$$\frac{\omega - \omega_0}{\omega_0} \approx -0.291 \frac{\epsilon_r - 1}{\epsilon_r + 2}\left(2.744 \frac{a}{b}\right)^3$$

where a is the radius of the small dielectric sphere and b is the radius of the conductor. This we can compare to the exact solution (Prob. 6-8), which is the same. The perturbational method used in conjunction with the quasi-static approximation gives excellent accuracy when properly used. This shift in resonant frequency caused by the introduction of a dielectric sample into a resonant cavity can be used to measure the constitutive parameters of matter.

7-4. Waveguide Perturbations. We shall now consider waveguides cylindrical in the general sense, that is, all $z = $ constant cross sections are identical. Figure 7-5a represents a cross section of the unperturbed waveguide, Fig. 7-5b represents a wall perturbation, and Fig. 7-5c represents a material perturbation. All perturbations must, of course, be independent of z. The guide boundary is taken as perfectly conducting in all cases.

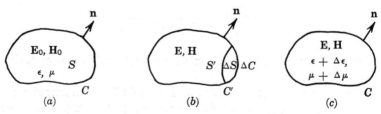

Fig. 7-5. Perturbations of cylindrical waveguides. (a) Original cross section; (b) wall perturbation; (c) material perturbation.

At the cutoff frequency a cylindrical waveguide is a two-dimensional resonator. We should therefore expect formulas similar to those for perturbations of cavities to apply to waveguides at cutoff. In fact, we can apply the cavity derivations directly to the region formed by the cylindrical waveguide bounded by two $z =$ constant planes, changing only some of the explanations. For example, in deriving Eq. (7-2), the left-hand side results from the integral

$$\oiint (\mathbf{H} \times \mathbf{E}_0^* + \mathbf{H}_0^* \times \mathbf{E}) \cdot d\mathbf{s}$$

taken over the perturbed surface. For a length of a cylindrical waveguide at cutoff, the fields are independent of z; so the surface integrals over the two $z =$ constant cross sections cancel each other. This leaves only the surface integral on the left-hand side of Eq. (7-2) taken over the wall of the waveguide. Following the derivation further, we find that Eq. (7-3) applies directly for calculating the change in waveguide cutoff frequency. But both numerator and denominator involve an integration with respect to z, which reduces to the length of the segment of the cylindrical waveguide. Hence, from Eq. (7-3) we obtain the change in cutoff frequency $\Delta\omega_c$ due to an inward perturbation of the waveguide wall as

$$\Delta\omega_c = \frac{j \oint_{\Delta C} \mathbf{H} \times \mathbf{E}_0^* \cdot \mathbf{n}\, dl}{\iint_{S'} (\epsilon \mathbf{E} \cdot \mathbf{E}_0^* + \mu \mathbf{H} \cdot \mathbf{H}_0^*)\, ds} \quad (7\text{-}21)$$

where ΔC is the contour about the volume of the perturbation and S' is the cross section of the perturbed waveguide (see Fig. 7-5b).

The crude approximation of replacing the perturbed fields $\mathbf{E}$, $\mathbf{H}$ by the unperturbed fields $\mathbf{E}_0$, $\mathbf{H}_0$ in Eq. (7-21) gives good results for smooth, shallow perturbations. This leads to

$$\frac{\Delta\omega_c}{\omega_c} \approx \frac{\iint_{\Delta S} (\mu|H_0|^2 - \epsilon|E_0|^2)\, ds}{\iint_{S} (\mu|H_0|^2 + \epsilon|E_0|^2)\, ds} \quad (7\text{-}22)$$

which is analogous to Eq. (7-4). Hence, an inward perturbation of the waveguide walls at a position of high E will lower the cutoff frequency, while one at a position of high H will raise the cutoff frequency. For perturbations not shallow and smooth, we can obtain a better approximation to $\Delta\omega_c$ by using a quasi-static approximation for $\mathbf{H}$ in the numerator of Eq. (7-21). An example of the perturbation of waveguide walls is the "ridge waveguide," formed from the rectangular waveguide by adding ridges along the center of the top and bottom walls.[1] Such ridges will lower the cutoff frequency of the dominant mode and will raise the cutoff frequency of the next higher mode (see Prob. 7-12). Hence, a greater range of single-mode operation can be obtained. The ridges also decrease the characteristic impedance of the guide; hence, they are used for impedance matching.

The formulas for material perturbations in cavities can also be specialized to the case of material perturbations in waveguides at cutoff. The reasoning is essentially the same as that used for the wall-perturbation case. Hence, from Eq. (7-10) we can obtain the exact formula for the change in cutoff frequency due to a change of matter with the waveguide. It is

$$\frac{\Delta \omega_c}{\omega_c} = - \frac{\iint (\Delta\epsilon \mathbf{E} \cdot \mathbf{E}_0^* + \Delta\mu \mathbf{H} \cdot \mathbf{H}_0^*)\, ds}{\iint (\epsilon \mathbf{E} \cdot \mathbf{E}_0^* + \mu \mathbf{H} \cdot \mathbf{H}_0^*)\, ds} \qquad (7\text{-}23)$$

where the integrals are taken over the guide cross section. Note that an increase in either ϵ or μ can only decrease the cutoff frequency of a waveguide. If $\Delta\epsilon$ and $\Delta\mu$ are small, we can replace $\mathbf{E}$, $\mathbf{H}$ by $\mathbf{E}_0$, $\mathbf{H}_0$ and obtain

$$\frac{\Delta \omega_c}{\omega_c} \approx - \frac{\iint (\Delta\epsilon |E_0|^2 + \Delta\mu |H_0|^2)\, ds}{\iint (\epsilon |E_0|^2 + \mu |H_0|^2)\, ds} \qquad (7\text{-}24)$$

This is analogous to Eq. (7-11). If $\Delta\epsilon$ and $\Delta\mu$ are large, but of small spatial extent, we can improve our approximation by using the quasi-static method of Sec. 7-3. For example, analogous to Eq. (7-18) we have in the nonmagnetic case

$$\frac{\Delta \omega_c}{\omega_c} \approx - \frac{\iint \Delta\epsilon \mathbf{E}_{\text{int}} \cdot \mathbf{E}_0^*\, ds}{2 \iint \epsilon |E_0|^2\, ds} \qquad (7\text{-}25)$$

where $\mathbf{E}_{\text{int}}$ is given by the appropriate one of Eqs. (7-14) to (7-16).

As long as the perturbed guide is homogeneous in ϵ and μ, we can determine the propagation constant at any frequency from the cutoff frequency according to

$$\gamma = \begin{cases} j\beta = jk\sqrt{1 - \left(\dfrac{\omega_c}{\omega}\right)^2} & \omega > \omega_c \\ \alpha = k_c \sqrt{1 - \left(\dfrac{\omega}{\omega_c}\right)^2} & \omega < \omega_c \end{cases} \qquad (7\text{-}26)$$

[1] S. B. Cohn, Properties of Ridge Waveguide, *Proc. IRE*, vol. 35, no. 8, pp. 783–788, August, 1947.

(This is proved in Sec. 8-1.) If the perturbed guide is inhomogeneous, no such simple relationship exists. In such cases we can obtain perturbational formulas for the change in γ. In the loss-free case we can express the unperturbed fields as

$$\begin{aligned} \mathbf{E}_0 &= \hat{\mathbf{E}}_0(x,y)e^{-j\beta_0 z} \\ \mathbf{H}_0 &= \hat{\mathbf{H}}_0(x,y)e^{-j\beta_0 z} \end{aligned} \quad (7\text{-}27)$$

and the perturbed fields as

$$\begin{aligned} \mathbf{E} &= \hat{\mathbf{E}}(x,y)e^{-j\beta z} \\ \mathbf{H} &= \hat{\mathbf{H}}(x,y)e^{-j\beta z} \end{aligned} \quad (7\text{-}28)$$

The perturbational formulas are then

$$\beta - \beta_0 = -j \frac{\oint_{\Delta C} (\hat{\mathbf{E}}_0^* \times \hat{\mathbf{H}}) \cdot \mathbf{n}\, dl}{\iint_S (\hat{\mathbf{E}}_0^* \times \hat{\mathbf{H}} + \hat{\mathbf{E}} \times \hat{\mathbf{H}}_0^*) \cdot \mathbf{u}_z\, ds} \quad (7\text{-}29)$$

in the case of a wall perturbation, and

$$\beta - \beta_0 = \omega \frac{\iint_S (\Delta\epsilon \hat{\mathbf{E}} \cdot \hat{\mathbf{E}}_0^* + \Delta\mu \hat{\mathbf{H}} \cdot \hat{\mathbf{H}}_0^*)\, ds}{\iint_S (\hat{\mathbf{E}}_0^* \times \hat{\mathbf{H}} + \hat{\mathbf{E}} \times \hat{\mathbf{H}}_0^*) \cdot \mathbf{u}_z\, ds} \quad (7\text{-}30)$$

in the case of a material perturbation. The perturbational formulas in the lossy case are given in Probs. 7-15 and 7-16.

To illustrate the derivation of the above formulas, consider a material perturbation. The unperturbed and perturbed fields satisfy Eqs. (7-9) with $\omega = \omega_0$, for the frequency is kept unchanged. The two equations following Eqs. (7-9) are still valid, and, with $\omega_0 = \omega$, their sum becomes

$$\nabla \cdot (\mathbf{H} \times \mathbf{E}_0^* + \mathbf{H}_0^* \times \mathbf{E}) = j\omega(\Delta\epsilon \mathbf{E} \cdot \mathbf{E}_0^* + \Delta\mu \mathbf{H} \cdot \mathbf{H}_0^*)$$

Integrating this equation throughout a region and applying the divergence theorem to the left-hand term, we obtain

$$\oint (\mathbf{H} \times \mathbf{E}_0^* + \mathbf{H}_0^* \times \mathbf{E}) \cdot d\mathbf{s} = j\omega \iiint (\Delta\epsilon \mathbf{E} \cdot \mathbf{E}_0^* + \Delta\mu \mathbf{H} \cdot \mathbf{H}_0^*)\, d\tau \quad (7\text{-}31)$$

This is an identity for any two fields of the same frequency in a region for which ϵ and μ are changed to $\epsilon + \Delta\epsilon$ and $\mu + \Delta\mu$. For material perturbations in a cylindrical waveguide, we express the fields according to Eqs. (7-27) and (7-28) and apply Eq. (7-31) to the differential slice of Fig. 7-6. On the

FIG. 7-6. Differential slice of a cylinder.

waveguide walls both **n** × **E** and **n** × **E**$_0$ vanish; so this part of the surface integral vanishes. Also, since the thickness of the slice is a differential distance,

$$\iint_{\text{top}} + \iint_{\text{bottom}} = dz \frac{\partial}{\partial z} \iint_S = -j(\beta - \beta_0) \, dz \iint_S$$

The right-hand side of Eq. (7-31) can be expressed as the integral over the cross section times dz; hence Eq. (7-31) reduces to

$$-j(\beta - \beta_0) \iint_S (\hat{\mathbf{H}} \times \hat{\mathbf{E}}_0^* + \hat{\mathbf{H}}_0^* \times \hat{\mathbf{E}}) \cdot \mathbf{u}_z \, ds$$

$$= j\omega \iint_S (\Delta\epsilon \hat{\mathbf{E}} \cdot \hat{\mathbf{E}}_0^* + \Delta\mu \hat{\mathbf{H}} \cdot \hat{\mathbf{H}}_0^*) \, ds$$

Rearrangement of this equation gives Eq. (7-30). In the derivation of Eq. (7-29), the right-hand side of Eq. (7-31) is zero, and the left-hand side equated to zero leads to the desired result.

Equations (7-29) and (7-30) as they stand are exact. To use them, we must make various approximations for $\hat{\mathbf{E}}$ and $\hat{\mathbf{H}}$, just as we did in the cavity problems of Secs. 7-3 and 7-4. For example, in the case of shallow, smooth deformations of waveguide walls, we can approximate $\hat{\mathbf{E}}$, $\hat{\mathbf{H}}$ by $\hat{\mathbf{E}}_0$, $\hat{\mathbf{H}}_0$ in Eq. (7-29). Using the conservation of complex power [Eq. (1-62)], we arrive at the result

$$\beta - \beta_0 \approx \omega \frac{\iint_{\Delta S} (\mu|\hat{H}_0|^2 - \epsilon|\hat{E}_0|^2) \, ds}{\iint_S (\hat{\mathbf{E}}_0^* \times \hat{\mathbf{H}}_0 + \hat{\mathbf{E}}_0 \times \hat{\mathbf{H}}_0^*) \cdot \mathbf{u}_z \, ds} \qquad (7\text{-}32)$$

(The denominator is twice the time-average power flow in the unperturbed guide.) If the perturbation is not shallow and smooth, better results can be obtained using a quasi-static modification. Similarly, for small $\Delta\epsilon$ and $\Delta\mu$ we have the approximation for material perturbations

$$\beta - \beta_0 \approx \omega \frac{\iint_S (\Delta\epsilon|\hat{E}_0|^2 + \Delta\mu|\hat{H}_0|^2) \, ds}{\iint_S (\hat{\mathbf{E}}_0^* \times \hat{\mathbf{H}}_0 + \hat{\mathbf{E}}_0 \times \hat{\mathbf{H}}_0^*) \cdot ds} \qquad (7\text{-}33)$$

For large $\Delta\epsilon$ and $\Delta\mu$ we can obtain better results by using the quasi-static approximation for the fields within $\Delta\epsilon$ and $\Delta\mu$.

As an example of the perturbational approach applied to a waveguide problem, consider a circular waveguide of radius b containing a concentric dielectric rod of radius a. The exact solution to this problem was

FIG. 7-7. Comparison of the perturbational solution with the exact solution for the partially filled circular waveguide, $\epsilon = 10\epsilon_0$, $b = 0.4\lambda_0$.

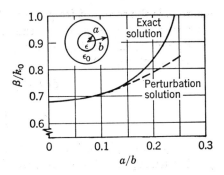

considered in Sec. 5-5, and a numerical example is shown in Fig. 5-11. For the perturbational solution we shall use Eq. (7-30) with $\Delta\mu = 0$. In the numerator we make the quasi-static approximation of Eq. (7-16), and in the denominator we approximate $\hat{\mathbf{E}}$, $\hat{\mathbf{H}}$ by $\hat{\mathbf{E}}_0$, $\hat{\mathbf{H}}_0$. The unperturbed field of the dominant TE_{11} mode for the circular waveguide is

$$E_\rho = \frac{1}{\rho} J_1\left(1.841 \frac{\rho}{b}\right) \sin\phi \qquad H_\rho = -\frac{E_\phi}{Z_0}$$

$$E_\phi = \frac{1.841}{b} J_1'\left(1.841 \frac{\rho}{b}\right) \cos\phi \qquad H_\phi = \frac{E_\rho}{Z_0}$$

where Z_0 is the characteristic impedance [Eq. (5-32)]. The denominator of Eq. (7-30) then becomes

$$\frac{2}{Z_0} \int_0^{2\pi} d\phi \int_0^b d\rho\, \rho(E_\rho^2 + E_\phi^2) = 0.7892 \frac{\pi}{\eta} \sqrt{1 - \left(\frac{\omega_c}{\omega}\right)^2}$$

where ω_c is the cutoff frequency. The numerator is easily evaluated as

$$\frac{\pi}{2} \frac{\epsilon_r - 1}{\epsilon_r + 1} \epsilon_0 \left(1.841 \frac{a}{b}\right)^2$$

and Eq. (7-30) reduces to

$$\frac{\beta - \beta_0}{k_0} = \frac{2.146}{\sqrt{1 - (\omega_c/\omega)^2}} \frac{\epsilon_r - 1}{\epsilon_r + 1} \left(\frac{a}{b}\right)^2 \qquad (7\text{-}34)$$

Figure 7-7 compares this solution to the exact solution of Fig. 5-11. Our approximations give good results for small a/b. At frequencies near the unperturbed cutoff frequency, the ω_c in Eq. (7-34) may be taken as that of the perturbed guide.

7-5. Stationary Formulas for Cavities. Suppose we have a resonant cavity formed by a perfect conductor enclosing a dielectric, possibly inhomogeneous. The "wave equations" are

$$\begin{aligned}\nabla \times \mu^{-1}\nabla \times \mathbf{E} - \omega_r^2 \epsilon \mathbf{E} &= 0 \\ \nabla \times \epsilon^{-1}\nabla \times \mathbf{H} - \omega_r^2 \mu \mathbf{H} &= 0\end{aligned} \qquad (7\text{-}35)$$

where ω_r is the resonant frequency. These reduce to the usual Helmholtz equations when ϵ and μ are constants. If the first of Eqs. (7-35) is scalarly multiplied by **E** and the resulting equation integrated throughout the cavity, we obtain

$$\omega_r^2 = \frac{\iiint \mathbf{E} \cdot \nabla \times \mu^{-1} \nabla \times \mathbf{E}\, d\tau}{\iiint \epsilon E^2\, d\tau} \qquad (7\text{-}36)$$

Similarly, multiplying the second of Eqs. (7-35) scalarly by **H** and integrating throughout the cavity, we obtain

$$\omega_r^2 = \frac{\iiint \mathbf{H} \cdot \nabla \times \epsilon^{-1} \nabla \times \mathbf{H}\, d\tau}{\iiint \mu H^2\, d\tau} \qquad (7\text{-}37)$$

Equations (7-36) and (7-37) are identities, but, even more important, they are useful for approximating ω_r by assuming field distributions in a cavity. They are particularly well-suited for this latter application because of their "stationary" character, which we shall now discuss.

We take Eq. (7-36) and substitute for the *true field* **E** a *trial field*

$$\mathbf{E}_{\text{trial}} = \mathbf{E} + \Delta\mathbf{E} = \mathbf{E} + p\mathbf{e} \qquad (7\text{-}38)$$

where p is an arbitrary parameter. This procedure gives

$$\omega^2(p) = \frac{\iiint (\mathbf{E} + p\mathbf{e}) \cdot \nabla \times \mu^{-1} \nabla \times (\mathbf{E} + p\mathbf{e})\, d\tau}{\iiint \epsilon(\mathbf{E} + p\mathbf{e}) \cdot (\mathbf{E} + p\mathbf{e})\, d\tau} \qquad (7\text{-}39)$$

where we show ω^2 as a function of p for fixed **e**. The Maclaurin expansion of ω^2 is

$$\omega^2(p) = \omega_r^2 + p \left.\frac{\partial \omega^2}{\partial p}\right|_{p=0} + \frac{p^2}{2!} \left.\frac{\partial^2 \omega^2}{\partial p^2}\right|_{p=0} + \cdots \qquad (7\text{-}40)$$

Note that the first term is the true resonant frequency, because $\omega^2(0) = \omega_r^2$. In the *variational notation*[1] the above expansion is written as

$$\omega^2(p) = \omega_r^2 + p\delta\omega^2 + \frac{p^2}{2!}\delta^2\omega^2 + \cdots \qquad (7\text{-}41)$$

By definition, each term of Eq. (7-41) equals the corresponding term of Eq. (7-40). The term $\delta\omega^2$ is called the *first variation* of ω^2, the term $\delta^2\omega^2$ is called the *second variation*, and so on. A formula for ω^2 is said to be

---

[1] F. B. Hildebrand, "Methods of Applied Mathematics," p. 130, Prentice-Hall, Inc., Englewood Cliffs, N.J., 1952.

stationary if the first variation of ω^2 vanishes. This is equivalent to

$$\frac{\partial \omega^2}{\partial p}\bigg|_{p=0} = 0 \tag{7-42}$$

The extension to more than one p parameter is straightforward.

We now wish to show that Eq. (7-39) is stationary. The derivative of the numerator $N(p)$ evaluated at $p = 0$ is

$$N'(0) = \iiint (\mathbf{E} \cdot \nabla \times \mu^{-1}\nabla \times \mathbf{e} + \mathbf{e} \cdot \nabla \times \mu^{-1}\nabla \times \mathbf{E}) \, d\tau$$

It is a vector identity that

$$\iiint \mathbf{E} \cdot \nabla \times \mu^{-1}\nabla \times \mathbf{e} \, d\tau = \iiint \mu^{-1}\nabla \times \mathbf{e} \cdot \nabla \times \mathbf{E} \, d\tau$$
$$+ \oint [(\mu^{-1}\nabla \times \mathbf{e}) \times \mathbf{E}] \cdot d\mathbf{s}$$

The last term vanishes, because $\mathbf{n} \times \mathbf{E} = 0$ on S. A similar identity states

$$\iiint \mu^{-1}\nabla \times \mathbf{e} \cdot \nabla \times \mathbf{E} \, d\tau = \iiint \mathbf{e} \cdot \nabla \times \mu^{-1}\nabla \times \mathbf{E} \, d\tau$$
$$- \oint [(\mu^{-1}\nabla \times \mathbf{E}) \times \mathbf{e}] \cdot d\mathbf{s}$$

Using these two identities and the first of Eqs. (7-35), we obtain

$$N'(0) = 2\omega_r^2 \iiint \epsilon \mathbf{e} \cdot \mathbf{E} \, d\tau - \oint [(\mu^{-1}\nabla \times \mathbf{E}) \times \mathbf{e}] \cdot d\mathbf{s}$$

The derivative of the denominator $D(p)$ of Eq. (7-39) is, for $p = 0$,

$$D'(0) = 2 \iiint \epsilon \mathbf{e} \cdot \mathbf{E} \, d\tau$$

We then obtain

$$\frac{\partial \omega^2}{\partial p}\bigg|_{p=0} = \frac{D(0)N'(0) - N(0)D'(0)}{D^2(0)}$$
$$= -\frac{\oint [(\mu^{-1}\nabla \times \mathbf{E}) \times \mathbf{e}] \cdot d\mathbf{s}}{\iiint \epsilon E^2 \, d\tau} \tag{7-43}$$

which has been simplified, using Eq. (7-36). The above equation vanishes if $\mathbf{n} \times \mathbf{e} = 0$ on S, which requires $\mathbf{n} \times \mathbf{E}_{\text{trial}} = 0$ on S. Hence, Eq. (7-36) is a stationary formula for the resonant frequency if the tangential components of the trial $\mathbf{E}$ vanish on the cavity walls.

Equation (7-36) can be put into a more symmetrical form by applying the identity

$$\iiint \mathbf{E} \cdot \nabla \times \mu^{-1} \nabla \times \mathbf{E} \, d\tau = \iiint \mu^{-1} \nabla \times \mathbf{E} \cdot \nabla \times \mathbf{E} \, d\tau \\ + \oiint [(\mu^{-1} \nabla \times \mathbf{E}) \times \mathbf{E}] \cdot d\mathbf{s}$$

The last term vanishes, because $\mathbf{n} \times \mathbf{E} = 0$ on S. Substituting this identity into Eq. (7-36), we obtain

$$\omega_r^2 = \frac{\iiint \mu^{-1} (\nabla \times \mathbf{E})^2 \, d\tau}{\iiint \epsilon E^2 \, d\tau} \qquad (7\text{-}44)$$

This formula proves to be stationary, provided $\mathbf{n} \times \mathbf{E}_{\text{trial}} = 0$ on S. If we look carefully at the first variation of Eq. (7-44), it is evident that the requirement $\mathbf{n} \times \mathbf{E}_{\text{trial}} = 0$ on S can be relaxed if the term

$$2 \oiint [(\mu^{-1} \nabla \times \mathbf{E}) \times \mathbf{E}] \cdot d\mathbf{s}$$

is added to the numerator. This gives

$$\omega_r^2 = \frac{\iiint \mu^{-1} (\nabla \times \mathbf{E})^2 \, d\tau + 2 \oiint [(\mu^{-1} \nabla \times \mathbf{E}) \times \mathbf{E}] \cdot d\mathbf{s}}{\iiint \epsilon E^2 \, d\tau} \qquad (7\text{-}45)$$

which is stationary, even if $\mathbf{n} \times \mathbf{E}_{\text{trial}} \neq 0$ on S. This is an important modification, because it is not always easy to find a trial field with vanishing tangential components on the cavity walls, especially if the geometry is complicated. Still further modifications in our formulas are required if $\mathbf{n} \times \mathbf{E}$ or $\mathbf{n} \times (\mu^{-1} \nabla \times \mathbf{E})$ are discontinuous over some surface within the cavity. All such modifications can be quite simply effected by the reaction concept of Sec. 7-7.

A similar procedure shows that Eq. (7-37) is a stationary formula in terms of $\mathbf{H}$, provided $\mathbf{n} \times (\epsilon^{-1} \nabla \times \mathbf{H}) = 0$ on S. The H-field formula corresponding to Eq. (7-44) is

$$\omega_r^2 = \frac{\iiint \epsilon^{-1} (\nabla \times \mathbf{H})^2 \, d\tau}{\iiint \mu H^2 \, d\tau} \qquad (7\text{-}46)$$

which turns out to be stationary subject to no boundary conditions on S. Further modifications to account for discontinuities in $\mathbf{n} \times \mathbf{H}$ or $\mathbf{n} \times (\epsilon^{-1} \nabla \times \mathbf{H})$ over surfaces within the cavity can be made. These modifications again follow directly from the methods of Sec. 7-7.

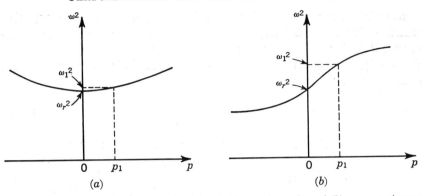

FIG. 7-8. Illustration of ω^2 versus p for (a) a stationary formula and (b) a nonstationary formula.

Let us now briefly consider the advantages of a stationary formula over a nonstationary one. Figure 7-8 shows pictorally the primary advantage. Given a class of trial fields of the form of Eq. (7-38), the parameter $\omega^2(p)$ determined from a stationary formula such as Eq. (7-39) will have a minimum or maximum at $p = 0$.[1] This is shown in Fig. 7-8a. The parameter ω^2 determined from a nonstationary formula must have some definite slope at $p = 0$, as shown in Fig. 7-8b. For a given error in the assumed field, say $\Delta E = p_1 e$, the corresponding error in the resonant frequency is $\omega_1{}^2 - \omega_r{}^2$. It is evident that for small p_1 the stationary formula gives a smaller error in ω^2 than does the nonstationary formula. This property is sometimes summarized as follows: "A parameter determined by a stationary formula is insensitive to small variations of the field about the true field." An error of the order of 10 per cent in the assumed field gives an error of the order of only 1 per cent in the parameter. In some cases the true field can be shown to yield an absolute minimum or maximum for the parameter. The stationary formula then gives upper or lower bounds to the parameter. Our formulas for ω^2 give upper bounds, as we shall show later.

We might also inquire about the general procedure of establishing stationary formulas. One characteristic of all such formulas is that the numerator and denominator contain squares of the trial field. This insures that amplitude of the trial field will have no effect on the calculation. Classically, the method of establishing stationary formulas is to construct formulas of the proper form and then separate the stationary ones from the nonstationary ones by testing the first variation. In Sec. 7-7 we shall give a general procedure which leads directly to the various stationary formulas.

[1] A complex parameter would have a saddle point at $p = 0$.

Now let us apply some of our stationary formulas to a problem for which we have an exact answer, so that we may get an idea of the accuracy obtainable. Consider the dominant mode of the circular cavity (Fig. 5-7), for the case $d < 2a$. The TM_{010} mode is dominant and the exact resonant frequency is

$$\omega_r = \frac{2.4048}{a\sqrt{\epsilon\mu}} \tag{7-47}$$

The field is sketched in Fig. 5-8 and is given mathematically by

$$E_z = \frac{\omega\mu}{j} J_0\left(2.405 \frac{\rho}{a}\right) \qquad H_\phi = \frac{2.405}{a} J_1\left(2.405 \frac{\rho}{a}\right)$$

Substitution of this true field into any of our stationary formulas must, of course, give us Eq. (7-47).

Suppose we first try a formula that requires no boundary conditions [Eq. (7-46)]. Assume as a trial field

$$\mathbf{H} = \mathbf{u}_\phi \rho \qquad \nabla \times \mathbf{H} = \mathbf{u}_z 2$$

Equation (7-46) then becomes

$$\omega^2 = \frac{\int_0^a 4\rho \, d\rho}{\epsilon\mu \int_0^a \rho^3 \, d\rho} = \frac{8}{\epsilon\mu a^2}$$

and our approximation is

$$\omega_r \approx \frac{2.818}{a\sqrt{\epsilon\mu}} \tag{7-48}$$

This is 16 per cent too high, which is a relatively poor result. This suggests that our trial field was too crude an approximation. We can improve our trial field by assuming

$$\mathbf{H} = \mathbf{u}_\phi \left(\rho - \frac{2\rho^2}{3a}\right) \qquad \nabla \times \mathbf{H} = \mathbf{u}_z 2\left(1 - \frac{\rho}{a}\right)$$

which is chosen to satisfy the condition $\mathbf{n} \times \mathbf{E} = 0$ on S. Equation (7-46) then yields

$$\omega^2 = \frac{\int_0^a 4\left(1 - \frac{\rho}{a}\right)^2 \rho \, d\rho}{\epsilon\mu \int_0^a \left(\rho - \frac{2\rho^2}{3a}\right)^2 \rho \, d\rho} = \frac{180}{\epsilon\mu 31 a^2}$$

and our approximation is now

$$\omega_r \approx \frac{2.410}{a\sqrt{\epsilon\mu}} \tag{7-49}$$

This is only 0.2 per cent in error. Even though a formula is stationary, we must use care in choosing trial fields. It is advisable to meet the physical boundary conditions as closely as possible, for this will help to obtain a trial field close to the true field. If the same trial field is used in Eq. (7-37), we again get Eq. (7-49), since $\mathbf{n} \times \mathbf{E} = 0$ on S.

Now consider a stationary **E**-field formula, say Eq. (7-44). This formula requires $\mathbf{n} \times \mathbf{E} = 0$ on S; hence we choose

$$\mathbf{E} = \mathbf{u}_z\left(1 - \frac{\rho}{a}\right) \qquad \nabla \times \mathbf{E} = \mathbf{u}_\phi \frac{1}{a}$$

Substituting this into Eq. (7-44), we obtain

$$\omega^2 = \frac{\int_0^a \frac{\rho}{a^2}\, d\rho}{\epsilon\mu \int_0^a \left(1 - \frac{\rho}{a}\right)^2 \rho\, d\rho} = \frac{6}{\epsilon\mu a^2}$$

Our approximation is therefore

$$\omega_r \approx \frac{2.449}{a\sqrt{\epsilon\mu}} \qquad (7\text{-}50)$$

which is 1.8 per cent too high. If we had chosen a trial **E** field not satisfying $\mathbf{n} \times \mathbf{E} = 0$ on S, we would have had to use Eq. (7-45).

Note that all our approximations are too high. This suggests that the true resonant frequency is an absolute minimum, which we shall now show. For example, take Eq. (7-39), and, by means of various identities, put it into the form

$$\omega^2 - \omega_r^2 = \frac{\iiint p\mathbf{e} \cdot (\nabla \times \mu^{-1}\nabla \times p\mathbf{e} - \omega_r^2 \epsilon p\mathbf{e})\, d\tau}{\iiint \epsilon(\mathbf{E} + p\mathbf{e})^2\, d\tau} \qquad (7\text{-}51)$$

It is known that the eigenfunctions, that is, the fields of the various modes, form a complete set of orthogonal functions in the cavity space.[1] Hence, the error field $p\mathbf{e}$ can be expanded in a series

$$p\mathbf{e} = \sum_i A_i \mathbf{E}_i$$

where the A_i are constants and the $\mathbf{E}_i$ are the various mode fields. Substituting the above equation into Eq. (7-51), making use of the wave

---

[1] Philip M. Morse and Herman Feshbach, "Methods of Theoretical Physics," part I, Chap. 6, McGraw-Hill Book Company, Inc., New York, 1953.

equation and the orthogonality relationships, we obtain

$$\omega^2 - \omega_r^2 = \frac{\sum_i (\omega_i^2 - \omega_r^2) A_i^2 \iiint \epsilon E_i^2 \, d\tau}{\iiint \epsilon (E_{\text{trial}})^2 \, d\tau} \tag{7-52}$$

where the ω_i are the resonant frequencies of the ith modes. Since we have chosen ω_r as the lowest eigenvalue, Eq. (7-52) is always positive. Hence, any ω calculated from Eq. (7-36) will be an upper bound to the true resonant frequency. Also, if we choose a trial field orthogonal to the field of the lowest mode, we have an upper bound to the next higher resonant frequency, and so on. This, of course, requires that the dominant mode be known exactly, which is seldom the case for complicated geometries.

Look now at Eq. (7-46). The trial field $\mathbf{H}$ = constant vector is a permissible trial field, since no boundary conditions are required. The result is $\omega_r = 0$, which is less than the true resonant frequency [Eq. (7-47)]. Why do we not have an upper bound in this case? The answer lies in the fact that we have overlooked the "static mode." A static magnetic field ($\omega_r = 0$) can exist in a cavity bounded by a perfect electric conductor. Fortunately, it is easy to insure that our trial field is orthogonal to all static fields, thereby obtaining an upper bound to the dominant a-c mode. Any trial field satisfying

$$\boldsymbol{\nabla} \cdot \mu \mathbf{H} = 0 \qquad \mu H_n = 0 \text{ on } S \tag{7-53}$$

is orthogonal to all static fields, as we shall now prove. The desired orthogonality is

$$\iiint \mu \mathbf{H} \cdot \mathbf{H}_{\text{static}} \, d\tau = 0$$

where, in general, $\mathbf{H}_{\text{static}} = -\boldsymbol{\nabla} U$. By virtue of the identity

$$\boldsymbol{\nabla} \cdot (U \mu \mathbf{H}) = \mu \mathbf{H} \cdot \boldsymbol{\nabla} U + U \boldsymbol{\nabla} \cdot \mu \mathbf{H}$$

the preceding equation becomes

$$\iiint U \boldsymbol{\nabla} \cdot \mu \mathbf{H} \, d\tau - \oiint U \mu \mathbf{H} \cdot d\mathbf{s} = 0$$

This requirement is met for all U by the conditions of Eq. (7-53). Our choices for $\mathbf{H}$ in the foregoing examples satisfied Eq. (7-53); so we obtained upper bounds to the dominant TM_{010} mode, as desired.

7-6. The Ritz Procedure. A further advantage of the variational formulation is that one can choose the best approximation to a stationary quantity obtainable from a given class of trial fields. This is done by including adjustable constants, or *variational parameters*, in the definition of the trial field and then choosing those parameters which will give a minimum or maximum of the stationary quantity. For example, if we choose

$$\mathbf{E}_{\text{trial}} = \mathbf{E}_{\text{trial}}(A_1, A_2, \ldots, A_n) \tag{7-54}$$

where the A_i are variational parameters, and substitute into the stationary formula Eq. (7-36), we obtain

$$\omega^2 = \omega^2(A_1, A_2, \ldots, A_n) \tag{7-55}$$

The best approximation to ω_r^2 will be the minimum value of ω^2, which can be chosen by requiring

$$\frac{\partial \omega^2}{\partial A_i} = 0 \qquad i = 1, 2, \ldots, n \tag{7-56}$$

This general method is known as the *Ritz procedure*.[1]

The most common way to include variational parameters is to express the trial field as a linear combination of functions

$$\mathbf{E}_{\text{trial}} = \mathbf{E}_0 + A_1 \mathbf{E}_1 + A_2 \mathbf{E}_2 + \cdots + A_n \mathbf{E}_n \tag{7-57}$$

Since the labor of the calculations increases approximately as the square of the number of terms in Eq. (7-57), it is desirable to keep n small. However, it is also necessary that some choice of the A_i will give a reasonably close approximation to the true field. When a complete set of functions $\mathbf{E}_i$ is used, the method may, in principle, lead to an exact solution. It is also sometimes convenient to choose the $\mathbf{E}_i$ as an orthogonal set.

For an example of the Ritz method, let us again consider the circular cavity of Fig. 5-7 and trial fields of the form

$$\mathbf{H} = \mathbf{u}_\phi (\rho + A\rho^2) \qquad \nabla \times \mathbf{H} = \mathbf{u}_z(2 + 3A\rho) \tag{7-58}$$

where A is a variational parameter. Note that $\mathbf{H}$ satisfies no boundary conditions on S; so we choose Eq. (7-46) as the stationary formula. Substituting the trial field into Eq. (7-46), we obtain

$$\omega^2 = \frac{\int_0^a (2 + 3A\rho)^2 \rho \, d\rho}{\epsilon\mu \int_0^a (\rho + A\rho^2)^2 \rho \, d\rho}$$
$$= \frac{15}{a^2 \epsilon \mu} \frac{8 + 16Aa + 9(Aa)^2}{15 + 24Aa + 10(Aa)^2} \tag{7-59}$$

[1] The method is also referred to as the "Rayleigh-Ritz procedure."

Note that the approximation of Eq. (7-49) is the special case $Aa = -\frac{2}{3}$. To determine A by the Ritz method, we set

$$\frac{\partial \omega^2}{\partial A} = 0$$

and obtain
$$24 + 55Aa + 28(Aa)^2 = 0$$

This can be solved for Aa as

$$Aa = \frac{-55 \pm \sqrt{337}}{56} = \begin{cases} -1.3100 \\ -0.6543 \end{cases} \quad (7\text{-}60)$$

A substitution of the second of these values into Eq. (7-59) gives

$$\omega = \frac{2.4087}{a\sqrt{\epsilon\mu}} \quad (7\text{-}61)$$

which is smaller than what the first of Eq. (7-60) gives. Hence, Eq. (7-61) is the desired "best" approximation to the true resonant frequency [Eq. (7-47)]. The solution $Aa = -1.31$ gives $ka = 7.191$, which is an approximation to the next higher eigenvalue 5.520. If the trial field has two variational parameters, we obtain approximations to the lowest three eigenvalues, and so on. The Ritz procedure also gives us an approximation to the true field, but it is difficult to establish the nature of the approximation.

7-7. The Reaction Concept.[1] A general procedure for establishing stationary formulas can be obtained, using the concept of reaction as defined in Sec. 3-8. To reiterate, the reaction of field a on source b is

$$\langle a,b \rangle = \int (\mathbf{E}^a \cdot d\mathbf{J}^b - \mathbf{H}^a \cdot d\mathbf{M}^b) \quad (7\text{-}62)$$

If all sources can be contained in a finite volume, the reciprocity theorem [Eq. (3-36)] is

$$\langle a,b \rangle = \langle b,a \rangle \quad (7\text{-}63)$$

The linearity of the field equations is reflected in the identities

$$\begin{aligned}\langle a, b+c \rangle &= \langle a,b \rangle + \langle a,c \rangle \\ \langle Aa,b \rangle &= A\langle a,b \rangle = \langle a,Ab \rangle\end{aligned} \quad (7\text{-}64)$$

where the notation Aa means the a field and source are multiplied by the number A.

Many of the parameters of interest in electromagnetic engineering are proportional to reactions. For example, the impedance parameters of a

---

[1] V. H. Rumsey, The Reaction Concept in Electromagnetic Theory, *Phys. Rev.*, ser. 2, vol. 94, no. 6, pp. 1483–1491, June 15, 1954.

multiport "network" are proportional to reactions, as shown by Eqs. (3-41). Approximations to the desired reactions can be obtained by assuming trial fields (or sources) to approximate the true fields (or sources). It is then argued that the best approximation to a desired reaction is that obtained by equating reactions between trial fields to the corresponding reactions between trial and true fields. To be specific, suppose we want an approximation to the reaction $\langle c_a, c_b \rangle$. (The symbol c stands for "correct.") The approximation $\langle a,b \rangle$ is then best if we subject it to

$$\langle a,b \rangle = \langle c_a, b \rangle = \langle a, c_b \rangle \tag{7-65}$$

because we have imposed all possible constraints. Equation (7-65) can be thought of as the statement that all trial sources look the same to themselves as to the correct sources.

The reaction $\langle a,b \rangle$ obtained from Eq. (7-65) is also stationary for small variations of a and b about c_a and c_b. This we can prove by letting

$$a = c_a + p_a e_a \qquad b = c_b + p_b e_b$$

and showing that

$$\left.\frac{\partial \langle a,b \rangle}{\partial p_a}\right|_{p_a = p_b = 0} = \left.\frac{\partial \langle a,b \rangle}{\partial p_b}\right|_{p_a = p_b = 0} = 0 \tag{7-66}$$

Substituting for a and b into Eqs. (7-65), we have the three relationships

$$\langle a,b \rangle = \langle c_a, c_b \rangle + p_a \langle e_a, c_b \rangle + p_b \langle c_a, e_b \rangle + p_a p_b \langle e_a, e_b \rangle$$
$$= \langle c_a, c_b \rangle + p_b \langle c_a, e_b \rangle$$
$$= \langle c_a, c_b \rangle + p_a \langle e_a, c_b \rangle$$

Using the last two equations in the first equation, we obtain

$$\langle a,b \rangle = \langle c_a, c_b \rangle - p_a p_b \langle e_a, e_b \rangle$$

It is now evident that Eqs. (7-66) are satisfied, proving the stationary character of $\langle a,b \rangle$.

We have a slightly different case when the reaction concept is used to determine resonant frequencies of cavities. The true field at resonance is a source-free field; so the reaction of any field with the true source is zero. Hence, if we let $a = b$ represent a trial field and associated source, Eq. (7-65) reduces to

$$\langle a,a \rangle = 0 \tag{7-67}$$

We can think of this as stating that the resonant frequencies are zeros of the input impedance.

To apply Eq. (7-67), we assume a trial field and determine its sources from the field equations. For example, an assumed **E** field can be sup-

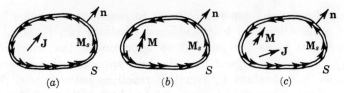

Fig. 7-9. Sources needed to support (a) a trial E field, (b) a trial H field, and (c) both a trial E field and a trial H field.

ported by the electric currents

$$\mathbf{J} = -j\omega\epsilon\mathbf{E} - \frac{1}{j\omega}\nabla \times (\mu^{-1}\nabla \times \mathbf{E}) \quad (7\text{-}68)$$

However, if the trial field does not satisfy $\mathbf{n} \times \mathbf{E} = 0$ on S, we need the additional magnetic surface currents

$$\mathbf{M}_s = \mathbf{n} \times \mathbf{E} \quad \text{on } S \quad (7\text{-}69)$$

to support the discontinuity in $\mathbf{E}$ at S. This is illustrated by Fig. 7-9a. We now substitute from Eqs. (7-68) and (7-69) into Eq. (7-67) and obtain

$$0 = \langle a,a \rangle = \iiint \mathbf{J} \cdot \mathbf{E}\, d\tau + \oiint \mathbf{M}_s \cdot \left(\frac{1}{j\omega\mu}\nabla \times \mathbf{E}\right) ds$$

$$= -j\omega \iiint \epsilon\mathbf{E} \cdot \mathbf{E}\, d\tau + \frac{j}{\omega}\iiint \mathbf{E} \cdot \nabla \times (\mu^{-1}\nabla \times \mathbf{E})\, d\tau$$

$$- \frac{j}{\omega}\oiint (\mathbf{n} \times \mathbf{E}) \cdot (\mu^{-1}\nabla \times \mathbf{E})\, ds$$

If $\mathbf{n} \times \mathbf{E} = 0$ on S, this reduces directly to Eq. (7-36). If $\mathbf{n} \times \mathbf{E} \neq 0$ on S, the above equation reduces to Eq. (7-45).

If a stationary formula in terms of the $\mathbf{H}$ field is desired, we consider the trial field to be supported by the sources

$$\begin{aligned}\mathbf{M} &= -j\omega\mu\mathbf{H} - \frac{1}{j\omega}\nabla \times (\epsilon^{-1}\nabla \times \mathbf{H}) \\ \mathbf{M}_s &= \mathbf{n} \times \left(\frac{1}{j\omega\epsilon}\nabla \times \mathbf{H}\right) \quad \text{on } S\end{aligned} \quad (7\text{-}70)$$

as represented by Fig. 7-9b. Application of Eq. (7-67) now leads to Eq. (7-46), or to Eq. (7-37) if $\mathbf{M}_s = 0$.

Stationary formulas in terms of *both* $\mathbf{E}$ and $\mathbf{H}$ are also possible. This time we consider both electric and magnetic currents, as shown in Fig. 7-9c. They are found from the trial fields according to

$$\begin{aligned}\mathbf{J} &= -j\omega\epsilon\mathbf{E} + \nabla \times \mathbf{H} \\ \mathbf{M} &= -j\omega\mu\mathbf{H} - \nabla \times \mathbf{E} \\ \mathbf{M}_s &= \mathbf{n} \times \mathbf{E} \quad \text{on } S\end{aligned} \quad (7\text{-}71)$$

Equation (7-67) then gives

$$0 = \iiint (\mathbf{E} \cdot \mathbf{J} - \mathbf{H} \cdot \mathbf{M}) \, d\tau - \oiint \mathbf{H} \cdot \mathbf{M}_s \, ds$$

$$= \iiint (-j\omega\epsilon E^2 + \mathbf{E} \cdot \nabla \times \mathbf{H} + \mathbf{H} \cdot \nabla \times \mathbf{E} + j\omega\mu H^2) \, d\tau$$

$$- \oiint \mathbf{E} \times \mathbf{H} \cdot ds$$

which can be rearranged to

$$\omega = j \frac{\iiint (\mathbf{E} \cdot \nabla \times \mathbf{H} + \mathbf{H} \cdot \nabla \times \mathbf{E}) \, d\tau - \oiint \mathbf{E} \times \mathbf{H} \cdot ds}{\iiint (\mu H^2 - \epsilon E^2) \, d\tau} \quad (7\text{-}72)$$

This is sometimes called a "mixed-field" stationary formula. The minus sign in the denominator might seem strange, but it is easily shown that E and H are 90° out of phase in the loss-free case (see Sec. 8-4). Hence, the denominator is twice the stored energy in the cavity.

Finally, if the trial fields have discontinuities in $\mathbf{n} \times \mathbf{E}$ or $\mathbf{n} \times \mathbf{H}$ over surfaces within the cavity, we must add the appropriate surface currents to support the discontinuities. This procedure leads to additional surface integrals in the stationary formulas, as shown in Probs. 7-27 and 7-28.

Earlier we showed that reactions constrained according to Eq. (7-65) were stationary. But in the above cavity formulas we calculated ω by forcing the reaction to vanish. We shall now prove that the ω so determined is stationary about the true resonant frequency. In the usual manner, we let the trial field be the true field plus a parameter times an error field, represented by

$$a = c + pe$$

For fixed e the reaction $\langle a,a \rangle$ is a function of both ω and p. Equation (7-67) constrains $\langle a,a \rangle$ to vanish; hence, as ω and p are varied, we have

$$\left.\frac{\partial \langle a,a \rangle}{\partial \omega}\right|_{\substack{\omega=\omega_r \\ p=0}} \delta\omega + \left.\frac{\partial \langle a,a \rangle}{\partial p}\right|_{\substack{\omega=\omega_r \\ p=0}} p = 0$$

The second term of this equation vanishes because $\langle a,a \rangle$ is stationary about $p = 0$. The coefficient of the first term is not in general zero; so

$$\delta\omega = 0$$

Thus, the first variation of ω vanishes, and all formulas for ω derived from Eq. (7-67) are stationary.

The reaction concept also provides us with an alternative way of viewing the Ritz procedure for improving the trial field or source. We assume the trial field or source to be a linear combination of functions, represented by

$$a = Uu + Vv + \cdots \qquad (7\text{-}73)$$

where $U, V, \ldots$ are numbers to be determined. According to the reaction concept, all trial fields should look the same to themselves as to the true source; hence we should enforce the conditions

$$\begin{aligned}\langle a,u\rangle &= \langle c,u\rangle \\ \langle a,v\rangle &= \langle c,v\rangle \\ &\cdots\cdots\cdots\end{aligned} \qquad (7\text{-}74)$$

Substituting from Eq. (7-73), we obtain the set of equations

$$\begin{aligned}U\langle u,u\rangle + V\langle v,u\rangle + \cdots &= \langle c,u\rangle \\ U\langle u,v\rangle + V\langle v,v\rangle + \cdots &= \langle c,v\rangle \\ &\cdots\cdots\cdots\cdots\end{aligned} \qquad (7\text{-}75)$$

which can be solved for the parameters $U, V, \ldots$. The solution so obtained is identical to that obtained by the Ritz procedure.

To illustrate, let us reconsider the example of Sec. 7-6, which was the Ritz procedure applied to the circular cavity (Fig. 5-7). Our trial field was Eq. (7-58); so for the same approximation by the reaction concept we choose

$$\mathbf{H}^u = \mathbf{u}_\phi \rho \qquad \mathbf{H}^v = \mathbf{u}_\phi \rho^2 \qquad (7\text{-}76)$$

The sources of these fields, according to Eq. (7-70), are

$$\begin{aligned}\mathbf{M}^u &= -\mathbf{u}_\phi j\omega\mu\rho & \mathbf{M}_s^u &= \frac{2j}{\omega\epsilon} \\ \mathbf{M}^v &= -\mathbf{u}_\phi \left(j\omega\mu\rho^2 + \frac{3j}{\omega\epsilon}\right) & \mathbf{M}_s^v &= \frac{3ja}{\omega\epsilon}\end{aligned} \qquad (7\text{-}77)$$

Calculating the various reactions according to Eq. (7-62), we obtain

$$\begin{aligned}\langle u,u\rangle &= 2\pi da^2 \left(j\omega\mu \frac{a^2}{4} + \frac{2}{j\omega\epsilon}\right) \\ \langle u,v\rangle &= \langle v,u\rangle = 2\pi da^3 \left(j\omega\mu \frac{a^2}{5} + \frac{2}{j\omega\epsilon}\right) \\ \langle v,v\rangle &= 2\pi da^4 \left(j\omega\mu \frac{a^2}{6} + \frac{9}{j\omega\epsilon 4}\right)\end{aligned} \qquad (7\text{-}78)$$

All reactions with the correct source are zero, because the true field is source-free. Hence, $\langle c,u\rangle = \langle c,v\rangle = 0$ and Eqs. (7-75) reduce to

$$\begin{aligned}U\langle u,u\rangle + V\langle v,u\rangle &= 0 \\ U\langle u,v\rangle + V\langle v,v\rangle &= 0\end{aligned}$$

These equations can have a nontrivial solution only if the determinant of the coefficients of U and V vanishes. Hence,

$$\langle u,u \rangle \langle v,v \rangle - \langle u,v \rangle^2 = 0 \tag{7-79}$$

is the equation from which ω is to be found. The solution of Eq. (7-79), with the reactions of Eqs. (7-78), yields Eq. (7-61).

7-8. Stationary Formulas for Waveguides. At cutoff, a waveguide is a two-dimensional resonator; so we should expect stationary formulas for the cutoff frequencies to be of the same form as those for the resonant frequencies of cavities. We must, of course, be careful in applying the reciprocity theorem, because the sources of our trial fields are not of finite extent. However, if we take a slice of the waveguide, as was done in Sec. 7-4, surface integrals over the top and bottom just cancel at resonance. The height of the slice is common to all terms and therefore cancels. Starting from Eq. (7-67), we arrive at stationary formulas differing from our cavity formulas only in that volume integrals are replaced by surface integrals and surface integrals by line integrals. Hence, the E-field formula corresponding to Eq. (7-45) is

$$\omega_c^2 = \frac{\iint \mu^{-1}(\nabla \times \mathbf{E})^2 \, ds + 2 \oint [(\mu^{-1}\nabla \times \mathbf{E}) \times \mathbf{E}] \cdot \mathbf{n} \, dl}{\iint \epsilon E^2 \, ds} \tag{7-80}$$

where $\mathbf{n}$ is the outward-pointing unit vector normal to the waveguide walls. The H-field formula corresponding to Eq. (7-46) is

$$\omega_c^2 = \frac{\iint \epsilon^{-1}(\nabla \times \mathbf{H})^2 \, ds}{\iint \mu H^2 \, ds} \tag{7-81}$$

and the mixed-field formula corresponding to Eq. (7-72) is

$$\omega_c = j \frac{\iint (\mathbf{E} \cdot \nabla \times \mathbf{H} + \mathbf{H} \cdot \nabla \times \mathbf{E}) \, ds - \oint \mathbf{E} \times \mathbf{H} \cdot \mathbf{n} \, dl}{\iint (\mu H^2 - \epsilon E^2) \, ds} \tag{7-82}$$

None of the above formulas require boundary conditions on the trial fields. Corrections for discontinuous trial fields can be made as outlined in the preceding section.

As an example, consider the partially filled rectangular waveguide of Fig. 4-8a. In Sec. 4-6 we obtained a transcendental equation for the cutoff frequency [Eq. (4-51)]. For a variational solution, let us use Eq. (7-80) and a trial field

$$\mathbf{E} = \mathbf{u}_y \sin \frac{\pi x}{a}$$

which is the empty-guide field. The result is[1]

$$\omega_c = \frac{\pi}{a\sqrt{\epsilon_2\mu_2}} \left[1 + \frac{\epsilon_1 - \epsilon_2}{\epsilon_2}\left(\frac{d}{a} - \frac{1}{2\pi}\sin\frac{2\pi d}{a}\right) \right]^{-\frac{1}{2}} \quad (7\text{-}83)$$

Note that this is an explicit formula for ω_c, in contrast to the exact equation, which is transcendental. Table 7-2 compares the above result with the exact solution for the case $\epsilon_1 = 2.45\epsilon_0$ and $\epsilon_2 = \epsilon_0$. We should expect the approximation to become worse as ϵ_1/ϵ_2 becomes larger, since the field then tends to concentrate more in the dielectric.

TABLE 7-2. RATIO OF WAVEGUIDE WIDTH TO CUTOFF WAVELENGTH FOR THE RECTANGULAR WAVEGUIDE WITH DIELECTRIC SLAB
("Exact" values read from curves by Frank)

d/a	a/λ_c (exact)	a/λ_c (approximate)
0	0.500	0.500
0.167	0.485	0.486
0.286	0.450	0.455
0.500	0.375	0.383
0.600	0.350	0.352
1.000	0.319	0.319

A knowledge of the cutoff frequency of a waveguide homogeneous in ϵ and μ is sufficient to determine the propagation constant at any other frequency according to Eq. (7-26). If the guide is inhomogeneously filled, as for example the above-treated rectangular waveguide with dielectric slab, there is no simple relationship between the cutoff frequency and the propagation constant. We therefore have need of stationary formulas for propagation constants.

In all of the previous examples, the field equations were given by an operator which was self-adjoint with respect to the desired integration.[2] For inhomogeneously filled waveguides, the field equations lead to an operator which is not self-adjoint. Hence, an appropriate adjoint operator must be found and the derivation of the stationary formulas suitably modified. It turns out that the operator for waves traveling in the $-z$ direction is the adjoint of the operator for waves traveling in the $+z$ direction, and the derivation proceeds as follows.

Define $+z$ traveling waves as

$$\begin{aligned} \mathbf{E}^+ &= \hat{\mathbf{E}}^+(x,y)e^{-j\beta z} = (\hat{\mathbf{E}}_t + \mathbf{u}_z\hat{E}_z)e^{-j\beta z} \\ \mathbf{H}^+ &= \hat{\mathbf{H}}^+(x,y)e^{-j\beta z} = (\hat{\mathbf{H}}_t + \mathbf{u}_z\hat{H}_z)e^{-j\beta z} \end{aligned} \quad (7\text{-}84)$$

[1] A. D. Berk, Variational Principles for Electromagnetic Resonators and Waveguides, *IRE Trans.*, vol. AP-4, no. 2, pp. 104–110, April, 1956.

[2] B. Friedman, "Principles and Techniques of Applied Mathematics," John Wiley and Sons, Inc., New York, 1956, p. 44.

Substituting these into the field equations, we obtain

$$\nabla \times \hat{\mathbf{E}}^+ + j\omega\mu\hat{\mathbf{H}}^+ = j\beta\mathbf{u}_z \times \hat{\mathbf{E}}^+$$
$$\nabla \times \hat{\mathbf{H}}^+ - j\omega\epsilon\hat{\mathbf{E}}^+ = j\beta\mathbf{u}_z \times \hat{\mathbf{H}}^+$$

Using analogous definitions for $-z$ traveling waves, we find

$$\nabla \times \hat{\mathbf{E}}^- + j\omega\mu\hat{\mathbf{H}}^- = -j\beta\mathbf{u}_z \times \hat{\mathbf{E}}^-$$
$$\nabla \times \hat{\mathbf{H}}^- - j\omega\epsilon\hat{\mathbf{E}}^- = -j\beta\mathbf{u}_z \times \hat{\mathbf{H}}^-$$

By direct substitution, it can be shown that for any $+z$ traveling wave solution there exists a $-z$ traveling wave solution given by

$$\mathbf{E}^- = \hat{\mathbf{E}}^-(x,y)e^{j\beta z} = (\hat{\mathbf{E}}_t - \mathbf{u}_z \hat{E}_z)e^{j\beta z}$$
$$\mathbf{H}^- = \hat{\mathbf{H}}^-(x,y)e^{j\beta z} = (-\hat{\mathbf{H}}_t + \mathbf{u}_z \hat{H}_z)e^{j\beta z} \qquad (7\text{-}85)$$

where the $\hat{\mathbf{E}}_t$, $\hat{\mathbf{H}}_t$, $\hat{E}_z$, and $\hat{H}_z$ of Eqs. (7-84) and (7-85) are the same functions.

Now multiply the first of the $+z$ wave equations scalarly by $\hat{\mathbf{H}}^-$, and the second of the $-z$ wave equations by $\hat{\mathbf{E}}^+$, and add the two resultant equations. This gives

$$\hat{\mathbf{H}}^- \cdot \nabla \times \hat{\mathbf{E}}^+ + \hat{\mathbf{E}}^+ \cdot \nabla \times \hat{\mathbf{H}}^- + j\omega\mu\hat{\mathbf{H}}^- \cdot \hat{\mathbf{H}}^+ - j\omega\epsilon\hat{\mathbf{E}}^+ \cdot \hat{\mathbf{E}}^-$$
$$= -2j\beta\hat{\mathbf{E}}_t \times \hat{\mathbf{H}}_t \cdot \mathbf{u}_z$$

which, when integrated over the guide cross section and rearranged, yields

$$\beta = \frac{\iint (\omega\epsilon\hat{\mathbf{E}}^+ \cdot \hat{\mathbf{E}}^- - \omega\mu\hat{\mathbf{H}}^+ \cdot \hat{\mathbf{H}}^- + j\hat{\mathbf{H}}^- \cdot \nabla \times \hat{\mathbf{E}}^+ + j\hat{\mathbf{E}}^+ \cdot \nabla \times \hat{\mathbf{H}}^-)\, ds}{2\iint \hat{\mathbf{E}}_t \times \hat{\mathbf{H}}_t \cdot \mathbf{u}_z \, ds}$$

(7-86)

This is a mixed-field formula, stationary if $\mathbf{n} \times \hat{\mathbf{E}} = 0$ on C.

For the E-field formulation, eliminate $\hat{\mathbf{H}}$ from the $+z$ and $-z$ wave equations, and proceed as in the derivation of Eq. (7-86). The resultant formula is

$$\beta^2 \iint \mu^{-1} E_t^2\, ds - j2\beta \iint \mu^{-1}\mathbf{E}_t \cdot \nabla E_z\, ds$$
$$+ \iint [\mu^{-1}(\nabla \times \hat{\mathbf{E}}^+) \cdot (\nabla \times \hat{\mathbf{E}}^-) - \omega^2\epsilon\hat{\mathbf{E}}^+ \cdot \hat{\mathbf{E}}^-]\, ds = 0 \quad (7\text{-}87)$$

stationary if $\mathbf{n} \times \mathbf{E} = 0$ on C. The H-field formula is given by Eq. (7-87) with ϵ, μ, $\mathbf{E}$ replaced by μ, ϵ, $\mathbf{H}$, and it is stationary with no boundary conditions on $\mathbf{H}$. Equations (7-86) and (7-87) remain stationary in the lossy case, for which $j\beta$ should be replaced by $\gamma = \alpha + j\beta$.

For an example of the calculation of propagation constants, consider the centered dielectric slab in a rectangular waveguide, as shown in the insert of Fig. 7-10. As a trial field, take

$$\hat{\mathbf{E}} = \mathbf{u}_y \sin\frac{\pi x}{a}$$

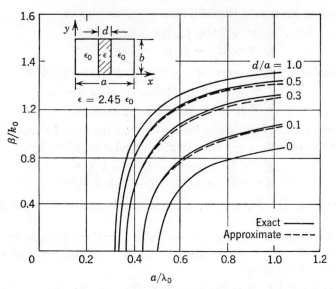

FIG. 7-10. Comparison of approximate and exact propagation constants for the rectangular waveguide with centered dielectric slab, $\epsilon = 2.45\epsilon_0$. (*After Berk.*)

and use Eq. (7-87). The result is[1]

$$\frac{\beta}{k_0} = \left[1 + \frac{\epsilon - \epsilon_0}{\epsilon_0}\left(\frac{d}{a} + \frac{1}{\pi}\sin\frac{\pi d}{a}\right) - \left(\frac{\pi}{k_0 a}\right)^2\right]^{1/2} \quad (7\text{-}88)$$

The exact solution is given in Prob. 4-19 and requires the solution of a transcendental equation. A comparison of a values obtained from Eq. (7-88) with the exact values for β/k_0 is shown in Fig. 7-10 for the case $\epsilon = 2.45\epsilon_0$.

7-9. Stationary Formulas for Impedance. A formula for impedance in terms of reaction is given by Eq. (3-41). Such a formula, when constrained according to Eq. (7-65), is a stationary formula for impedance.

Figure 7-11 represents a perfectly conducting antenna excited by a current source. The resultant current on the antenna will distribute itself so that tangential components of the total electric field vanish on the conductor. The antenna terminals are close together; so the reaction of any field with the current source is of the form $-VI$. If a trial-current distribution $\mathbf{J}_s{}^a$ is assumed on the antenna, the formula for input impedance [Eq. (3-41)] is

$$Z_{\text{in}} = -\frac{\langle a,a \rangle}{I^2} = -\frac{1}{I^2}\oint \mathbf{E}^a \cdot \mathbf{J}_s{}^a\, ds \quad (7\text{-}89)$$

[1] Berk, *op. cit.*

where I is the input current. The impedance as calculated by Eq. (7-89) is stationary about the true current, as we shall now show. On the antenna surface, the tangential components of the true field $\mathbf{E}^c$ are zero except at the input; hence

$$\langle c,a \rangle = -V_c I = -I^2 Z_{\text{in}} = \langle a,a \rangle$$

Also, $\langle c,a \rangle = \langle a,c \rangle$ by reciprocity; so the constraints of Eq. (7-65) have been met, and Eq. (7-89) is a stationary formula.

Equation (7-89) was used to calculate impedance before its stationary character was noticed.[1] This method should not be confused with the *induced emf method*

$$Z_{\text{in}} = -\frac{1}{|I|^2} \oint \mathbf{E} \cdot \mathbf{J}_s^* \, ds \qquad (7\text{-}90)$$

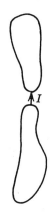

FIG. 7-11. An antenna excited by a current source.

which is based on the conservation of complex power. Equation (7-90) is not stationary unless both the true current and the trial current are real. When the trial current is assumed real, we get the same answer from Eqs. (7-89) and (7-90). Hence, the input impedances for waveguide feeds calculated in Sec. 4-10 are also variational solutions to the same problems.

If we have two sets of input terminals, as, for example, in the case of the two linear antennas shown in Fig. 7-12, the variational formula for mutual impedance is

$$Z_{ab} = -\frac{\langle a,b \rangle}{I_a I_b} = \frac{-1}{I_a I_b} \oint \mathbf{E}^a \cdot \mathbf{J}_s^b \, ds \qquad (7\text{-}91)$$

where I_a and I_b are the input currents at terminals a and b, respectively. The demonstration that the constraints of Eq. (7-65) are met is similar to that for self-impedance. Note that Eq. (7-91) involves the assumption of currents due to *both* sources, since $\mathbf{E}^a$ is the field of $\mathbf{J}_s^a$. The extension to N sets of terminals is straightforward.

The calculation of mutual impedance is usually simpler than the calculation of self-impedance because the source and field points are separated. Let us therefore take a mutual-impedance problem as our first example. Consider the parallel linear antennas of length $\lambda/2$ as shown in the insert of Fig. 7-12. No appreciable error will be incurred by assuming the currents as filamentary, as long as the antenna diameters are small compared to wavelength and compared to antenna separation. Let the z axis lie

---

[1] P. S. Carter, Circuit Relations in Radiating Systems and Applications to Antenna Problems. *Proc. IRE*, vol. 20, no. 6, pp. 1004–1041, June, 1932.

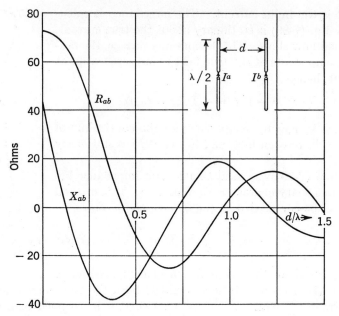

FIG. 7-12. Mutual impedance $Z_{ab} = R_{ab} + jX_{ab}$ between parallel $\lambda/2$ linear antennas in free space.

along antenna a, and assume

$$I^a = I_a \cos \frac{2\pi z}{\lambda} \qquad I^b = I_b \cos \frac{2\pi z}{\lambda} \tag{7-92}$$

Our formula for mutual impedance [Eq. (7-91)] becomes, in this case,

$$Z_{ab} = -\frac{1}{I_a I_b} \int_{-\lambda/4}^{\lambda/4} E_z^a I^b \, dz$$

By the usual vector-potential method we have

$$E_z^a = \frac{1}{j\omega\epsilon}\left(\frac{\partial^2}{\partial z^2} + k^2\right) A_z^a$$

where, at antenna b,

$$A_z^a = \frac{1}{4\pi}\int_{-\lambda/4}^{\lambda/4} I^a(z') \frac{e^{-jk\sqrt{d^2+(z-z')^2}}}{\sqrt{d^2+(z-z')^2}} \, dz'$$

Substituting for E_z^a and I^b in our expression for Z_{ab}, we obtain

$$Z_{ab} = -\int_{-\lambda/4}^{\lambda/4} dz \int_{-\lambda/4}^{\lambda/4} dz' \cos\frac{2\pi z}{\lambda} \cos\frac{2\pi z'}{\lambda} G(z,z') \tag{7-93}$$

where

$$G(z,z') = \frac{1}{4\pi j\omega\epsilon}\left(\frac{\partial^2}{\partial z^2} + k^2\right)\frac{e^{-jk\sqrt{d^2+(z-z')^2}}}{\sqrt{d^2+(z-z')^2}} \tag{7-94}$$

The integrations of Eq. (7-93) can be expressed in terms of sine integrals and cosine integrals. The details of the integration can be found in the literature.[1] Letting

$$Z_{ab} = R_{ab} + jX_{ab}$$

we obtain for the result

$$R_{ab} = \frac{\eta}{4\pi} \{2 \text{ Ci}(kd) - \text{Ci}[\sqrt{(kd)^2 + \pi^2} + \pi^2] - \text{Ci}[\sqrt{(kd)^2 + \pi^2} - \pi^2]\}$$

$$X_{ab} = \frac{-\eta}{4\pi} \{2 \text{ Si}(kd) - \text{Si}[\sqrt{(kd)^2 + \pi^2} + \pi^2] - \text{Si}[\sqrt{(kd)^2 + \pi^2} - \pi^2]\}$$

(7-95)

where $\text{Ci}(x)$ and $\text{Si}(x)$ are as defined in Prob. 2-44. Figure 7-12 shows a plot of Eqs. (7-95). The mutual impedance between linear antennas of other lengths and orientations can be found in the literature.[1,2]

The evaluation of the self-impedance of a linear antenna is more difficult because of the singular integrands encountered. Let us use this problem to illustrate the use of adjustable parameters in the trial current. The geometry of the center-fed linear antenna is shown in the insert of Fig. 7-13. Let the current on the antenna be represented by two functions, according to Eq. (7-73). Our trial current is then a surface current of the form

$$\mathbf{J}_s = U\mathbf{J}_s^u + V\mathbf{J}_s^v \tag{7-96}$$

where U and V are adjustable parameters. According to the reaction concept, the trial functions should look the same to the assumed current as to the true current; hence we enforce the conditions

$$\langle a,u \rangle = U\langle u,u \rangle + V\langle v,u \rangle = \langle c,u \rangle$$
$$\langle a,v \rangle = U\langle u,v \rangle + V\langle v,v \rangle = \langle c,v \rangle$$

where $\langle c,u \rangle$ and $\langle c,v \rangle$ can be calculated, as we shall later show. Solving for U and V, we have in matrix notation

$$\begin{bmatrix} U \\ V \end{bmatrix} = \begin{bmatrix} \langle u,u \rangle & \langle v,u \rangle \\ \langle u,v \rangle & \langle v,v \rangle \end{bmatrix}^{-1} \begin{bmatrix} \langle c,u \rangle \\ \langle c,v \rangle \end{bmatrix} \tag{7-97}$$

Substituting for U and V into Eq. (7-96) and calculating the self-reaction, we obtain

$$\langle a,a \rangle = [\langle c,u \rangle \; \langle c,v \rangle] \begin{bmatrix} \langle u,u \rangle & \langle v,u \rangle \\ \langle u,v \rangle & \langle v,v \rangle \end{bmatrix}^{-1} \begin{bmatrix} \langle c,u \rangle \\ \langle c,v \rangle \end{bmatrix} \tag{7-98}$$

---

[1] P. S. Carter, Circuit Relations in Radiating Systems and Applications to Antenna Problems, *Proc. IRE*, vol. 20, no. 6, pp. 1004–1041, June, 1932.

[2] G. Brown and R. King, High Frequency Models in Antenna Investigations, *Proc. IRE*, vol. 22, no. 4, pp. 457–480, April, 1934.

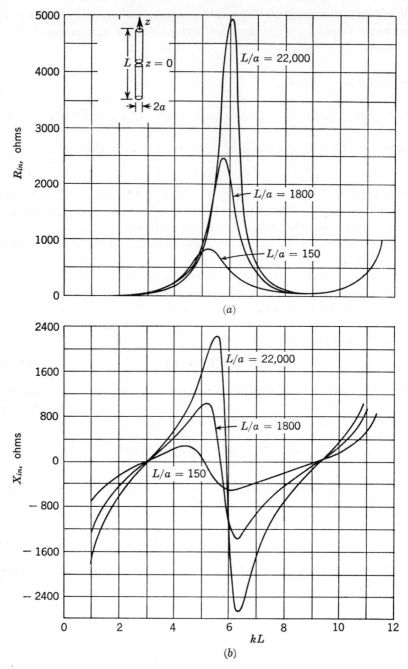

Fig. 7-13. Variational solution for the input impedance of the symmetrical cylindrical antenna. (*After Y. Y. Hu.*) (*a*) Input resistance; (*b*) input reactance.

Equations (7-97) and (7-98) also apply to the case of N adjustable constants if the various matrices are extended to N rows and/or columns. Expanding Eq. (7-98), using the reciprocity condition $\langle u,v\rangle = \langle v,u\rangle$, we obtain

$$\langle a,a\rangle = \frac{\langle c,u\rangle^2\langle v,v\rangle - 2\langle c,u\rangle\langle c,v\rangle\langle u,v\rangle + \langle c,v\rangle^2\langle u,u\rangle}{\langle u,u\rangle\langle v,v\rangle - \langle u,v\rangle^2}$$

Now note that $\mathbf{n} \times \mathbf{E}^c = 0$ on the antenna surface except at the feed; so

$$\langle c,x\rangle = -V_{\text{in}}I_x$$

for any x, where V_{in} is the input voltage and I_x is the x current at the input. Using the above two relationships in Eq. (7-89), we obtain

$$Z_{\text{in}} = Z_{\text{in}}^2 \frac{I_v^2\langle u,u\rangle - 2I_uI_v\langle u,v\rangle + I_u^2\langle v,v\rangle}{\langle u,v\rangle^2 - \langle u,u\rangle\langle v,v\rangle}$$

which can be rearranged to read

$$Z_{\text{in}} = \frac{\langle u,v\rangle^2 - \langle u,u\rangle\langle v,v\rangle}{I_v^2\langle u,u\rangle - 2I_uI_v\langle u,v\rangle + I_u^2\langle v,v\rangle} \tag{7-99}$$

where I_u and I_v are the values of the u and v trial currents at the input.

Let us now look at the form of the reactions. The currents will be rotationally symmetric z-directed surface currents on the cylinder $\rho = a$, where a is the antenna radius. These currents can be expressed as

$$\mathbf{J}_s^x = \frac{1}{2\pi a} I^x(z) \mathbf{u}_z \tag{7-100}$$

where I^x is the total current and $x = u, v$. By the potential integral method we can calculate the field of the current $\mathbf{J}_s^x$ as

$$E_z^x = \frac{1}{8\pi^2 j\omega\epsilon}\left(k^2 + \frac{\partial^2}{\partial z^2}\right)\int_{-L/2}^{L/2} dz' \int_0^{2\pi} d\phi' J_s^x G \tag{7-101}$$

where
$$G = \frac{e^{-jk\sqrt{\rho^2+a^2-2\rho a\cos(\phi-\phi')+(z-z')^2}}}{\sqrt{\rho^2+a^2-2\rho a\cos(\phi-\phi')+(z-z')^2}} \tag{7-102}$$

The various reactions of Eq. (7-99) are then given by

$$\langle x,y\rangle = \int_{-L/2}^{L/2} dz \int_0^{2\pi} a\,d\phi\, E_z^x J_s^y \tag{7-103}$$

where E_z^x is given by Eq. (7-101) with $\rho = a$. Note the singular nature of the Green's function [Eq. (7-102)] at $\rho = a$.

A precise evaluation of Eq. (7-103) would be difficult; so the following approximation is usually used. The field of the current is approximated by the field of a filamentary current of the same magnitude. This is equivalent to replacing Eq. (7-102) for $\rho = a$ by

$$G = \frac{e^{-jk\sqrt{a^2+(z-z')^2}}}{\sqrt{a^2 + (z - z')^2}} \quad (7\text{-}104)$$

For thin antennas, the error introduced by Eq. (7-104) is negligible, as can be shown by the following argument. The field of the filament of current is a source-free field in the region external to the linear antenna. We can therefore assume that this field exists and calculate the equivalent currents on the surface of the antenna. As long as the equivalent magnetic currents are negligible, as they will be for thin antennas, we can take the equivalent electric currents for our trial currents. The resultant current is essentially that of Eq. (7-100). Using the above approximation for G, we obtain from Eq. (7-103)

$$\langle x,y \rangle = \frac{1}{4\pi j\omega\epsilon} \int_{-L/2}^{L/2} dz \int_{-L/2}^{L/2} dz' \, I^x(z')I^y(z)\left(k^2 + \frac{\partial^2}{\partial z^2}\right)G \quad (7\text{-}105)$$

where G is given by Eq. (7-104). Note that, to this approximation, the self-reaction is equal to the mutual reaction between two identical antennas fed in phase and separated by a distance a. Hence, Eqs. (7-95) with d replaced by a give the first-order (one trial function) variational solution for the input impedance of a $\lambda/2$ linear antenna. In particular, note that for very small $a = d$, Eqs. (7-95) reduce to

$$R_{\text{in}} \approx 73.1 \qquad X_{\text{in}} \approx 42.5 \quad (7\text{-}106)$$

as is evident from Fig. 7-12. Resonance ($X = 0$) occurs for L slightly less than $\lambda/2$.

For trial functions in the second-order solution,

$$\begin{aligned} I^u &= \sin k\left(\frac{L}{2} - |z|\right) \\ I^v &= 1 - \cos k\left(\frac{L}{2} - |z|\right) \end{aligned} \quad (7\text{-}107)$$

have been used in the literature. The evaluation of Eq. (7-105) for $\langle x,y \rangle = \langle u,u \rangle$, $\langle u,v \rangle$, and $\langle v,v \rangle$ is long and involved, and formulas in terms of sine integrals and cosine integrals have been given by Storer[1] and Hu.[2] Numerical values of the input impedance are given in Fig. 7-13. The antenna is said to be resonant when X is zero and $kL \approx n\pi$, n odd. It is said to be antiresonant when X is zero and $kL \approx n\pi$, n even. Note that,

[1] J. E. Storer, Variational Solution to the Problem of the Symmetrical Cylindrical Antenna, *Cruft Lab. Rep.* TR 101, Cambridge, Mass., 1952.

[2] Y. Y. Hu, Back-scattering Cross Sections of a Center-loaded Cylindrical Antenna, *IRE Trans.*, vol AP-6, no. 1, pp. 140–148, January, 1958.

in the vicinity of resonance, R is insensitive to antenna thickness. It is in these regions that the analysis of Sec. 2-10 gives good results. Both trial currents of Eqs. (7-107) are zero at the input for $kL = 4\pi$. Hence, the input impedance calculated therefrom cannot be valid in the vicinity of $kL = 4\pi$. Perhaps a better choice for the v current would be

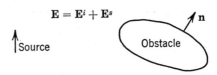

FIG. 7-14. Wave scattering by an obstacle.

$$I^v = \frac{L}{2} - |z|$$

which is finite at $z = 0$ for all $L > 0$. However, calculations have not been made for this choice.

7-10. Stationary Formulas for Scattering. Let us first treat the backscattering, or radar echo, type of problem by the variational method. The problem is represented by Fig. 7-14. It consists of a source and one or more obstacles, and we wish to determine the field scattered back to the source. For simplicity, the obstacle will be considered a perfect conductor and the source a current element Il. The more general case of dielectric obstacles is considered in Sec. 7-11.

Let the incident field, that is, the free-space field of the source alone, be denoted by $\mathbf{E}^i$. The total field $\mathbf{E}$ with the obstacle present is then the sum of the incident field $\mathbf{E}^i$ plus the scattered field $\mathbf{E}^s$. The reaction of the scattered field on the current element is

$$\langle s,i \rangle = IlE_l^s = -IV^s \tag{7-108}$$

where V^s is the scattered voltage appearing across l. Let the *echo* be defined as the ratio of E_l^s to Il. Then, using reciprocity, we have

$$\text{Echo} = \frac{E_l^s}{Il} = \frac{\langle s,i \rangle}{(Il)^2} = \frac{\langle i,s \rangle}{(Il)^2}$$

$$= \frac{1}{(Il)^2} \oint \mathbf{E}^i \cdot \mathbf{J}_s \, ds \tag{7-109}$$

where $\mathbf{J}_s$ is the current induced on the perfectly conducting obstacle. The boundary condition at the obstacle is $\mathbf{n} \times \mathbf{E} = 0$, or

$$\mathbf{n} \times \mathbf{E}^i = -\mathbf{n} \times \mathbf{E}^s \qquad \text{on } S \tag{7-110}$$

Hence, Eq. (7-109) can be written as

$$\text{Echo} = \frac{-1}{(Il)^2} \oint \mathbf{E}^s \cdot \mathbf{J}_s \, ds = -\frac{\langle c,c \rangle}{(Il)^2} \tag{7-111}$$

where $\langle c,c \rangle$ stands for the self-reaction of the "correct" currents induced on the obstacle by the source.

For a stationary formula, we assume a current $\mathbf{J}^a$ on S and approximate $\langle c,c \rangle$ by $\langle a,a \rangle$, subject to the constraint

$$\langle a,a \rangle = \langle c,a \rangle = -\langle i,a \rangle \tag{7-112}$$

The last equality results from Eq. (7-110). To express this constraint in a form for which $\langle a,a \rangle$ is insensitive to the amplitude of $\mathbf{J}^a$, we take

$$\langle a,a \rangle = \frac{\langle i,a \rangle^2}{\langle a,a \rangle}$$

and, replacing $\langle c,c \rangle$ by $\langle a,a \rangle$ in Eq. (7-111), we have

$$\text{Echo} \approx \frac{-\langle i,a \rangle^2}{(Il)^2 \langle a,a \rangle} = -\frac{\left(\oint \mathbf{E}^i \cdot \mathbf{J}^a \, ds \right)^2}{(Il)^2 \oint \mathbf{E}^a \cdot \mathbf{J}^a \, ds} \tag{7-113}$$

where $\mathbf{E}^a$ is the field produced by the assumed currents $\mathbf{J}^a$. This is the variational formulation of the problem. Note the close similarity of the echo problem to the impedance problem of the preceding section. The impedance problem is essentially an echo problem for which the source is at the obstacle. A more general formulation of the echo problem can be made by replacing Il with an arbitrary source.

The tensor Green's functions of Sec. 3-10 can be used to put Eq. (7-113) into a more descriptive form. Define $[\Gamma(\mathbf{r},\mathbf{r}')]$ as the tensor of proportionality between a current element $d\mathbf{J}^a$ at $\mathbf{r}'$ and the field $d\mathbf{E}^a$ that it produces at $\mathbf{r}$, that is,

$$d\mathbf{E}^a(\mathbf{r}) = [\Gamma(\mathbf{r},\mathbf{r}')] \, d\mathbf{J}^a(\mathbf{r}')$$

Then Eq. (7-113) can be written as

$$\text{Echo} \approx \frac{-\left[\frac{1}{Il} \oint \mathbf{E}^i(\mathbf{r}) \cdot \mathbf{J}^a(\mathbf{r}) \, ds \right]^2}{\oint ds \oint ds' \, \mathbf{J}^a(\mathbf{r}) \cdot [\Gamma(\mathbf{r},\mathbf{r}')] \, \mathbf{J}^a(\mathbf{r}')}$$

This equation is in a form characteristic of variational solutions in general.

A commonly calculated parameter is the echo area, defined by Eq. (3-30). For linearly polarized fields, the echo area is given by

$$A_e = \lim_{r \to \infty} \left(4\pi r^2 \left| \frac{E^s}{E^i} \right|^2 \right) \tag{7-114}$$

If, in Fig. 7-14, we let Il be z-directed and located on the x axis, and then let $r = x \to \infty$, we have, in the vicinity of the obstacle,

$$\mathbf{E}^i = \mathbf{u}_z \frac{j\eta Il}{2\lambda r} e^{jkx} = \mathbf{u}_z E_0 e^{jkx}$$

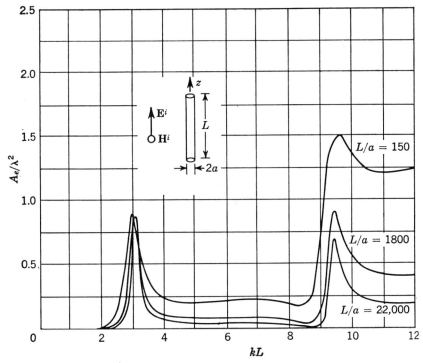

FIG. 7-15. Broadside echo area A_e of a wire. (*After Y. Y. Hu.*)

Also, by definition, we have echo $= E_z^s/Il$; hence from Eq. (7-113)

$$E_z^s = \frac{\eta E_0 \left(\oint \mathbf{u}_z \cdot \mathbf{J}^a e^{jkx}\, ds\right)^2}{j2\lambda r \oint \mathbf{E}^a \cdot \mathbf{J}^a\, ds}$$

Therefore, by Eq. (7-114), our stationary formula for echo area is

$$A_e = \pi \left|\frac{\eta}{\lambda} \frac{\left(\oint J_z^a e^{jkx}\, ds\right)^2}{\oint \mathbf{E}^a \cdot \mathbf{J}^a\, ds}\right|^2 \qquad (7\text{-}115)$$

when the incident plane wave is z-polarized and $-x$ traveling.

As an example, consider the scattering of a plane wave by a thin conducting wire, as represented by the insert of Fig. 7-15. The integral in the denominator of Eq. (7-115) is just the self-reaction of the assumed current on the wire. This is the same type of reaction that we encountered in the linear-antenna problem, approximated by Eq. (7-105). Defining A as the self-reaction, we have

$$A = \oint \mathbf{E}^a \cdot \mathbf{J}^a\, ds \approx \frac{1}{4\pi j\omega\epsilon} \int_{-L/2}^{L/2} dz \int_{-L/2}^{L/2} dz'\, I^a(z) I^a(z') \left(k^2 + \frac{\partial^2}{\partial z^2}\right) G \qquad (7\text{-}116)$$

where G is given by Eq. (7-104). For the current on the wire we should expect a constant current "forced" by the incident field plus a "natural-mode" sinusoidal current. At the ends of the wire, the current should be practically zero; hence we assume for our trial current

$$I^a = \cos kz - \cos k \frac{L}{2} \qquad (7\text{-}117)$$

Equation (7-116) can then be evaluated as

$$\text{Re}(A) = \frac{-\eta}{4\pi} [(kL + kL \cos kL - 2 \sin kL) \, \text{Si}(kL) \\ + \log 2\gamma kL - \text{Ci}(2kL) - \sin^2(kL)] \qquad (7\text{-}118)$$

$$\text{Im}(A) = \frac{-\eta}{4\pi} \left\{ (kL + kL \cos kL - 2 \sin kL) \left[\text{Ci}(kL) + \log \frac{2}{\gamma ka} \right] \right. \\ \left. + \text{Si}(2kL) - (1 + \cos kL) \sin kL \right\}$$

where $\gamma = 1.781$. The integral in the numerator of Eq. (7-115) evaluates to

$$\int_{-L/2}^{L/2} I^a(z) \, dz = \frac{1}{k} \left(2 \sin \frac{kL}{2} - kL \cos \frac{kL}{2} \right) = \frac{B}{k} \qquad (7\text{-}119)$$

which defines B. Hence, the echo area is

$$A_e = \frac{\lambda^2}{16\pi^3} \left| \frac{B^2}{A/\eta} \right|^2 \qquad (7\text{-}120)$$

with A and B given by Eqs. (7-118) and (7-119). This solution gives good accuracy out to about $kL = 8$. Figure 7-15 shows a plot of A_e/λ^2 for the second-order solution (two trial functions), as calculated by Y. Y. Hu.[1] The results for plane waves incident at an arbitrary angle are given by Tai.[2] He also shows the effect of choosing different trial functions.

In two-dimensional problems, the quantity *echo width* L_e corresponds to the echo area of the three-dimensional problems. The echo width is defined as the width of incident wave which carries sufficient power to produce, by cylindrically omnidirectional radiation, the same back-scattered power density. In equation form, the echo width is

$$L_e = \lim_{\rho \to \infty} \left(2\pi\rho \frac{\overline{\mathcal{S}}^s}{\overline{\mathcal{S}}^i} \right) \qquad (7\text{-}121)$$

---

[1] Y. Y. Hu, Back-scattering Cross Section of a Center-loaded Cylindrical Antenna, *IRE Trans.*, vol. AP-6, no. 1, pp. 140–148, January, 1958.

[2] C. T. Tai, Electromagnetic Back-scattering from Cylindrical Wires, *J. Appl. Phy.*, vol. 23, no. 8, pp. 909–916, August, 1952.

or, for linear polarization,

$$L_e = \lim_{\rho \to \infty} \left(2\pi\rho \left| \frac{E^s}{E^i} \right|^2 \right) \quad (7\text{-}122)$$

where superscripts s and i stand for "scattered" and "incident," respectively. Going through a development similar to that used for Eq. (7-115), except that a line source is used, we obtain

$$L_e = \frac{\pi}{2\lambda} \left| \eta \frac{\left(\oint J_z e^{jkx} \, dl \right)^2}{\oint E_z{}^a J_z{}^a \, dl} \right|^2 \quad (7\text{-}123)$$

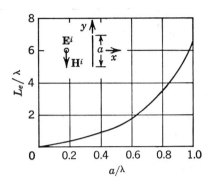

Fig. 7-16. Echo width L_e of a conducting ribbon of width a.

if the incident field is z-polarized and $-x$ traveling. Similarly,

$$L_e = \frac{\pi}{2\lambda} \left| \eta \frac{\left(\oint J_y{}^a e^{jkx} \, dl \right)^2}{\oint \mathbf{E}^a \cdot \mathbf{J}^a \, dl} \right|^2 \quad (7\text{-}124)$$

if the incident field is y-polarized and $-x$ traveling. From symmetry, $\mathbf{J}^a$ in Eq. (7-124) should have no z component. In both Eqs. (7-123) and (7-124), it is assumed that the scatterers are cylinders generated by elements parallel to the z axis and the line integrals are in a transverse ($z = $ constant) plane.

For an example of a two-dimensional problem, consider a z-polarized plane wave normally incident on a conducting ribbon of width a. This is illustrated by the insert of Fig. 7-16. Assume that the current induced on the ribbon is uniform, that is,

$$J_z{}^a = 1 \quad (7\text{-}125)$$

Because the current is real, the integral in the denominator of Eq. (7-123) is

$$\int_{-a/2}^{a/2} E_z{}^a J_z{}^a \, dy = \int_{-a/2}^{a/2} E_z{}^a J_z{}^{a*} \, dy = -P$$

where P is the complex power per unit length supplied by $J_z{}^a$. But we have already analyzed the ribbon of uniform current in Sec. 4-12, the result being

$$P = |I|^2 Z = a^2 \frac{\eta^2}{2} Y_{\text{apert}}$$

where Y_{apert} is plotted in Fig. 4-22. The echo width, according to Eq.

Fig. 7-17. Differential scattering.

(7-123), is then

$$L_e \approx \frac{\pi}{2\lambda} \left| \frac{\eta a^2}{a^2(\eta^2/2)Y_{\text{apert}}} \right|^2 = \frac{2\pi}{\lambda} \left| \frac{1}{\eta Y_{\text{apert}}} \right|^2 \quad (7\text{-}126)$$

A plot of this is shown in Fig. 7-16. For large a we can use Eq. (4-107) and obtain

$$L_e \xrightarrow[a/\lambda \to \infty]{} \frac{2\pi a^2}{\lambda} \quad (7\text{-}127)$$

which is also the physical optics approximation (see Fig. 3-21).

The more general case of differential scattering, or transmission,[1] is represented by Fig. 7-17. The problem consists of a transmitter, which illuminates the obstacle, and a receiver at which we wish to evaluate the scattered signal. For simplicity, let us consider both the source and receiver to be unit electric currents. Then, according to Eq. (3-39), the voltage across the receiving current due to the transmitting current is

$$V_r = -\langle t, r \rangle_{\text{obstacle present}} \quad (7\text{-}128)$$

where t and r refer to the source or field of the transmitter and receiver, respectively. The total signal received is the superposition of the incident field, due to the transmitter alone, plus the scattered field, due to the currents c on the obstacle. Hence,

$$V_r = V_r^i + V_r^s = -\langle t, r \rangle - \langle c, r \rangle \quad (7\text{-}129)$$

where $\langle t, r \rangle$ is calculated with the obstacle absent and $\langle c, r \rangle$ involves the free-space field of the currents on the obstacle. The transmitter and receiver currents are assumed to be known (they are current elements in our simplified case); so V_r^i can, in principle, be calculated exactly. Our problem is to obtain the variational formula for V_r^s.

We shall here consider only the simple case of a perfectly conducting obstacle, the general case being considered in Sec. 7-11. Applying reci-

---

[1] A transmission problem involves the evaluation of the total field at the receiver, while a scattering problem involves the evaluation of only the scattered field.

procity, we have, for the scattered voltage at the receiver,

$$-V_r^s = \langle c,r \rangle = \langle r,c \rangle = \oiint (\mathbf{E}^i)^r \cdot (\mathbf{J}_s^c)^t \, ds \tag{7-130}$$

where $(\mathbf{J}_s^c)^t$ is the surface current induced on the obstacle by the transmitter and $(\mathbf{E}^i)^r$ is the field of the receiver current calculated with the obstacle absent (the incident field). Our boundary conditions on the various true fields are $\mathbf{n} \times \mathbf{E} = 0$ at the obstacle boundary; hence

$$\begin{aligned} \mathbf{n} \times (\mathbf{E}^i)^r &= -\mathbf{n} \times (\mathbf{E}^s)^r \\ \mathbf{n} \times (\mathbf{E}^i)^t &= -\mathbf{n} \times (\mathbf{E}^s)^t \end{aligned} \tag{7-131}$$

where superscripts i and s refer to incident and scattered components, and t and r refer to transmitter and receiver sources. Hence, by Eqs. (7-130) and (7-131), we have

$$V_r^s = \oiint (\mathbf{E}^s)^r \cdot (\mathbf{J}_s^c)^t \, ds = \langle c_r, c_t \rangle \tag{7-132}$$

where $\langle c_r, c_t \rangle$ stands for the reaction between the field of the "correct" currents induced on the obstacle by the receiver and the "correct" currents induced by the transmitter. For our stationary formula, we approximate $\langle c_r, c_t \rangle$ by $\langle a_r, a_t \rangle$, where the a's denote assumed currents on the obstacle, and constrain the latter according to Eq. (7-65), which is

$$\langle a_r, a_t \rangle = \langle c_r, a_t \rangle = \langle a_r, c_t \rangle \tag{7-133}$$

In the language of the reaction concept, Eq. (7-133) says that the assumed currents look the same to each other as to their respective true currents. By Eqs. (7-131) and reciprocity, Eqs. (7-133) become

$$\begin{aligned} \langle a_r, a_t \rangle &= \langle c_r, a_t \rangle = -\langle r, a_t \rangle \\ \langle a_r, a_t \rangle &= \langle a_r, c_t \rangle = \langle c_t, a_r \rangle = -\langle t, a_r \rangle \end{aligned} \tag{7-134}$$

Substituting from Eqs. (7-134) into Eq. (7-132), we have for our variational formula

$$\begin{aligned} V_r^s \approx \langle a_r, a_t \rangle &= \frac{\langle r, a_t \rangle \langle t, a_r \rangle}{\langle a_r, a_t \rangle} \\ &= \frac{\left[\oiint (\mathbf{E}^i)^r \cdot (\mathbf{J}_s^a)^t \, ds \right] \left[\oiint (\mathbf{E}^i)^t \cdot (\mathbf{J}_s^a)^r \, ds \right]}{\oiint (\mathbf{E}^a)^r \cdot (\mathbf{J}_s^a)^t \, ds} \end{aligned} \tag{7-135}$$

where $(\mathbf{E}^a)^r$ is the field due to the assumed currents $(\mathbf{J}_s^a)^r$, which approximate the currents induced by the receiver. Note that Eq. (7-135) involves the assumption of currents on the obstacle due to sources at both the transmitter and receiver. Note also that Eq. (7-135) reduces to the formula for back-scattering [Eq. (7-113)] when the transmitter and receiver coincide.

7-11. Scattering by Dielectric Obstacles.[1]

The problem of differential scattering by a dielectric obstacle is represented by Fig. 7-17 if the obstacle is now considered as a dielectric body. We shall assume it to be nonmagnetic ($\mu = \mu_0$), but it may be lossy if ϵ is complex. The extension to magnetic obstacles is given in Prob. 7-42.

When the obstacle is excited by a source, there will be induced in it polarization currents given by

$$\mathbf{J}^c = j\omega(\epsilon - \epsilon_0)\mathbf{E} = \kappa\mathbf{E} = \kappa(\mathbf{E}^i + \mathbf{E}^s) \qquad (7\text{-}136)$$

Superscripts t or r will be added to the various quantities to indicate that the exciting source is at the transmitter or receiver, respectively. The treatment of differential scattering of the preceding section made no assumptions about the nature of the obstacle in the derivation of Eq. (7-130); hence for unit currents at t and r

$$-V_r{}^s = \langle r,c \rangle = \iiint (\mathbf{E}^i)^r \cdot (\mathbf{J}^c)^t \, d\tau \qquad (7\text{-}137)$$

where the notation is the same as in the preceding section. Using the relationship $\mathbf{E}^i = \mathbf{E} - \mathbf{E}^s$ and Eq. (7-136), we can rewrite Eq. (7-137) as

$$\begin{aligned}-V_r{}^s &= \iiint \kappa^{-1}(\mathbf{J}^c)^r \cdot (\mathbf{J}^c)^t \, d\tau - \iiint (\mathbf{E}^s)^r \cdot (\mathbf{J}^c)^t \, d\tau \\ &= F(c_r,c_t) - \langle c_r,c_t \rangle \end{aligned} \qquad (7\text{-}138)$$

which defines the functional F. Note that F is symmetrical in c_r and c_t and is actually the reaction between $\mathbf{E}^r$ and $(\mathbf{J}^c)^t$ with the obstacle present.

To obtain a stationary formula for the scattered voltage at the receiver, we approximate the true currents c by trial currents a and set

$$-V_r{}^s \approx F(a_r,a_t) - \langle a_r,a_t \rangle = G(a_r,a_t) \qquad (7\text{-}139)$$

subject to constraints of the form of Eq. (7-65) applied to G. Such constraints are

$$G(a_r,a_t) = G(c_r,a_t) = G(a_r,c_t) \qquad (7\text{-}140)$$

and we find

$$\begin{aligned} G(c_r,a_t) &= \langle r,a_t \rangle = \iiint (\mathbf{E}^i)^r \cdot (\mathbf{J}^a)^t \, d\tau \\ G(a_r,c_t) &= \langle t,a_r \rangle = \iiint (\mathbf{E}^i)^t \cdot (\mathbf{J}^a)^r \, d\tau \end{aligned} \qquad (7\text{-}141)$$

Combining the preceding equations to render $V_r{}^s$ insensitive to the ampli-

---

[1] M. H. Cohen, Application of the Reaction Concept to Scattering Problems, *IRE Trans.*, vol. AP-3, no. 4, pp. 193–199, October, 1955.

tudes of the trial functions, we have the variational formula

$$-V_r{}^s = \frac{\langle r, a_t \rangle \langle t, a_r \rangle}{F(a_t, a_t) - \langle a_r, a_t \rangle}$$

$$= \frac{\left[\iiint (\mathbf{E}^i)^r \cdot (\mathbf{J}^a)^t \, d\tau \right] \left[\iiint (\mathbf{E}^i)^t \cdot (\mathbf{J}^a)^r \, d\tau \right]}{\iiint \kappa^{-1} (\mathbf{J}^a)^r \cdot (\mathbf{J}^a)^t \, d\tau - \iiint (\mathbf{E}^a)^r \cdot (\mathbf{J}^a)^t \, d\tau} \quad (7\text{-}142)$$

For the lossy case, $\kappa = j\omega\epsilon + \sigma$. For a perfectly conducting obstacle, $\sigma \to \infty$; hence $\kappa^{-1} \to 0$ and Eq. (7-142) reduces to Eq. (7-135).

When the transmitter and receiver are represented by the same source, we have the back-scattering problem. Using the definition of Eq. (7-109) for echo, when the source is a unit current, we have

$$\text{Echo} = \frac{-V_r{}^s}{l^2} = \frac{(\langle i, a \rangle / l)^2}{F(a,a) - \langle a, a \rangle}$$

$$= \frac{\left(\frac{1}{l} \iiint \mathbf{E}^i \cdot \mathbf{J}^a \, d\tau \right)^2}{\iiint \kappa^{-1} (\mathbf{J}^a)^2 \, d\tau - \iiint \mathbf{E}^a \cdot \mathbf{J}^a \, d\tau} \quad (7\text{-}143)$$

The echo area, defined by Eq. (7-114), can be obtained from Eq. (7-143) by letting the source recede to infinity. The steps parallel those used to obtain Eq. (7-115). For a z-polarized, $-x$ traveling incident wave, we obtain

$$A_e = \pi \left| \frac{\eta}{\lambda} \frac{\left(\iiint J_z{}^a e^{jkx} \, d\tau \right)^2}{\iiint \kappa^{-1} (J^a)^2 \, d\tau - \iiint \mathbf{E}^a \cdot \mathbf{J}^a \, d\tau} \right|^2 \quad (7\text{-}144)$$

In two-dimensional problems, the echo width, defined by Eq. (7-122), is found to be

$$L_e = \frac{\pi}{2\lambda} \left| \frac{\eta \left(\iint J_z{}^a e^{jkx} \, ds \right)^2}{\iint \kappa^{-1} (J_z{}^a)^2 \, ds - \iint E_z{}^a J_z{}^a \, ds} \right|^2 \quad (7\text{-}145)$$

if the incident wave is $-x$ traveling and z-polarized, and

$$L_e = \frac{\pi}{2\lambda} \left| \frac{\eta \left(\iint J_y{}^a e^{jkx} \, ds \right)^2}{\iint \kappa^{-1} (J^a)^2 \, ds - \iint \mathbf{E}^a \cdot \mathbf{J}^a \, ds} \right|^2 \quad (7\text{-}146)$$

if the incident wave is $-x$ traveling and y-polarized. The surface integrals in Eqs. (7-145) and (7-146) are over the cross section of the obstacle in a $z = $ constant plane.

To illustrate the accuracy that we might expect from the variational formulas, let us consider a problem for which the exact solution is available, the circular dielectric cylinder. The incident wave is z-polarized, and the cylinder is defined by $\rho = a = \lambda_0/2$, as shown in the insert of Fig. 7-18. For our first approximation, let us take

$$\mathbf{J}^a = \mathbf{u}_z e^{jkx} \tag{7-147}$$

where $k = \omega \sqrt{\epsilon \mu_0}$ is the wave number of the dielectric. This very crude assumption yields curve (b) of Fig. 7-18. For a better approximation, which yields curve (c) of Fig. 7-18, take

$$\mathbf{J}^a = \mathbf{u}_z(e^{jkx} + Ae^{-jkx}) \tag{7-148}$$

where A is a variational parameter to be determined either by the Ritz procedure or by the reaction concept. While Eq. (7-148) is a better approximation than Eq. (7-147), it is still crude. The integrations occurring in the various reactions were accomplished by expressing the exponentials and Hankel functions as Bessel function series, according to Sec. 5-8. The resulting series converged fairly rapidly.

An alternative procedure for treating dielectric obstacles can be given

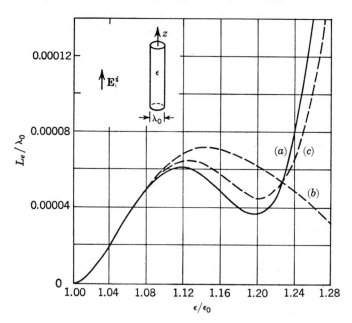

FIG. 7-18. Scattering by a dielectric cylinder (a) exact solution, (b) first-order variational solution, and (c) second-order variational solution. (*After Cohen.*)

in terms of equivalent currents over the surface of the obstacle.[1] This method leads to more than one formula for the desired parameter, and Rumsey discusses how to choose the best approximation according to the reaction concept.

7-12. Transmission through Apertures. The problem of transmission through apertures in an infinitely thin, perfectly conducting plane is closely related to the problem of scattering by plane obstacles. The precise interrelationship is shown by the following extension of Babinet's principle for optics.

Consider the three cases of a given source (a) radiating in free space, (b) radiating in the presence of an electrically conducting screen, and (c) radiating in the presence of a magnetically conducting screen, as shown in Fig. 7-19. The electric and magnetic screens are said to be *complementary* if the two screens superimposed cover the entire $y = 0$ plane with no overlapping. (The aperture of one is identical to the obstacle of the other.) Let the fields $y > 0$ be designated $(\mathbf{E}^i,\mathbf{H}^i)$, $(\mathbf{E}^e,\mathbf{H}^e)$, and $(\mathbf{E}^m,\mathbf{H}^m)$ for the cases (a), (b), and (c), respectively. Then *Babinet's principle* for complementary screens states that

$$\mathbf{E}^e + \mathbf{E}^m = \mathbf{E}^i \qquad \mathbf{H}^e + \mathbf{H}^m = \mathbf{H}^i \qquad (7\text{-}149)$$

proved as follows. Let S_s be the screen surface of Fig. 7-19b, and S_a be the aperture surface of Fig. 7-19b. The total field in each case is the incident field $\mathbf{E}^i$ plus the scattered field $\mathbf{E}^s$ produced by the currents on the screen. An element of electric current produces no components of $\mathbf{H}$ tangential to any plane containing the element (see Sec. 2-9). The currents induced on the screen thus produce no tangential $\mathbf{H}$ over the $y = 0$ plane; hence

$$\mathbf{n} \times \mathbf{H}^e = \mathbf{n} \times \mathbf{H}^i \qquad \text{over } S_a$$

On the screen itself we have the boundary condition

$$\mathbf{n} \times \mathbf{E}^e = 0 \qquad \text{over } S_s$$

For the complementary magnetic screen, following similar reasoning, we find

$$\mathbf{n} \times \mathbf{E}^m = \mathbf{n} \times \mathbf{E}^i \qquad \text{over } S_s$$
$$\mathbf{n} \times \mathbf{H}^m = 0 \qquad \text{over } S_a$$

By the above four equations, the sum $\mathbf{E}^e + \mathbf{E}^m$, $\mathbf{H}^e + \mathbf{H}^m$ satisfies

$$\mathbf{n} \times (\mathbf{E}^e + \mathbf{E}^m) = \mathbf{n} \times \mathbf{E}^i \qquad \text{over } S_s$$
$$\mathbf{n} \times (\mathbf{H}^e + \mathbf{H}^m) = \mathbf{n} \times \mathbf{H}^i \qquad \text{over } S_a$$

[1] V. H. Rumsey, The Reaction Concept in Electromagnetic Theory, *Phys. Rev.*, 2 ser., vol. 94, no. 6, pp. 1485–1491, June 15, 1954.

Hence, the $e + m$ field has the same $\mathbf{n} \times \mathbf{E}$ as the incident field over part of the $y = 0$ plane and the same $\mathbf{n} \times \mathbf{H}$ over the rest of the $y = 0$ plane. These conditions are sufficient to determine $\mathbf{E}$, $\mathbf{H}$ in the region $y > 0$ according to the uniqueness theorem (Sec. 3-3); so Babinet's principle [Eq. (7-149)] follows.

An alternative statement of Babinet's principle can be given in terms of the dual problem to Fig. 7-19c, shown in Fig. 7-19d. If the original source is replaced by its dual ($\mathbf{J}$ replaced by $\mathbf{K}$), the magnetic screen replaced by an electric screen, and the medium replaced by its "reciprocal" (η by $1/\eta$), then $\mathbf{E}$ will be numerically equal to $-\mathbf{H}^m$ and $\mathbf{H}$ numerically equal to $\mathbf{E}^m$ (see Table 3-2). If the field of this dual problem is

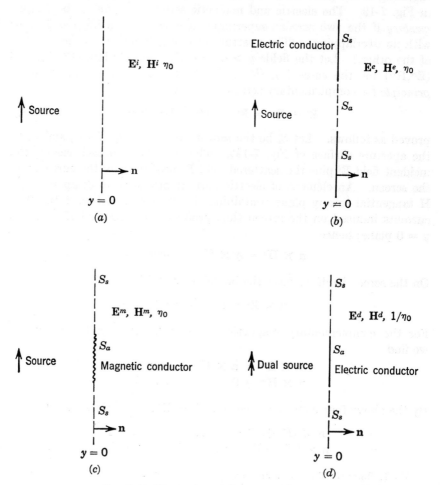

FIG. 7-19. Illustration of Babinet's principle.

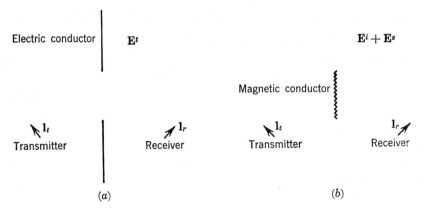

Fig. 7-20. The transmitted field E^t of (a) is equal to the scattered field E^s of (b).

denoted by E^d, H^d, Babinet's principle [Eq. (7-149)] becomes

$$E^e + H^d = E^i \qquad H^e - E^d = H^i \qquad (7\text{-}150)$$

The problem of Fig. 7-19d is more easily approximated physically than is the problem of Fig. 7-19c.

The direct application of Babinet's principle to the problems of Fig. 7-20a and b shows that the field transmitted by an aperture in a plane-conducting screen is equal to the negative of the field scattered by the complementary obstacle. Hence, stationary formulas for the signal at a receiver on the shadow side of a screen are of the same form as the stationary formulas for the scattered signal at a receiver in the complementary problem. In Fig. 7-20b, let the sources at the transmitter and receiver be magnetic currents across the "terminals" 1_t and 1_r. Then, dual to Eq. (7-135), we have at the receiver

$$H^s \cdot 1_r = - \frac{\left[\iint (H^i)^r \cdot (M_s^a)^t \, ds \right] \left[\iint (H^i)^t \cdot (M_s^a)^r \, ds \right]}{\iint (H^a)^r \cdot (M_s^a)^t \, ds} \qquad (7\text{-}151)$$

where M_s^a denotes the assumed magnetic current on the obstacle. It approximates the true magnetic current

$$M_s = (E^+ - E^-) \times n = 2E^s \times n \qquad (7\text{-}152)$$

where E^+ and E^- denote E in the regions $y > 0$ and $y < 0$, respectively, and $n = u_y$. The interrelationships between Fig. 7-20a and b can be expressed as

$$H^t = -H^s \qquad E^t = -E^s$$

Hence, from Eq. (7-151), we obtain for the aperture problem

$$\mathbf{H}^t \cdot \mathbf{l}_r = \frac{\left[\iint (\mathbf{H}^i)^r \cdot (\mathbf{n} \times \mathbf{E}^a)^t \, ds\right] \left[\iint (\mathbf{H}^i)^t \cdot (\mathbf{n} \times \mathbf{E}^a)^r \, ds\right]}{\iint (\mathbf{H}^a)^r \cdot (\mathbf{n} \times \mathbf{E}^a)^t \, ds} \quad (7\text{-}153)$$

where $\mathbf{E}^a$ is an assumed field in the aperture and $\mathbf{H}^a$ is the magnetic field calculated from the $\mathbf{E}^a$. The sources of $\mathbf{H}^i$ are magnetic current elements across $\mathbf{l}_t$ and $\mathbf{l}_r$, and, to apply Eq. (7-153), we must assume an $\mathbf{n} \times \mathbf{E}$ in the aperture due to $(\mathbf{H}^i)^t$ alone and due to $(\mathbf{H}^i)^r$ alone. If $\mathbf{l}_t$ and $\mathbf{l}_r$ are images of each other, as they appear in Fig. 7-20, then the aperture problem becomes the same as an echo problem, because of the symmetry of the plane screens about $y = 0$.

Sometimes it is the total power transmitted through the aperture that is of interest. We define the *transmission coefficient* T of an aperture as the ratio of the power transmitted through an aperture to the power incident on the aperture, that is,

$$T = \frac{\operatorname{Re} \iint_{\text{apert}} \mathbf{E}^t \times \mathbf{H}^{t*} \cdot d\mathbf{s}}{\operatorname{Re} \iint_{\text{apert}} \mathbf{E}^i \times \mathbf{H}^{i*} \cdot d\mathbf{s}} = \frac{\bar{\mathcal{O}}_t}{\bar{\mathcal{O}}_i} \quad (7\text{-}154)$$

Note that T depends on both the nature of the source and the geometry of the aperture. Another quantity sometimes defined is the *transmission area*, which is the transmission coefficient times the area of the aperture.

We shall explicitly consider uniform plane waves normally incident on an aperture in a plane screen, as shown in Fig. 7-21a. Let the incident

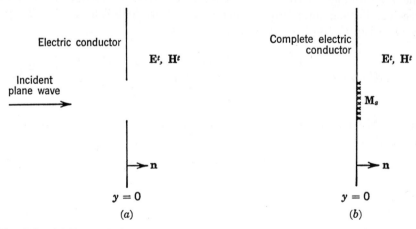

FIG. 7-21. (a) Transmission through an aperture, and (b) equivalent problem for the region $y > 0$.

wave be specified by

$$\mathbf{H}^i = \mathbf{u}e^{-jky} \qquad \mathbf{E}^i = \eta \mathbf{H}^i \times \mathbf{u}_y \qquad (7\text{-}155)$$

where $\mathbf{u}$ is any unit vector orthogonal to $\mathbf{u}_y$. In the proof of Babinet's principle, we noted that in the aperture

$$\mathbf{n} \times \mathbf{H}^t = \mathbf{n} \times \mathbf{H}^i \qquad (7\text{-}156)$$

because the currents on the conducting screen produce no tangential components of $\mathbf{H}$ in the $y = 0$ plane. Equation (7-155) chooses $\mathbf{H}^i$ to be real in the $y = 0$ plane; so by Eq. (7-156) $\mathbf{n} \times \mathbf{H}^i$ is real in the aperture. Hence,

$$\bar{\mathcal{P}}_t = \mathrm{Re} \iint_{\mathrm{apert}} \mathbf{E}^t \times \mathbf{H}^{t*} \cdot d\mathbf{s} = \mathrm{Re} \iint_{\mathrm{apert}} \mathbf{E}^t \times \mathbf{H}^t \cdot d\mathbf{s} \qquad (7\text{-}157)$$

Now consider the problem of Fig. 7-21b, which for

$$\mathbf{M}_s = \mathbf{E}^t \times \mathbf{n} \qquad (7\text{-}158)$$

is equivalent to Fig. 7-21a in the region $y > 0$. Hence, in the equivalent problem,

$$\bar{\mathcal{P}}_t = - \mathrm{Re} \iint \mathbf{M}_s \cdot \mathbf{H}^t \cdot d\mathbf{s} = \mathrm{Re} \langle c,c \rangle \qquad (7\text{-}159)$$

where $\langle c,c \rangle$ is the self-reaction of the correct magnetic currents *radiating in the presence of an electric conductor covering the entire $y = 0$ plane*.

For a variational formulation, we approximate $\langle c,c \rangle$ by $\langle a,a \rangle$ and constrain $\langle a,a \rangle$ according to Eqs. (7-65), that is,

$$\langle c,c \rangle \approx \langle a,a \rangle = \langle c,a \rangle = \langle a,c \rangle$$

where all sources radiate in the presence of the conducting plane. We have $\langle a,c \rangle = \langle c,a \rangle$ by reciprocity, and $\langle c,a \rangle$ can be calculated because we know $\mathbf{n} \times \mathbf{H}^c = \mathbf{n} \times \mathbf{H}^i$. Hence, our stationary formula for $\langle c,c \rangle$ is

$$\langle c,c \rangle \approx \frac{\langle c,a \rangle^2}{\langle a,a \rangle} = - \frac{\left(\iint \mathbf{H}^i \cdot \mathbf{M}_s^a \, ds \right)^2}{\iint \mathbf{H}^a \cdot \mathbf{M}_s^a \, ds} \qquad (7\text{-}160)$$

where $\mathbf{H}^a$ is the field of the assumed current $\mathbf{M}_s^a$. For the incident field of Eq. (7-155), we have the power incident on the aperture given by

$$\bar{\mathcal{P}}_i = \eta A \qquad (7\text{-}161)$$

where A is the area of the aperture. Hence, combining Eqs. (7-154) to (7-161), we have

$$T = \frac{1}{\eta A} \mathrm{Re} \left[\frac{\left(\iint \mathbf{u} \cdot \mathbf{n} \times \mathbf{E}^a \, ds \right)^2}{\iint \mathbf{H}^a \cdot \mathbf{n} \times \mathbf{E}^a \, ds} \right] \qquad (7\text{-}162)$$

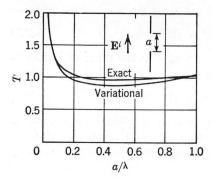

FIG. 7-22. Transmission coefficient for a slotted conductor, incident wave polarized transverse to slot axis. (*After Miles.*)

where $\mathbf{E}^a$ is the assumed tangential electric field in the aperture and $\mathbf{H}^a$ is the magnetic field calculated from $\mathbf{E}^a$ by the methods of Sec. 3-6.

As an example, let us consider the two-dimensional problem of transmission through a slot, as shown in the insert of Fig. 7-22. If we assume $\mathbf{E}^a$ in the slot to be real, then

$$\mathbf{E}^a \times \mathbf{H}^{a*} = (\mathbf{E}^a \times \mathbf{H}^a)^*$$

and the denominator of Eq. (7-162) is

$$\iint \mathbf{H}^a \cdot \mathbf{n} \times \mathbf{E}^a \, ds = \left(\iint \mathbf{E}^a \times \mathbf{H}^{a*} \cdot d\mathbf{s} \right)^*$$

In Sec. 4-11 we defined the admittance of an aperture as

$$Y_{\text{apert}} = \frac{1}{|V|^2} \iint \mathbf{E} \times \mathbf{H}^* \cdot d\mathbf{s}$$

and calculated it for a slot for particular assumed $\mathbf{E}$'s. Hence, applying Eq. (7-162) to a unit length of our two-dimensional slot, we have

$$T = \frac{1}{\eta a} \operatorname{Re} \left[\frac{\left(\int \mathbf{u} \cdot \mathbf{E}^a \times d\mathbf{l} \right)^2}{|V|^2 Y^*_{\text{apert}}} \right] \quad (7\text{-}163)$$

where a is the width of the slot. When the incident wave is polarized transverse to the slot, we have the case of Fig. 4-22; hence we take

$$E^a = 1 \quad (7\text{-}164)$$

in the slot. Now Eq. (7-163) reduces to

$$T = \frac{1}{\eta a} \operatorname{Re} \left(\frac{1}{Y^*_{\text{apert}}} \right) \quad (7\text{-}165)$$

where $Y_{\text{apert}} = G_a + jB_a$ is shown in Fig. 4-22. From Eqs. (4-106) we have for small a

$$T \xrightarrow[ka \to 0]{} \frac{\pi^2}{ka \log ka} \quad (7\text{-}166)$$

and from Eqs. (4-107) we have for large a

$$T \xrightarrow[ka \to \infty]{} 1 \quad (7\text{-}167)$$

This last result is the geometrical optics approximation. The variational solution is compared to the exact solution, which can be obtained by solving the wave equation in elliptic coordinates[1] (Fig. 7-22). The case of a plane wave at an arbitrary angle of incidence is considered by Miles.[2]

If the incident wave is polarized parallel to the axis of the slot, we have the case of Fig. 4-23; so to make use of the analysis of Sec. 4-11 we would assume

$$E^a = \cos\frac{\pi x}{a} \tag{7-168}$$

in the slot. Equation (7-163) then reduces to

$$T = \frac{4a}{\pi^2 \eta} \operatorname{Re}\left(\frac{1}{Y^*_{\text{apert}}}\right) \tag{7-169}$$

where $Y_{\text{apert}} = G_a + jB_a$ is shown in Fig. 4-23. From Eqs. (4-115), we have for small a

$$T \xrightarrow[ka \to 0]{} 6.85 \left(\frac{a}{\lambda}\right)^3 \tag{7-170}$$

For large a we should expect the field in the aperture to be uniform. Hence, we should not expect the trial field of Eq. (7-168) to give good results for large a, say $a > \lambda$. Equation (7-169) actually approaches 0.81 for large a, instead of the expected value 1.

PROBLEMS

7-1. Suppose the cavities of Fig. 7-1 contain lossy material characterized by σ, ϵ, and μ. Show that the perturbational formula corresponding to Eq. (7-3) is

$$\omega - \omega_0 = \frac{j \oiint_{\Delta S} \mathbf{H} \times \mathbf{E}_0 \cdot d\mathbf{s}}{\iiint [\epsilon \mathbf{E} \cdot \mathbf{E}_0 - \mu \mathbf{H} \cdot \mathbf{H}_0] \, d\tau}$$

Note that both ω and ω_0 must be complex. A complex resonance in the low-loss case can be interpreted according to

$$\omega = \omega_r \left(1 + \frac{j}{2Q}\right)$$

where ω_r is the real resonant frequency and Q is the quality factor (see Sec. 8-14).

7-2. Consider the perturbation of a cavity (say Fig. 7-1a) from one having perfectly conducting walls to one having a wall impedance Z, defined by

$$\mathbf{n} \times \mathbf{E} = Z\mathbf{H}_t$$

---

[1] Morse and Rubenstein, The Diffraction of Waves by Ribbons and Slits, *Phys. Rev.*, vol. 54, no. 11, pp. 895–898, December, 1938.

[2] J. W. Miles, On the Diffraction of an Electromagnetic Wave through a Plane Screen, *J. Appl. Phy.*, vol. 20, no. 8, pp. 760–770, August, 1949.

at the walls. Show that the exact perturbational formula is

$$\omega - \omega_0 = \frac{-j \oiint Z\mathbf{H} \cdot \mathbf{H}_0 \, ds}{\iiint (\epsilon \mathbf{E} \cdot \mathbf{E}_0 - \mu \mathbf{H} \cdot \mathbf{H}_0) \, d\tau}$$

where the subscript 0 denotes unperturbed quantities. Note that ω_0 is real but ω is complex if Z has a real part.

7-3. Use the results of Prob. 7-2 and the approximations

$$\mathbf{E} \approx \mathbf{E}_0 = |\mathbf{E}_0| \qquad \mathbf{H} \approx \mathbf{H}_0 = j|\mathbf{H}_0|$$

to show that

$$\omega - \omega_0 \approx \frac{j \oiint Z|H_0|^2 \, ds}{\iiint (\epsilon |E_0|^2 + \mu |H_0|^2) \, d\tau}$$

Use the relationships

$$\omega = \omega_r \left(1 + \frac{j}{2Q}\right) \qquad Z = \mathcal{R} + j\mathcal{X}$$

and show that the perturbational formula gives

$$\omega_r - \omega_0 \approx -\frac{\oiint \mathcal{X}|H_0|^2 \, ds}{2 \iiint \mu|H_0|^2 \, d\tau} \qquad Q \approx \frac{\omega_0 \iiint \mu|H_0|^2 \, d\tau}{\oiint \mathcal{R}|H_0|^2 \, ds}$$

Note that the formula for Q is identical to the one that we have been using if $\mathcal{R} = \text{Re}(\eta)$, where η is the intrinsic impedance of the conducting walls.

7-4. Use the results of Prob. 7-3, and show that the fractional change in resonance due to metal walls is

$$\frac{\omega_r - \omega_0}{\omega_0} \approx -\frac{1}{2Q}$$

where ω_0 is the resonant frequency of the cavity with perfectly conducting walls.

7-5. Suppose the cavities of Fig. 7-2 are characterized by σ and $\sigma + \Delta\sigma$ in addition to ϵ, μ and $\epsilon + \Delta\epsilon$, $\mu + \Delta\mu$. Show that the perturbational formula corresponding to Eq. (7-10) is then

$$\frac{\omega - \omega_0}{\omega} = -\frac{\iiint [(\Delta\epsilon - j\Delta\sigma/\omega)\mathbf{E} \cdot \mathbf{E}_0 - \Delta\mu \mathbf{H} \cdot \mathbf{H}_0] \, d\tau}{\iiint [(\epsilon - j\sigma/\omega)\mathbf{E} \cdot \mathbf{E}_0 - \mu \mathbf{H} \cdot \mathbf{H}_0] \, d\tau}$$

Both ω_0 and ω are complex if σ and $\sigma + \Delta\sigma$ are not identically zero.

7-6. Use the result of Prob. 7-5 for the case $\sigma = 0$, and let $\mathbf{E} \approx \mathbf{E}_0 = |\mathbf{E}_0|$, $\mathbf{H} \approx \mathbf{H}_0 = j|\mathbf{H}_0|$, $\omega \approx \omega_r + j\omega_0/2Q$, to show that

$$Q \approx \frac{\omega_0 \iiint \epsilon|E_0|^2 \, d\tau}{\iiint \Delta\sigma |E_0|^2 \, d\tau}$$

and that Eq. (7-11) still applies with ω changed to ω_r.

7-7. Suppose that a small sample of lossy dielectric is introduced into a cavity whose unperturbed resonant frequency is ω_0. Show that

$$\frac{\epsilon' - \epsilon_0}{\epsilon''} \approx 2Q \frac{\omega_0 - \omega_r}{\omega_0}$$

where $\hat{\epsilon} = \epsilon' - j\epsilon''$ is the complex permittivity of the sample and ω_r is the perturbed resonant frequency. If the losses of the unperturbed cavity are significant, then

$$\frac{1}{Q} \approx \frac{1}{Q_s} - \frac{1}{Q_0}$$

where Q_s and Q_0 are the Q's of the cavity with and without the sample, respectively.

7-8. Consider a rectangular cavity with a small centered dielectric cylinder, as shown in Fig. 7-23a. Show that the change in the resonant frequency of the dominant mode due to the introduction of the dielectric is

$$\frac{\omega - \omega_0}{\omega_0} \approx \frac{2A}{bc}(1 - \epsilon_r)$$

where A is the cross-sectional area of the cylinder. Use a quasi-static approximation.

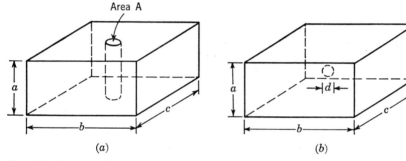

FIG. 7-23. Rectangular cavity with (a) dielectric cylinder and (b) dielectric sphere.

7-9. Consider the rectangular cavity with a small centered dielectric sphere, as shown in Fig. 7-23b. Show that the change in the resonant frequency of the dominant mode due to the introduction of the dielectric is

$$\frac{\omega - \omega_0}{\omega} \approx -\frac{\pi d^3}{abc}\frac{\epsilon_r - 1}{\epsilon_r + 2}$$

where d is the diameter of the sphere. Use a quasi-static approximation.

7-10. Consider the circular waveguide of Fig. 5-2. Suppose the wall is slightly flattened at the point $\phi = 90°$. Show that the change in cutoff frequency for the x-polarized (**E** in the center points in the x direction) dominant mode is

$$\frac{\Delta\omega_c}{\omega_c} \approx -0.418 \frac{A}{\pi a^2}$$

where A is the cross-sectional area of the deformation and $\omega_c = 1.841/a\sqrt{\epsilon\mu}$ is the unperturbed cutoff frequency. For the y-polarized dominant mode,

$$\frac{\Delta\omega_c}{\omega_c} \approx 1.42 \frac{A}{\pi a^2}$$

Hence, the mode degeneracy has been removed.

7-11. Figure 7-24a shows a small centered dielectric cylinder in a rectangular waveguide. Show that the change in cutoff frequency of the dominant mode from that for the empty guide is

$$\frac{\Delta \omega_c}{\omega_c} \approx -\frac{\pi d^2}{2ab}\frac{\epsilon_r - 1}{\epsilon_r + 1}$$

where $\omega_c = \pi/b \sqrt{\epsilon\mu}$. Use a perturbational method and a quasi-static approximation.

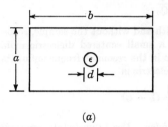
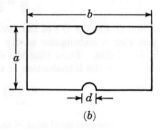

(a) (b)

Fig. 7-24. Rectangular waveguide with (a) dielectric cylinder and (b) conducting ridges.

7-12. Consider the rectangular waveguide with small semicircular ridges, as shown in Fig. 7-24b. Use a perturbational method and a quasi-static approximation to show that the dominant-mode cutoff frequency differs from the TE_{01} rectangular guide cutoff, according to

$$\frac{\Delta \omega_c}{\omega_c} \approx -\frac{\pi d^2}{2ab}$$

where $\omega_c = \pi/b \sqrt{\epsilon\mu}$. Show that the next higher mode ($b \leq 2a$) cutoff frequency differs from the TE_{02} rectangular guide cutoff, according to

$$\frac{\Delta \omega_c}{\omega_c} \approx \frac{\pi d^2}{4ab}$$

where $\omega_c = 2\pi/b \sqrt{\epsilon\mu}$. Hence, the mode separation is increased.

7-13. Consider the rectangular waveguide with the bottom covered by a thin dielectric slab (Fig. 4-6 with $d \leq a$). Use a perturbational method and quasi-static approximation to show that the phase constant is

$$\beta \approx \beta_0 + \frac{\epsilon_2}{\epsilon_1}\frac{k_1^2 - k_2^2}{2\beta_0}\frac{d}{a}$$

where $\beta_0 = k_2 \sqrt{1 - (f_c/f)^2}$ is the empty-guide phase constant. Note that this is the same as the first term of an expansion of the exact characteristic equation, as given in Prob. 4-14.

7-14. Consider the rectangular waveguide with a centered dielectric cylinder, as shown in Fig. 7-24a. Use a perturbational method and quasi-static approximation to show that

$$\frac{\beta - \beta_0}{k_0} \approx \frac{\pi d^2}{2ab}\frac{\epsilon_r - 1}{\epsilon_r + 1}\frac{1}{\sqrt{1 - (\omega_c/\omega)^2}}$$

where ω_c can be taken as the cutoff frequency of the perturbed guide, given in Prob.

7-11, if ω is close to ω_c. Show that at the unperturbed TE_{01} cutoff frequency

$$\beta \approx k_0 \sqrt{\frac{\pi d^2}{4ab} \frac{\epsilon_r + 1}{\epsilon_r - 1}}$$

7-15. Suppose that a waveguide is filled with lossy material, and consider a perturbation of its perfectly conducting walls. Represent the unperturbed fields (subscript 0) and the perturbed fields (no subscript) by

$$\mathbf{E}_0 = \hat{\mathbf{E}}_0 e^{-\gamma_0 z} \qquad \mathbf{E} = \hat{\mathbf{E}} e^{\gamma z}$$
$$\mathbf{H}_0 = \hat{\mathbf{H}}_0 e^{-\gamma_0 z} \qquad \mathbf{H} = \hat{\mathbf{H}} e^{\gamma z}$$

Note the opposite directions of propagation. Show that the formula corresponding to Eq. (7-29) is

$$\gamma - \gamma_0 = \frac{\oint_{\Delta C} \hat{\mathbf{E}}_0 \times \hat{\mathbf{H}} \cdot \mathbf{n}\, dl}{\iint_{S'} (\hat{\mathbf{E}}_0 \times \hat{\mathbf{H}} - \hat{\mathbf{E}} \times \hat{\mathbf{H}}_0) \cdot \mathbf{u}_z\, ds}$$

Show that this reduces to Eq. (7-29) in the loss-free case.

7-16. Consider the perturbation of material in a lossy waveguide from ϵ, μ, σ to $\epsilon + \Delta\epsilon$, $\mu + \Delta\mu$, $\sigma + \Delta\sigma$. Represent the fields as in Prob. 7-15, and show that the formula corresponding to Eq. (7-30) is

$$\gamma - \gamma_0 = -j \frac{\iint [(\omega\,\Delta\epsilon - j\,\Delta\sigma)\hat{\mathbf{E}} \cdot \hat{\mathbf{E}}_0 - \omega\,\Delta\mu \hat{\mathbf{H}} \cdot \hat{\mathbf{H}}_0]\, ds}{\iint (\hat{\mathbf{E}}_0 \times \hat{\mathbf{H}} - \hat{\mathbf{E}} \times \hat{\mathbf{H}}_0) \cdot \mathbf{u}_z\, ds}$$

Show that this reduces to Eq. (7-30) in the loss-free case.

7-17. Use the results of Prob. 7-16, and let the unperturbed guide be loss-free. Denote the propagation constant of the perturbed guide by $\gamma = \alpha + j\beta$, and let $\mathbf{E} \approx \mathbf{E}_0^*$ and $\mathbf{H} \approx -\mathbf{H}_0^*$. Show that the resultant approximation for β is Eq. (7-33) and

$$\alpha \approx \frac{\iint \Delta\sigma\, |\hat{\mathbf{E}}_0|^2\, ds}{2\,\mathrm{Re} \iint \hat{\mathbf{E}}_0 \times \hat{\mathbf{H}}_0^* \cdot \mathbf{u}_z\, ds}$$

Note that this is an approximate form of Eq. (2-76).

7-18. Consider the perturbation of the walls of a waveguide from a perfect conductor to an impedance sheet Z such that

$$\mathbf{n} \times \mathbf{E} = Z\mathbf{H}$$

Represent the unperturbed and perturbed fields as in Prob. 7-15, and show that

$$\gamma - \gamma_0 = \frac{\oint Z\hat{\mathbf{H}} \cdot \hat{\mathbf{H}}_0\, dl}{\iint (\hat{\mathbf{E}}_0 \times \hat{\mathbf{H}} - \hat{\mathbf{E}} \times \hat{\mathbf{H}}_0) \cdot \mathbf{u}_z\, ds}$$

7-19. Use the results of Prob. 7-18 and let the unperturbed guide be loss-free, so that $\gamma_0 = j\beta_0$. In the perturbed guide, let $Z = \mathcal{R} + j\mathcal{X}$, $\gamma = \alpha + j\beta$, $\mathbf{E} = \mathbf{E}_0^*$,

$\mathbf{H} = -\mathbf{H}_0^*$, and show that

$$\beta - \beta_0 \approx \frac{\oint \mathfrak{X} |\hat{H}_0|^2 \, dl}{2 \operatorname{Re} \iint \hat{\mathbf{E}}_0 \times \hat{\mathbf{H}}_0^* \cdot \mathbf{u}_z \, ds}$$

$$\alpha \approx \frac{\oint \mathfrak{R} |\hat{H}_0|^2 \, dl}{2 \operatorname{Re} \iint \hat{\mathbf{E}}_0 \times \hat{\mathbf{H}}_0^* \cdot \mathbf{u}_z \, ds}$$

If $Z = \eta$, the intrinsic impedance of metal walls, the above formula for α is the approximation that we have been using to calculate attenuation in metal waveguides.

7-20. Show that

$$\omega_r{}^2 = \frac{\iiint \mu^{-1} |\nabla \times \mathbf{E}|^2 \, d\tau}{\iiint \epsilon |E|^2 \, d\tau}$$

is a stationary formula for the resonant frequency of a *loss-free* cavity, provided $\mathbf{n} \times \mathbf{E} = 0$ on S, but is not stationary if losses are present.

7-21. Show that Eq. (7-46) is a stationary formula for $\omega_r{}^2$, with no boundary conditions required on $\mathbf{H}$.

7-22. Consider the rectangular cavity (Fig. 2-19) and the stationary formula [Eq. (7-44)]. Use a trial field

$$\mathbf{E} = \mathbf{u}_x y z (y - b)(z - c)$$

and show that Eq. (7-44) gives

$$\omega_r \approx \frac{\sqrt{10}}{bc} \sqrt{\frac{b^2 + c^2}{\epsilon \mu}}$$

In the exact solution [Eq. (2-95)], the numerical factor is π instead of $\sqrt{10}$.

7-23. Consider a small deformation of the walls of a cavity, such as represented by Fig. 7-1. Take the variational formula [Eq. (7-45)], which requires no boundary conditions on $\mathbf{E}$, and take the unperturbed cavity field $\mathbf{E}_0$ as a trial field. Show that Eq. (7-45) reduces to

$$\frac{\omega^2 - \omega_0^2}{\omega_0^2} \approx \frac{\iiint_{\Delta\tau} (\mu |H_0|^2 - \epsilon |E_0|^2) \, d\tau}{\iiint_{\tau} \epsilon |E_0|^2 \, d\tau}$$

Show that this formula is essentially the same as Eq. (7-4).

7-24. Figure 5-31b shows a partially filled circular cavity. Use Eq. (7-46) and a trial field

$$\mathbf{H} = \mathbf{u}_\phi J_1 \left(2.405 \frac{\rho}{a} \right)$$

to show that the dominant mode resonance is

$$\omega_r \approx \frac{2.405}{a \sqrt{\epsilon_0 \mu_0}} \sqrt{1 - \frac{b}{d}\left(1 - \frac{1}{\epsilon_r}\right)}$$

Compare with the results of Prob. 5-24.

7-25. Consider a waveguide whose cross section is an equilateral triangle of side length a. Use variational formulas to approximate the lowest cutoff frequency. The exact solution is

$$\omega_c = \frac{4\pi}{3a\sqrt{\epsilon\mu}}$$

7-26. Consider the rectangular cavity (Fig. 2-19) and the mixed-field variational formula [Eq. (7-72)]. Choose a trial field

$$\mathbf{E} = \mathbf{u}_x \sin\frac{\pi y}{b}\sin\frac{\pi z}{c}$$

$$\mathbf{H} = \mathbf{u}_y A_1 \sin\frac{\pi y}{b}\cos\frac{\pi z}{c} + \mathbf{u}_z A_2 \cos\frac{\pi y}{b}\sin\frac{\pi z}{c}$$

where A_1 and A_2 are variational parameters. Determine A_1 and A_2 by the Ritz method, and show that the resultant formula for ω_r is the exact formula [Eq. (2-95)]. Why do we get an exact solution in this case?

7-27. In Fig. 7-25, the surface S represents a perfect electric conductor enclosing a cavity. A variational solution is desired in terms of a trial field satisfying $\mathbf{n} \times \mathbf{E} = 0$

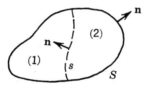

Fig. 7-25. Trial fields are discontinuous over s.

on S and $\mathbf{n} \times (\mu^{-1}\nabla \times \mathbf{E})$ continuous at s, but with $\mathbf{n} \times \mathbf{E}$ discontinuous at s. Show that the stationary E-field formula is

$$\omega^2 = \frac{\iiint \frac{1}{\mu}(\nabla \times \mathbf{E})^2\,d\tau + 2\iint_s (\mathbf{E}_1 - \mathbf{E}_2) \times \frac{1}{\mu}\nabla \times \mathbf{E}\cdot d\mathbf{s}}{\iiint \epsilon E^2\,d\tau}$$

where subscripts 1 and 2 refer to regions 1 and 2 (Fig. 7-25). Show also that a variational solution in terms of trial fields satisfying $\mathbf{n} \times \mathbf{E} = 0$ on S and $\mathbf{n} \times \mathbf{E}$ continuous at s, but with $\mathbf{n} \times (\mu^{-1}\nabla \times \mathbf{E})$ discontinuous at s, is given by Eq. (7-44).

7-28. Show that the variational H-field formula for Prob. 7-27 is of the same form as the above E-field formula, given by replacing $\mathbf{E}$ by $\mathbf{H}$, ϵ by μ, and μ by ϵ. Show that no boundary conditions at S are required in the H-field formula.

7-29. Consider a perturbation of material in a cavity, such as represented by Fig. 7-2. Take the mixed-field variational formula [Eq. (7-72)], and take the unperturbed cavity field $\mathbf{E}_0$, $\mathbf{H}_0$ as a trial field. Show that Eq. (7-72) then reduces to Eq. (7-11).

7-30. Repeat Prob. 7-26, using the reaction concept of Sec. 7-7.

7-31. Consider the partially filled rectangular waveguide of Fig. 4-8a. Use the E-field variational formula [Eq. (7-8)], and the trial field

$$\mathbf{E} = \mathbf{u}_y \sin\frac{\pi x}{a}$$

and show that

$$\omega_c \approx \frac{\pi}{a}\left[1 + \frac{\epsilon_1 - \epsilon_2}{\epsilon_2}\left(\frac{d}{a} - \frac{1}{2\pi}\sin\frac{2\pi d}{a}\right)\right]^{-\frac{1}{2}}$$

Compare some calculated points with the exact solution (Fig. 4-9).

7-32. Use the reaction concept to derive the mixed-field variational formula for waveguide phase constants

$$\beta = \frac{\omega \iint (\mu \hat{H}^2 + \epsilon \hat{E}^2)\, ds - j \oint \hat{\mathbf{E}} \times \hat{\mathbf{H}} \cdot \mathbf{n}\, dl}{2 \iint \hat{\mathbf{E}} \times \hat{\mathbf{H}} \cdot \mathbf{u}_z\, ds}$$

which corresponds to Eq. (7-85) if $\mathbf{n} \times \mathbf{E} = 0$ on C. No boundary conditions are required in the above formula.

7-33. Consider the variational formula of Prob. 7-32 and a perturbation of waveguide walls, as illustrated by Fig. 7-5a and b. Use the unperturbed field $\mathbf{E}_0$, $\mathbf{H}_0$ as a trial field, and show that the formula of Prob. 7-32 reduces to Eq. (7-32).

7-34. Consider the variational formula of Eq. (7-85) and a perturbation of matter in a waveguide, represented by Fig. 7-5a and c. Use the unperturbed field $\mathbf{E}_0$, $\mathbf{H}_0$ as a trial field, and show that Eq. (7-85) reduces to Eq. (7-33).

7-35. Figure 7-26 shows a coaxial stub to parallel-plate waveguide feed system. Assume $a \ll \lambda$ so that a reasonable trial current is a uniform current. Show by the variational method that the impedance seen by the coax is

$$Z \approx \frac{\eta}{4} ka \left(1 - j\frac{2}{\pi}\log\frac{\gamma k d}{4}\right)$$

where $\gamma = 1.781$.

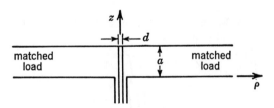

Fig. 7-26. Coax to parallel-plate feed.

7-36. In Prob. 7-35, remove the restriction on a and assume a trial current on the stub

$$I = \cos k(a - z)$$

Obtain the input impedance seen by the coax by the variational method.

7-37. Repeat Prob. 7-36 for the second-order variational solution, assuming trial currents

$$I^u = \cos k(a - z) \qquad I^v = 1$$

Note that only one new reaction is needed in addition to those obtained in Probs. 7-35 and 7-36. Specialize the result to $a = \lambda/4$.

7-38. Consider the two-dimensional problem of plane-wave scattering by a conducting ribbon, shown in the insert of Fig. 7-16, but with the opposite polarization.

PERTURBATIONAL AND VARIATIONAL TECHNIQUES

In other words, H^i is parallel to the axis of the ribbon. Use the trial current

$$J_s = u_y \cos \frac{\pi y}{a}$$

and show that the variational solution is

$$L_e \approx \frac{32a^4}{\pi^3 \lambda} \left| \frac{1}{\eta Y_{\text{apert}}} \right|^2$$

where ηY_{apert} is given in Fig. 4-23. Show that as $ka \to \infty$ this answer reduces to 0.66 times the physical optics solution. Why should we expect the above formula to be inaccurate for large ka?

7-39. Consider plane-wave scattering by a wire, represented by Fig. 7-15. At the first resonance ($L \approx \lambda/2$), the current is

$$I^a \approx \cos kz$$

and we know that (see Fig. 2-24)

$$\langle a,a \rangle \approx 73$$

The imaginary part of $\langle a,a \rangle$ is zero because the length is adjusted for resonance. Using Eq. (7-115), show that at resonance the echo area is

$$A_e \approx 0.86\lambda^2$$

This is relatively insensitive to the diameter of the wire.

7-40. Figure 7-27 represents a resonant length of wire illuminated by a uniform plane wave at the angle θ, polarized in the r-z plane. Using the approximations of Prob. 7-39, show that the back-scattering area is

$$A_e \approx 0.86\lambda^2 \left[\frac{\cos\left(\frac{\pi}{2} \cos \theta\right)}{\sin \theta} \right]^4$$

Again this is relatively insensitive to the diameter of the wire.

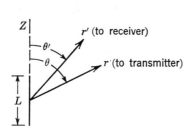

FIG. 7-27. Scattering by a resonant wire ($L \approx \lambda/2$).

7-41. Repeat Prob. 7-40 for the case of differential scattering, showing that the differential echo area is

$$A_e \approx 0.86\lambda^2 \left[\frac{\cos\left(\frac{\pi}{2} \cos \theta\right)}{\sin \theta} \frac{\cos\left(\frac{\pi}{2} \cos \theta'\right)}{\sin \theta'} \right]^2$$

where A_e is defined by Eq. (7-114) with E^s evaluated in the θ' direction.

7-42. Consider differential scattering by a magnetic obstacle (Fig. 7-17) and define

$$\kappa_e = j\omega(\epsilon - \epsilon_0) \qquad \kappa_m = j\omega(\mu - \mu_0)$$

Show that, instead of Eq. (7-143), we have

$$\text{Echo} = \frac{(\langle i,a \rangle / l)^2}{F(a,a) - \langle a,a \rangle}$$

where
$$\langle i,a \rangle = \iiint (\mathbf{E}^i \cdot \mathbf{J}^a - \mathbf{H}^i \cdot \mathbf{M}^a) \, d\tau$$

$$F(a,a) = \iiint [\kappa_e^{-1}(J^a)^2 - \kappa_m^{-1}(M^a)^2] \, d\tau$$

$$\langle a,a \rangle = \iiint (\mathbf{E}^a \cdot \mathbf{J}^a - \mathbf{H}^a \cdot \mathbf{M}^a) \, d\tau$$

In the above formulas, $\mathbf{E}^i$, $\mathbf{H}^i$ is the incident field, $\mathbf{J}^a$ and $\mathbf{M}^a$ are the assumed electric and magnetic polarization currents on the obstacle, and $\mathbf{E}^a$, $\mathbf{H}^a$ is the field from $\mathbf{J}^a$, $\mathbf{M}^a$.

7-43. Figure 7-28a represents a metal antenna cut from a plane conductor and fed across the slot ab. Figure 7-28b represents the aperture formed by the remainder of the metal plane left after the metal antenna was cut. The aperture antenna, fed

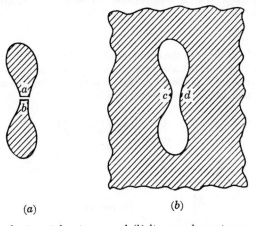

Fig. 7-28. (a) A sheet-metal antenna and (b) its complementary aperture antenna.

across cd, is said to be complementary to the metal antenna. Let Z_m be the input impedance of the metal antenna and Y_s be the input admittance to the slot antenna, and show that

$$\frac{Z_m}{Y_s} = \frac{\eta^2}{4}$$

Hint: Consider line integrals of $\mathbf{E}$ and $\mathbf{H}$ from a to b and c to d, and use duality.

7-44. Consider a narrow resonant slot of approximate length $\lambda/2$ in a conducting screen. Show that the transmission coefficient is

$$T \approx 0.52 \frac{\lambda}{w}$$

where w is the width of the slot. *Hint:* Use the result of Prob. 7-43 and assumptions similar to those of Prob. 7-39.

CHAPTER 8

MICROWAVE NETWORKS

8-1. Cylindrical Waveguides. Several special cases of the cylindrical waveguide, such as the rectangular and circular guides, already have been considered. We now wish to give a general treatment of cylindrical (cross section independent of z) waveguides consisting of a homogeneous isotropic dielectric bounded by a perfect electric conductor. Figure 8-1 represents the cross section of one such waveguide.

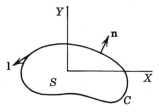

FIG. 8-1. Cross section of a cylindrical waveguide.

Our formulation of the problem will be similar to that given by Marcuvitz.[1]

As shown in Sec. 3-12, general solutions for the field in a homogeneous region can be constructed from solutions to the Helmholtz equation

$$\nabla^2 \psi + k^2 \psi = 0 \qquad (8\text{-}1)$$

In cylindrical coordinates, this equation can be partially separated by taking

$$\psi = \Psi(x,y) Z(z) \qquad (8\text{-}2)$$

The resultant pair of equations are

$$\nabla_t^2 \Psi + k_c^2 \Psi = 0 \qquad (8\text{-}3)$$

$$\frac{d^2 Z}{dz^2} + k_z^2 Z = 0 \qquad (8\text{-}4)$$

where the separation constants k_z and k_c are related by

$$k_c^2 + k_z^2 = k^2 \qquad (8\text{-}5)$$

and ∇_t is the two-dimensional (transverse to z) del operator

$$\nabla_t = \nabla - \mathbf{u}_z \frac{\partial}{\partial z} \qquad (8\text{-}6)$$

---

[1] N. Marcuvitz, "Waveguide Handbook," MIT Radiation Laboratory Series, vol. 10, sec. 1-2, McGraw-Hill Book Company, Inc., New York, 1951.

Solutions to Eq. (8-4) are of the general form

$$Z(z) = Ae^{-jk_z z} + Be^{jk_z z} \qquad (8\text{-}7)$$

which, for k_z real, is a superposition of $+z$ and $-z$ traveling waves. The k_z are determined from Eq. (8-5) after the k_c (cutoff wave numbers) are found by solving the boundary-value problem.

For TE modes, we take $\mathbf{F} = \mathbf{u}_z \psi^e$ (superscript e denotes TE) and determine

$$\mathbf{E}^e = -\mathbf{u}_x \frac{\partial \psi^e}{\partial y} + \mathbf{u}_y \frac{\partial \psi^e}{\partial x} = (\mathbf{u}_z \times \nabla_t \Psi^e) Z^e \qquad (8\text{-}8)$$

The component of $\mathbf{E}$ tangential to the waveguide boundary C is

$$E_l^e = \mathbf{l} \cdot (\mathbf{u}_z \times \nabla_t \Psi^e) = (\mathbf{n} \cdot \nabla_t \Psi^e) Z^e$$

where $\mathbf{l}$ is the unit tangent to C and $\mathbf{n}$ is the unit normal to C (see Fig. 8-1). The boundary is perfectly conducting; hence $E_l = 0$ on C and

$$\frac{\partial \Psi^e}{\partial n} = 0 \qquad \text{on } C \qquad (8\text{-}9)$$

The associated magnetic field is given by

$$\mathbf{H}^e = -\frac{1}{j\omega\mu} \nabla \times \mathbf{E}^e = \frac{1}{j\omega\mu} \left(\mathbf{u}_x \frac{\partial^2 \psi^e}{\partial x \partial z} + \mathbf{u}_y \frac{\partial^2 \psi^e}{\partial y \partial z} + \mathbf{u}_z k_c^2 \psi^e \right)$$

For more concise notation, we define a *transverse field vector* as

$$\mathbf{H}_t = \mathbf{H} - \mathbf{u}_z H_z \qquad (8\text{-}10)$$

and rewrite the above as

$$\mathbf{H}_t^e = \frac{1}{j\omega\mu} (\nabla_t \Psi^e) \frac{dZ^e}{dz} \qquad H_z^e = \frac{k_c^2}{j\omega\mu} \Psi^e Z^e \qquad (8\text{-}11)$$

It is evident from Eqs. (8-8) and (8-11) that lines of $\mathcal{E}$ and $\mathcal{H}_t$ are everywhere perpendicular to each other.

For TM modes, we take $\mathbf{A} = \mathbf{u}_z \psi^m$ (superscript m denotes TM) and, dual to Eq. (8-8), we determine

$$\mathbf{H}^m = -(\mathbf{u}_z \times \nabla_t \Psi^m) Z^m \qquad (8\text{-}12)$$

Defining the transverse electric field vector $\mathbf{E}_t$ by Eq. (8-10) with $\mathbf{H}$ replaced by $\mathbf{E}$, we have, dual to Eq. (8-11),

$$\mathbf{E}_t^m = \frac{1}{j\omega\epsilon} (\nabla_t \Psi^m) \frac{dZ^m}{dz} \qquad E_z^m = \frac{k_c^2}{j\omega\epsilon} \Psi^m Z^m \qquad (8\text{-}13)$$

From the second of these equations, it is evident that for E_z to vanish on C we must meet the boundary condition

$$\Psi^m = 0 \qquad \text{on } C \qquad (8\text{-}14)$$

provided $k_c \neq 0$. Note that Eq. (8-14) also satisfies the condition $\mathbf{1} \cdot \mathbf{E}_t = 0$ on C. When the waveguide cross section is multiply connected, such as in coaxial lines, it is possible to have $k_c = 0$. In this case, the necessary boundary condition is $\Psi^m =$ constant on each conductor. The corresponding field is TEM to z and is a transmission-line mode.

It should be kept in mind that Eq. (8-3) subject to boundary conditions is an eigenvalue problem, giving rise to a discrete set of modes. These modes can be suitably ordered, and the various equations of this section then apply to each mode. It is convenient to introduce *mode functions* $\mathbf{e}(x,y)$ and $\mathbf{h}(x,y)$, *mode voltages* $V(z)$, and *mode currents* $I(z)$ according to

$$\begin{aligned} \mathbf{E}^e = \mathbf{e}^e V^e & \qquad \mathbf{E}_t^m = \mathbf{e}^m V^m \\ \mathbf{H}_t^e = \mathbf{h}^e I^e & \qquad \mathbf{H}^m = \mathbf{h}^m I^m \end{aligned} \qquad (8\text{-}15)$$

Comparing Eqs. (8-15) with Eqs. (8-8) and (8-11), we see that we may choose

$$\begin{aligned} \mathbf{e}^e = \mathbf{u}_z \times \boldsymbol{\nabla}_t \Psi^e = \mathbf{h}^e \times \mathbf{u}_z & \qquad V^e = Z^e \\ \mathbf{h}^e = -\boldsymbol{\nabla}_t \Psi^e = \mathbf{u}_z \times \mathbf{e}^e & \qquad I^e = -\frac{1}{j\omega\mu}\frac{dZ^e}{dz} \end{aligned} \qquad (8\text{-}16)$$

for TE modes, and, comparing Eqs. (8-15) with Eqs. (8-12) and (8-13),

$$\begin{aligned} \mathbf{e}^m = -\boldsymbol{\nabla}_t \Psi^m = \mathbf{h}^m \times \mathbf{u}_z & \qquad V^m = -\frac{1}{j\omega\epsilon}\frac{dZ^m}{dz} \\ \mathbf{h}^m = -\mathbf{u}_z \times \boldsymbol{\nabla}_t \Psi^m = \mathbf{u}_z \times \mathbf{e}^m & \qquad I^m = Z^m \end{aligned} \qquad (8\text{-}17)$$

for TM modes. Furthermore, we *normalize* the mode vectors according to

$$\begin{aligned} \iint (\mathbf{e}^e)^2 \, ds = \iint (\mathbf{h}^e)^2 \, ds = 1 \\ \iint (\mathbf{e}^m)^2 \, ds = \iint (\mathbf{h}^m)^2 \, ds = 1 \end{aligned} \qquad (8\text{-}18)$$

where the integration extends over the guide cross section. Hence, all amplitude factors are included in the V's and I's.

We shall now show that *all eigenvalues are real*. Consider the two-dimensional divergence theorem

$$\iint \boldsymbol{\nabla}_t \cdot \mathbf{A} \, ds = \oint \mathbf{A} \cdot \mathbf{n} \, dl$$

and let $\mathbf{A} = \Psi^* \boldsymbol{\nabla}_t \Psi$. Then,

$$\boldsymbol{\nabla}_t \cdot \mathbf{A} = \boldsymbol{\nabla}_t \Psi^* \cdot \boldsymbol{\nabla}_t \Psi + \Psi^* \boldsymbol{\nabla}_t^2 \Psi = |\boldsymbol{\nabla}_t \Psi|^2 - k_c^2 |\Psi|^2$$

and the divergence theorem becomes

$$\iint (|\boldsymbol{\nabla}_t \Psi|^2 - k_c^2 |\Psi|^2) \, ds = \oint \Psi^* \frac{\partial \Psi}{\partial n} \, dl$$

But the boundary conditions on the eigenfunction Ψ are either $\Psi = 0$ or $\partial\Psi/\partial n = 0$ on C. Hence, the right-hand term vanishes and

$$k_c^2 = \frac{\iint |\nabla_t \Psi|^2 \, ds}{\iint |\Psi|^2 \, ds} \tag{8-19}$$

The eigenvalue k_c^2 is therefore positive real. There is also no loss of generality if we take *all eigenfunctions Ψ to be real*. To justify this statement, suppose Ψ is not real, and let $\Psi = u + jv$. Then the Helmholtz equation is

$$\nabla_t^2 \Psi + k_c^2 \Psi = \nabla_t^2 u + k_c^2 u + j(\nabla_t^2 v + k_c^2 v) = 0$$

which, since k_c^2 is real, represents two Helmholtz equations for the real functions u and v. The boundary conditions, either

$$\Psi = u + jv = 0 \qquad \text{on } C$$

or

$$\frac{\partial \Psi}{\partial n} = \frac{\partial u}{\partial n} + j \frac{\partial v}{\partial n} = 0 \qquad \text{on } C$$

are satisfied independently by u and v; so u and v are solutions to the same boundary-value problem. Hence, u and v for a particular k_c can differ only by a constant, and Ψ is in phase over a guide cross section. We can take it to be real and include any phase in the V and I functions.

Let us now look at the propagation constant $\gamma = jk_z$. For ϵ and μ real, we have a cutoff wavelength

$$\lambda_c = \frac{2\pi}{k_c} \tag{8-20}$$

and a cutoff frequency

$$f_c = \frac{k_c}{2\pi \sqrt{\epsilon\mu}} \tag{8-21}$$

Then, from Eq. (8-5), we have the propagation constant given by

$$\gamma = jk_z = \begin{cases} j\beta = jk \sqrt{1 - \left(\frac{f_c}{f}\right)^2} & f > f_c \\ \alpha = k_c \sqrt{1 - \left(\frac{f}{f_c}\right)^2} & f < f_c \end{cases} \tag{8-22}$$

These are, of course, just the relationships that we previously established for the rectangular and circular waveguides. Figure 2-18 illustrates the behavior of α and β versus f. When the mode is propagating ($f > f_c$), the concepts of *guide wavelength*,

$$\lambda_g = \frac{2\pi}{\beta} = \frac{\lambda}{\sqrt{1 - (f_c/f)^2}} \tag{8-23}$$

where λ is the intrinsic wavelength in the dielectric, and *guide phase velocity*,

$$v_g = \frac{\omega}{\beta} = \frac{v_p}{\sqrt{1 - (f_c/f)^2}} \qquad (8\text{-}24)$$

where v_p is the intrinsic phase velocity, are useful. These parameters are discussed in Sec. 2-7.

Turning now to the mode voltages and currents, we see from their definitions [Eqs. (8-16) and (8-17)] that V and I satisfy Eq. (8-4). Hence, in general they are of the form of Eq. (8-7), or

$$\begin{aligned} V(z) &= V^+ e^{-\gamma z} + V^- e^{\gamma z} \\ I(z) &= I^+ e^{-\gamma z} + I^- e^{\gamma z} \end{aligned} \qquad (8\text{-}25)$$

where superscripts $+$ and $-$ denote positively and negatively traveling (or attenuating) wave components. Also, from Eqs. (8-4), (8-16), and (8-17) it is apparent that

$$\frac{V^+}{I^+} = Z_0 \qquad \frac{V^-}{I^-} = -Z_0 \qquad (8\text{-}26)$$

where the *characteristic impedance* Z_0 is, for TE modes,

$$Z_0^e = \frac{j\omega\mu}{\gamma} = \begin{cases} \dfrac{\omega\mu}{\beta} = \dfrac{\eta}{\sqrt{1 - (f_c/f)^2}} & f > f_c \\ \dfrac{j\omega\mu}{\alpha} = \dfrac{j\omega\mu}{k_c \sqrt{1 - (f/f_c)^2}} & f < f_c \end{cases} \qquad (8\text{-}27)$$

and, for TM modes,

$$Z_0^m = \frac{\gamma}{j\omega\epsilon} = \begin{cases} \dfrac{\beta}{\omega\epsilon} = \eta \sqrt{1 - \left(\dfrac{f_c}{f}\right)^2} & f > f_c \\ \dfrac{\alpha}{j\omega\epsilon} = \dfrac{k_c}{j\omega\epsilon} \sqrt{1 - \left(\dfrac{f}{f_c}\right)^2} & f < f_c \end{cases} \qquad (8\text{-}28)$$

Note that these are just the characteristic wave impedances that we previously defined for rectangular and circular waveguides. Figure 4-3 illustrates the behavior of the Z_0's versus frequency. Finally, from Eqs. (8-4), (8-16), and (8-17), we can show that V and I also satisfy the *transmission-line equations*

$$\begin{aligned} \frac{dV}{dz} &= -\gamma Z_0 I \\ \frac{dI}{dz} &= -\gamma Y_0 V \end{aligned} \qquad (8\text{-}29)$$

where $Y_0 = 1/Z_0$ is the *characteristic admittance*. Hence, the analogy

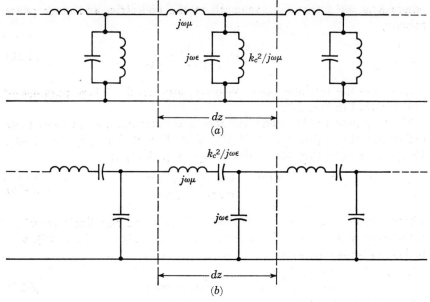

FIG. 8-2. Equivalent transmission lines for waveguide modes (series elements labeled in ohms, shunt elements in mhos). (a) TE modes, (b) TM modes.

with transmission lines is complete, and all of the techniques for analyzing transmission lines can be applied to each waveguide mode.[1]

We may define an *equivalent transmission line* for each waveguide mode as one for which γ and Z_0 are the same as those of the waveguide mode. Such an equivalent circuit may help us to visualize waveguide behavior by presenting it in terms of the more familiar transmission-line behavior. For a dissipationless transmission line, we have

$$Z_0 = \sqrt{\frac{Z}{Y}} = \sqrt{\frac{X}{B}}$$
$$\gamma = \sqrt{ZY} = j\sqrt{XB}$$

(see Sec. 2-6). Equating the above Z_0 and γ to those of a TE waveguide mode, we obtain

$$jX = j\omega\mu \qquad jB = j\omega\epsilon + \frac{k_c^2}{j\omega\mu} \qquad (8\text{-}30)$$

Thus, the transmission line equivalent to a TE mode is as shown in Fig. 8-2a. Similarly, for a TM mode we obtain

$$jX = j\omega\mu + \frac{k_c^2}{j\omega\epsilon} \qquad jB = j\omega\epsilon \qquad (8\text{-}31)$$

[1] For example, see Wilbur LePage and Samuel Seely, "General Network Analysis," Chaps. 9 and 10, McGraw-Hill Book Company, Inc., New York, 1952.

The transmission line equivalent to a TM mode is therefore as shown in Fig. 8-2b. If the dielectric is lossy, the equivalent transmission will also have resistances, obtained by replacing $j\omega\epsilon$ by $\sigma + j\omega\epsilon$ in Eqs. (8-30) and (8-31). In the light of filter theory, we can recognize the equivalent transmission lines as high-pass filters.

The power transmitted along the waveguide is, of course, obtained by integrating the Poynting vector over the guide cross section. Hence, for the $+z$ direction,

$$P_z = \iint \mathbf{E} \times \mathbf{H}^* \cdot \mathbf{u}_z \, ds = VI^* \iint \mathbf{e} \times \mathbf{h}^* \cdot \mathbf{u}_z \, ds$$
$$= VI^* \iint e^2 \, ds = VI^* \tag{8-32}$$

and the time-average power transmitted is

$$\bar{\mathcal{P}}_z = \text{Re}\,(VI^*) \tag{8-33}$$

Hence, in terms of the mode voltage and current, power is calculated by the usual circuit-theory formulas.

It is also worthwhile to note that the mode patterns, that is, pictures of lines of $\mathcal{E}$ and $\mathcal{K}$ at some instant, can be obtained directly from the Ψ's. For TE modes, $\mathbf{H}_t$ is proportional to $\nabla_t \Psi^e$, and $\mathbf{E}$ is perpendicular to $\mathbf{H}_t$. Hence, *lines of constant Ψ^e are also lines of instantaneous $\mathcal{E}$.* Lines of instantaneous $\mathcal{K}_t$ are everywhere perpendicular to lines of instantaneous $\mathcal{E}$. Similarly, for TM modes, *lines of constant Ψ^m are also lines of instantaneous $\mathcal{K}$,* and lines of instantaneous $\mathcal{E}_t$ are everywhere perpendicular to lines of instantaneous $\mathcal{K}$. It is therefore quite easy to sketch the mode patterns directly from the eigenfunctions Ψ.

Recognizing that the general exposition of cylindrical waveguides has been quite lengthy, let us summarize the results. Table 8-1 lists the more important relationships that we have derived. Those equations common to both TE and TM modes are written centered in the table. Keep in mind that all of the equations apply to *each mode* and that many modes may exist simultaneously in any given waveguide.

Finally, for future reference, let us tabulate the normalized eigenfunctions for the special cases already treated. For the rectangular waveguide of Fig. 2-16, we can pick the Ψ's from Eqs. (4-19) and (4-21) and normalize them according to Eq. (8-18). The result is

$$\Psi_{mn}{}^e = \frac{1}{\pi}\sqrt{\frac{ab\epsilon_m\epsilon_n}{(mb)^2 + (na)^2}} \cos\left(\frac{m\pi}{a}x\right)\cos\left(\frac{n\pi}{b}y\right)$$
$$\Psi_{mn}{}^m = \frac{2}{\pi}\sqrt{\frac{ab}{(mb)^2 + (na)^2}} \sin\left(\frac{m\pi}{a}x\right)\sin\left(\frac{n\pi}{b}y\right) \tag{8-34}$$

where $m, n = 0, 1, 2, \ldots, (m = n = 0$ excepted). Similarly, for the

TABLE 8-1. SUMMARY OF EQUATIONS FOR THE CYLINDRICAL WAVEGUIDE
(TEM MODES NOT INCLUDED)

	TE modes	TM modes
Transverse Helmholtz equation	$\nabla_t^2 \Psi + k_c^2 \Psi = 0$	
Boundary relations	$\dfrac{\partial \Psi^e}{\partial n} = 0 \quad \text{on } C$	$\Psi^m = 0 \quad \text{on } C$
Mode vectors	$\mathbf{e}^e = \mathbf{u}_z \times \nabla_t \Psi^e$ $\mathbf{h}^e = -\nabla_t \Psi^e$	$\mathbf{e}^m = -\nabla_t \Psi^m$ $\mathbf{h}^m = -\mathbf{u}_z \times \nabla_t \Psi^m$
	$\mathbf{e} = \mathbf{h} \times \mathbf{u}_z$ $\mathbf{h} = \mathbf{u}_z \times \mathbf{e}$	
Normalization	$\iint e^2 \, ds = \iint h^2 \, ds = 1$	
Propagation constant	$\gamma = jk_z = \begin{cases} j\beta = jk\sqrt{1-(f_c/f)^2} & f > f_c \\ \alpha = k_c\sqrt{1-(f/f_c)^2} & f < f_c \end{cases}$	
Characteristic Z and Y	$Z_0^e = \dfrac{j\omega\mu}{\gamma} = \dfrac{1}{Y_0^e}$	$Z_0^m = \dfrac{\gamma}{j\omega\epsilon} = \dfrac{1}{Y_0^m}$
Transmission-line equations	$\dfrac{dV}{dz} + \gamma Z_0 I = 0$ $\dfrac{dI}{dz} + \gamma Y_0 V = 0$	
Mode voltage and current	$V = V^+ e^{-\gamma z} + V^- e^{\gamma z}$ $I = \dfrac{1}{Z_0}(V^+ e^{-\gamma z} - V^- e^{\gamma z})$	
Transverse field	$\mathbf{E}_t = \mathbf{e}V$ $\mathbf{H}_t = \mathbf{h}I$	
Longitudinal field	$H_z^e = \dfrac{k_c^2}{j\omega\mu} \Psi^e V^e$	$E_z^m = \dfrac{k_c^2}{j\omega\epsilon} \Psi^m I^m$
z-directed power	$P_z = VI^*$	

circular waveguide of Fig. 5-2, we can pick the Ψ's from Eqs. (5-23) and (5-27) and normalize them. The result is

$$\Psi_{np}{}^e = \sqrt{\frac{\epsilon_n}{\pi[(x'_{np})^2 - n^2]}} \frac{J_n(x'_{np}\rho/a)}{J_n(x'_{np})} \begin{Bmatrix} \sin n\phi \\ \cos n\phi \end{Bmatrix}$$

$$\Psi_{np}{}^m = \sqrt{\frac{\epsilon_n}{\pi}} \frac{J_n(x_{np}\rho/a)}{x_{np} J_{n+1}(x_{np})} \begin{Bmatrix} \sin n\phi \\ \cos n\phi \end{Bmatrix}$$

(8-35)

where $n = 0, 1, 2, \ldots$, and $p = 1, 2, 3, \ldots$. The x_{np} are given by Table 5-2, and the x'_{np} are given by Table 5-3. Normalized eigenfunctions for the parallel-plate guide are given in Prob. 8-1. Normalized eigenfunctions for the coaxial and elliptic waveguides are given by Marcuvitz.[1]

8-2. Modal Expansions in Waveguides. An arbitrary field inside a section of waveguide can be expanded as a sum over all possible modes. This concept was used in Sec. 4-4 for the special case of the rectangular waveguide. We now wish to consider such expansions for cylindrical waveguides in general. The equations in Sec. 8-1 apply to each mode. Henceforth, to identify a particular mode, we shall use the subscript i to denote the mode number.

Let us first show that each mode vector $\mathbf{e}_i$ is orthogonal to all other mode vectors. For this, we shall use the divergence theorem in two dimensions,

$$\iint \nabla_t \cdot \mathbf{A} \, ds = \oint \mathbf{A} \cdot \mathbf{n} \, dl$$

Green's first identity in two dimensions,

$$\iint (\nabla_t \psi \cdot \nabla_t \phi + \psi \nabla_t{}^2 \phi) \, ds = \oint \psi \frac{\partial \phi}{\partial n} \, dl$$

and Green's second identity in two dimensions,

$$\iint (\psi \nabla_t{}^2 \phi - \phi \nabla_t{}^2 \psi) \, ds = \oint \left(\psi \frac{\partial \phi}{\partial n} - \phi \frac{\partial \psi}{\partial n} \right) dl$$

First, consider two TE modes and form the product

$$\mathbf{e}_i{}^e \cdot \mathbf{e}_j{}^e = \mathbf{h}_i{}^e \cdot \mathbf{h}_j{}^e = \nabla_t \Psi_i{}^e \cdot \nabla_t \Psi_j{}^e$$

Letting $\psi = \Psi_i{}^e$ and $\phi = \Psi_j{}^e$ in Green's first identity, we obtain

$$\iint \mathbf{e}_i{}^e \cdot \mathbf{e}_j{}^e \, ds = -(k_{cj}{}^e)^2 \iint \Psi_i{}^e \Psi_j{}^e \, ds$$

Using the same substitution in Green's second identity, we have

$$[(k_{ci}{}^e)^2 - (k_{cj}{}^e)^2] \iint \Psi_i{}^e \Psi_j{}^e \, ds = 0$$

[1] N. Marcuvitz, "Waveguide Handbook," MIT Radiation Laboratory Series, vol. 10, chap. 2, McGraw-Hill Book Company, Inc., New York, 1951.

Hence, if $k_{ci}^e \neq k_{cj}^e$, the integral must vanish, and the preceding equation becomes[1]

$$\iint \mathbf{e}_i^e \cdot \mathbf{e}_j^e \, ds = 0 \qquad i \neq j \tag{8-36}$$

A dual analysis applies to the TM modes, and we have

$$\iint \mathbf{e}_i^m \cdot \mathbf{e}_j^m \, ds = 0 \qquad i \neq j \tag{8-37}$$

Finally, we must consider the TE-TM cross products

$$\mathbf{e}_i^e \cdot \mathbf{e}_j^m = \mathbf{h}_i^e \cdot \mathbf{h}_j^m = -(\mathbf{u}_z \times \nabla_t \Psi_i^e) \cdot \nabla_t \Psi_j^m$$

If we let $\mathbf{A} = \Psi_j^m \mathbf{u}_z \times \nabla_t \Psi_i^e$ in the divergence theorem, the contour integral vanishes because of the boundary conditions, and we obtain

$$\iint \nabla_t \Psi_j^m \cdot \mathbf{u}_z \times \nabla_t \Psi_i^e \, ds = 0$$

Comparing the preceding two equations, we see that

$$\iint \mathbf{e}_i^e \cdot \mathbf{e}_j^m \, ds = 0 \qquad \text{for all } i, j \tag{8-38}$$

The orthogonality relationships [Eqs. (8-36) to (8-38)] also are valid for the **e**'s replaced by the **h**'s.

At any cross section along a cylindrical waveguide, the field can be expressed as a summation over all possible modes:

$$\begin{aligned} \mathbf{E}_t &= \sum_i \mathbf{e}_i^e V_i^e + \mathbf{e}_i^m V_i^m \\ \mathbf{H}_t &= \sum_i \mathbf{h}_i^e I_i^e + \mathbf{h}_i^m I_i^m \end{aligned} \tag{8-39}$$

Because of the orthogonality of the mode vectors, we can determine the mode voltages and/or mode currents at any cross section by multiplying each side of Eqs. (8-39) by an arbitrary mode vector and integrating over the guide cross section. Noting that the mode vectors are normalized, we obtain

$$\begin{aligned} \iint \mathbf{E}_t \cdot \mathbf{e}_i^p \, ds &= V_i^p \\ \iint \mathbf{H}_t \cdot \mathbf{h}_i^p \, ds &= I_i^p \end{aligned} \tag{8-40}$$

where $p = e$ or m. Since there are two independent constants in V and I for each mode, as shown by Eqs. (8-25) and (8-26), we need two "cross-

[1] A discrete spectrum of eigenvalues is assumed. However, orthogonal sets of mode functions for degenerate cases can also be found.

sectional" boundary conditions. These may be (1) matched waveguide and E_t over one cross section, (2) matched waveguide and H_t over one cross section, (3) E_t over two cross sections, (4) H_t over two cross sections, and (5) E_t over one cross section and H_t over another cross section. The solutions of Sec. 4-9 are examples of case (1). Furthermore, when we have currents in a waveguide, we can obtain additional cases involving discontinuities in E_t and/or H_t over waveguide cross sections. The solutions of Sec. 4-10 are examples of this situation.

It is also of interest to note that, when many modes exist simultaneously in a cylindrical waveguide, *each mode propagates energy as if it exists alone.* Hence, the equivalent circuit of a section of waveguide in which N modes exist is N separate transmission lines of the form of Fig. 8-2. To show this power orthogonality, we calculate the z-directed complex power

$$P_z = \iint \mathbf{E} \times \mathbf{H}^* \cdot \mathbf{u}_z \, ds = \iint \left(\sum_i \mathbf{e}_i V_i \right) \times \left(\sum_j \mathbf{h}_j I_j^* \right) \cdot \mathbf{u}_z \, ds$$

$$= \sum_{i,j} V_i I_j^* \iint \mathbf{e}_i \cdot \mathbf{e}_j \, ds = \sum_i V_i I_i^* \qquad (8\text{-}41)$$

which is a summation of the powers carried by each mode. (We have used the indices i and j to order both TE and TM modes in the above proof.) The energy stored per unit length in a waveguide is also the sum of the energies stored in each mode (see Prob. 8-3).

8-3. The Network Concept. In Sec. 3-8, we saw that, given N sets of "circuit" terminals, the voltages at the terminals were related to the currents by an impedance matrix. This impedance matrix was shown to be symmetrical, that is, the usual circuit-theory reciprocity applied if the medium was isotropic. We shall now show that the same network formulation applies if, instead of circuit voltages and currents, the modal voltages and currents of waveguide "ports" are used.

Let Fig. 8-3 represent a general "microwave network," that is, a system for which a closed surface separating the network from the rest of space can be found such that $\mathbf{n} \times \mathbf{E} = 0$ on the surface except over one or more waveguide cross sections. Suppose that only one mode propagates

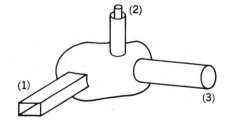

Fig. 8-3. A microwave network.

in each waveguide.[1] Then, assuming we are far enough along each waveguide for higher-order modes to die out, only the dominant mode exists in each guide. A knowledge of the mode V or I in the guide is equivalent to a knowledge of $\mathbf{E}_t$ or $\mathbf{H}_t$, respectively, since the mode vectors depend only on the geometry. Hence, according to the uniqueness concepts of Sec. 3-3, a knowledge of V (or I) in all guides is sufficient to determine I (or V) in all guides. Furthermore, the relationship must be linear if the medium is linear, and an impedance matrix $[z]$ is defined by

$$\begin{bmatrix} V_1 \\ V_2 \\ V_3 \end{bmatrix} = \begin{bmatrix} z_{11} & z_{12} & z_{13} \\ z_{21} & z_{22} & z_{23} \\ z_{31} & z_{32} & z_{33} \end{bmatrix} \begin{bmatrix} I_1 \\ I_2 \\ I_3 \end{bmatrix} \qquad (8\text{-}42)$$

where V_n and I_n are the mode voltage and current in the nth waveguide. The inverse relationship to Eq. (8-42) defines an admittance matrix $[y]$ according to

$$\begin{bmatrix} I_1 \\ I_2 \\ I_3 \end{bmatrix} = \begin{bmatrix} y_{11} & y_{12} & y_{13} \\ y_{21} & y_{22} & y_{23} \\ y_{31} & y_{32} & y_{33} \end{bmatrix} \begin{bmatrix} V_1 \\ V_2 \\ V_3 \end{bmatrix} \qquad (8\text{-}43)$$

Equations (8-42) and (8-43) have been written explicitly for the three-port network of Fig. 8-3 but, of course, can be similarly written for any N-port network. Now that we have established these linear sets of equations, we can use all the usual techniques for solving linear equations. The electrical engineer knows these techniques by the name of "network theory."[2]

It is also of interest to show that, for isotropic media,

$$z_{ij} = z_{ji} \qquad y_{ij} = y_{ji} \qquad (8\text{-}44)$$

that is, microwave networks are reciprocal in the same sense as are the usual lumped-element networks. To prove this, let us apply the Lorentz reciprocity theorem [Eq. (3-34)]. It states that

$$\oint \mathbf{E}^a \times \mathbf{H}^b \cdot d\mathbf{s} = \oint \mathbf{E}^b \times \mathbf{H}^a \cdot d\mathbf{s}$$

for two fields $\mathbf{E}^a$, $\mathbf{H}^a$ and $\mathbf{E}^b$, $\mathbf{H}^b$ in linear, isotropic media. We visualize a surface surrounding an N-port microwave network such that $\mathbf{E}_t = 0$ on S except over the waveguide cross sections, where

$$(\mathbf{E}_t)_n = \mathbf{e}_n V_n \qquad (\mathbf{H}_t)_n = \mathbf{h}_n I_n$$

---

[1] If N modes propagate in a single waveguide, then that guide will be represented by N ports on the equivalent network.

[2] For example, see C. D. Montgomery, R. H. Dicke, and E. M. Purcell (eds.), "Principles of Microwave Circuits," Chap. 4, MIT Radiation Laboratory Series, vol. 8, McGraw-Hill Book Company, Inc., New York, 1948.

(The n here refers to the nth waveguide, not the nth mode.) Hence, the desired surface integrals become

$$\oint \mathbf{E}^a \times \mathbf{H}^b \cdot d\mathbf{s} = \sum_{n=1}^{N} V_n{}^a I_n{}^b \oint \mathbf{e}_n \times \mathbf{h}_n \cdot d\mathbf{s} = \sum_{n=1}^{N} V_n{}^a I_n{}^b$$

and the Lorentz reciprocity theorem reduces to

$$\sum_{n=1}^{N} V_n{}^a I_n{}^b = \sum_{n=1}^{N} V_n{}^b I_n{}^a \tag{8-45}$$

To show that Eq. (8-45) is equivalent to Eqs. (8-44), it is merely necessary to consider the special cases (1) all $I_n{}^a = 0$ except $I_i{}^a$ and (2) all $I_n{}^b = 0$ except $I_j{}^b$. Then $V_j{}^a = z_{ji} I_i{}^a$ and $V_i{}^b = z_{ij} I_j{}^b$, and Eq. (8-45) reduces to $z_{ij} = z_{ji}$. Similarly, taking all $V_n{}^a = 0$ except $V_i{}^a$, and all $V_n{}^b = 0$ except $V_j{}^b$ in Eq. (8-45) establishes $y_{ij} = y_{ji}$.

8-4. One-port Networks. A one-port network is characterized by a single impedance or admittance element. Visualize a surface enclosing the network such that the field is zero on the surface except where it crosses the input guide, as shown in Fig. 8-4. We then have

$$P_{\text{in}} = -\oint \mathbf{E} \times \mathbf{H}^* \cdot d\mathbf{s} = -VI^* \oint \mathbf{e} \times \mathbf{h} \cdot d\mathbf{s} = VI^*$$

where V and I are the mode voltage and current at the "reference plane," that is, at the cross section cut by the surface enclosing the network. Because of the conservation of complex power [Eq. (1-62)], we have

$$VI^* = P_{\text{in}} = \bar{\mathcal{P}}_d + j2\omega(\bar{\mathcal{W}}_m - \bar{\mathcal{W}}_e) \tag{8-46}$$

where $\bar{\mathcal{P}}_d$ is the power dissipated, $\bar{\mathcal{W}}_m$ is the magnetic energy stored, and $\bar{\mathcal{W}}_e$ is the electric energy stored in the network. The input impedance to the network is therefore

$$Z = \frac{P_{\text{in}}}{|I|^2} = \frac{1}{|I|^2}[\bar{\mathcal{P}}_d + j2\omega(\bar{\mathcal{W}}_m - \bar{\mathcal{W}}_e)] \tag{8-47}$$

which is well known for lumped-element network theory. Similarly, the

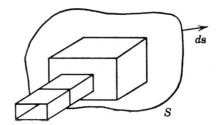

Fig. 8-4. A one-port network and a surface enclosing it.

input admittance is

$$Y = \frac{P_{in}^*}{|V|^2} = \frac{1}{|V|^2}[\bar{\mathcal{P}}_d - j2\omega(\bar{\mathcal{W}}_m - \bar{\mathcal{W}}_e)] \qquad (8\text{-}48)$$

As usual, we define the real and imaginary parts of Z to be resistance and reactance, and the real and imaginary parts of Y to be conductance and susceptance, respectively.

$$Z = R + jX \qquad Y = G + jB \qquad (8\text{-}49)$$

From Eqs. (8-47) to (8-49) we can draw the following conclusions. (1) A dissipationless network has $R = G = 0$. (2) The R and G cannot be negative in the lossy case. (3) At resonance ($X = B = 0$) the electric and magnetic energies are equal. (4) The Z and Y satisfy

$$Z^*(-\omega) = Z(\omega) \qquad Y^*(-\omega) = Y(\omega)$$

and hence R and G are even functions of ω and X and B are odd functions of ω.

In the lossless case, VI^* is imaginary, and hence V must be 90° out of phase with I. We shall now show that everywhere within the network **E** is in phase with V and **H** is in phase with I. Hence, **E** is 90° out of phase with **H**. Suppose we choose our reference plane such that V is real. Then **n** × **E** is real over the reference cross section of the input guide and zero over the rest of the enclosing surface (see Fig. 8-4). These boundary conditions, as well as the field equations

$$\nabla \times \mathbf{E} = -j\omega\mu\mathbf{H} \qquad \nabla \times \mathbf{H} = j\omega\epsilon\mathbf{E} \qquad (8\text{-}50)$$

can be satisfied by assuming **E** real and **H** imaginary. This is therefore a possible solution, and, assuming uniqueness,[1] it must be the only solution.

Let us now consider the effect of a change in frequency. The frequency derivatives of Eqs. (8-50) are

$$\nabla \times \frac{\partial \mathbf{E}}{\partial \omega} = -j\mu\mathbf{H} - j\omega\mu \frac{\partial \mathbf{H}}{\partial \omega}$$
$$\nabla \times \frac{\partial \mathbf{H}}{\partial \omega} = j\epsilon\mathbf{E} + j\omega\epsilon \frac{\partial \mathbf{E}}{\partial \omega} \qquad (8\text{-}51)$$

If we scalarly multiply the first of these by **H*** and the conjugate of the

---

[1] It may be recalled that the uniqueness theorem of Sec. 3-3 required some dissipation for its proof. Hence, our conclusions apply only if we visualize some slight loss. However, even in the loss-free case, any field having **n** × **E** = 0 over the entire boundary would be uncoupled to the input ports, and would have no influence on the external behavior of the network.

second of Eqs. (8-50) by $\partial \mathbf{E}/\partial \omega$, and subtract, we obtain

$$\nabla \cdot \left(\frac{\partial \mathbf{E}}{\partial \omega} \times \mathbf{H}^*\right) = -j\mu |H|^2 - j\omega\mu \frac{\partial \mathbf{H}}{\partial \omega} \cdot \mathbf{H}^* + j\omega\epsilon \mathbf{E}^* \cdot \frac{\partial \mathbf{E}}{\partial \omega}$$

Similarly, if we scalarly multiply the second of Eqs. (8-51) by $\mathbf{E}^*$ and the conjugate of the first of Eqs. (8-50) by $\partial \mathbf{H}/\partial \omega$, and subtract, we obtain

$$\nabla \cdot \left(\frac{\partial \mathbf{H}}{\partial \omega} \times \mathbf{E}^*\right) = j\epsilon |E|^2 + j\omega\epsilon \frac{\partial \mathbf{E}}{\partial \omega} \cdot \mathbf{E}^* - j\omega\mu \mathbf{H}^* \cdot \frac{\partial \mathbf{H}}{\partial \omega}$$

We now subtract the above equation from the preceding one and obtain

$$\nabla \cdot \left(\frac{\partial \mathbf{E}}{\partial \omega} \times \mathbf{H}^* - \frac{\partial \mathbf{H}}{\partial \omega} \times \mathbf{E}^*\right) = -j\mu |H|^2 - j\epsilon |E|^2 \qquad (8\text{-}52)$$

Finally, this equation is integrated throughout a region of space, and the divergence theorem applied to the left-hand term.

$$\oint \left(\frac{\partial \mathbf{E}}{\partial \omega} \times \mathbf{H}^* - \frac{\partial \mathbf{H}}{\partial \omega} \times \mathbf{E}^*\right) \cdot d\mathbf{s} = -j \iiint (\mu |H|^2 + \epsilon |E|^2) \, d\tau \qquad (8\text{-}53)$$

Note that the right-hand side is proportional to the total electromagnetic energy contained within the region.

Equation (8-53) is now applied to the one-port network (Fig. 8-4). The field vanishes over the enclosing surface except where it crosses the input port, and the left-hand side of Eq. (8-53) becomes

$$\iint \left(\frac{\partial V}{\partial \omega} I^* + \frac{\partial I}{\partial \omega} V^*\right) \mathbf{e} \times \mathbf{h} \cdot d\mathbf{s} = -\left(I^* \frac{\partial V}{\partial \omega} + V^* \frac{\partial I}{\partial \omega}\right)$$

where V and I are the mode voltage and current at the input reference plane. Hence, we can write Eq. (8-53) as

$$I^* \frac{\partial V}{\partial \omega} + V^* \frac{\partial I}{\partial \omega} = j \iiint (\mu |H|^2 + \epsilon |E|^2) \, d\tau$$
$$= 2j(\overline{\mathcal{W}}_m + \overline{\mathcal{W}}_e) \qquad (8\text{-}54)$$

The input reactance X and susceptance B are given by

$$jX = \frac{V}{I} = -\frac{j}{B}$$

Their frequency derivatives are therefore

$$\frac{dX}{d\omega} = -\frac{j}{I} \frac{\partial V}{\partial \omega}\bigg|_{I \text{ constant}}$$
$$\frac{dB}{d\omega} = -\frac{j}{V} \frac{\partial I}{\partial \omega}\bigg|_{V \text{ constant}} \qquad (8\text{-}55)$$

Hence, from Eq. (8-54), it follows that

$$\frac{dX}{d\omega} = \frac{2}{|I|^2}(\overline{\mathcal{W}}_m + \overline{\mathcal{W}}_e)$$
$$\frac{dB}{d\omega} = \frac{2}{|V|^2}(\overline{\mathcal{W}}_m + \overline{\mathcal{W}}_e) \qquad (8\text{-}56)$$

Equations (8-56) state that the *slope of the reactance or susceptance for a loss-free one-port network is always positive*. This is known as *Foster's reactance theorem*. From Eqs. (8-47) and (8-48) we also have for loss-free networks

$$X = \frac{2\omega}{|I|^2}(\overline{\mathcal{W}}_m - \overline{\mathcal{W}}_e)$$
$$B = \frac{2\omega}{|V|^2}(\overline{\mathcal{W}}_e - \overline{\mathcal{W}}_m) \qquad (8\text{-}57)$$

Solving Eqs. (8-56) and (8-57) for the energies, we obtain

$$\overline{\mathcal{W}}_e = \frac{|I|^2}{4}\left(\frac{dX}{d\omega} - \frac{X}{\omega}\right) = \frac{|V|^2}{4}\left(\frac{dB}{d\omega} + \frac{B}{\omega}\right)$$
$$\overline{\mathcal{W}}_m = \frac{|I|^2}{4}\left(\frac{dX}{d\omega} + \frac{X}{\omega}\right) = \frac{|V|^2}{4}\left(\frac{dB}{d\omega} - \frac{B}{\omega}\right) \qquad (8\text{-}58)$$

Because the energies are positive, it follows that

$$\frac{dX}{d\omega} > \frac{X}{\omega} \qquad \frac{dB}{d\omega} > \frac{B}{\omega} \qquad (8\text{-}59)$$

that is, the slope of the reactance or susceptance is always greater than the slope of a straight line from the origin to the point of consideration. Relationships (8-56) to (8-59) were first established in lumped-element network theory.[1]

An important consequence of Eqs. (8-56) and (8-57) is that *all poles and zeros of the reactance or susceptance function for a loss-free one-port network are simple*. To prove this, suppose X vanishes at a resonant frequency ω_0. The Taylor series about ω_0 is then

$$X(\omega) = a_1(\omega - \omega_0) + a_2(\omega - \omega_0)^2 + \cdots$$

and $X'(\omega_0) = a_1$, which must be positive by Foster's reactance theorem. Hence, X has a simple zero at ω_0 and $B = 1/X$ has a simple pole at ω_0. Similar reasoning shows that the zeros of B are simple; hence the poles of X are simple. Furthermore, the poles and zeros for the reactance or susceptance function of a loss-free one-port network must alternate along

---

[1] R. M. Foster, A Reactance Theorem, *Bell System Tech. J.*, vol. 3, pp. 259–267, April, 1924.

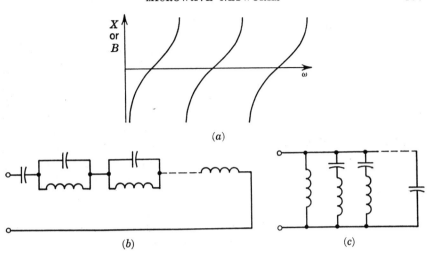

FIG. 8-5. (a) Typical reactance or susceptance function, (b) a Foster equivalent network of the first type, and (c) a Foster equivalent network of the second type.

the ω axis; else $X'(\omega)$ will not always be greater than zero. Figure 8-5a illustrates the general behavior of a reactance or susceptance function. Equivalent circuits for reactance functions of the Foster type[1] are illustrated by Fig. 8-5b and c. Other equivalent circuits of the Cauer type,[1] or of mixed Foster-Cauer type, can be found. An important difference between microwave networks (distributed elements) and lumped-element networks is that the former have infinitely many resonances, while the latter have a finite number of resonances.

The loss-free network is, of course, only an approximation to physical networks. It is therefore desirable to know how the behavior of networks with small losses differs from the behavior of loss-free networks. It is known from the usual network theory that a slight amount of dissipation shifts the poles and zeros of the impedance function from the ω axis to points above it. Hence, the reactance (imaginary part of Z) of a slightly dissipative network would not become infinite for any real ω but would be somewhat like that shown in Fig. 8-6. Also, since $Z(\omega)$ is an analytic function of ω, the resistance (real part of Z) is not independent of X. A study of the resistance corresponding to the reactance of Fig. 8-6 reveals that it would behave somewhat like the dashed curve of Fig. 8-6. An example of a lossy one-port network is the linear antenna of Fig. 7-13, for which the power "loss" is actually radiated power. The effect of small losses can be shown in the equivalent circuits by adding

[1] For example, see M. Van Valkenburg, "Network Analysis," Chap. 12, Prentice-Hall, Inc., Englewood Cliffs, N.J., 1955.

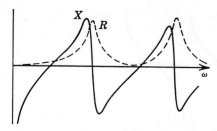

FIG. 8-6. The effect of small losses on the impedance of a microwave network.

large resistances in parallel with the LC resonators of Fig. 8-5b and by adding small resistances in series with the LC resonators of Fig. 8-5c.

8-5. Two-port Networks. The primary uses of two-port networks in microwave theory are (1) transmission of energy from one place to another and (2) filtering of signals from one another. While much of the theory can be presented in terms of the impedance matrix $[z]$, defined by

$$\begin{bmatrix} V_1 \\ V_2 \end{bmatrix} = \begin{bmatrix} z_{11} & z_{12} \\ z_{21} & z_{22} \end{bmatrix} \begin{bmatrix} I_1 \\ I_2 \end{bmatrix} \qquad (8\text{-}60)$$

or in terms of the admittance matrix

$$[y] = [z]^{-1} \qquad (8\text{-}61)$$

it is often more convenient to use other matrices which emphasize the waveguide character of the ports. The port voltages and currents can be considered to be the superposition of incident and reflected components. Hence, for port 1,

$$V_1 = V_1{}^i + V_1{}^r$$
$$I_1 = I_1{}^i + I_1{}^r = \frac{1}{Z_{01}}(V_1{}^i - V_1{}^r) \qquad (8\text{-}62)$$

and similar equations apply to port 2. Figure 8-7 suggests this traveling-wave concept. Mathematically, Eqs. (8-62) are merely a linear transformation from the two quantities V_1, I_1 to $V_1{}^i$, $V_1{}^r$, and it is apparent that Z_{01} can be arbitrarily chosen. However, it is usually convenient to make the natural choice that Z_{01} is the characteristic impedance of the waveguide connected to port 1. Another choice, convenient from a mathematical viewpoint, is to normalize the characteristic impedance by choosing all Z_0's equal to unity. We shall make the former choice.

From the traveling-wave viewpoint, a possible matrix for describing

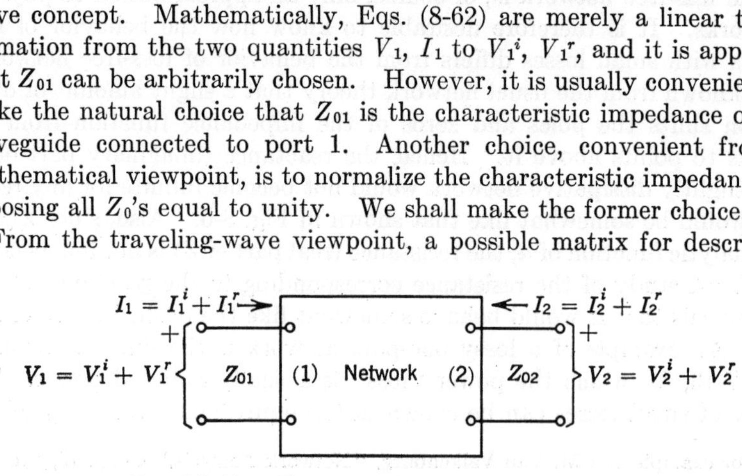

FIG. 8-7. Traveling waves for a two-port network.

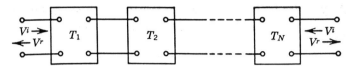

FIG. 8-8. N two-port networks cascaded.

two-port microwave networks is the *transmission matrix* $[T]$, defined by

$$\begin{bmatrix} V_2^r \\ V_2^i \end{bmatrix} = \begin{bmatrix} T_{11} & T_{12} \\ T_{21} & T_{22} \end{bmatrix} \begin{bmatrix} V_1^i \\ V_1^r \end{bmatrix} \tag{8-63}$$

This matrix is particularly convenient when microwave networks are cascaded, as illustrated by Fig. 8-8. The incident and reflected waves at the input of network $n + 1$ are the reflected and incident waves, respectively, at the output of network n. Hence, the T matrix of the over-all network is the product of the T matrices of the individual networks, that is,

$$[T] = [T_N][T_{N-1}] \cdots [T_2][T_1] \tag{8-64}$$

Another matrix commonly used to describe microwave networks is the *scattering matrix* $[S]$ defined by

$$\begin{bmatrix} V_1^r \\ V_2^r \end{bmatrix} = \begin{bmatrix} S_{11} & S_{12} \\ S_{21} & S_{22} \end{bmatrix} \begin{bmatrix} V_1^i \\ V_2^i \end{bmatrix} \tag{8-65}$$

This matrix is convenient for considerations of impedance matching. It can also be easily extended to the case of multiport networks. Note that S_{11} is the reflection coefficient seen at port 1 when port 2 is matched and S_{22} is the reflection coefficient seen at port 2 when port 1 is matched.

The various matrices defined for a two-port network are, of course, related to one another. For example, $[y]$ is the inverse of $[z]$, as stated by Eq. (8-61). The relationship of $[S]$ to $[z]$ is more complicated. Defining the matrix

$$[z_0] = \begin{bmatrix} Z_{01} & 0 \\ 0 & Z_{02} \end{bmatrix}$$

we have $\qquad [S] = [z - z_0][z + z_0]^{-1} \tag{8-66}$

Similarly, the transmission matrix is related to the scattering matrix by

$$[T] = \begin{bmatrix} S_{21} - \dfrac{S_{22}S_{11}}{S_{12}} & \dfrac{S_{22}}{S_{12}} \\ -\dfrac{S_{11}}{S_{12}} & \dfrac{1}{S_{12}} \end{bmatrix} \tag{8-67}$$

The derivation of Eqs. (8-66) and (8-67), along with other relationships among the various matrices, can be found in vol. 8 of the Radiation

Laboratory Series.[1] For networks constructed of linear isotropic matter, the reciprocity relationships [Eqs. (8-44)] apply. From Eq. (8-66), it is evident that reciprocity requires

$$S_{ij} = S_{ji} \tag{8-68}$$

in the scattering matrix. From Eq. (8-67), it follows that reciprocity requires

$$T_{11}T_{22} - T_{12}T_{21} = \frac{Z_{02}}{Z_{01}} \tag{8-69}$$

in the transmission matrix. Equations (8-66) and (8-68) also apply to multiport networks.

There are realizability conditions imposed on the matrices by the conservation of energy theorem. These conditions can be obtained from the corresponding one-port conditions by terminating the two-port network in various ways to form a one-port. For example, if port 2 is open-circuited ($I_2 = 0$), then z_{11} is the input impedance. Similarly, when port 1 is open-circuited, z_{22} is the input impedance looking from port 2. Hence, by Eq. (8-47) we know

$$\operatorname{Re}(z_{11}) \geq 0 \qquad \operatorname{Re}(z_{22}) \geq 0 \tag{8-70}$$

Similarly, using the y matrix and short circuits on the ports, we obtain from Eq. (8-48) that

$$\operatorname{Re}(y_{11}) \geq 0 \qquad \operatorname{Re}(y_{22}) \geq 0 \tag{8-71}$$

More generally, since Eqs. (8-47) and (8-48) must be valid for *any* passive termination, we can show that

$$\begin{aligned} \operatorname{Re}(z_{11})\operatorname{Re}(z_{22}) - \operatorname{Re}(z_{12})\operatorname{Re}(z_{21}) \geq 0 \\ \operatorname{Re}(y_{11})\operatorname{Re}(y_{22}) - \operatorname{Re}(y_{12})\operatorname{Re}(y_{21}) \geq 0 \end{aligned} \tag{8-72}$$

Finally, when the network is loss-free, the elements of the impedance and admittance matrices become imaginary, and restrictions on them can be obtained from the corresponding restrictions in the one-port case. Such considerations are particularly useful in the theory of filters.[2]

Our principal concern for the remainder of this chapter will be to obtain equivalent circuits for microwave networks. For any particular network, an infinite number of equivalent circuits will exist. One of our tasks will be to choose a "natural" equivalent circuit, that is, one which suggests the physical nature of the network. For example, a section of

---

[1] C. D. Montgomery, R. H. Dicke, and E. M. Purcell (eds.), "Principles of Microwave Circuits," Chap. 4, MIT Radiation Laboratory Series, vol. 8, McGraw-Hill Book Company, Inc., New York, 1948.

[2] M. Van Valkenburg, "Network Analysis," Chap. 13, Prentice-Hall, Inc., Englewood Cliffs, N.J., 1955.

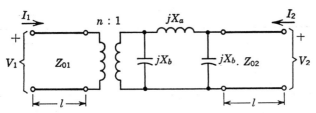

FIG. 8-9. A typical equivalent circuit for a loss-free two-port microwave network.

waveguide would not be represented by an equivalent tee or pi circuit, since this would hide the transmission-line character of the guide. For loss-free networks, we shall use the symbolism of Table 8-2 in equivalent circuits. It should be emphasized that it is only the *sign* of a reactance or susceptance that dictates whether an inductor or capacitor is chosen. The reactance or susceptance does *not*, in general, have the simple frequency dependence of a lumped-element inductor or capacitor. Figure 8-9 illustrates a typical equivalent circuit for a loss-free two-port network.

TABLE 8-2. SYMBOLISM USED IN EQUIVALENT CIRCUITS OF LOSS-FREE NETWORKS

Element	Symbol	Represents
Inductor	—⟿— jX	Positive reactance
	—⟿— jB	Negative susceptance
Capacitor	—⊣⊢— jX	Negative reactance
	—⊣⊢— jB	Positive susceptance
Ideal transformer	$n:1$	Change in impedance level
Transmission line	Z_0, l	Waveguide section

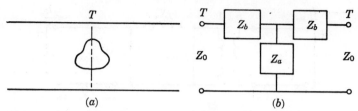

Fig. 8-10. (a) A symmetrical obstacle in a cylindrical waveguide, and (b) an equivalent circuit.

In the case of dissipative networks, resistors in series with X or in parallel with B can be used to represent the losses. Similarly, the characteristic impedances and propagation constants of the equivalent transmission lines can be assumed complex to account for losses. Most of the networks used in microwave practice are only slightly lossy, and the small losses introduce only second-order corrections to the reactances calculated on a loss-free basis.

8-6. Obstacles in Waveguides. An object in a cylindrical waveguide can be represented as a two-port network. Figure 8-10a shows an obstacle, symmetric about the cross section T, in a waveguide. Figure 8-10b shows a possible equivalent circuit. In the more general case of an unsymmetrical object, the two Z_b's would probably be different from each other, and it might even be desirable to choose two reference planes T. In the loss-free case, the Z's will all be jX's.

Before considering the obstacle problem, let us consider "dominant-mode sources" in cylindrical waveguides. Figure 8-11 shows the electric source $\mathbf{J}_s$ in a waveguide terminated at $z = 0$ by a magnetic conductor and matched as $z \to -\infty$. The method of treating this problem is that used in Sec. 3-1 for rectangular guides, as, for example, Fig. 3-2. Let superscripts (1) denote the region $-l < z < 0$, and superscripts (2) denote the region $z < -l$. Then in region 1 there will be an incident wave plus a reflected wave such that $\mathbf{H}_t = 0$ at $z = 0$. Hence,

$$\mathbf{E}_t^{(1)} = A(e^{-j\beta z} + e^{j\beta z})\mathbf{e} = 2A \cos(\beta z)\,\mathbf{e}$$
$$\mathbf{H}_t^{(1)} = \frac{A}{Z_0}(e^{-j\beta z} - e^{j\beta z})\mathbf{h} = \frac{2A}{jZ_0} \sin(\beta z)\,\mathbf{h} \qquad (8\text{-}73)$$

where $\mathbf{e}$ and $\mathbf{h}$ are the mode vectors, β is the phase constant, and Z_0 is the characteristic impedance, all of the dominant mode (see Table 8-1). In region (2) there will be only a wave in the $-z$ direction; hence

$$\mathbf{E}_t^{(2)} = Be^{j\beta z}\,\mathbf{e}$$
$$\mathbf{H}_t^{(2)} = \frac{-B}{Z_0} e^{j\beta z}\,\mathbf{h}$$

Continuity of $\mathbf{E}_t$ at $z = -l$ requires that

$$2A \cos \beta l = Be^{-j\beta l}$$

which determines B in terms of A. The boundary condition on $\mathbf{H}$ at $z = -l$ is

$$\mathbf{u}_z \times [\mathbf{H}^{(1)} - \mathbf{H}^{(2)}] = \mathbf{J}_s$$

which leads to

$$\mathbf{J}_s = -\frac{2A}{Z_0} e^{j\beta l} \mathbf{e}$$

A quantity of interest to us is the self-reaction of the current sheet

$$\langle s,s \rangle = \iint \mathbf{E} \cdot \mathbf{J}_s \, ds = -\frac{2A^2}{Z_0}(1 + e^{j2\beta l}) \tag{8-74}$$

We shall use dominant-mode current sheets as mathematical "waveguide probes" to determine the equivalent circuit impedances.

Now return to the original problem, Fig. 8-10a. We define *even excitation* of the waveguide as the case of equal incident waves from both $z < 0$ and $z > 0$, phased so that $\mathbf{E}_t$ is maximum and $\mathbf{H}_t$ is zero at $z = 0$. By symmetry arguments, the $\mathbf{H}_t$ scattered by the obstacle will also be zero in the $z = 0$ cross section; so a magnetic conductor can be placed over the $z = 0$ plane without changing the field. This divides the problem into two isolated parts, one of which is shown in Fig. 8-12a. The excitation is provided by the dominant-mode source $\mathbf{J}_s$, which we have just analyzed. The equivalent circuit of Fig. 8-12a is shown in Fig. 8-12b. (The magnetic conductor is equivalent to an open circuit, and the $\mathbf{J}_s$ is equivalent to a shunt current source I.)

We now further restrict the problem to the loss-free case. Then the dominant mode will be a pure standing wave in the region $-l < z < 0$ of Fig. 8-12a. If $\mathbf{J}_s$ is located where $\mathbf{E}_t = 0$, then by the usual transmission-line formulas

$$\frac{Z}{Z_0} = \frac{Z_b + 2Z_a}{Z_0} = -j \tan \beta l \tag{8-75}$$

For the source of arbitrary l, the total reaction on $\mathbf{J}_s$ is

$$\text{Reaction} = \iint \mathbf{E} \cdot \mathbf{J}_s \, ds = \iint (\mathbf{E}^s + \mathbf{E}^c) \cdot \mathbf{J}_s \, ds$$
$$= \langle s,s \rangle + \langle c,s \rangle$$

where $\mathbf{E}^s$ is the field of $\mathbf{J}_s$ alone, and $\mathbf{E}^c$ is the field of the current on the

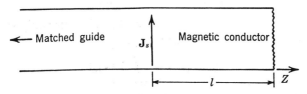

FIG. 8-11. A dominant-mode source in a waveguide terminated by a magnetic conductor.

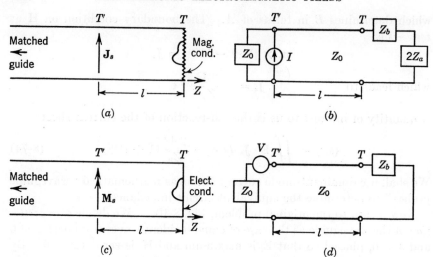

Fig. 8-12. Even excitation of Fig. 8-10a is represented by (a), which has an equivalent network (b). Odd excitation of Fig. 8-10a is represented by (c), which has an equivalent network (d).

obstacle alone, both radiating in the waveguide terminated by the magnetic conductor at $z = 0$. If l is adjusted to a cross section for which $\mathbf{E}_t = 0$, then the reaction vanishes and the above equation becomes

$$\langle c,s \rangle = -\langle s,s \rangle = \frac{2A^2}{Z_0}(1 + e^{j2\beta l})$$

where the last equality is Eq. (8-74). Taking A as real, we have

$$\operatorname{Re}\langle c,s\rangle = \frac{2A^2}{Z_0}(1 + \cos 2\beta l)$$

$$\operatorname{Im}\langle c,s\rangle = \frac{2A^2}{Z_0}(\sin 2\beta l)$$

and, using the identity

$$\tan\frac{\alpha}{2} = \frac{\sin\alpha}{1 + \cos\alpha}$$

Eq. (8-75) becomes

$$\frac{X_b + 2X_a}{Z_0} = -\frac{\operatorname{Im}\langle c,s\rangle}{\operatorname{Re}\langle c,s\rangle} \qquad (8\text{-}76)$$

We have replaced the Z_a and Z_b by jX_a and jX_b because only the loss-free case is being considered. By reciprocity,

$$\langle c,s\rangle = \langle s,c\rangle = \int_{\text{obst}} \mathbf{E}^s \cdot d\mathbf{J}^c \qquad (8\text{-}77)$$

where $\mathbf{E}^s$ is the incident field, given by Eq. (8-73), and $\mathbf{J}^c$ is the current on the obstacle.[1] Note that the problem is now identical to the echo problems of Secs. 7-10 and 7-11, except that all currents radiate in the environment of the waveguide plus the magnetic conductor.

For the case of a perfectly conducting object, the obstacle current is a surface current $\mathbf{J}_s^c$, and $\mathbf{n} \times \mathbf{E} = 0$ on its boundary. Hence,

$$\mathbf{n} \times \mathbf{E}^s = -\mathbf{n} \times \mathbf{E}^c$$

and
$$\langle s,c \rangle = -\langle c,c \rangle = -\iint \mathbf{E}^c \cdot \mathbf{J}_s^c \, ds \qquad (8\text{-}78)$$

where $\langle c,c \rangle$ represents the self-reaction of the currents induced on the obstacle. By Eqs. (8-76) to (8-78), we therefore have

$$\frac{X_b + 2X_a}{Z_0} = -\frac{\operatorname{Im} \langle c,c \rangle}{\operatorname{Re} \langle c,c \rangle} \qquad (8\text{-}79)$$

Our problem is now one of finding the self-reaction of the currents induced by the incident field of Eq. (8-73) with A real.

For a stationary formula, we assume currents $\mathbf{J}_s^a$ on the obstacle and calculate $\langle a,a \rangle$ subject to the constraints

$$\langle a,a \rangle = \langle c,a \rangle = \langle a,c \rangle$$

(see Sec. 7-7). The last equality is met by reciprocity, and, since $\mathbf{n} \times \mathbf{E}^s = -\mathbf{n} \times \mathbf{E}^c$ on the obstacle surface,

$$\langle c,a \rangle = -\langle s,a \rangle$$

Hence, our stationary formula for $\langle c,c \rangle$ is

$$\langle c,c \rangle \approx \frac{\langle s,a \rangle^2}{\langle a,a \rangle} \qquad (8\text{-}80)$$

This, coupled with Eq. (8-79), represents the variational solution to the problem. If the trial current is taken as real, then $\langle s,a \rangle$ is real because $\mathbf{E}^s$ is real. Equation (8-80) can then be written as

$$\langle c,c \rangle \approx \left| \frac{\langle s,a \rangle}{\langle a,a \rangle} \right|^2 \langle a,a \rangle^*$$

and Eq. (8-79) becomes

$$\frac{X_b + 2X_a}{Z_0} \approx \frac{\operatorname{Im} \langle a,a \rangle}{\operatorname{Re} \langle a,a \rangle} \qquad (8\text{-}81)$$

This formula applies only when $\mathbf{J}_s^a$ is real, which is usually the case. The change of sign in going from Eq. (8-79) to Eq. (8-81) can be explained by noting that $\mathbf{J}_s^c$ is *not* real for the given $\mathbf{E}^s$, but is usually at some constant phase.

---

[1] The obstacle may be a conductor, a nonmagnetic dielectric, or a magnetic dielectric ($\mu \neq \mu_0$). In the latter case the term $-\int \mathbf{H}^s \cdot d\mathbf{M}^c$ must be added to the right-hand side of Eq. (8-77).

We define *odd excitation* of the waveguide (Fig. 8-10a) as the case of equal incident waves from both $z < 0$ and $z > 0$, phased so that $\mathbf{E}_t = 0$ and $\mathbf{H}_t$ is maximum at $z = 0$. By symmetry, the $\mathbf{E}_t$ scattered by the obstacle must also be zero in the $z = 0$ cross section, and so an electric conductor can be placed over the $z = 0$ plane without changing the field. This divides the problem into two isolated parts, one of which is shown in Fig. 8-12c. The excitation is provided by a dominant-mode magnetic source $\mathbf{M}_s$, which, together with the electric conductor covering the $z = 0$ plane, is dual to Fig. 8-11. The equivalent circuit of Fig. 8-12c is shown in Fig. 8-12d. (The electric conductor at $z = 0$ is equivalent to a short circuit, and the $\mathbf{M}_s$ is equivalent to a series voltage source V.)

The analysis of Fig. 8-12c is dual to that used for Fig. 8-12a. Hence, dual to Eqs. (8-73), in the region $-l < z < 0$ we have a source field

$$\mathbf{H}_t^s = 2C \cos(\beta z)\, \mathbf{h}$$
$$\mathbf{E}_t^s = -\frac{2C}{jY_0} \sin(\beta z)\, \mathbf{e} \qquad (8\text{-}82)$$

where $Y_0 = 1/Z_0$ is the characteristic admittance of the dominant mode. Dual to Eq. (8-79) we have

$$\frac{j}{Y_0 Z_b} = \frac{1}{Y_0 X_b} = \frac{\operatorname{Im}\langle c,c\rangle}{\operatorname{Re}\langle c,c\rangle} \qquad (8\text{-}83)$$

where $\langle c,c\rangle$ is the self-reaction of the obstacle currents radiating in the presence of an *electric* conductor over the $z = 0$ cross section (see Fig. 8-12c). Finally, for a variational solution, currents $\mathbf{J}_s^a$ are assumed on the obstacle, and their self-reaction $\langle a,a\rangle$ is calculated. If the $\mathbf{J}_s^a$ is *real*, then dual to Eq. (8-81) we have

$$\frac{1}{Y_0 X_b} \approx -\frac{\operatorname{Im}\langle a,a\rangle}{\operatorname{Re}\langle a,a\rangle} \qquad (8\text{-}84)$$

where $\langle a,a\rangle$ is calculated with an electric conductor over the $z = 0$ plane.

8-7. Posts in Waveguides. Some variational solutions for circular posts in rectangular waveguides can be carried out relatively simply. Figure 8-13 illustrates three classes of obstacles: (1) those cylindrical to y,

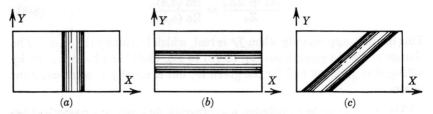

Fig. 8-13. Posts in a rectangular waveguide, (a) cylindrical to y, (b) cylindrical to x, and (c) otherwise.

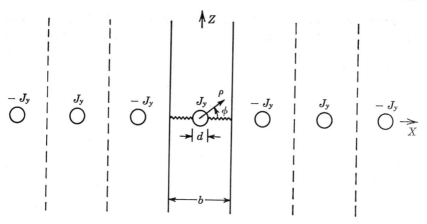

FIG. 8-14. Image system for the circular post in a rectangular waveguide.

(2) those cylindrical to x, and (3) all other cases. [The cylinders are not necessarily circular, and case (1) is different from case (2) only because of the excitation.] It is assumed that the incident wave in each case is the dominant mode with $\mathbf{E}$ parallel to y and $\mathbf{H}_t$ parallel to x. Then the field of case (1) will be TM to y, expressible in terms of a single wave function $A_y = \psi$ (see Sec. 4-4). The field of case (2) will be TE to x, expressible in terms of a single wave function $F_x = \psi$. Type (3) problems require two scalar wave functions to express the field (see Sec. 3-12).

We shall consider only the centered circular post, as shown in the insert of Fig. 8-15. For even excitation (Fig. 8-12a), assume a constant current on the post

$$\mathbf{J}_s{}^a = \mathbf{u}_y \frac{I}{\pi d} \tag{8-85}$$

The field produced by $\mathbf{J}_s{}^a$ in the waveguide closed by the magnetic conductor will be the same as the free-space field from the image system of Fig. 8-14. Hence, we can write

$$E_y{}^a = E_y{}^{\text{post}} + E_y{}^{\text{images}}$$

where the first term is the free-space field of $\mathbf{J}_s{}^a$ and the second term is the free-space field from all its images. The self-reaction of $\mathbf{J}_s{}^a$ in the waveguide with magnetic conductor is one-half that for the complete post in a waveguide; hence

$$\langle a,a \rangle = \frac{1}{2} \int_0^a dy \int_0^{2\pi} \frac{d}{2} d\phi \, (J_y E_y)$$
$$= \frac{aI}{4\pi} \int_0^{2\pi} (E_y{}^{\text{post}} + E_y{}^{\text{images}}) \, d\phi \tag{8-86}$$

Now the "post" term is independent of ϕ since the $J_s{}^a$ is independent of ϕ. The "image" term is a source-free field in the vicinity of the post and can therefore be expressed as

$$E_y{}^{\text{images}} = \sum_{n=-\infty}^{\infty} A_n J_n(k\rho) e^{jn\phi}$$

(see Sec. 5-8). Thus,

$$\int_0^{2\pi} E_y{}^{\text{images}} d\phi = 2\pi A_0 J_0\left(k\frac{d}{2}\right) = 2\pi J_0\left(k\frac{d}{2}\right) E_y{}^{\text{images}}\bigg|_{\rho=0}$$

and Eq. (8-86) reduces to

$$\langle a,a \rangle = \frac{aI}{2}\left[E_y{}^{\text{post}}\bigg|_{\rho=\frac{d}{2}} + J_0\left(k\frac{d}{2}\right) E_y{}^{\text{images}}\bigg|_{\rho=0} \right] \qquad (8\text{-}87)$$

The field of a single cylinder of constant current was calculated in Sec. 5-6. Abstracting from Eq. (5-92), we have

$$E_y{}^{\text{post}} = -\frac{\eta}{4} kIJ_0\left(k\frac{d}{2}\right) H_0{}^{(2)}(k\rho) \qquad \rho \geq \frac{d}{2}$$

The field from each image is also of the above form, with ρ replaced by the distance to the image. Hence, Eq. (8-87) becomes

$$\langle a,a \rangle = K\left[H_0{}^{(2)}\left(k\frac{d}{2}\right) + J_0\left(k\frac{d}{2}\right) 2\sum_{n=1}^{\infty} (-1)^n H_0{}^{(2)}(nkb) \right] \qquad (8\text{-}88)$$

where $\quad K = -\dfrac{\eta}{8} kaI^2 J_0\left(k\dfrac{d}{2}\right)$

is an unimportant constant. Equation (8-88) is an exact evaluation of $\langle a,a \rangle$ for the assumed current of Eq. (8-85).

Unfortunately, the Hankel function summation in Eq. (8-88) converges slowly and is not convenient for computation. However, we shall now show that it can be transformed to

$$\sum_{n=1}^{\infty} (-1)^n H_0{}^{(2)}(nkb) = \frac{2}{\pi}\left[\frac{1}{\sqrt{(2b/\lambda)^2 - 1}} - \frac{\pi}{4} \right.$$
$$\left. + j\left(\frac{1}{2}\log\frac{2\gamma b}{\lambda} - 1 + S \right) \right] \qquad (8\text{-}89)$$

where $\gamma = 1.781$ and S is the rapidly convergent summation

$$S\left(\frac{b}{\lambda}\right) = \sum_{n=3,5,7,\ldots}^{\infty} \left[\frac{1}{\sqrt{n^2 - (2b/\lambda)^2}} - \frac{1}{n} \right] \qquad (8\text{-}90)$$

The free-space field of a filament of current is given by Eq. (5-84). Hence, the left-hand side of Eq. (8-89) is the E_y from all images of the filament

$$I = \frac{-2}{\eta k}$$

across the center of the original waveguide. (This problem is Fig. 8-14 with $J_s{}^a$ replaced by the above I.) Then, by the method of Sec. 4-10, we can find the *total* field in the $z = 0$ cross section due to the above I. It is

$$E_y{}^{\text{total}} = \frac{2}{\pi}\left[\frac{\sin(\pi x/b)}{\sqrt{(2b/\lambda)^2 - 1}} + j\sum_{n=2}^{\infty}\frac{\sin(n\pi/2)\sin(n\pi x/b)}{\sqrt{n^2 - (2b/\lambda)^2}}\right] \quad (8\text{-}91)$$

where only the first term is real because it is assumed that $1 < (2b/\lambda) < 2$. For large n, the above summation has terms equal to those of

$$\sum_{n=1}^{\infty}\frac{1}{n}\sin\left(\frac{n\pi}{2}\right)\sin\left[n\left(\frac{\pi}{2}+\delta\right)\right] = \sum_{n=1,3,5,\ldots}^{\infty}\frac{\cos n\delta}{n}$$

$$= \operatorname{Re}\sum_{n=1,3,5,\ldots}^{\infty}\frac{(e^{j\delta})^n}{n} = \operatorname{Re}\left(\frac{1}{2}\log\frac{1+e^{j\delta}}{1-e^{j\delta}}\right) = \operatorname{Re}\left(\frac{1}{2}\log\frac{j\sin\delta}{1-\cos\delta}\right)$$

$$= \operatorname{Re}\left(\frac{1}{2}\log\frac{j}{\tan(\delta/2)}\right) = -\frac{1}{2}\log\tan\frac{\delta}{2}$$

Hence, letting $x = (b/2) + \rho$ in Eq. (8-91) and $\delta = \pi\rho/b$ in the above identity, we can add and subtract the latter from the former and obtain

$$E_y{}^{\text{total}} \xrightarrow[\rho\to 0]{} \frac{2}{\pi}\left[\frac{1}{\sqrt{(2b/\lambda)^2 - 1}} + j\left(\frac{1}{2}\log\frac{2b}{\pi\rho} - 1 + S\right)\right]$$

The free-space E_y from the same filament I is

$$E_y = \tfrac{1}{2}H_0^{(2)}(k\rho) \xrightarrow[\rho\to 0]{} \tfrac{1}{2} + j\frac{1}{\pi}\log\frac{2}{\gamma k\rho}$$

When this is subtracted from the total E_y, and ρ set equal to zero, we have the right-hand side of Eq. (8-89).

Returning now to the self-reaction, we substitute Eq. (8-89) into Eq. (8-88) and obtain

$$\operatorname{Re}\langle a,a\rangle = C\frac{2}{\sqrt{(2b/\lambda)^2 - 1}} = C\frac{\lambda_g}{b}$$

$$\operatorname{Im}\langle a,a\rangle = C\left[-\frac{\pi N_0(kd/2)}{2J_0(kd/2)} + \log\frac{2\gamma b}{\lambda} - 2 + 2S\right] \quad (8\text{-}92)$$

where C is the unimportant constant,

$$C = -\frac{\eta k a}{4\pi} I^2 J_0{}^2\left(k\frac{d}{2}\right)$$

Equation (8-92) is still exact for the current assumed in Eq. (8-85). However, because of the crudeness of our initial trial current, we can expect our result to be valid only for small d/λ. Hence, we use small-argument formulas for the Bessel functions and obtain

$$\operatorname{Im}\langle a,a\rangle \approx C\left(\log\frac{4b}{\pi d} - 2 + 2S\right) \qquad (8\text{-}93)$$

Now, substituting from Eqs. (8-92) and (8-93) into Eq. (8-81), we have

$$\frac{X_b + 2X_a}{Z_0} \approx \frac{b}{\lambda_g}\left[\log\frac{4b}{\pi d} - 2 + 2S\left(\frac{b}{\lambda}\right)\right] \qquad (8\text{-}94)$$

where S is given by Eq. (8-90).

For odd excitation (Fig. 8-12c), we assume a current

$$\mathbf{J}_s{}^a = \mathbf{u}_y \sin\phi \qquad (8\text{-}95)$$

induced on the post. The appropriate variational formula is Eq. (8-84),

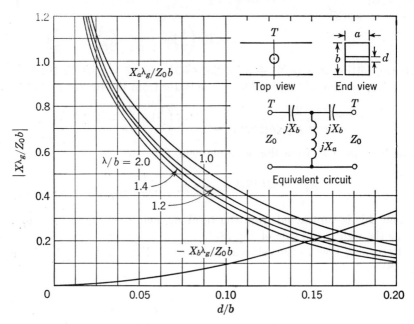

Fig. 8-15. The centered circular inductive post in a rectangular waveguide. (*After Marcuvitz.*)

the exact evaluation of which follows steps similar to those used to derive Eq. (8-94). The result is

$$\frac{X_b}{Z_0} \approx -\frac{b}{\lambda_g}\left(\frac{\pi d}{b}\right)^2 \qquad (8\text{-}96)$$

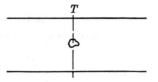

Fig. 8-16. A small obstacle in a waveguide.

Figure 8-15 shows X_a and X_b as calculated from a second-order variational solution.[1] Our solution [Eqs. (8-94) and (8-96)] is accurate for small d/b, the error being of the order of 10 per cent for $d/b = 0.15$. Formulas and calculations for off-centered posts are also available.[1] A solution for the circular capacitive post (Fig. 8-13b) is given in Prob. 8-12.

8-8. Small Obstacles in Waveguides. Figure 8-16 represents a small obstacle in a waveguide of arbitrary cross section. If the obstacle is symmetrical about a transverse plane, the equivalent circuit is as shown in Fig. 8-10b. If the obstacle is loss-free, the Z's are jX's. The formulation of the problem for a conducting obstacle is that of Sec. 8-6. An approximate evaluation of the reactions, made possible because the obstacles are small and not too near the guide walls, will now be discussed.

Consider even excitation of the guide (Fig. 8-12a). The effect of a small obstacle is small; hence Z_b is small and Z_a is large. Equation (8-81) is then

$$\frac{X_a}{Z_0} \approx \frac{1}{2} \frac{\text{Im}\,\langle a,a \rangle}{\text{Re}\,\langle a,a \rangle} \qquad (8\text{-}97)$$

where $\langle a,a \rangle$ is the self-reaction of the assumed currents in the waveguide.

Let us first make some qualitative observations. In a rectangular waveguide, the reaction $\langle a,a \rangle$ is the free-space self-reaction of the obstacle plus the mutual reaction with all its images. For real current, the imaginary part of the free-space self-reaction becomes extremely large as the obstacle becomes small. Hence, for sufficiently small obstacles, we can let

$$\text{Im}\,\langle a,a \rangle \approx \text{Im}\,\langle a,a \rangle_{\text{free space}} \qquad (8\text{-}98)$$

In contrast to this, the real part of the free-space reaction approaches a constant, independent of the size of the obstacle, as the obstacle becomes small. The mutual reaction between the obstacle and its images therefore cannot be neglected. However, because the real part of the reaction is independent of the size and shape of the obstacle, we can calculate the dipole moment Il of the free-space obstacle and let

$$\text{Re}\,\langle a,a \rangle \approx \text{Re}\,\langle Il,Il \rangle \qquad (8\text{-}99)$$

---

[1] N. Marcuvitz, "Waveguide Handbook," MIT Radiation Laboratory Series, vol. 10, pp. 257–263, McGraw-Hill Book Company, Inc., New York, 1951.

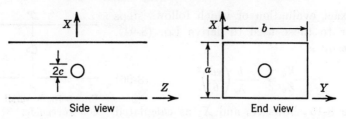

FIG. 8-17. A small conducting sphere centered in a rectangular waveguide.

The right-hand term represents the self-reaction of a current element Il in the waveguide.

As an example, consider the small sphere of radius c in the center of a rectangular waveguide, as shown in Fig. 8-17. As our trial current, assume $\mathbf{J}_s{}^a$ is that which produces the dipole field external to the sphere. This current, even though we shall not need it explicitly, is approximately

$$\mathbf{J}_s{}^a = -\mathbf{u}_\theta \frac{Il}{2\pi c^2} \sin\theta \qquad (8\text{-}100)$$

where θ is measured from the x direction. Because the above current produces the same field as an x-directed element of moment Il, the imaginary part of the free-space self-reaction is the imaginary part of Eq. (2-115) evaluated at $r = c$. Hence,

$$\operatorname{Im}\langle a,a\rangle_{\text{free space}} = -\eta \frac{2\pi}{3}\left(\frac{Il}{\lambda}\right)^2\left(\frac{1}{kc}\right)^3$$

Equation (8-98) is therefore

$$\operatorname{Im}\langle a,a\rangle \approx \frac{\eta\lambda(Il)^2}{12\pi^2 c^3} \qquad (8\text{-}101)$$

For the real part of $\langle a,a\rangle$, we can use the analysis of Sec. 4-10 for a current sheet

$$J_x = Il\,\delta\!\left(x - \frac{a}{2}\right)\delta\!\left(y - \frac{b}{2}\right)$$

Because the current is real, we can set $\operatorname{Re}\langle Il,Il\rangle = -\operatorname{Re}(P)$ of Eq. (4-87) and obtain

$$\operatorname{Re}\langle Il,Il\rangle = -\frac{ab}{4} Z_0 (J_{01})^2$$

where, from Eq. (4-86),

$$J_{01} = \frac{2}{ab} Il$$

Hence, Eq. (8-99) becomes

$$\operatorname{Re}\langle a,a\rangle \approx -\frac{Z_0}{ab}(Il)^2 = -\frac{\eta\lambda_g}{ab\lambda}(Il)^2 \qquad (8\text{-}102)$$

Substituting from Eqs. (8-101) and (8-102) into Eq. (8-97), we have

$$\frac{X_a}{Z_0} \approx - \frac{\lambda^2 ab}{24\pi^2 \lambda_g c^3} \tag{8-103}$$

This is the small-obstacle approximation for a centered sphere in a rectangular waveguide. Our free-space reaction is the Rayleigh approximation [Eq. (6-106)], which is valid for $c/\lambda < 0.1$. Hence, we should expect Eq. (8-103) to be accurate when $c/\lambda < 0.1$ and $c \ll a/2$.

Now consider odd excitation of the guide (Fig. 8-12c). The evaluation of X_b can then be made according to Eq. (8-84). Taking the current as real, we evaluate the imaginary part of $\langle a,a \rangle$ according to the free-space approximation [Eq. (8-98)]. However, because of the symmetry of the obstacle and of the excitation, there can be no net electric dipole moment, and Eq. (8-99) does not apply. There will be a magnetic moment Kl (unless the obstacle has zero axial thickness), which can be calculated from the assumed current. Then, analogous to Eq. (8-99), we use the approximation

$$\mathrm{Re}\,\langle a,a \rangle \approx \mathrm{Re}\,\langle Kl, Kl \rangle \tag{8-104}$$

where the right-hand term represents the self reaction of a magnetic current element Kl in the waveguide.

Return now to the specific problem of a conducting sphere in a rectangular guide (Fig. 8-17). It is evident from symmetry that, for odd excitation, the resultant magnetic dipole will be y-directed. For the trial current, assume that which produces the magnetic dipole field external to the sphere. The free-space self-reaction of this current is then just the dual of that for the electric dipole, given by Eq. (8-101). Hence,

$$\mathrm{Im}\,\langle a,a \rangle \approx \mathrm{Im}\,\langle a,a \rangle_{\text{free space}} = \frac{\lambda (Kl)^2}{\eta 12\pi^2 c^3} \tag{8-105}$$

For the real part of $\langle a,a \rangle$, we evaluate the right-hand side of Eq. (8-104) by methods dual to those used to establish Eq. (8-102). For the centered y-directed magnetic current element in the rectangular guide, we obtain

$$\mathrm{Re}\,\langle a,a \rangle \approx \mathrm{Re}\,\langle Kl, Kl \rangle = \frac{Y_0}{ab}(Kl)^2 = \frac{\lambda}{ab\eta\lambda_g}(Kl)^2$$

Substituting from this and from Eq. (8-105) into Eq. (8-84), we have

$$\frac{Z_0}{X_b} \approx - \frac{ab\lambda_g}{12\pi^2 c^3} \tag{8-106}$$

The accuracy of this formula is at least as good as that of Eq. (8-103). The evaluation of other small-obstacle equivalent circuits can be found in the literature.[1]

[1] A. A. Oliner, Equivalent Circuits for Small Symmetrical Longitudinal Apertures and Obstacles, *IRE Trans.*, vol. MTT-8, no. 1, January, 1960.

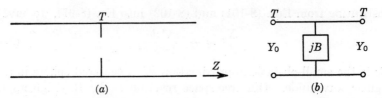

FIG. 8-18. (a) A diaphragm in a waveguide, and (b) an equivalent circuit.

8-9. Diaphragms in Waveguides. Figure 8-18a represents a cylindrical waveguide of arbitrary cross section with an infinitely thin electric conductor covering part of the $z = 0$ plane. This conductor is called a *diaphragm*, and the opening in it is called a *window*. The diaphragm plus the window cover the entire $z = 0$ cross section. The exact equivalent circuit is just a shunt element, as shown in Fig. 8-18b. Depending upon the shape of the diaphragm or window, the susceptance may be positive (capacitive), negative (inductive), or change from positive to negative as the frequency is varied (resonant when $B = 0$).

To evaluate the shunt susceptance, we can use the method of Sec. 8-7. Taking the case of even excitation (Fig. 8-12a), the diaphragm problem reduces to Fig. 8-19a. The equivalent circuit is shown in Fig. 8-19b. The appropriate stationary formula is Eq. (8-81), which reduces to

$$\frac{2Y_0}{B} \approx -\frac{\operatorname{Im}\langle a,a\rangle}{\operatorname{Re}\langle a,a\rangle} \qquad (8\text{-}107)$$

where $\langle a,a\rangle$ is the self-reaction of the assumed current $\mathbf{J}_s{}^a$ on the dia-

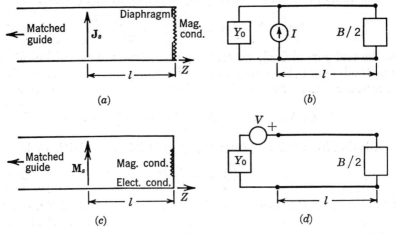

FIG. 8-19. Symmetrical excitation of Fig. 8-10a is represented by (a), which has an equivalent circuit (b). Symmetrical excitation of Fig. 8-10a is also represented by (c), which has an equivalent circuit (d).

phragm. We can think of Fig. 8-19a as being constructed by placing pieces of electric conductor on top of a magnetic conductor.

Because the diaphragm problem is self-dual, we have the alternative representation of Fig. 8-19c. This can be viewed as a construction of the window by placing pieces of magnetic conductor on top of an electric conductor. The source has been changed to a magnetic current sheet, instead of the electric current sheet of Fig. 8-19a, so that complete duality is preserved. Then, dual to Eq. (8-107), we have

$$\frac{B}{2Y_0} \approx \frac{\text{Im}\,\langle a,a\rangle_m}{\text{Re}\,\langle a,a\rangle_m} \tag{8-108}$$

where the subscripts m are added to emphasize that $\langle a,a\rangle_m$ is the self-reaction of assumed magnetic currents $\mathbf{M}_s^a$ on the window, that is,

$$\langle a,a\rangle_m = -\iint \mathbf{H}^a \cdot \mathbf{M}_s^a\, ds \tag{8-109}$$

Because the $\mathbf{M}_s^a$ is related to the tangential $\mathbf{E}$ in the window of the original problem according to

$$\mathbf{M}_s^a = \mathbf{u}_z \times \mathbf{E} \tag{8-110}$$

Eq. (8-108) is known as an *aperture-field formulation* of the problem. This is in contrast to Eq. (8-107) which is an *obstacle-current formulation*. Note that Eq. (8-108) can also be viewed as a specialization of Eq. (8-84).

To illustrate the theory, consider a capacitive diaphragm in a rectangular waveguide (Fig. 8-20). (Note that it must be capacitive, because it is a special case of Fig. 8-13b.) Take the E-field formula [Eq. (8-108)] and note that

$$\langle a,a\rangle_m = -\iint \mathbf{H}^a \cdot \mathbf{M}_s^a\, ds = -\iint \mathbf{E} \times \mathbf{H} \cdot \mathbf{u}_z\, ds$$
$$= \left(-\iint \mathbf{E} \times \mathbf{H}^* \cdot \mathbf{u}_z\, ds\right)^* = P^*$$

because $\mathbf{E}$ is real. Hence, the problem is the same as those treated in

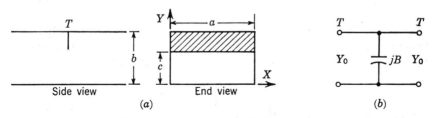

FIG. 8-20. (a) Capacitive diaphragm, and (b) an equivalent circuit.

Sec. 4-9. In particular, if we assume

$$E_y^a\Big|_{z=0} = \begin{cases} \sin\dfrac{\pi x}{a} & y < c \\ 0 & y > c \end{cases} \tag{8-111}$$

in the window, we then have precisely the problem of Fig. 4-17. Hence, from Eq. (4-77), we have

$$\langle a,a \rangle_m = P^* = |V|^2 Y_a = |V|^2 \left(Y_0 \frac{a}{2b} + jB_a \right)$$

where Y_a is the aperture admittance. Finally, substituting from the above into Eq. (8-108), we obtain

$$\frac{B}{Y_0} \approx \frac{4b}{a}\frac{B_a}{Y_0} = \frac{8b}{\lambda_g}\left(\frac{\lambda_g B_a}{2aY_0}\right) \tag{8-112}$$

where the quantity in parentheses is plotted in Fig. 4-17.

A more general treatment of the problem proceeds as follows. We know from the discussion of Fig. 8-13b that the field must be TE to x, and so the most general form for the tangential field in the window is

$$E_y^a\Big|_{z=0} = \begin{cases} f(y)\sin\dfrac{\pi x}{a} & y < c \\ 0 & y > c \end{cases} \tag{8-113}$$

Then, by the methods of Sec. 4-9, we calculate

$$\langle a,a \rangle_m = P^* = \frac{ab}{2}\sum_{n=0}^{\infty}\frac{1}{\epsilon_n}(Y_0)_{1n}|E_{1n}|^2$$

where, by Eq. (4-73), the Fourier coefficients E_{1n} are

$$E_{1n} = \frac{\epsilon_n}{b}\int_0^c f(y)\cos\frac{n\pi y}{b}\,dy$$

and the characteristic admittances of the TEx_{1n} modes are

$$(Y_0)_{1n} = \frac{j2bY_0}{\lambda_g\sqrt{n^2 - (2b/\lambda_g)^2}} \tag{8-114}$$

The Y_0 and λ_g pertain to the dominant mode, which is the only mode having real characteristic impedance, because of our assumption that only the dominant mode propagates. Hence, Eq. (8-108) becomes

$$\frac{B}{2Y_0} \approx \frac{\sum_{n=1}^{\infty}|Y_0|_{1n}|E_{1n}|^2}{2Y_0|E_{10}|^2}$$

which, upon substitution from the preceding equations, becomes

$$\frac{B}{Y_0} \approx \frac{8b}{\lambda_g} \frac{\sum_{n=1}^{\infty} \frac{1}{\sqrt{n^2 - (2b/\lambda_g)^2}} \left[\int_0^c f(y) \cos \frac{n\pi y}{b} dy\right]^2}{\left[\int_0^c f(y) dy\right]^2} \qquad (8\text{-}115)$$

Equation (8-112) represents the special case $f(y) = 1$. Better approximations to B/Y_0 can be obtained by using a better choice for $f(y)$, or by applying the Ritz procedure.

The stationary formula in terms of obstacle current [Eq. (8-107)] is specialized to the capacitive diaphragm as follows. The field is TE to x, given by Eqs. (4-32) with

$$\psi = \sin \frac{\pi x}{a} \sum_{n=0}^{\infty} A_n \cos \frac{n\pi y}{b} e^{\gamma_n z}$$

where

$$\gamma_n = \sqrt{\left(\frac{\pi}{a}\right)^2 + \left(\frac{n\pi}{b}\right)^2 - k^2}$$

The current on a diaphragm backed by a magnetic conductor (Fig. 8-19a) is then

$$J_x = H_y\Big|_{z=0} = \frac{j\pi^2}{ab\omega\mu} \cos \frac{\pi x}{a} \sum_{n=0}^{\infty} n A_n \sin \frac{n\pi y}{b}$$

$$J_y = -H_x\Big|_{z=0} = \frac{(\pi/a)^2 - k^2}{j\omega\mu} \sin \frac{\pi x}{a} \sum_{n=0}^{\infty} A_n \cos \frac{n\pi y}{b}$$

Hence, the current has both x and y components, but the A_n can be determined from the y component alone. The x component then adjusts itself to make the field TE to x. If we assume a current

$$J_y{}^a = g(y) \sin \frac{\pi x}{a} \qquad (8\text{-}116)$$

and define Fourier coefficients

$$J_n = \frac{\epsilon_n}{b} \int_c^b g(y) \cos \frac{n\pi y}{b} dy \qquad (8\text{-}117)$$

then

$$A_n = \frac{j\omega\mu}{(\pi/a)^2 - k^2} J_n = \frac{Z_0}{\gamma} J_n$$

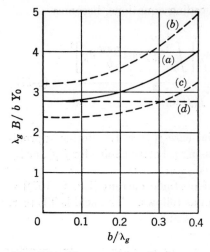

FIG. 8-21. The capacitive diaphragm with $c = 2b$. (a) Exact solution, (b) crude aperture-field variational solution, (c) crude obstacle-current variational solution, and (d) crude quasi-static solution.

Also, at $z = 0$, the tangential electric intensity is given by $E_x = 0$, and

$$E_y = -\sin\frac{\pi x}{a} \sum_{n=0}^{\infty} \gamma_n A_n \cos\frac{n\pi y}{b}$$

Hence, in the same manner as Eq. (4-74) was derived, we find the self-reaction of $\mathbf{J}_s{}^a$ as

$$\langle a,a \rangle = \frac{ab}{2} \sum_{n=0}^{\infty} \frac{1}{\epsilon_n} (Z_0)_n J_n{}^2$$

where the characteristic impedances $(Z_0)_n$ are the reciprocals of Eqs. (8-114). Because only the dominant mode propagates, only the $n = 0$ term of the summation is real, and Eq. (8-107) reduces to

$$\frac{2Y_0}{B} \approx \frac{\sum_{n=1}^{\infty} (Z_0)_n J_n{}^2}{2Z_0 J_0{}^2}$$

Substituting for J_n from Eq. (8-117) and for $(Z_0)_n = 1/(Y_0)_{1n}$ from Eq. (8-114), we finally have

$$\frac{Y_0}{B} \approx \frac{\lambda_g}{8b} \frac{\sum_{n=1}^{\infty} \sqrt{n^2 - \left(\frac{2b}{\lambda_g}\right)^2} \left[\int_c^b g(y) \cos\frac{n\pi y}{b} dy\right]^2}{\left[\int_c^b g(y) dy\right]^2} \qquad (8\text{-}118)$$

This is the stationary formula in terms of obstacle current for the capacitive diaphragm of Fig. 8-20.

Figure 8-21 compares various solutions to the capacitive diaphragm problem for the case of a diaphragm covering half the guide cross section. Curve (a) is called the exact solution because the estimated error is less than the accuracy of the graph. This solution is obtained by finding a quasi-static field and then using it in the variational formula, Eq. (8-115).[1] Curve (b) is the crude aperture-field variational solution, Eq. (8-112), which is also Eq. (8-115) with $f(y) = 1$. Curve (c) is a crude

[1] N. Marcuvitz, "Waveguide Handbook," MIT Radiation Laboratory Series, vol. 10, secs. 3-5 and 5-1, McGraw-Hill Book Company, Inc., New York, 1951.

obstacle-current variational solution, Eq. (8-118), with

$$g(y) = \sin \frac{\pi(y-c)}{2(b-c)} \tag{8-119}$$

(If the case $g = 1$ is tried, the solution diverges, because the boundary condition that the current vanishes at $y = c$ is violated.) Curve (d) is a first-order quasi-static solution to the problem[1]

$$\frac{B}{Y_0} \approx \frac{8b}{\lambda_g} \log \csc \frac{\pi c}{2b} \tag{8-120}$$

In practice, waveguides are usually operated with $b/\lambda_g < 0.25$; so this last solution is a good approximation for most purposes.

Note that the aperture-field variational solution, curve (b), is above the true solution, and the obstacle-current variational solution, curve (c), is below the true solution. That this is so for any trial functions $f(y)$ and $g(y)$ follows from the fact that Eqs. (8-115) and (8-118) are positive definite and hence are an absolute minimum for the true fields. Since Eq. (8-115) gives B/Y_0 and Eq. (8-118) gives Y_0/B, the former yields upper bounds and the latter yields lower bounds to the true B/Y_0. The existence of variational formulas for both upper and lower bounds is not very common and is a consequence of the self-duality of the problem plus the positive-definite nature of the resulting variational formulas.

Our crude variational solutions give an error of the order of 20 per cent, but it is remarkable that they are as close as that. A quasi-static solution to the problem is

$$f(y) = \frac{\cos(\pi y/2b)}{\sqrt{\sin^2(\pi c/2b) - \sin^2(\pi y/2b)}} \tag{8-121}$$

which actually has a singularity at $y = c$. Hence, our approximation $f(y) = 1$ was an exceedingly crude choice, yet it led to usable results. Our approximation to $g(y)$ [Eq. (8-119)] is equally crude. If we were to use Eq. (8-121) in Eq. (8-115), the result would be very close to the true solution.

It is interesting to note that the three diaphragms shown in Fig. 8-22 all have the same equivalent circuits. This is evident, because the image systems for all three cases are identical.

The treatment of the inductive diaphragm (Fig. 8-23) is similar to that of the capacitive diaphragm. The general variational formulas for upper and lower bounds are given in Probs. 8-14 and 8-15. For a crude aperture-field solution, we can assume Eq. (4-75) for $E_y{}^a$ in the aperture.

[1] W. R. Smythe, "Static and Dynamic Electricity," 2d ed., Sec. 15-10, McGraw-Hill Book Company, Inc., New York, 1950.

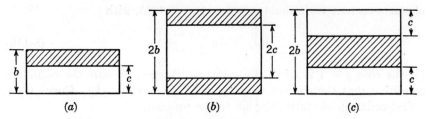

FIG. 8-22. These three diaphragms give rise to the same shunt capacitance.

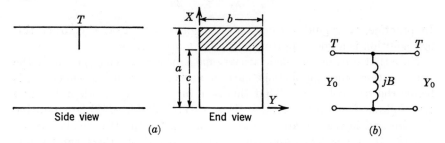

FIG. 8-23. (a) Inductive diaphragm, and (b) an equivalent circuit.

This procedure gives

$$\frac{B}{Y_0} \approx -\frac{\lambda_g}{a}\left[\frac{\pi a}{c}\frac{1-(c/a)^2}{\sin(\pi c/a)}\right]^2 \left(\frac{\eta b}{\lambda} B_a\right) \qquad (8\text{-}122)$$

where B_a is the aperture susceptance plotted in Fig. 4-19. The values of $-B/Y_0$ calculated from Eq. (8-122) will be higher than the true values (of the order of 20 per cent higher). The problem can also be treated by quasi-static methods, a first-order solution being[1]

$$\frac{B}{Y_0} \approx -\frac{\lambda_g}{a}\left(1 + \csc^2\frac{\pi c}{2a}\right)\cot^2\frac{\pi c}{2a} \qquad (8\text{-}123)$$

A combination of the quasi-static and variational methods can be used to obtain solutions of high accuracy.[2]

8-10. Waveguide Junctions. We shall now consider waveguide junctions formed by butting two cylindrical guides together, possibly with a diaphragm covering part of the $z = 0$ cross section. Figure 8-24 represents the general problem. No longer is there symmetry about the $z = 0$ cross section; so the methods of Sec. 8-6 do not apply directly. We there-

---

[1] W. R. Smythe, "Static and Dynamic Electricity," 2d ed., p. 555, McGraw-Hill Book Company, Inc., New York, 1950.

[2] N. Marcuvitz, "Waveguide Handbook," MIT Radiation Laboratory Series, vol. 10, sec. 5-2, McGraw-Hill Book Company, Inc., New York, 1951.

MICROWAVE NETWORKS 421

fore take the more fundamental approach of constructing complete solutions in each region and enforcing

$$\iint_{z=0} \mathbf{E}^+ \times \mathbf{H}^+ \cdot d\mathbf{s} = \iint_{z=0} \mathbf{E}^- \times \mathbf{H}^- \cdot d\mathbf{s} \qquad (8\text{-}124)$$

where superscripts $+$ and $-$ refer to regions $z > 0$ and $z < 0$, respectively. In terms of the reaction concept, we can think of Eq. (8-124) as stating that the reaction is conserved at the junction.

An equivalent network for the junction is shown in Fig. 8-24b. It is evident that only a shunt element is required to represent the junction, because an electric conductor placed across the entire $z = 0$ cross section presents a short circuit to both waveguides. The characteristic admittances of the equivalent transmission lines are taken to be the characteristic wave admittances of the guides, and the ideal transformer represents the change in admittance level. If the characteristic admittance of the right-hand transmission line were chosen as n^2 times the characteristic wave admittance of the guide, then the transformer would not be needed. We shall use Eq. (8-124) to obtain stationary formulas for B and n^2.

It is assumed that the excitation is at $z = -\infty$; hence in the region $z < 0$

$$\begin{aligned}
\mathbf{E}_t^- &= (e^{-j\beta z} + \Gamma e^{j\beta z})\frac{V_0}{1+\Gamma}\mathbf{e}_0 + \sum_i V_i e^{\alpha_i z}\mathbf{e}_i \\
\mathbf{H}_t^- &= Y_0^-(e^{-j\beta z} - \Gamma e^{j\beta z})\frac{V_0}{1+\Gamma}\mathbf{h}_0 - \sum_i Y_i V_i e^{\alpha_i z}\mathbf{h}_i
\end{aligned} \qquad (8\text{-}125)$$

where $\mathbf{e}_i$, $\mathbf{h}_i$ are the mode vectors, α_i are the cutoff mode-attenuation constants, Y_i are the characteristic admittances, and Γ is the reflection coefficient for the dominant mode. The subscripts 0 denote dominant-mode parameters. Matched conditions are assumed at $z = \infty$; hence in

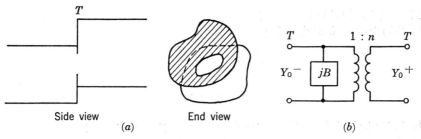

Fig. 8-24. (a) A waveguide junction, and (b) an equivalent circuit.

the region $z > 0$

$$\mathbf{E}_t^+ = \hat{V}_0 e^{-j\hat{\beta}z}\hat{\mathbf{e}}_0 + \sum_i \hat{V}_i e^{-\hat{\alpha}_i z}\hat{\mathbf{e}}_i$$

$$\mathbf{H}_t^+ = Y_0^+\hat{V}_0 e^{-j\hat{\beta}z}\hat{\mathbf{h}}_0 + \sum_i \hat{Y}_i\hat{V}_i e^{-\hat{\alpha}_i z}\hat{\mathbf{h}}_i \qquad (8\text{-}126)$$

where the carets distinguish the various parameters from their $z < 0$ counterparts. The application of Eq. (8-124) to the above field expressions yields

$$Y_0^+\hat{V}_0^2 + \sum_i \hat{Y}_i\hat{V}_i^2 = \frac{1-\Gamma}{1+\Gamma} Y_0^- V_0^2 - \sum_i Y_i V_i^2 \qquad (8\text{-}127)$$

Now the relative admittance seen from the left-hand guide is

$$\frac{1-\Gamma}{1+\Gamma} = \frac{Y}{Y_0^-} = \frac{G}{Y_0^-} + j\frac{B}{Y_0^-} \qquad (8\text{-}128)$$

Remembering that the Y_0 are real and the Y_i, $i \neq 0$, are imaginary, for real V_i and $\hat{V}_i$ we have

$$\frac{jB}{Y_0^-} = \frac{\sum_i Y_i V_i^2 + \sum_i \hat{Y}_i \hat{V}_i^2}{Y_0^- V_0^2} \qquad (8\text{-}129)$$

$$\frac{G}{Y_0^-} = \frac{Y_0^+ \hat{V}_0^2}{Y_0^- V_0^2}$$

From our equivalent circuit, with matched conditions at $z = \infty$, it is evident that

$$\frac{G}{Y_0^-} = n^2 \frac{Y_0^+}{Y_0^-}$$

hence
$$n^2 = \frac{\hat{V}_0^2}{V_0^2} \qquad (8\text{-}130)$$

Finally, to obtain the V_i and $\hat{V}_i$, we need only specialize Eqs. (8-125) and (8-126) to $z = 0$ and, using the methods of Sec. 8-2, obtain

$$V_i = \iint_{\text{apert}} \mathbf{E}_t \cdot \mathbf{e}_i \, ds$$

$$\hat{V}_i = \iint_{\text{apert}} \mathbf{E}_t \cdot \hat{\mathbf{e}}_i \, ds \qquad (8\text{-}131)$$

Note that the integration extends only over the aperture, because $\mathbf{E}_t = 0$ on the conductor. Equations (8-129) and (8-130), with V_i and $\hat{V}_i$ given by Eq. (8-131), are formulas stationary with respect to small variations in the aperture $\mathbf{E}_t$ about the correct field. Alternative stationary formulas in terms of current on the conducting wall at $z = 0$ can also be obtained (see Prob. 8-18). Note that Eq. (8-129) specialized to the case of two identical guides is the diaphragm solution of the preceding section.

To illustrate the theory, consider the rectangular waveguide junctions of Sec. 4-9. For the capacitive junction (Fig. 4-16), the dominant-mode vectors are

$$\mathbf{e}_0 = \mathbf{u}_y \sqrt{\frac{2}{ac}} \sin \frac{\pi x}{a} \qquad \hat{\mathbf{e}}_0 = \mathbf{u}_y \sqrt{\frac{2}{ab}} \sin \frac{\pi x}{a}$$

Hence, regardless of our assumed tangential $\mathbf{E}$ in the aperture

$$\mathbf{E}_t{}^a = \mathbf{u}_y f(y) \sin \frac{\pi x}{a} \qquad (8\text{-}132)$$

we have by Eqs. (8-130) and (8-131)

$$n^2 = \frac{c}{b} \qquad (8\text{-}133)$$

This is therefore the exact transformation ratio of the ideal transformer. In Sec. 4-9, we calculated the aperture susceptance corresponding to the crude choice $f(y) = 1$. The first summation in the numerator of Eq. (8-129) then vanishes, and the second summation is related to the aperture susceptance of Eq. (4-78) by

$$\sum_i \hat{Y}_i \hat{V}_i{}^2 = j|V|^2 B_a = jc^2 B_a$$

But, for $f(y) = 1$, we have $V_0{}^2 = ac/2$; hence, by Eq. (8-129),

$$\frac{B}{Y_0} \approx \frac{2c^2 B_a}{acY_0} = \frac{4c}{\lambda_g} \left(\frac{\lambda_g Z_0}{2a} B_a \right) \qquad (8\text{-}134)$$

where the quantity in parentheses is plotted in Fig. 4-17. The general expression [Eq. (8-129)] is positive definite in our particular case; so Eq. (8-134) gives values of B/Y_0 higher than the true values. However, because the field in the aperture is less singular at the edge of a step than at a knife edge, we should expect the assumption $f(y) = 1$ to give better results in the junction problem than in the corresponding diaphragm problem. Our approximate answer [Eq. (8-134)] gives an accuracy of the order of 10 per cent, as illustrated by Table 8-3. This can be compared to the 20 per cent accuracy in the corresponding diaphragm problem, illustrated by Fig. 8-21.

The inductive junction of Fig. 4-18 is treated in a similar manner. In general, the field in the aperture is of the form $E_y = f(x)$, and for the

TABLE 8-3. COMPARISON OF EQ. (8-134) TO THE EXACT SOLUTION[1] FOR THE CASE $c/b = 0.5$

$\dfrac{b}{\lambda_g}$	$\dfrac{\lambda_g B}{c Y_0}$	
	Exact	Approximate
0	1.57	1.63
0.2	1.69	1.84
0.3	1.93	2.10
0.4	2.44	2.67

[1] N. Marcuvitz, "Waveguide Handbook," MIT Radiation Laboratory Series, vol. 10, sec. 5-24, McGraw-Hill Book Company, Inc., New York, 1951.

solution of Sec. 4-9 we assumed

$$\mathbf{E}_t{}^a = \mathbf{u}_y f(x) = \mathbf{u}_y \sin \frac{\pi x}{c} \qquad (8\text{-}135)$$

By Eq. (8-130), we then find the transformation ratio of the ideal transformer as

$$n^2 \approx \frac{4c}{\pi^2 a} \left[\frac{\sin(\pi c/a)}{1 - (c/a)^2} \right]^2 \qquad (8\text{-}136)$$

and, by Eq. (8-129), the normalized shunt susceptance as

$$\frac{B}{Y_0{}^-} \approx -\frac{2\lambda_g}{c} \left(-\frac{\eta b}{\lambda} B_a \right) \qquad (8\text{-}137)$$

where the quantity in parentheses is plotted in Fig. 4-19. Note that, in contrast to Eq. (8-133), the transformation ratio [Eq. (8-136)] depends on the assumed aperture field and is therefore approximate. Note also that the characteristic wave impedances of the two guides, $z < 0$ and $z > 0$, are now different; so the superscript $-$ has been retained on $Y_0{}^-$ in Eq. (8-137). Finally, the value of $-B/Y_0{}^-$ obtained from Eq. (8-137) will be larger than the true solution, because of the positive definiteness of the variational formula.

The alternative equivalent circuit of Fig. 8-25 illustrates a very useful way of viewing the waveguide junction of Fig. 8-24a. We have separated the shunt susceptance into two parts, which, by Eq. (8-129), can be identified as

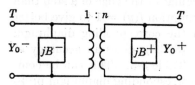

FIG. 8-25. Alternative equivalent circuit for Fig. 8-24a.

$$\frac{jB^-}{Y_0{}^-} = \frac{\sum_i Y_i V_i^2}{Y_0{}^- V_0^2} \qquad \frac{jB^+}{Y_0{}^+} = \frac{\sum_i \hat{Y}_i \hat{V}_i^2}{Y_0{}^+ \hat{V}_0^2} \qquad (8\text{-}138)$$

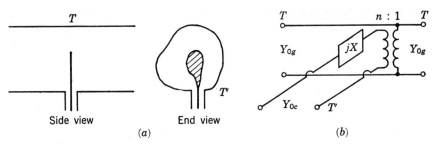

FIG. 8-26. (a) A thin coax-to-waveguide feed, and (b) an equivalent circuit.

where the V_i and $\hat{V}_i$ are given by Eq. (8-131). Note that B^- depends only on guide $z < 0$, and in particular is one-half the shunt susceptance of a diaphragm, assuming $\mathbf{E}_t$ in the aperture is unchanged. This assumption is, of course, incorrect, but our formulas are stationary; so B^- in the junction problem is approximately $B/2$ in the corresponding diaphragm problem. Similarly, B^+ is approximately $B/2$ for the diaphragm problem corresponding to the guide $z > 0$. Hence, by defining aperture susceptances according to Eqs. (8-138), we effectively divide the problem into two parts, each part relatively insensitive to the other. An aperture susceptance calculated for the aperture and one guide, such as Figs. 4-17 and 4-19, thereby becomes useful for a wide variety of problems.

8-11. Waveguide Feeds. We shall now consider thin coax-to-waveguide feeds, as illustrated by Fig. 8-26a. By thin, we mean that the dimension in the axial (z) direction is small. The analysis will be exact only for zero-thickness junctions. An equivalent circuit when only one mode propagates is shown in Fig. 8-26b. When more than one mode propagates, say N modes, there will be N ideal transformers in series, each coupling to one mode. The justification for this equivalent circuit will be found in the analysis.

Let the feed be viewed as a sheet of current $\mathbf{J}_s$ in the $z = 0$ cross section. (This neglects the effect of the gap, which is usually small.) Then, in the region $z > 0$, we have

$$\mathbf{E}_t{}^+ = \sum_i \frac{V_i}{1 + \Gamma_i{}^+} (e^{-\gamma_i z} + \Gamma_i{}^+ e^{\gamma_i z}) \mathbf{e}_i$$
$$\mathbf{H}_t{}^+ = \sum_i \frac{V_i Y_i}{1 + \Gamma_i{}^+} (e^{-\gamma_i z} - \Gamma_i{}^+ e^{\gamma_i z}) \mathbf{h}_i$$
(8-139)

where $\Gamma_i{}^+$ is the $+z$ reflection coefficient of the ith mode referred to

$z = 0$. Similarly, for $z < 0$,

$$\mathbf{E}_t^- = \sum_i \frac{V_i}{1 + \Gamma_i^-} (e^{\gamma_i z} + \Gamma_i^- e^{-\gamma_i z}) \mathbf{e}_i$$

$$\mathbf{H}_t^- = -\sum_i \frac{V_i Y_i}{1 + \Gamma_i^-} (e^{\gamma_i z} - \Gamma_i^- e^{-\gamma_i z}) \mathbf{h}_i \tag{8-140}$$

where Γ_i^- is the $-z$ reflection coefficient of the ith mode referred to $z = 0$. We have ensured continuity of $\mathbf{E}_t$ at $z = 0$ by choosing coefficients V_i the same in both Eqs. (8-139) and (8-140). The boundary condition on $\mathbf{H}$ at $z = 0$ is

$$\mathbf{J}_s = \mathbf{u}_z \times (\mathbf{H}_t^+ - \mathbf{H}_t^-)\bigg|_{z=0}$$

$$= \sum_i V_i Y_i \left(\frac{1 - \Gamma_i^-}{1 + \Gamma_i^-} + \frac{1 - \Gamma_i^+}{1 + \Gamma_i^+} \right) \mathbf{u}_z \times \mathbf{h}_i \tag{8-141}$$

Multiplying each side by $\mathbf{e}_i$ and integrating over the guide cross section, we have

$$V_i Y_i \left(\frac{1 - \Gamma_i^-}{1 + \Gamma_i^-} + \frac{1 - \Gamma_i^+}{1 + \Gamma_i^+} \right) = -\iint \mathbf{J}_s \cdot \mathbf{e}_i \, ds \tag{8-142}$$

The field is then completely determined if the Γ's and $\mathbf{J}_s$ are known.

We now use the stationary formula of Eq. (7-89) to determine the impedance seen by the coax. This formula is

$$Z_{\text{in}} = -\frac{1}{I_{\text{in}}^2} \iint \mathbf{E} \cdot \mathbf{J}_s \, ds$$

where the integration extends over the $z = 0$ guide cross section and I_{in} is the current at the reference plane T'. Using the first of Eqs. (8-139) for $\mathbf{E}_t$ and Eq. (8-141) for $\mathbf{J}_s$, we obtain

$$Z_{\text{in}} = \frac{1}{I_{\text{in}}^2} \sum_i V_i^2 Y_i \left(\frac{1 - \Gamma_i^-}{1 + \Gamma_i^-} + \frac{1 - \Gamma_i^+}{1 + \Gamma_i^+} \right)$$

Finally, substituting for V_i from Eq. (8-142), we have

$$Z_{\text{in}} = \frac{1}{I_{\text{in}}^2} \sum_i \frac{Z_i \left(\iint \mathbf{J}_s \cdot \mathbf{e}_i \, ds \right)^2}{(1 - \Gamma_i^-)(1 + \Gamma_i^-)^{-1} + (1 - \Gamma_i^+)(1 + \Gamma_i^+)^{-1}} \tag{8-143}$$

where Z_i is the characteristic impedance of the ith mode. This is a stationary formula for the input impedance of a zero-thickness coax-to-waveguide feed. We can put it into a slightly different form by noting that the wave impedance of an ith mode referred to $z = 0$ is

$$\hat{Z}_i = Z_i \frac{1 + \Gamma_i}{1 - \Gamma_i} \qquad (8\text{-}144)$$

Hence, Eq. (8-143) can also be written as

$$Z_{in} = \frac{1}{I_{in}^2} \sum_i \left(\iint \mathbf{J}_s \cdot \mathbf{e}_i \, ds \right)^2 \frac{\hat{Z}_i^+ \hat{Z}_i^-}{\hat{Z}_i^+ + \hat{Z}_i^-} \qquad (8\text{-}145)$$

This shows that the guides $z > 0$ and $z < 0$ appear in parallel for each mode. Nonpropagating modes decay exponentially from the junction and their Γ_i may be taken as zero unless some obstacle is close to the feed.

If we assume that only one mode propagates, then all Z_i are imaginary except $i = 0$, and all $\Gamma_i = 0$ except $i = 0$, provided the terminations are not too close to the feed. Equation (8-143) or (8-145) can then be written as

$$\begin{aligned} Z_{in} &= n^2 Z_0 \left(\frac{1 - \Gamma_0^-}{1 + \Gamma_0^-} + \frac{1 - \Gamma_0^+}{1 + \Gamma_0^+} \right)^{-1} + jX \\ &= n^2 \frac{\hat{Z}_0^+ \hat{Z}_0^-}{\hat{Z}_0^+ + \hat{Z}_0^-} + jX \end{aligned} \qquad (8\text{-}146)$$

where

$$n^2 = \frac{1}{I_{in}^2} \left(\iint \mathbf{J}_s \cdot \mathbf{e}_0 \, ds \right)^2 \qquad (8\text{-}147)$$

$$jX = \frac{1}{2 I_{in}^2} \sum_{i \neq 0} Z_i \left(\iint \mathbf{J}_s \cdot \mathbf{e}_i \, ds \right)^2 \qquad (8\text{-}148)$$

Equation (8-146) is, of course, just that for the equivalent circuit of Fig. 8-26b.

As an example, consider a probe in a rectangular guide (Fig. 8-27). Assume

$$\mathbf{J}_s = \begin{cases} \mathbf{u}_x \sin k(d - x) \, \delta(y - c) & x < d \\ 0 & x > d \end{cases} \qquad (8\text{-}149)$$

where $k = 2\pi/\lambda$ is the wave number of free space. The dominant-mode vector is

$$\mathbf{e}_0 = \mathbf{u}_x \sqrt{\frac{2}{ab}} \sin \frac{\pi y}{b}$$

Equation (8-147) is therefore

$$n = \frac{\sqrt{2/ab}}{\sin kd} \int_0^d dx \int_0^b dy \, \sin k(d - x) \, \delta(y - c) \sin \frac{\pi y}{b}$$

giving

$$n^2 = \frac{2}{k^2 ab} \sin^2 \frac{\pi c}{b} \tan^2 \left(k \frac{d}{2} \right) \qquad (8\text{-}150)$$

The summation for X [Eq. (8-148)] diverges, because the current was

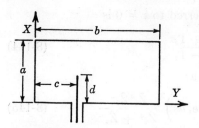

Fig. 8-27. Probe in a rectangular waveguide.

taken as filamentary. If the probe is taken as circular in cross section, the reactance can be evaluated by methods similar to those used in Sec. 8-7. However, if the probe is very thick, we shall have to modify the equivalent circuit of Fig. 8-26b. The reactance of a short probe can be estimated by the small-obstacle approximation of Sec. 8-8. It is evident from the small-obstacle analysis that X is capacitive (negative) for a short probe and is of the order of magnitude of X for a probe over a conducting ground plane.

Note that our present solution [Eqs. (8-146) to (8-148)], specialized to a rectangular waveguide matched in both directions, is the same problem treated in Sec. 4-10. From our equivalent circuit (Fig. 8-26), it is evident that the coax sees

$$R_{in} = n^2 \frac{Z_0}{2}$$

under matched conditions. Hence,

$$n^2 = \frac{2R_{in}}{Z_0} \qquad (8\text{-}151)$$

where R_{in} is the quantity calculated in Sec. 4-10. For example, when the probe is connected to the opposite wall of the waveguide, as in Fig. 4-20, we have from Eq. (4-91)

$$n^2 = \frac{2a}{b}\left(\frac{\tan ka}{ka}\right)^2 \sin^2 \frac{\pi c}{b} \qquad (8\text{-}152)$$

Other possible feeds are shown in Fig. 4-28.

8-12. Excitation of Apertures. We now wish to consider conducting bodies containing apertures excited by waveguides. The general problem is represented by Fig. 8-28a. As far as the waveguide is concerned, the aperture appears simply as a load across the reference plane T. A variational solution to the problem can be obtained by assuming tangential **E** in the aperture, calculating the resultant fields on each side of the aperture, and then conserving the flux of reaction by

$$\iint_{\text{apert}} (\mathbf{E} \times \mathbf{H} \cdot d\mathbf{s})_{\text{ext}} = \iint_{\text{apert}} (\mathbf{E} \times \mathbf{H} \cdot d\mathbf{s})_{\text{int}} \qquad (8\text{-}153)$$

This is the same approach that we took in Sec. 8-10 for the waveguide junction. Indeed, we can think of our present problem as a junction between the waveguide and external space.

Once the tangential **E** in the aperture is assumed, the problem separates into two parts, external and internal. We have anticipated this separation by taking the equivalent circuit as shown in Fig. 8-28b, where jB represents the internal susceptance of the diaphragm and Y_apert the external admittance of the aperture. The ideal transformer accounts for possible differences of impedance reference in the internal and external problems. The internal problem is identical to one-half of the waveguide-junction problem. Let us therefore abstract from Eq. (8-138)

$$\frac{jB}{Y_0} = \frac{\sum_i Y_i V_i^2}{Y_0 V_0^2} \qquad (8\text{-}154)$$

where
$$V_i = \iint_\text{apert} \mathbf{E}_t^a \cdot \mathbf{e}_i \, ds \qquad (8\text{-}155)$$

These formulas give the internal shunt susceptance B in terms of an assumed $\mathbf{E}_t^a$ in the aperture. For the external problem, we define an aperture admittance as

$$Y_\text{apert} = \frac{1}{V^2} \iint_\text{apert} \mathbf{E}_t^a \times \mathbf{H}^a \cdot d\mathbf{s} \qquad (8\text{-}156)$$

where V is some reference voltage and $\mathbf{H}^a$ is the external magnetic field calculated from the assumed $\mathbf{E}_t^a$. Examples of some aperture-admittance calculations are given in Sec. 4-11. (These calculations were made on a conservation of power basis, but, because $\mathbf{E}^a$ was assumed real, they are the same as variational solutions.) To determine n^2 we note that the dominant-mode voltage coupled to the aperture is V_0, but we have referred the aperture admittance to V; hence

$$n^2 = \frac{V^2}{V_0^2} \qquad (8\text{-}157)$$

where V_0 is given by Eq. (8-155) applied to the dominant mode.

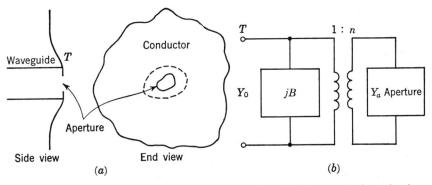

FIG. 8-28. (a) An aperture excited by a waveguide, and (b) an equivalent circuit.

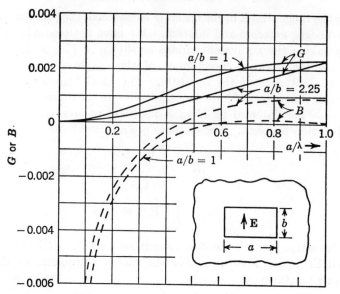

FIG. 8-29. Aperture admittance for rectangular apertures in ground planes, referred to the dominant-mode voltage of a rectangular waveguide of the same dimensions. (*After Cohen, Crowley, and Levis.*)

An aperture of practical importance is the rectangular aperture in a conducting ground plane, as shown in the insert of Fig. 8-29. The aperture admittance has been calculated for the assumed field

$$\mathbf{E}_t^a = \mathbf{u}_y \sin \frac{\pi x}{a} \qquad (8\text{-}158)$$

in the aperture, referred to the voltage

$$V = \sqrt{\frac{ab}{2}} \qquad (8\text{-}159)$$

which is the dominant-mode voltage for a waveguide of the same dimensions as the aperture. Hence, when the aperture is simply the flanged open end of a rectangular waveguide, then $n = 1$. The field due to $\mathbf{E}_t^a$ in the aperture can be found by the methods of Sec. 3-6, and the aperture **admittance** calculated by Eq. (8-156). The mathematical details are tedious but can be found in the literature.[1] Figure 8-29 shows the aperture admittances for a square aperture and for a rectangular aperture with sides in the ratio 1 to 1 and 2.25 to 1.[2]

---

[1] Cohen, Crowley, and Levis, The Aperture Admittance of a Rectangular Waveguide Radiating into Half-space, *Ohio State Univ. Antenna Lab. Rept.* ac 21114 SR no. 22, 1953.

[2] Additional calculations have been made by R. J. Tector, The Cavity-backed Slot Antenna, *Univ. Illinois Antenna Lab. Rept.* 26, 1957.

As an example, suppose we have a square waveguide of height and width a feeding a rectangular aperture with sides in the ratio $a/b = 2.25$, as shown in Fig. 8-30. The waveguide is excited in the dominant y-polarized mode, for which

$$\mathbf{e}_0 = \mathbf{u}_y \frac{\sqrt{2}}{a} \sin \frac{\pi x}{a}$$

Hence, by Eqs. (8-155) and (8-158), we have

$$V_0 = \frac{\sqrt{2}}{a} \int_0^a dx \int_0^b dy \sin^2 \frac{\pi x}{a} = \frac{b}{\sqrt{2}}$$

and so, by Eqs. (8-157) and (8-159),

$$n^2 = \frac{a}{b} = 2.25$$

The shunt susceptance B is one-half that for the diaphragm of Fig. 8-22b. An approximation to B is therefore given by Eq. (8-120) with B replaced by $B/2$, b by $a/2$, and c by $b/2$, giving

$$\frac{B}{Y_0} \approx \frac{8a}{\lambda_g} \log \csc \frac{\pi b}{2a} = 3.54 \frac{a}{\lambda_g}$$

Hence, the terminating admittance seen by the waveguide is

$$Y \approx j3.54 \frac{a}{\lambda_g} + 2.25 Y_{\text{apert}}$$

where Y_{apert} is given by the $a/b = 2.25$ curves of Fig. 8-29.

8-13. Modal Expansions in Cavities. Consider a cavity formed by a perfect conductor enclosing a dielectric medium. Each mode must

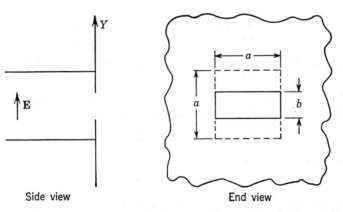

Fig. 8-30. A square waveguide feeding a rectangular aperture in a ground plane.

satisfy the field equations

$$\nabla \times \mathbf{E}_i = -j\omega_i\mu\mathbf{H}_i \qquad \nabla \times \mathbf{H}_i = j\omega_i\epsilon\mathbf{E}_i \qquad (8\text{-}160)$$

where i is a mode index. Either $\mathbf{E}_i$ or $\mathbf{H}_i$ may be eliminated from the above pair of equations, giving the wave equations

$$\begin{aligned}\nabla \times (\mu^{-1}\nabla \times \mathbf{E}_i) - \omega_i^2\epsilon\mathbf{E}_i = 0 \\ \nabla \times (\epsilon^{-1}\nabla \times \mathbf{H}_i) - \omega_i^2\mu\mathbf{H}_i = 0\end{aligned} \qquad (8\text{-}161)$$

valid even if ϵ and μ are functions of position. Each of these wave equations, coupled with the boundary condition

$$\mathbf{n} \times \mathbf{E}_i = \mathbf{n} \times (\epsilon^{-1}\nabla \times \mathbf{H}_i) = 0 \qquad \text{on } S \qquad (8\text{-}162)$$

where $\mathbf{n}$ is the unit normal directed outward from the cavity boundary S, is an eigenvalue problem in the classical sense.[1] Hence, for ϵ and μ real (no dissipation), the eigenvalues ω_i (resonant frequencies) are real, and the eigenfunctions $\mathbf{E}_i$, $\mathbf{H}_i$ form a complete orthogonal set in the Hermitian sense. Furthermore, we wish to normalize the mode vectors, so that the orthogonality relationships are

$$\iiint \epsilon \mathbf{E}_i \cdot \mathbf{E}_j^* \, d\tau = \begin{cases} 0 & i \neq j \\ 1 & i = j \end{cases} \qquad (8\text{-}163)$$

which can be derived from Eqs. (8-160) in the usual manner. Normalization of the $\mathbf{E}_i$ also normalizes the $\mathbf{H}_i$, because

$$\iiint \epsilon |E_i|^2 \, d\tau = \iiint \mu |H_i|^2 \, d\tau$$

that is, the time-average electric and magnetic energies are equal. Hence, the orthogonality relationships for the $\mathbf{H}_i$ corresponding to the orthonormal $\mathbf{E}_i$ are

$$\iiint \mu \mathbf{H}_i \cdot \mathbf{H}_j^* \, d\tau = \begin{cases} 0 & i \neq j \\ 1 & i = j \end{cases} \qquad (8\text{-}164)$$

We have already shown in Sec. 8-4 that if $\mathbf{E}_i$ is chosen real, then the corresponding $\mathbf{H}_i$ is imaginary, and vice versa.

Now suppose that electric sources exist within the cavity, as suggested by Fig. 8-31a. The field equations are then

$$\nabla \times \mathbf{E} = -j\omega\mu\mathbf{H} \qquad \nabla \times \mathbf{H} = j\omega\epsilon\mathbf{E} + \mathbf{J}$$

and the wave equation is

$$\nabla \times (\mu^{-1}\nabla \times \mathbf{E}) - \omega^2\epsilon\mathbf{E} = -j\omega\mathbf{J} \qquad (8\text{-}165)$$

---

[1] Philip M. Morse and Herman Feshbach, "Methods of Theoretical Physics," chap. 6, part I, McGraw-Hill Book Company, New York, 1953.

FIG. 8-31. A cavity containing (a) electric sources, and (b) magnetic sources.

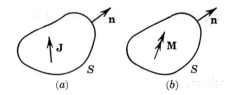

Because the E_i are a complete set, we can let

$$\mathbf{E} = \sum_i A_i \mathbf{E}_i \qquad (8\text{-}166)$$

Substituting this into Eq. (8-165), we have

$$\sum_i A_i [\nabla \times (\mu^{-1} \nabla \times \mathbf{E}_i) - \omega^2 \epsilon \mathbf{E}_i] = -j\omega \mathbf{J}$$

which, by Eq. (8-161), can be written as

$$\sum_i A_i (\omega_i^2 - \omega^2) \epsilon \mathbf{E}_i = -j\omega \mathbf{J}$$

If each side is now multiplied scalarly by $\mathbf{E}_j^*$ and integrated over the volume of the cavity, all terms except $i = j$ vanish because of orthogonality [Eq. (8-163)], and we have

$$(\omega_i^2 - \omega^2) A_i = -j\omega \iiint \mathbf{J} \cdot \mathbf{E}_i^* \, d\tau \qquad (8\text{-}167)$$

which determines the A_i. Hence, Eq. (8-166) becomes

$$\mathbf{E} = \sum_i \frac{j\omega \mathbf{E}_i}{\omega^2 - \omega_i^2} \iiint \mathbf{J} \cdot \mathbf{E}_i^* \, d\tau \qquad (8\text{-}168)$$

and the corresponding $\mathbf{H}$, obtained from the field equations, is

$$\mathbf{H} = \sum_i \frac{j\omega_i \mathbf{H}_i}{\omega^2 - \omega_i^2} \iiint \mathbf{J} \cdot \mathbf{E}_i^* \, d\tau \qquad (8\text{-}169)$$

Note that the field becomes extremely large as ω approaches some resonant frequency. In fact, the field becomes infinite at a resonant frequency in the loss-free case, which is to be expected. Actually, in any physical problem there will always be some dissipation; so the ω_i are complex. Hence, the field is large, but finite, at all real resonant frequencies.

The dual problem is that of magnetic sources in a cavity, represented by Fig. 8-31b. In this case, the wave equation in $\mathbf{H}$ is

$$\nabla \times (\epsilon^{-1} \nabla \times \mathbf{H}) - \omega^2 \mu \mathbf{H} = -j\omega \mathbf{M} \qquad (8\text{-}170)$$

We then expand **H** in terms of the orthonormal mode vectors $\mathbf{H}_i$ as

$$\mathbf{H} = \sum_i B_i \mathbf{H}_i \qquad (8\text{-}171)$$

where, dual to Eq. (8-167), the B_i are given by

$$(\omega_i^2 - \omega^2) B_i = -j\omega \iiint \mathbf{M} \cdot \mathbf{H}_i^* \, d\tau \qquad (8\text{-}172)$$

Hence, the expansion of **H** due to magnetic currents **M** is[1]

$$\mathbf{H} = \sum_i \frac{j\omega \mathbf{H}_i}{\omega^2 - \omega_i^2} \iiint \mathbf{M} \cdot \mathbf{H}_i^* \, d\tau \qquad (8\text{-}173)$$

and the corresponding **E** field is

$$\mathbf{E} = \sum_i \frac{j\omega_i \mathbf{E}_i}{\omega^2 - \omega_i^2} \iiint \mathbf{M} \cdot \mathbf{H}_i^* \, d\tau \qquad (8\text{-}174)$$

If both electric and magnetic sources exist within the cavity, we can superimpose Eqs. (8-168) and (8-174) for a solution.

8-14. Probes in Cavities. Mathematically, we can represent a probe in a cavity in terms of electric currents in the cavity, as shown in Fig. 8-31a. The impedance seen at the input terminals to the probe can then be calculated by the variational formula

$$Z_{\text{in}} = -\frac{1}{I^2} \iiint \mathbf{E} \cdot \mathbf{J}^a \, d\tau \qquad (8\text{-}175)$$

where $\mathbf{J}^a$ is the assumed current distribution on the probe, and I is the corresponding input current. All mode vectors $\mathbf{E}_i$ will be chosen real; so the field produced by $\mathbf{J}^a$ is given by Eq. (8-168) with the * dropped. Substituting this equation into Eq. (8-175), we obtain

$$Z_{\text{in}} = -\frac{j\omega}{I^2} \sum_i \frac{a_i^2}{\omega^2 - \omega_i^2} \qquad (8\text{-}176)$$

where

$$a_i = \iiint \mathbf{E}_i \cdot \mathbf{J}^a \, d\tau \qquad (8\text{-}177)$$

The analysis neglects the effect of the aperture through which the probe is fed. This effect is usually negligible and can be taken into account by the methods of the next section.

As long as there is no dissipation, the input impedance will be purely reactive. However, if the cavity is lossy but high Q, the effect of dissipa-

---

[1] The eigenvalue $\omega_i = 0$ must be included in both Eqs. (8-168) and (8-173). The modes associated with $\omega_i = 0$ account for the irrotational parts of **E** and **H**. See, for example, Teichmann and Wigner, *J. Appl. Phy.*, vol. 24, March, 1953.

tion can be taken into account by letting the resonant frequencies be complex, according to[1]

$$\omega_i^2 = \omega_r^2 \left(1 + \frac{j}{Q}\right) \quad (8\text{-}178)$$

where Q is the quality factor. In the vicinity of a resonant frequency, say ω_0 (not necessarily the dominant resonant frequency), we can approximate Eq. (8-176) by

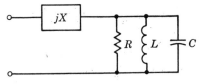

FIG. 8-32. An equivalent circuit for a probe-fed cavity in the vicinity of resonance.

$$Z_{in} \approx jX - \frac{j\omega(a_0/I)^2}{\omega^2 - \omega_0^2(1 + j/Q)} \quad (8\text{-}179)$$

where X is the reactance due to all modes except the $i = 0$ mode

$$X = -\frac{\omega}{I^2} \sum_{i \neq 0} \frac{a_i^2}{\omega^2 - \omega_i^2} \quad (8\text{-}180)$$

The effect of dissipation in modes not near resonance is negligible; hence, it is not included in Eq. (8-180). An equivalent circuit which represents Eq. (8-179) is shown in Fig. 8-32. To determine the values of R, L, and C, we need only compare the formula for the impedance of the parallel RLC circuit

$$Z = -\frac{j\omega/C}{\omega^2 - \omega_0^2(1 + j/Q)} \qquad \begin{aligned} \omega_0^2 &= \frac{1}{LC} \\ Q &= \frac{R}{\omega L} \approx \frac{R}{\omega_0 L} \end{aligned}$$

with the last term of Eq. (8-179). It is then evident that

$$R = \frac{Q}{\omega_0}\left(\frac{a_0}{I}\right)^2 \qquad L = \left(\frac{a_0}{I\omega_0}\right)^2 \qquad C = \left(\frac{I}{a_0}\right)^2 \quad (8\text{-}181)$$

where a_0 is obtained from Eq. (8-177).

To illustrate the theory, consider a probe in a rectangular cavity (Fig. 8-33). The normalized mode vector of the dominant mode is

$$\mathbf{E}_0 = \mathbf{u}_x \frac{2}{\sqrt{\epsilon abc}} \sin\frac{\pi y}{b} \sin\frac{\pi z}{c} \quad (8\text{-}182)$$

where the normalization factor was obtained from Eq. (2-97). For the current on the probe, we assume

$$J_z^a = \begin{cases} I\dfrac{\sin k(d-x)}{\sin kd}\delta(y-b')\delta(y-c') & x < d \\ 0 & x > d \end{cases} \quad (8\text{-}183)$$

[1] M. E. Van Valkenberg, "Network Analysis," p. 364, Prentice-Hall, Inc., Englewood Cliffs, N.J., 1955.

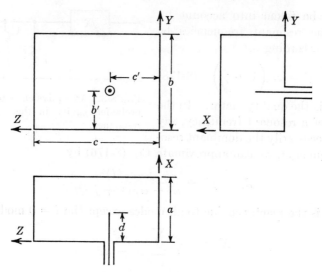

Fig. 8-33. Probe in a rectangular cavity.

Then, by Eq. (8-177), we have

$$\frac{a_0}{I} = \frac{2}{k\sqrt{\epsilon abc}} \tan\left(k\frac{d}{2}\right) \sin\left(\pi \frac{b'}{b}\right) \sin\left(\frac{\pi c'}{c}\right) \qquad (8\text{-}184)$$

The other parameters needed to evaluate R, L, and C are the resonant frequency $f_r = \omega_0/2\pi$, given by Eq. (2-95), and the quality factor Q, given by Eq. (2-101). The evaluation of the series reactance X is a much more difficult problem. We cannot, of course, use the filamentary current of Eq. (8-183) to evaluate X, since the resulting reactance would be infinite. The actual diameter of the stub must be considered. To a very rough approximation, X will be of the same order of magnitude as for a stub over a ground plane. Hence, for short stubs, the reactance is capacitive.

When the stub is bent into a small loop and joined to the cavity wall, the system is often called a *loop feed*. The treatment of loops in cavities is essentially the same as the treatment of stubs, once a current is assumed on the loop. The series reactance X for small loops is inductive, in contrast to the small-stub case, for which it is capacitive. Some explicit loop feeds are considered in Probs. 8-24 and 8-26.

8-15. Aperture Coupling to Cavities. The general problem of coupling a cavity to a waveguide through an aperture is represented by Fig. 8-34a. For a variational treatment of the problem, we assume an aperture field $\mathbf{E}_t{}^a$ and conserve reaction according to

$$\iint_{\text{apert}} (\mathbf{E}_t{}^a \times \mathbf{H}^a \cdot d\mathbf{s})_{\text{guide}} = \iint_{\text{apert}} (\mathbf{E}_t{}^a \times \mathbf{H}^a \cdot d\mathbf{s})_{\text{cavity}} \qquad (8\text{-}185)$$

For a given $\mathbf{E}_t{}^a$, each side of this equation can be considered separately, which amounts to dividing the original problem into two parts, as shown in Fig. 8-34b and c. The equivalent current

$$\mathbf{M}_s{}^a = \mathbf{n} \times \mathbf{E}_t{}^a \tag{8-186}$$

in the cavity part is the negative of the terminating current in the waveguide part. The waveguide part of the problem is identical to the problems treated in Secs. 8-10 and 8-11, and is therefore of the form

$$\iint_{\text{apert}} (\mathbf{E}_t{}^a \times \mathbf{H}^a \cdot d\mathbf{s})_{\text{guide}} = -YV_0{}^2 + \sum_{n \neq 0} Y_n V_n{}^2$$

where the V_n are the various mode voltages, the Y_n are the mode-characteristic admittances, and Y is the admittance seen by the dominant mode. Hence, we can rewrite Eq. (8-185) as

$$\frac{Y}{Y_0} = jB_g - \frac{1}{Y_0 V_0{}^2} \iint_{\text{apert}} (\mathbf{E}_t{}^a \times \mathbf{H}^a \cdot d\mathbf{s})_{\text{cavity}} \tag{8-187}$$

where Y_0 is the characteristic admittance of the dominant mode and

$$B_g = -j \sum_{n \neq 0} \frac{Y_n}{Y_0} \left(\frac{V_n}{V_0}\right)^2 \tag{8-188}$$

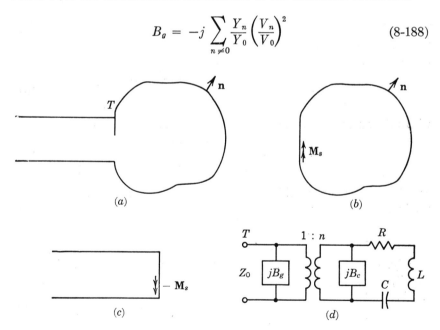

FIG. 8-34. (a) Aperture coupling from a waveguide to a cavity can be divided into two parts, (b) the cavity, and (c) the waveguide. An equivalent circuit in the vicinity of resonance is shown in (d).

is the shunt susceptance introduced by the waveguide part of the problem. The calculation of B_g was treated in Sec. 8-10.

For the cavity part of the problem, we can determine the field by Eq. (8-173) with the current given by Eq. (8-186). Taking the mode vectors $\mathbf{H}_i$ as real, we obtain

$$\mathbf{H}^a = \sum_i \frac{j\omega \mathbf{H}_i}{\omega^2 - \omega_i^2} \iint \mathbf{E}_t^a \times \mathbf{H}_i \cdot d\mathbf{s}$$

The right-hand side of Eq. (8-185) is then given by

$$\iint_{\text{apert}} (\mathbf{E}_t^a \times \mathbf{H}^a \cdot d\mathbf{s})_{\text{cavity}} = \sum_i \frac{j\omega b_i^2}{\omega^2 - \omega_i^2} \qquad (8\text{-}189)$$

where

$$b_i = \iint \mathbf{E}_t^a \times \mathbf{H}_i \cdot d\mathbf{s} \qquad (8\text{-}190)$$

In the vicinity of a resonant frequency, we again take losses into account by Eq. (8-178), and Eq. (8-187) can be written as

$$\frac{Y}{Y_0} \approx jB_g - \frac{j\omega}{Y_0 V_0^2} \left[\sum_{i \neq 0} \frac{b_i^2}{\omega^2 - \omega_i^2} + \frac{b_0^2}{\omega^2 - \omega_0^2(1 + j/Q)} \right]$$

The first term in the brackets represents the susceptance due to all non-resonant modes in the cavity, and the second term gives the resonant-mode effect. The above equation can therefore be written as

$$\frac{Y}{Y_0} \approx jB_g + \frac{n^2}{Y_0} \left[jB_c - \frac{j\omega(b_0/V)^2}{\omega^2 - \omega_0^2(1 + j/Q)} \right] \qquad (8\text{-}191)$$

where the susceptance due to nonresonant cavity modes is

$$B_c = -\frac{\omega}{V^2} \sum_{i \neq 0} \frac{b_i^2}{\omega^2 - \omega_i^2} \qquad (8\text{-}192)$$

and, to account for an arbitrary reference-voltage V, we have introduced the ideal transformer

$$n^2 = \left(\frac{V}{V_0}\right)^2 \qquad (8\text{-}193)$$

Finally, we can represent the last term of Eq. (8-191) as a series RLC circuit, as shown in Fig. 8-34d. The formula for admittance of a series RLC circuit is

$$Y = \frac{-j\omega/L}{\omega^2 - \omega_0^2(1 + j/Q)} \qquad \omega_0^2 = \frac{1}{LC}$$

$$Q = \frac{1}{\omega CR} \approx \frac{1}{\omega_0 CR}$$

MICROWAVE NETWORKS

Comparing this with the last term of Eq. (8-191), we see that

$$\frac{1}{R} = \frac{Q}{\omega_0}\left(\frac{b_0}{V}\right)^2 \qquad C = \left(\frac{b_0}{V\omega_0}\right)^2 \qquad L = \left(\frac{V}{b_0}\right)^2 \qquad (8\text{-}194)$$

where b_0 is obtained from Eq. (8-190).

Let us illustrate the above theory by a treatment of the rectangular waveguide to rectangular cavity junction, shown in Fig. 8-35. The waveguide part of the problem is identical to problems previously considered. In particular, B_g will be approximately one-half of Eq. (8-120) with the appropriate interchange of symbols, or

$$\frac{B_g}{Y_0} \approx \frac{4a'}{\lambda_g} \log \csc \frac{\pi d}{2a'} \qquad (8\text{-}195)$$

For the cavity part of the problem, let us make our often-used assumption

$$\mathbf{E}_t{}^a = \mathbf{u}_x \sin \frac{\pi y}{b} \qquad (8\text{-}196)$$

in the aperture. Also, let us refer the cavity admittances to

$$V = \sqrt{\frac{bd}{2}} \qquad (8\text{-}197)$$

which is the waveguide dominant-mode voltage that would be excited by Eq. (8-196) if the waveguide were the same height as the aperture (n^2 would be 1 in that case). In our particular problem, the waveguide dominant-mode voltage is $V_0 = \sqrt{ba'/2}$; hence

$$n^2 = \frac{d}{a'} \qquad (8\text{-}198)$$

Rather than calculating Eq. (8-192) directly, let us view the aperture as the junction between two waveguides of height a' and a. The suscept-

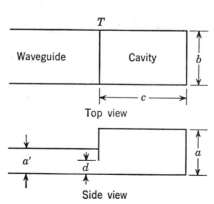

Fig. 8-35. Aperture coupling from a rectangular waveguide to a rectangular cavity.

ance B_c referred to the mode voltage of a waveguide of height a could then be approximated by Eq. (8-195) with a' replaced by a. But we wish to refer it to the V of Eq. (8-197); so we should multiply by d/a and obtain

$$B_c \approx \frac{4d}{\lambda_g} \log \csc \frac{\pi d}{2a} \qquad (8\text{-}199)$$

Finally, to determine the R, L, and C, we need the normalized dominant-mode vector, which is

$$\mathbf{H}_0 = \frac{2}{\sqrt{\mu abc(b^2 + c^2)}} \left(\mathbf{u}_y b \sin \frac{\pi y}{b} \cos \frac{\pi z}{c} - \mathbf{u}_z c \cos \frac{\pi y}{b} \sin \frac{\pi z}{c} \right) \qquad (8\text{-}200)$$

Hence, from Eqs. (8-190), (8-196), and (8-197), we obtain

$$\left(\frac{b_0}{V}\right)^2 = \frac{2d}{\mu ac(1 + c/b)^2} \qquad (8\text{-}201)$$

The resonant frequency $f_r = \omega_0/2\pi$ is given by Eq. (2-95) and the quality factor Q by Eq. (2-101). Hence, all parameters of the equivalent circuit (Fig. 8-34d) have been evaluated.

PROBLEMS

8-1. Consider the parallel-plate waveguide formed by conductors covering the $y = 0$ and $y = b$ planes. Show that the eigenfunctions, normalized on a per unit width basis, are

$$\Psi_0{}^m = \frac{y}{\sqrt{b}}$$

$$\Psi_n{}^m = \frac{\sqrt{2b}}{n\pi} \sin \frac{n\pi y}{b}$$

$$\Psi_n{}^e = \frac{\sqrt{2b}}{n\pi} \cos \frac{n\pi y}{b}$$

where $n = 1, 2, 3, \ldots$.

8-2. Consider an x-directed current element Il at the point x', y', z' in a rectangular waveguide (Fig. 2-16). Show that the field is given by formulas of Table 8-1 where Ψ's are given by Eqs. (8-34) and, for $n, m \neq 0$,

$$V_{mn}{}^m = \sqrt{\frac{b}{a}} f_{mn} \qquad V_{mn}{}^e = -\sqrt{\frac{a}{b}} f_{mn}$$

where

$$f_{mn} = Il(Z_0)_{mn} \frac{\sqrt{(mb)^2 + (na)^2}}{mb^2 + na^2} \cos \frac{m\pi x'}{a} \sin \frac{n\pi y'}{b} e^{-\gamma_{mn}|z-z'|}$$

and, for $m = 0$,

$$V_{0n}{}^e = -Il(Z_0)_{0n} \sqrt{\frac{\epsilon_n}{2ab}} \sin \frac{n\pi y'}{b} e^{-\gamma_{0n}|z-z'|}$$

8-3. For the general cylindrical waveguide (Fig. 8-1), show that the time-average electric energy per unit length of guide is

$$\bar{W}_e = \frac{1}{2}\sum_i \epsilon|V_i^e|^2 + \epsilon|V_i^m|^2 + \epsilon\left(\frac{k_{ci}^m}{\omega\epsilon}\right)^2 |I_i^m|^2$$

and the time-average magnetic energy per unit length of guide is

$$\bar{W}_m = \frac{1}{2}\sum_i \mu|I_i^m|^2 + \mu|I_i^e|^2 + \mu\left(\frac{k_{ci}^e}{\omega\mu}\right)^2 |V_i^e|^2$$

Note that these are just the sum of the energies in each mode alone.

8-4. Let the T equivalent circuit of Fig. 8-10b represent a section of waveguide of length l, propagation constant $j\beta$, and characteristic impedance Z_0. Show that

$$Z_a = -jZ_0 \csc \beta l$$
$$Z_b = jZ_0 \tan \beta l/2$$

8-5. Using the usual perturbational method, show that, for general cylindrical waveguides, the attenuation constant due to conductor loss is

$$\alpha_c = \frac{1}{2}\frac{\mathfrak{R}}{\eta}\frac{k}{\beta} k_c^2 \oint \left(\frac{\partial \Psi^m}{\partial n}\right)^2 dl$$

for TM modes, and

$$\alpha_c = \frac{1}{2}\frac{\mathfrak{R}}{\eta}\frac{\beta}{k}\left[\oint \left(\frac{\partial \Psi^e}{\partial l}\right)^2 dl + \frac{k_c^4}{\beta^2}\oint (\Psi^e)^2 dl\right]$$

for TE modes. $\mathfrak{R}$ denotes intrinsic resistance of the metal walls, η intrinsic impedance of the dielectric, and the other symbols correspond to their usage in Table 8-1.

8-6. Use the above formulas to determine the attenuation in rectangular waveguides (Prob. 4-4) and in circular waveguides (Prob. 5-9).

8-7. Consider a one-port network, and define the reflection coefficient $\Gamma = V^r/V^i$. Show that, for Z_0 real,

$$\bar{\mathcal{P}}_d = (1 - |\Gamma|^2)|V^i|^2/Z_0$$

and

$$\bar{W}_m - \bar{W}_e = \frac{1}{\omega}|V^i|^2 \text{ Im }(\Gamma)/Z_0$$

Hence, in a source-free network, $|\Gamma|^2 \leq 1$, and, at resonance, Γ is real.

8-8. Derive Eqs. (8-72).

8-9. Let the characteristic impedances of ports (1) and (2) of Fig. 8-7 be normalized to unity. Show that the transmission matrix $[T]$ is related to the impedance matrix $[z]$ by

$$2T_{11} = z_{21} + \frac{1}{z_{12}}(1 - z_{11})(z_{22} - 1)$$

$$2T_{12} = -z_{21} + \frac{1}{z_{12}}(1 + z_{11})(z_{22} - 1)$$

$$2T_{21} = z_{21} + \frac{1}{z_{12}}(1 - z_{11})(z_{22} + 1)$$

$$2T_{22} = -z_{21} + \frac{1}{z_{12}}(1 + z_{11})(z_{22} + 1)$$

Show that in the loss-free case $T_{11} = T_{22}^*$ and $T_{12} = T_{21}^*$.

8-10. Add a magnetic current sheet $\mathbf{M}_s$ coincident with the electric current sheet $\mathbf{J}_s$ of Fig. 8-11. Determine $\mathbf{M}_s$ and $\mathbf{J}_s$ such that they are a unidirectional dominant-mode source, sending waves in the $+z$ direction only. Determine the self-reaction of this source in the presence of the magnetic conductor terminating the guide.

8-11. Derive Eq. (8-96).

8-12. Consider the centered capacitive post in a rectangular waveguide, shown in Fig. 8-36. Show that the equivalent network parameters are

$$\frac{B_a}{Y_0} \approx \frac{Y_0}{B_b} \approx \frac{\pi^2 d^2}{2a\lambda_g}$$

The approximations are good for $d/a < 0.3$ and $a/\lambda_g < 0.2$.

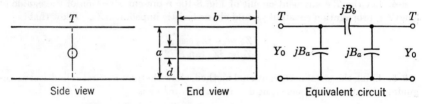

Fig. 8-36. Centered capacitive post in a rectangular waveguide.

8-13. Consider the inductive diaphragm of Fig. 8-23. Approximating $\mathbf{E}_t$ in the aperture by

$$\mathbf{E}_t{}^a = \mathbf{u}_y \sin \frac{\pi x}{c}$$

show that Eq. (8-122) is a crude variational solution for the shunt susceptance.

8-14. The inductive diaphragm (Fig. 8-23) has boundaries cylindrical to y. The incident mode is TM to y; hence, the entire field must be TM to y. Express the field as $\mathbf{H} = \nabla \times \mathbf{u}_y \psi$ where

$$\psi = \sum_{n=1}^{\infty} A_n \sin \frac{n\pi x}{a} e^{\gamma_n z}$$

In the aperture, tangential $\mathbf{E}$ must be of the form

$$\mathbf{E}_t = \mathbf{u}_y f(x)$$

Show that

$$-\frac{B}{Y_0} \approx \frac{2\lambda_g}{a} \frac{\sum_{n=2}^{\infty} \sqrt{\left(\frac{n}{2}\right)^2 - \left(\frac{a}{\lambda}\right)^2} \left[\int_0^c f(x) \sin \frac{n\pi x}{a}\, dx\right]^2}{\left[\int_0^c f(x) \sin \frac{\pi x}{a}\, dx\right]^2}$$

is a variational formula for the shunt susceptance. Note that it gives upper bounds to $-B/Y_0$. Problem 8-13 is the special case $f(x) = \sin(\pi x/c)$.

8-15. Consider the inductive diaphragm (Fig. 8-23) and the variational formula in terms of obstacle current [Eq. (8-107)]. On the diaphragm, the current is of the form

$$\mathbf{J}_s = \mathbf{u}_y g(x)$$

Show that

$$-\frac{Y_0}{B} \approx \frac{a}{2\lambda_g} \sum_{n=2}^{\infty} \frac{1}{\sqrt{(n/2)^2 - (a/\lambda)^2}} \frac{\left[\int_c^a g(x) \sin \frac{n\pi x}{a} dx\right]^2}{\left[\int_c^a g(x) \sin \frac{\pi x}{a} dx\right]^2}$$

is the variational formula for lower bounds to $-B/Y_0$.

8-16. Show that the shunt susceptance of the capacitive diaphragm of Fig. 8-37 is given by the same formula as applies to Fig. 8-22a.

FIG. 8-37. A capacitive diaphragm (metal shown dashed).

8-17. Consider the capacitive junction of Fig. 8-38. Show that the parameters of the equivalent circuit are

$$\frac{B^+}{Y_0} \approx \frac{4b^+}{\lambda_g} \log \csc \frac{\pi c}{2b^+}$$

$$\frac{B^-}{Y_0} \approx \frac{4b^-}{\lambda_g} \log \csc \frac{\pi c}{2b^-}$$

$$n^2 = \frac{b^-}{b^+}$$

Use the approximation of Eq. (8-120).

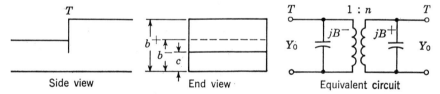

FIG. 8-38. A capacitive junction.

8-18. Consider the waveguide junction of Fig. 8-24a and the equivalent circuit of Fig. 8-25. Show that, analogous to Eqs. (8-138),

$$\frac{Y_0^-}{jB^-} = \frac{\sum_i Z_i I_i^2}{Z_0^- I_0^2} \qquad \frac{Y_0^+}{jB^+} = \frac{\sum_i \hat{Z}_i \hat{I}_i^2}{Z_0^+ \hat{I}_0^2}$$

and $n^2 = I_0^2/\hat{I}_0^2$. The mode currents are given by

$$I_i = \iint \mathbf{H}_t^- \cdot \mathbf{h}_i \, ds \qquad \hat{I}_i = \iint \mathbf{H}_t^+ \cdot \hat{\mathbf{h}}_i \, ds$$

where $\mathbf{H}_t^+$ and $\mathbf{H}_t^-$ denote tangential $\mathbf{H}$ on the $+z$ and $-z$ sides of the junction, respectively. Variational formulas are obtained by assuming $\mathbf{H}_t^+$ and $\mathbf{H}_t^-$ subject to the restriction $\mathbf{H}_t^+ = \mathbf{H}_t^-$ in the aperture.

8-19. Let $\psi(x,y) = f(\rho,\phi)$ be a solution to the two-dimensional source-free Helmholtz equation $\rho < a$. Prove that

$$\int_0^{2\pi} f(a,\phi)e^{jn\phi}\,d\phi = \frac{j^n}{2\pi} J_n(ka)[e^{jnD}\psi(0)]$$

where e^{jnD} is an operator defined by

$$\sin D = \frac{1}{jk}\frac{\partial}{\partial x} \qquad \cos D = \frac{1}{jk}\frac{\partial}{\partial z}$$

and $e^{jnD}\psi(0)$ means $e^{jnD}\psi(x,y)$ evaluated at $x = 0$, $y = 0$. This is a kind of mean-value theorem.

8-20. Consider the coax to waveguide feed of Fig. 4-20. Let d denote the diameter of the coaxial stub, and let $a \ll \lambda$. Show that, for the equivalent circuit of Fig. 8-26b,

$$n^2 \approx \frac{2a}{b}\sin^2\frac{\pi c}{b}$$

$$X \approx -\eta\frac{a}{\lambda}\log\frac{\gamma k d}{2}$$

where $\gamma = 1.781$.

8-21. Let the rectangular aperture of Fig. 8-29 be thin ($b \ll \lambda$) and of resonant length ($a \approx \lambda/2$). Show that

$$Y_{\text{apert}} \approx 0.004\frac{b}{\lambda}$$

Hint: Use the duality concept of Prob. 7-43 and the approximations of Prob. 7-39. Note that the aperture radiates only into half-space.

8-22. Figure 8-39 represents a parallel-plate transmission line radiating through a slot into half-space. Let Fig. 8-28b represent the equivalent circuit, and evaluate the parameters, using the aperture admittance of Fig. 4-22.

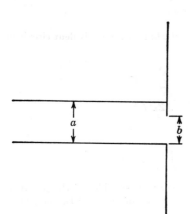

FIG. 8-39. A parallel-plate transmission line radiating into half-space.

MICROWAVE NETWORKS 445

8-23. Figure 8-40 represents a rectangular waveguide having sides a, b radiating into half-space through a narrow resonant slot. Using the results of Prob. 8-21, show that reflectionless transmission through the slot occurs when

$$\frac{a}{\lambda} \approx \frac{0.54 \cos^2(\pi\lambda/4b)}{\frac{b}{\lambda}\left[1 - \left(\frac{\lambda}{2b}\right)^2\right]^{3/2}}$$

When $b/\lambda < 0.7$, the above formula can be approximated by

$$\frac{a}{\lambda} \approx \frac{2}{3}\sqrt{1 - \left(\frac{\lambda}{2b}\right)^2}$$

The waveguide is excited in the dominant mode.

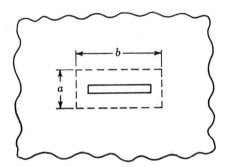

Fig. 8-40. A rectangular waveguide radiating into half-space through a resonant slot.

8-24. Consider the loop-fed rectangular cavity of Fig. 8-41. Assume that the loop is small, so that the current on it may be assumed constant. Show that the elements of the equivalent circuit (Fig. 8-32) are given by Eqs. (8-181) where

$$\frac{a_0}{I} = \frac{2d}{\sqrt{\epsilon abc}} \sin\left(\pi \frac{b'}{b}\right) \sin\left(\pi \frac{c'}{c}\right)$$

When $c' \ll c$, this reduces to

$$\frac{a_0}{I} \approx \frac{2\pi A}{c \sqrt{\epsilon abc}} \sin\left(\pi \frac{b'}{b}\right)$$

where $A = c'd$ is the area of the loop.

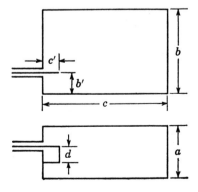

Fig. 8-41. A loop-fed rectangular cavity.

8-25. Show that the normalized mode vector for the dominant mode of the circular cavity (Fig. 8-42) is

$$\mathbf{E}_0 \approx \mathbf{u}_z \frac{1}{a\sqrt{\epsilon\pi b}\, J_1(x_{01})} J_0\left(x_{01}\frac{\rho}{a}\right)$$

where $x_{01} = 2.405$.

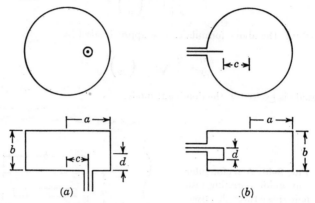

FIG. 8-42. A circular cavity with (a) probe feed, and (b) loop feed.

8-26. Figure 8-42a represents a probe-fed circular cavity. Assume sinusoidal distribution of current on the probe, and show that the elements of the equivalent circuit (Fig. 8-32) are given by Eqs. (8-181) where

$$\frac{a_0}{I} = \frac{1}{ka\sqrt{\epsilon\pi b}\, J_1(x_{01})} \tan\left(k\frac{d}{2}\right) J_0\left(x_{01}\frac{c}{a}\right)$$

and $x_{01} = 2.405$.

8-27. Figure 8-42b represents a loop-fed circular cavity. Assume uniform current on the loop, and show that the elements of the equivalent circuit (Fig. 8-32) are given by Eqs. (8-181) where

$$\frac{a_0}{I} = \frac{d}{a\sqrt{\epsilon\pi b}\, J_1(x_{01})} J_0\left(x_{01}\frac{c}{a}\right)$$

Show that, when $c \approx a$, this reduces to

$$\frac{a_0}{I} \approx \frac{Ax_{01}}{a^2\sqrt{\epsilon\pi b}}$$

where $A = (a - c)d$ is the area of the loop.

8-28. Reconsider Fig. 8-41 for the case of a small loop. Represent the loop by a magnetic-current element Kl, according to Fig. 3-3, and evaluate

$$R = -\frac{\text{Re}\,\langle a,a\rangle}{I^2} = \frac{Kl \cdot \mathbf{H}}{I^2}$$

The result is the same as the small-loop answer in Prob. 8-24.

8-29. Reconsider Fig. 8-42b by the method outlined in Prob. 8-28. Show that the result is the same as the small-loop answer of Prob. 8-26.

8-30. Show that the normalized $\mathbf{H}$ mode vector for the dominant mode of the spherical cavity (Fig. 6-2) is

$$\mathbf{H} = \mathbf{u}_\phi \frac{0.536}{r\sqrt{a\mu}} \hat{J}_1\left(2.744\frac{r}{a}\right) \sin\theta$$

APPENDIX A

VECTOR ANALYSIS

We shall normally orient rectangular (x,y,z), cylindrical (ρ,ϕ,z), and spherical (r,θ,ϕ) coordinates as shown in Fig. A-1. Coordinate transformations are then given by

$$
\begin{aligned}
x &= \rho \cos \phi = r \sin \theta \cos \phi \\
y &= \rho \sin \phi = r \sin \theta \sin \phi \\
z &= r \cos \theta \\
\rho &= \sqrt{x^2 + y^2} = r \sin \theta \\
\phi &= \tan^{-1} \frac{y}{x} \\
r &= \sqrt{x^2 + y^2 + z^2} = \sqrt{\rho^2 + z^2} \\
\theta &= \tan^{-1} \frac{\sqrt{x^2 + y^2}}{z} = \tan^{-1} \frac{\rho}{z}
\end{aligned}
\qquad \text{(A-1)}
$$

Transformations of the coordinate components of a vector among the three coordinate systems are given by

$$
\begin{aligned}
A_x &= A_\rho \cos \phi - A_\phi \sin \phi \\
&= A_r \sin \theta \cos \phi + A_\theta \cos \theta \cos \phi - A_\phi \sin \phi \\
A_y &= A_\rho \sin \phi + A_\phi \cos \phi \\
&= A_r \sin \theta \sin \phi + A_\theta \cos \theta \sin \phi + A_\phi \cos \phi \\
A_z &= A_r \cos \theta - A_\theta \sin \theta \\
A_\rho &= A_x \cos \phi + A_y \sin \phi = A_r \sin \theta + A_\theta \cos \theta \\
A_\phi &= - A_x \sin \phi + A_y \cos \phi \\
A_r &= A_x \sin \theta \cos \phi + A_y \sin \theta \sin \phi + A_z \cos \theta \\
&= A_\rho \sin \theta + A_z \cos \theta \\
A_\theta &= A_x \cos \theta \cos \phi + A_y \cos \theta \sin \phi - A_z \sin \theta \\
&= A_\rho \cos \theta - A_z \sin \theta
\end{aligned}
\qquad \text{(A-2)}
$$

The coordinate-unit vectors in the three systems are denoted by $(\mathbf{u}_x,\mathbf{u}_y,\mathbf{u}_z)$, $(\mathbf{u}_\rho,\mathbf{u}_\phi,\mathbf{u}_z)$, and $(\mathbf{u}_r,\mathbf{u}_\theta,\mathbf{u}_\phi)$. Differential elements of volume are

$$
d\tau = dx\, dy\, dz = \rho\, d\rho\, d\phi\, dz = r^2 \sin \theta\, dr\, d\theta\, d\phi \qquad \text{(A-3)}
$$

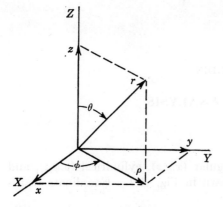

Fig. A-1. Normal coordinate orientation.

differential elements of vector area are

$$ds = \mathbf{u}_x\,dy\,dz + \mathbf{u}_y\,dx\,dz + \mathbf{u}_z\,dx\,dy$$
$$= \mathbf{u}_\rho \rho\,d\phi\,dz + \mathbf{u}_\phi\,d\rho\,dz + \mathbf{u}_z \rho\,d\rho\,d\phi \qquad (A\text{-}4)$$
$$= \mathbf{u}_r r^2 \sin\theta\,d\theta\,d\phi + \mathbf{u}_\theta r \sin\theta\,dr\,d\phi + \mathbf{u}_\phi r\,dr\,d\theta$$

and differential elements of vector length are

$$d\mathbf{l} = \mathbf{u}_x\,dx + \mathbf{u}_y\,dy + \mathbf{u}_z\,dz$$
$$= \mathbf{u}_\rho\,d\rho + \mathbf{u}_\phi \rho\,d\phi + \mathbf{u}_z\,dz \qquad (A\text{-}5)$$
$$= \mathbf{u}_r\,dr + \mathbf{u}_\theta r\,d\theta + \mathbf{u}_\phi r \sin\theta\,d\phi$$

The elementary algebraic operations are the same in all right-handed orthogonal coordinate systems. Letting $(\mathbf{u}_1, \mathbf{u}_2, \mathbf{u}_3)$ denote the unit vectors and (A_1, A_2, A_3) the corresponding vector components, we have addition defined by

$$\mathbf{A} + \mathbf{B} = \mathbf{u}_1(A_1 + B_1) + \mathbf{u}_2(A_2 + B_2) + \mathbf{u}_3(A_3 + B_3) \qquad (A\text{-}6)$$

scalar multiplication defined by

$$\mathbf{A} \cdot \mathbf{B} = A_1 B_1 + A_2 B_2 + A_3 B_3 \qquad (A\text{-}7)$$

and vector multiplication defined by

$$\mathbf{A} \times \mathbf{B} = \begin{vmatrix} \mathbf{u}_1 & \mathbf{u}_2 & \mathbf{u}_3 \\ A_1 & A_2 & A_3 \\ B_1 & B_2 & B_3 \end{vmatrix} \qquad (A\text{-}8)$$

The above formula is a determinant, to be expanded in the usual manner.

The differential operators that we have occasion to use are the gradient (∇w), divergence ($\nabla \cdot \mathbf{A}$), curl ($\nabla \times \mathbf{A}$), and Laplacian ($\nabla^2 w$). In rectangular coordinates we can think of del (∇) as the vector operator

$$\nabla = \mathbf{u}_x \frac{\partial}{\partial x} + \mathbf{u}_y \frac{\partial}{\partial y} + \mathbf{u}_z \frac{\partial}{\partial z} \qquad (A\text{-}9)$$

VECTOR ANALYSIS

and the various operations are

$$\nabla w = \mathbf{u}_x \frac{\partial w}{\partial x} + \mathbf{u}_y \frac{\partial w}{\partial y} + \mathbf{u}_z \frac{\partial w}{\partial z}$$

$$\nabla \cdot \mathbf{A} = \frac{\partial A_x}{\partial x} + \frac{\partial A_y}{\partial y} + \frac{\partial A_z}{\partial z}$$

$$\nabla \times \mathbf{A} = \begin{vmatrix} \mathbf{u}_x & \mathbf{u}_y & \mathbf{u}_z \\ \frac{\partial}{\partial x} & \frac{\partial}{\partial y} & \frac{\partial}{\partial z} \\ A_x & A_y & A_z \end{vmatrix} \quad (A\text{-}10)$$

$$\nabla^2 w = \frac{\partial^2 w}{\partial x^2} + \frac{\partial^2 w}{\partial y^2} + \frac{\partial^2 w}{\partial z^2}$$

In cylindrical coordinates we have

$$\nabla w = \mathbf{u}_\rho \frac{\partial w}{\partial \rho} + \mathbf{u}_\phi \frac{1}{\rho} \frac{\partial w}{\partial \phi} + \mathbf{u}_z \frac{\partial w}{\partial z}$$

$$\nabla \cdot \mathbf{A} = \frac{1}{\rho} \frac{\partial}{\partial \rho} (\rho A_\rho) + \frac{1}{\rho} \frac{\partial A_\phi}{\partial \phi} + \frac{\partial A_z}{\partial z}$$

$$\nabla \times \mathbf{A} = \mathbf{u}_\rho \left(\frac{1}{\rho} \frac{\partial A_z}{\partial \phi} - \frac{\partial A_\phi}{\partial z} \right) + \mathbf{u}_\phi \left(\frac{\partial A_\rho}{\partial z} - \frac{\partial A_z}{\partial \rho} \right) \quad (A\text{-}11)$$

$$+ \mathbf{u}_z \left[\frac{1}{\rho} \frac{\partial}{\partial \rho} (\rho A_\phi) - \frac{1}{\rho} \frac{\partial A_\rho}{\partial \phi} \right]$$

$$\nabla^2 w = \frac{1}{\rho} \frac{\partial}{\partial \rho} \left(\rho \frac{\partial w}{\partial \rho} \right) + \frac{1}{\rho^2} \frac{\partial^2 w}{\partial \phi^2} + \frac{\partial^2 w}{\partial z^2}$$

In spherical coordinates we have

$$\nabla w = \mathbf{u}_r \frac{\partial w}{\partial r} + \mathbf{u}_\theta \frac{1}{r} \frac{\partial w}{\partial \theta} + \mathbf{u}_\phi \frac{1}{r \sin \theta} \frac{\partial w}{\partial \phi}$$

$$\nabla \cdot \mathbf{A} = \frac{1}{r^2} \frac{\partial}{\partial r} (r^2 A_r) + \frac{1}{r \sin \theta} \frac{\partial}{\partial \theta} (A_\theta \sin \theta) + \frac{1}{r \sin \theta} \frac{\partial A_\phi}{\partial \phi}$$

$$\nabla \times \mathbf{A} = \mathbf{u}_r \frac{1}{r \sin \theta} \left[\frac{\partial}{\partial \theta} (A_\phi \sin \theta) - \frac{\partial A_\theta}{\partial \phi} \right]$$

$$+ \mathbf{u}_\theta \frac{1}{r} \left[\frac{1}{\sin \theta} \frac{\partial A_r}{\partial \phi} - \frac{\partial}{\partial r} (r A_\phi) \right] \quad (A\text{-}12)$$

$$+ \mathbf{u}_\phi \frac{1}{r} \left[\frac{\partial}{\partial r} (r A_\theta) - \frac{\partial A_r}{\partial \theta} \right]$$

$$\nabla^2 w = \frac{1}{r^2} \frac{\partial}{\partial r} \left(r^2 \frac{\partial w}{\partial r} \right) + \frac{1}{r^2 \sin \theta} \frac{\partial}{\partial \theta} \left(\sin \theta \frac{\partial w}{\partial \theta} \right) + \frac{1}{r^2 \sin^2 \theta} \frac{\partial^2 w}{\partial \phi^2}$$

A number of useful vector identities, which are independent of the choice of coordinate system, are as follows. For addition and multiplication we have

$$A^2 = \mathbf{A} \cdot \mathbf{A}$$
$$|A|^2 = \mathbf{A} \cdot \mathbf{A}^*$$
$$\mathbf{A} + \mathbf{B} = \mathbf{B} + \mathbf{A}$$
$$\mathbf{A} \cdot \mathbf{B} = \mathbf{B} \cdot \mathbf{A}$$
$$\mathbf{A} \times \mathbf{B} = -\mathbf{B} \times \mathbf{A} \quad \text{(A-13)}$$
$$(\mathbf{A} + \mathbf{B}) \cdot \mathbf{C} = \mathbf{A} \cdot \mathbf{C} + \mathbf{B} \cdot \mathbf{C}$$
$$(\mathbf{A} + \mathbf{B}) \times \mathbf{C} = \mathbf{A} \times \mathbf{C} + \mathbf{B} \times \mathbf{C}$$
$$\mathbf{A} \cdot \mathbf{B} \times \mathbf{C} = \mathbf{B} \cdot \mathbf{C} \times \mathbf{A} = \mathbf{C} \cdot \mathbf{A} \times \mathbf{B}$$
$$\mathbf{A} \times (\mathbf{B} \times \mathbf{C}) = (\mathbf{A} \cdot \mathbf{C})\mathbf{B} - (\mathbf{A} \cdot \mathbf{B})\mathbf{C}$$

For differentiation we have

$$\nabla(v + w) = \nabla v + \nabla w$$
$$\nabla \cdot (\mathbf{A} + \mathbf{B}) = \nabla \cdot \mathbf{A} + \nabla \cdot \mathbf{B}$$
$$\nabla \times (\mathbf{A} + \mathbf{B}) = \nabla \times \mathbf{A} + \nabla \times \mathbf{B}$$
$$\nabla(vw) = v\,\nabla w + w\,\nabla v$$
$$\nabla \cdot (w\mathbf{A}) = w\nabla \cdot \mathbf{A} + \mathbf{A} \cdot \nabla w$$
$$\nabla \times (w\mathbf{A}) = w\nabla \times \mathbf{A} - \mathbf{A} \times \nabla w \quad \text{(A-14)}$$
$$\nabla \cdot (\mathbf{A} \times \mathbf{B}) = \mathbf{B} \cdot \nabla \times \mathbf{A} - \mathbf{A} \cdot \nabla \times \mathbf{B}$$
$$\nabla^2 \mathbf{A} = \nabla(\nabla \cdot \mathbf{A}) - \nabla \times \nabla \times \mathbf{A}$$
$$\nabla \times (v\,\nabla w) = \nabla v \times \nabla w$$
$$\nabla \times \nabla w = 0$$
$$\nabla \cdot \nabla \times \mathbf{A} = 0$$

For integration we have

$$\iiint \nabla \cdot \mathbf{A}\,d\tau = \oiint \mathbf{A} \cdot d\mathbf{s}$$
$$\iint \nabla \times \mathbf{A} \cdot d\mathbf{s} = \oint \mathbf{A} \cdot d\mathbf{l}$$
$$\iiint \nabla \times \mathbf{A}\,d\tau = -\oiint \mathbf{A} \times d\mathbf{s} \quad \text{(A-15)}$$
$$\iiint \nabla w\,d\tau = \oiint w\,d\mathbf{s}$$
$$\iint \mathbf{n} \times \nabla w\,d\mathbf{s} = \oint w\,d\mathbf{l}$$

Finally, we have the Helmholtz identity

$$4\pi \mathbf{A} = -\nabla \iiint \frac{\nabla' \cdot \mathbf{A}}{|\mathbf{r} - \mathbf{r}'|}\,d\tau' + \nabla \times \iiint \frac{\nabla' \times \mathbf{A}}{|\mathbf{r} - \mathbf{r}'|}\,d\tau' \quad \text{(A-16)}$$

valid if **A** is well-behaved in all space and vanishes at least as rapidly as r^{-2} at infinity.

APPENDIX B

COMPLEX PERMITTIVITIES

The following is a table of relative a-c capacitivities ϵ_r' and relative dielectric loss factors ϵ_r'' where

$$\hat{\epsilon}_r = \frac{\hat{\epsilon}}{\epsilon_0} = \frac{\epsilon'}{\epsilon_0} - j\frac{\epsilon''}{\epsilon_0} = \epsilon_r' - j\epsilon_r''$$

is the relative complex permittivity. The measurements, along with many others, were reported in "Tables of Dielectric Materials" (vol. IV, *Mass. Inst. Technol., Research Lab. Insulation, Tech. Rept.*). They also appear in Part V of "Dielectric Materials and Applications," Technology Press, M.I.T., Cambridge, Mass., 1954.

Material	T°C		Frequency, cycles per second									
			10^2	10^3	10^4	10^5	10^6	10^7	10^8	3×10^8	3×10^9	10^{10}
Amber (fossil resin)	25	ϵ_r' $10^4\epsilon_r''$	2.7 34	2.7 49	2.7 84	2.7 116	2.65 148	2.65 180	....	2.6 223	2.6 234	
Bakelite (no filler)	24	ϵ_r' $10^3\epsilon_r''$	8.2 1100	7.15 585	6.5 410	5.9 330	5.4 320	4.9 360	4.4 340	....	3.64 190	3.52 130
Beeswax (white)	23	ϵ_r' $10^4\epsilon_r''$	2.65 360	2.63 310	2.56 680	2.48 470	2.43 205	2.41 165	2.39 145	....	2.35 120	2.35 113
Carbon tetrachloride	25	ϵ_r' $10^4\epsilon_r''$	2.17 130	2.17 17	2.17 0.9	2.17 1	2.17 1	2.17 5	2.17 5	2.17 3	2.17 8	2.17 35
Clay soil (dry)	25	ϵ_r' $10^3\epsilon_r''$	4.73 570	3.94 470	3.27 390	2.79 280	2.57 170	2.44 98	....	2.38 48	2.27 34	2.16 28
Ethyl alcohol (absolute)	25	ϵ_r' $10^2\epsilon_r''$	....	....	....	....	24.5 220	24.1 80	23.7 150	22.3 600	6.5 165	1.7 10
Fiberglas BK 174 (laminated)	24	ϵ_r' $10^2\epsilon_r''$	14.2 365	9.8 255	7.2 115	5.9 52	5.3 24	5.0 17	4.8 12.5	4.54 10	4.40 13	4.37 16
Glass, phosphate (2 per cent iron oxide)	25	ϵ_r' $10^4\epsilon_r''$	5.25 115	5.25 95	5.25 85	5.25 80	5.25 75	5.25 85	5.24 105	5.23 130	5.17 240	5.00 210
Glass, lead-barium	25	ϵ_r' $10^4\epsilon_r''$	6.78 160	6.77 120	6.76 100	6.75 85	6.73 85	6.72 95	6.70 115	6.69 130	....	6.64 470

Material	T (°C)											
Gutta-percha	25	ϵ_r' $10^4\epsilon_r''$	2.61 13	2.60 10	2.58 23	2.55 54	2.53 105	2.50 200	2.47 300	2.45 270	2.40 145	2.38 120
Loamy soil (dry)	25	ϵ_r' $10^4\epsilon_r''$	3.06 2100	2.83 1400	2.69 950	2.60 780	2.53 460	2.48 360	....	2.47 160	2.44 27	2.44 34
Lucite HM-119	23	ϵ_r' $10^4\epsilon_r''$	3.20 2000	2.84 1250	2.75 865	2.68 580	2.63 380	2.60 260	2.58 175	....	2.57 126	2.57 82
Mycalex 400 (mica, glass)	25	ϵ_r' $10^4\epsilon_r''$	7.47 220	7.45 140	7.42 120	7.40 105	7.39 95	7.38 95	....	....	....	7.12 235
Neoprene compound (38 per cent GN)	24	ϵ_r' $10^4\epsilon_r''$	6.70 1070	6.60 730	6.54 750	6.47 970	6.26 2400	5.54 6600	4.5 4050	4.24 2700	4.00 1350	4.00 1050
Nylon 66	25	ϵ_r' $10^4\epsilon_r''$	3.88 560	3.75 725	3.60 840	3.45 880	3.33 860	3.24 790	3.16 660	....	3.03 390	
Paper (Royalgrey)	25	ϵ_r' $10^4\epsilon_r''$	3.30 190	3.29 250	3.22 380	3.10 620	2.99 1150	2.86 1600	2.77 1800	2.75 1800	2.70 1500	2.62 1050
Paraffin 132° ASTM	25	ϵ_r' $10^4\epsilon_r''$	2.25 5	2.25 5	2.25 5	2.25 5	2.25 5	2.25 5	2.25 5	....	2.25 4.5	2.24 5
	81	ϵ_r' $10^4\epsilon_r''$	2.02 10	2.02 2.4	2.02 1	2.02 4	2.02 4	2.02 6	....	....	2.00 10.4	
Plexiglas	27	ϵ_r' $10^4\epsilon_r''$	3.40 2050	3.12 1450	2.95 885	2.84 570	2.76 385	2.71 270	....	2.66 165	2.60 150	2.59 175
Polyethylene (pure)	24	ϵ_r' $10^4\epsilon_r''$	2.25 11	2.25 7	2.25 7	2.25 11	2.25 9	2.25 7	....	....	2.25 7	2.25 9

Material	T°C		Frequency, cycles per second									
			10^2	10^3	10^4	10^5	10^6	10^7	10^8	3×10^8	3×10^9	10^{10}
Polystyrene (sheet stock)	25	ϵ_r' $10^4 \epsilon_r''$	2.56 1.3	2.56 1.3	2.56 1.3	2.56 1.3	2.56 1.8	2.56 5	2.55 3	2.55 9	2.55 8.5	2.54 11
Porcelain, wet process	25	ϵ_r' $10^4 \epsilon_r''$	6.47 1800	6.24 1100	6.08 800	5.98 630	5.87 530	5.82 670	5.80 780	5.75 805	....	5.51 850
Porcelain, dry process	25	ϵ_r' $10^4 \epsilon_r''$	5.50 1200	5.36 750	5.23 550	5.14 440	5.08 380	5.04 350	5.04 390	5.02 490	....	4.74 740
Pyranol 1478	26	ϵ_r' $10^4 \epsilon_r''$	4.55 680	4.53 64	4.53 9	4.53 23	4.53 9	4.53 55	....	4.50 1700	3.80 8800	
Quartz, fused	25	ϵ_r' $10^4 \epsilon_r''$	3.78 32	3.78 28	3.78 23	3.78 15	3.78 7.5	3.78 4	3.78 4	....	3.78 2.3	3.78 4
Resin No. 90S	25	ϵ_r' $10^4 \epsilon_r''$	3.25 3500	2.94 1450	2.80 770	2.72 450	2.64 300	2.61 240	2.58 215	....	2.54 160	2.53 145
Rubber, pale crepe (Hevea)	25	ϵ_r' $10^4 \epsilon_r''$	2.4 67	2.4 43	2.4 34	2.4 34	2.4 43	2.4 77	2.4 120	....	2.15 65	
Sandy soil (dry)	25	ϵ_r' $10^4 \epsilon_r''$	3.42 6700	2.91 2300	2.75 940	2.65 530	2.59 440	2.55 410	....	2.55 250	2.55 160	2.53 92
Sealing wax (Red Empress)	25	ϵ_r' $10^4 \epsilon_r''$	3.68 920	3.52 530	3.40 340	3.32 260	3.29 260	3.27 330	3.2 380	....	3.09 380	

Material	T (°C)											
Shellac, natural XL (3.5 per cent wax)	28	ϵ'_r / $10^4 \epsilon''_r$	3.86 / 250	3.81 / 280	3.75 / 480	3.66 / 825	3.47 / 1100	3.26 / 1150	3.10 / 930	… / …	2.86 / 730	… / …
Styrofoam 103.7	70	ϵ'_r / $10^4 \epsilon''_r$	6.50 / 6800	5.65 / 4850	5.10 / 3300	4.60 / 2100	4.33 / 1700	4.00 / 2200	3.80 / 2700	… / …	3.45 / 2500	… / …
Sulfur, sublimed	25	ϵ'_r / $10^4 \epsilon''_r$	1.03 / 2	1.03 / 1	1.03 / 1	1.03 / 1	1.03 / 2	1.03 / 2	… / …	… / …	1.03 / 1	1.03 / 1.5
Teflon	25	ϵ'_r / $10^4 \epsilon''_r$	3.69 / 11	3.69 / 8	3.69 / 8	3.69 / 8	3.69 / 8	3.69 / 8	… / …	… / …	3.62 / 1.5	3.58 / 5.5
Vaseline	22	ϵ'_r / $10^4 \epsilon''_r$	2.1 / 11	2.1 / 7	2.1 / 7	2.1 / 7	2.1 / 4	2.1 / 4	2.1 / 4	2.1 / 3	2.1 / 3	2.08 / 8
	25	ϵ'_r / $10^4 \epsilon''_r$	2.16 / 6.5	2.16 / 4.3	2.16 / 4	2.16 / 2	2.16 / 2	2.16 / 7	2.16 / 9	… / …	2.16 / 14	2.16 / 22
	80	ϵ'_r / $10^4 \epsilon''_r$	2.10 / 34	2.10 / 7.5	2.10 / 2	2.10 / 2	2.10 / 2	… / …	… / …	… / …	2.10 / 19	2.10 / 46
Water	1.5	ϵ'_r / $10^2 \epsilon''_r$	… / …	… / …	… / …	87.0 / 1650	87.0 / 165	87 / 17	87 / 61	86.5 / 280	80.5 / 2500	38 / 3900
	25	ϵ'_r / $10^2 \epsilon''_r$	… / …	… / …	… / …	78.2 / 3100	78.2 / 310	78.2 / 36	78 / 39	77.5 / 125	76.7 / 1200	55 / 3000
	55	ϵ'_r / $10^2 \epsilon''_r$	… / …	… / …	… / …	… / …	68.2 / 490	… / …	… / …	68 / 63	67.5 / 600	60 / 2200
	85	ϵ'_r / $10^2 \epsilon''_r$	… / …	… / …	… / …	58 / 7200	58 / 720	58 / 73	58 / 18	57 / 42	56.5 / 310	54 / 1400

APPENDIX C

FOURIER SERIES AND INTEGRALS

A periodic function $f(x)$ with period a and satisfying the Dirichlet conditions can be expanded in a Fourier series

$$f(x) = \frac{a_0}{2} + \sum_{n=1}^{\infty} \left[a_n \cos\left(\frac{2n\pi}{a} x\right) + b_n \sin\left(\frac{2n\pi}{a} x\right) \right] \quad \text{(C-1)}$$

where
$$a_n = \frac{2}{a} \int_0^a f(x) \cos\left(\frac{2n\pi}{a} x\right) dx$$
$$b_n = \frac{2}{a} \int_0^a f(x) \sin\left(\frac{2n\pi}{a} x\right) dx \quad \text{(C-2)}$$

Such a series converges to $f(x)$ at each point of continuity and to the mid-point of each discontinuity. Also, a finite Fourier series ($n \leq N$) is a least-mean-square error approximation to $f(x)$. Alternatively, the Fourier series can be written as

$$f(x) = \sum_{n=-\infty}^{\infty} c_n e^{j(2n\pi/a)x} \quad \text{(C-3)}$$

where
$$c_n = \frac{1}{a} \int_0^a f(x) e^{-j(2n\pi/a)x} dx \quad \text{(C-4)}$$

A comparison of Eq. (C-1) with Eq. (C-3) reveals that

$$2c_n = a_n - jb_n \quad \text{(C-5)}$$

Equation (C-1) is called the *trigonometric form*, and Eq. (C-3) the *exponential form* of the Fourier series.

Now consider a nonperiodic function, as represented by Fig. C-1a. In a given interval, say $0 < x < a$, the function can be represented by Eq. (C-1). However, outside the given interval, the series does not equal $f(x)$, but instead the series gives a periodic extension of $f(x)$, as represented by Fig. C-1b. Moreover, we can represent $f(x)$ in the interval $0 < x < a$ in terms of a Fourier series of arbitrary period $b \geq a$, but the series will not be unique until we specify the manner of extending the function beyond $x = a$. In particular, if we choose a period $2a$ and take

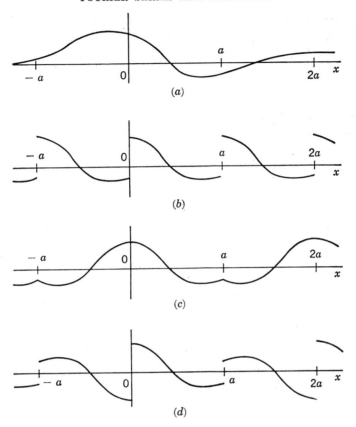

FIG. C-1. (a) A function can be represented in the interval $0 < x < a$ by (b) a "complete" Fourier series, (c) a Fourier cosine series, and (d) a Fourier sine series.

the even extension of $f(x)$ from a to $2a$, as shown in Fig. C-1c, we have the Fourier cosine series

$$f(x) = \frac{A_0}{2} + \sum_{n=1}^{\infty} A_n \cos\left(\frac{n\pi}{a} x\right) \tag{C-6}$$

where

$$A_n = \frac{1}{a} \int_0^a f(x) \cos\left(\frac{n\pi}{a} x\right) dx \tag{C-7}$$

Similarly, if we choose a period $2a$ and take the odd extension of $f(x)$ from a to $2a$, as shown in Fig. C-1d, we have the Fourier sine series

$$f(x) = \sum_{n=1}^{\infty} B_n \sin\left(\frac{n\pi}{a} x\right) \tag{C-8}$$

where

$$B_n = \frac{1}{a} \int_0^a f(x) \sin\left(\frac{n\pi}{a} x\right) dx \tag{C-9}$$

The representation of Eq. (C-6) converges to $f(x)$ on the closed interval $0 \leq x \leq a$, while Eq. (C-8) converges to $f(x)$ on the open interval $0 < x < a$. At $x = 0$ and $x = a$, Eq. (C-8) converges to zero, which is the mid-point of the discontinuity in the extended function (see Fig. C-1d).

A function $f(x)$ can also be represented as a superposition of sinusoidal functions in an infinite interval, say $-\infty < x < \infty$. In this case, the summation must be replaced by an integration, and we have

$$f(x) = \frac{1}{2\pi} \int_{-\infty}^{\infty} \tilde{f}(w) e^{jwx} \, dw \tag{C-10}$$

where
$$\tilde{f}(w) = \int_{-\infty}^{\infty} f(x) e^{-jwx} \, dx \tag{C-11}$$

The $\tilde{f}(w)$ is called the *Fourier transform* of $f(x)$. Equation (C-11) is called the *direct transformation*, and Eq. (C-10) is called the *inverse transformation*. Sufficient conditions on $f(x)$ for $\tilde{f}(w)$ to exist are

$$\int_{-\infty}^{\infty} |f(x)| \, dx < \infty \tag{C-12}$$

and $f(x)$ satisfies the Dirichlet conditions. The inversion [Eq. (C-10)] then converges to $f(x)$ at all points of continuity and to the mid-point of points of discontinuity. Fourier integrals corresponding to the trigonometric series of Eq. (C-1) also exist, but we shall not consider them here.

A useful relationship between the Fourier coefficients a_n, b_n, c_n and the integral of $|f(x)|^2$ over its period, known as *Parseval's theorem*, is

$$\frac{1}{a} \int_0^a |f(x)|^2 \, dx = |a_0|^2 + \frac{1}{2} \sum_{n=1}^{\infty} (|a_n|^2 + |b_n|^2)$$
$$= \sum_{n=-\infty}^{\infty} |c_n|^2 \tag{C-13}$$

This is readily proved by substituting for $f(x)$ in the left-hand term from Eq. (C-1) or (C-3) and interchanging summation and integration. All cross-product terms drop out because of orthogonality. Similarly, for the Fourier integral, we have a Parseval theorem

$$\int_{-\infty}^{\infty} |f(x)|^2 \, dx = \frac{1}{2\pi} \int_{-\infty}^{\infty} |\tilde{f}(w)|^2 \, dw \tag{C-14}$$

or, more generally,

$$\int_{-\infty}^{\infty} f(x) g^*(x) \, dx = \frac{1}{2\pi} \int_{-\infty}^{\infty} \tilde{f}(w) \tilde{g}^*(w) \, dw \tag{C-15}$$

The proof of Eq. (C-15) is summarized as follows

$$\int_{-\infty}^{\infty} f(x)g^*(x)\,dx = \int_{-\infty}^{\infty} \left[\frac{1}{2\pi}\int_{-\infty}^{\infty} \bar{f}(w)e^{jwx}\,dx\right] g^*(x)\,dx$$

$$= \frac{1}{2\pi}\int_{-\infty}^{\infty} \bar{f}(w)\left[\int_{-\infty}^{\infty} g^*(x)e^{jwx}\,dx\right] dw$$

A similar generalization of Eq. (C-13) can also be given.

Finally, the *impulse function* (delta function) is useful in Fourier analysis. By definition, the impulse function $\delta(x)$ satisfies the integral equation

$$\int_a^b f(x)\,\delta(x - x')\,dx = \begin{cases} f(x') & a < x' < b \\ 0 & x' < a \text{ or } x' > b \end{cases} \quad \text{(C-16)}$$

for all $f(x)$. It is evident that $\delta(x)$ is not a function in the usual sense, but its use can be justified by rigorous means.[1] It is helpful to visualize the impulse function as

$$\delta(x) = \begin{cases} \dfrac{1}{c} & -\dfrac{c}{2} < x < \dfrac{c}{2} \\ 0 & |x| > \dfrac{c}{2} \end{cases} \quad \text{(C-17)}$$

where c is an appropriately small number. Such a picture gives an intuitive justification of Eq. (C-16). From Eqs. (C-11) and (C-16), it follows that

$$\bar{\delta}(w) = \int_{-\infty}^{\infty} \delta(x)e^{-jwx}\,dx = 1 \quad \text{(C-18)}$$

that is, the transform of the δ function contains all frequencies in equal amounts. The inverse of Eq. (C-18) is

$$\frac{1}{2\pi}\int_{-\infty}^{\infty} e^{jwx}\,dw = \delta(x) \quad \text{(C-19)}$$

which is a particularly simple and useful result. Our use of $\delta(x)$ will be primarily as shorthand notation for Eq. (C-17).

[1] L. Schwartz, "Théorie des distributions," *Actualities scientifiques et industrielles,* nos. 1091 and 1122, Hermann & Cie., Paris, 1950–1951.

APPENDIX D

BESSEL FUNCTIONS

Bessel's equation of order v is

$$x \frac{d}{dx}\left(x \frac{dy}{dx}\right) + (x^2 - v^2)y = 0 \qquad \text{(D-1)}$$

Solutions may be obtained by the method of Frobenius, the result being

$$\begin{aligned} J_v(x) &= \sum_{m=0}^{\infty} \frac{(-1)^m (x)^{2m+v}}{m!(m+v)!\, 2^{2m+v}} \\ J_{-v}(x) &= \sum_{m=0}^{\infty} \frac{(-1)^m (x)^{2m-v}}{m!(m-v)!\, 2^{2m-v}} \end{aligned} \qquad \text{(D-2)}$$

where $m! = \Gamma(m+1)$ in general. As long as v is not an integer, these are two independent solutions to Bessel's equation. However, when $v = n$ is an integer, we have

$$J_{-n}(x) = (-1)^n J_n(x) \qquad \text{(D-3)}$$

and Eqs. (D-2) are no longer two independent solutions. In this case a second solution may be obtained by a limiting procedure. It is conventional to define another solution to Bessel's equation as

$$N_v(x) = \frac{J_v(x) \cos v\pi - J_{-v}(x)}{\sin v\pi} \qquad \text{(D-4)}$$

where, for integral $v = n$,

$$N_n(x) = \lim_{v \to n} N_v(x) \qquad \text{(D-5)}$$

This limit exists and establishes a second solution to Bessel's equation of integral order. The $J_v(x)$ are called Bessel functions of the first kind of order v, and the $N_v(x)$ are called Bessel functions of the second kind of order v.

BESSEL FUNCTIONS

Of particular interest are integral orders of Bessel functions. From Eq. (D-2) and (D-5), one can determine

$$J_0(x) = \sum_{m=0}^{\infty} \frac{(-1)^m}{(m!)^2} \left(\frac{x}{2}\right)^{2m}$$

$$N_0(x) = \frac{2}{\pi} \log \frac{\gamma x}{2} J_0(x) + \frac{2}{\pi} \sum_{m=1}^{\infty} \frac{(-1)^{m+1}}{(m!)^2} \left(\frac{x}{2}\right)^{2m} \phi(m)$$

(D-6)

for the zero-order functions, and

$$J_n(x) = \sum_{m=0}^{\infty} \frac{(-1)^m}{m!(m+n)!} \left(\frac{x}{2}\right)^{2m+n}$$

$$N_n(x) = \frac{2}{\pi} \log \frac{\gamma x}{2} J_n(x) - \frac{1}{\pi} \sum_{m=0}^{n-1} \frac{(n-m-1)!}{m!} \left(\frac{2}{x}\right)^{n-2m}$$

$$- \frac{1}{\pi} \sum_{m=0}^{\infty} \frac{(-1)^m}{m!(m+n)!} \left(\frac{x}{2}\right)^{n+2m} [\phi(m) + \phi(m+n)]$$

(D-7)

for $n > 0$, where $\log \gamma = 0.5772$ (Euler's constant)

$$\gamma = 1.781$$

$$\phi(m) = 1 + \tfrac{1}{2} + \tfrac{1}{3} + \cdots + \frac{1}{m}$$

(D-8)

The Bessel functions have been tabulated over an extensive range of orders and arguments, and tables are available. Figure D-1 shows

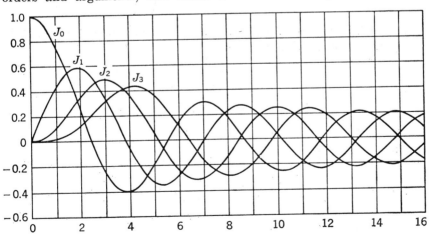

FIG. D-1. Bessel functions of the first kind.

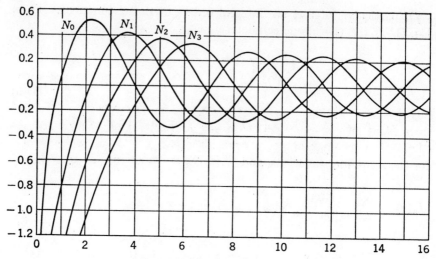

Fig. D-2. Bessel functions of the second kind.

curves for the lowest-order functions of the first kind, and Fig. D-2 shows those for the second kind. For small arguments, we have from the series

$$J_0(x) \xrightarrow[x \to 0]{} 1$$
$$N_0(x) \xrightarrow[x \to 0]{} \frac{2}{\pi} \log \frac{\gamma x}{2} \qquad \text{(D-9)}$$

and, for $v > 0$,

$$J_v(x) \xrightarrow[x \to 0]{} \frac{1}{v!} \left(\frac{x}{2}\right)^v$$
$$N_v(x) \xrightarrow[x \to 0]{} -\frac{(v-1)!}{\pi} \left(\frac{2}{x}\right)^v \qquad \text{(D-10)}$$

provided Re $(v) > 0$. For large arguments, asymptotic series exist, the leading terms of which are

$$J_v(x) \xrightarrow[x \to \infty]{} \sqrt{\frac{2}{\pi x}} \cos\left(x - \frac{\pi}{4} - \frac{v\pi}{2}\right)$$
$$N_v(x) \xrightarrow[x \to \infty]{} \sqrt{\frac{2}{\pi x}} \sin\left(x - \frac{\pi}{4} - \frac{v\pi}{2}\right) \qquad \text{(D-11)}$$

provided $|\text{phase}(x)| < \pi$.

For the expression of wave phenomena, it is convenient to define linear combinations of the Bessel functions

$$H_v^{(1)}(x) = J_v(x) + jN_v(x)$$
$$H_v^{(2)}(x) = J_v(x) - jN_v(x) \qquad \text{(D-12)}$$

called Hankel functions of the first and second kinds. Small-argument and large-argument formulas are obtained from those for J_v and N_v. In particular, the large-argument formulas become

$$H_v^{(1)}(x) \xrightarrow[x \to \infty]{} \sqrt{\frac{2}{j\pi x}} j^{-v} e^{jx}$$

$$H_v^{(2)}(x) \xrightarrow[x \to \infty]{} \sqrt{\frac{2j}{\pi x}} j^v e^{-jx} \quad \text{(D-13)}$$

which place into evidence the wave character of the Hankel functions.

Derivative formulas and recurrence formulas can be obtained by differentiation of Eqs. (D-2). Letting $B_v(x)$ denote an arbitrary solution to Bessel's equation, we have

$$B_v'(x) = B_{v-1} - \frac{v}{x} B_v$$

$$B_v'(x) = -B_{v+1} + \frac{v}{x} B_v \quad \text{(D-14)}$$

which, in the special case $v = 0$, become

$$B_0'(x) = -B_1(x) \quad \text{(D-15)}$$

The difference of Eqs. (D-14) yields the recurrence formula

$$B_v(x) = \frac{2(v-1)}{x} B_{v-1} - B_{v-2} \quad \text{(D-16)}$$

which is useful for calculating $B_n(x)$, $n > 1$, from a knowledge of $B_0(x)$ and $B_1(x)$. The Wronskian of Bessel's equation is often encountered in problem solving. This is

$$J_v(x) N_v'(x) - N_v(x) J_v'(x) = \frac{2}{\pi x} \quad \text{(D-17)}$$

from which Wronskians for other pairs of solutions can be easily obtained.

When $x = ju$ is imaginary, modified Bessel functions of the first and second kind can be defined as

$$I_v(u) = j^v J_v(-ju)$$

$$K_v(u) = \frac{\pi}{2} (-j)^{v+1} H_v^{(2)}(-ju) \quad \text{(D-18)}$$

These are real functions for real u. General formulas for I_v and K_v can be obtained from the corresponding formulas for J_v and $H_v^{(2)}$. Figure D-3 shows curves of the zero- and first-order modified Bessel functions. The large-argument formulas, obtained from Eqs. (D-11) and (D-12),

$$I_v(u) \xrightarrow[u \to \infty]{} \frac{e^u}{\sqrt{2\pi u}}$$

$$K_v(u) \xrightarrow[u \to \infty]{} \sqrt{\frac{\pi}{2u}} e^{-u} \quad \text{(D-19)}$$

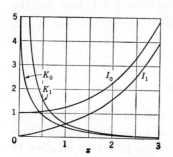

FIG. D-3. Modified Bessel functions.

illustrate the evanescent character of the modified Bessel functions. Derivative formulas and recurrence formulas can be readily obtained from Eqs. (D-14) to (D-16).

Bessel functions of order $n + \frac{1}{2}$ are used in the solution of the Helmholtz equation in spherical coordinates. In scalar-wave problems, it is conventional to define spherical Bessel functions as

$$b_n(x) = \sqrt{\frac{\pi}{2x}}\, B_{n+\frac{1}{2}}(x) \qquad \text{(D-20)}$$

The b_n are given the name and letter as the corresponding $B_{n+\frac{1}{2}}$. (For example, j_n is the spherical Bessel function of the first kind, $h_n{}^{(2)}$ is the spherical Hankel function of the second kind, etc.) In a-c electromagnetic field problems, it is convenient to define the alternative spherical Bessel functions

$$\hat{B}_n(x) = \sqrt{\frac{\pi x}{2}}\, B_{n+\frac{1}{2}}(x) \qquad \text{(D-21)}$$

where $\hat{B}_n$ is given the same name and symbol as the corresponding $B_{n+\frac{1}{2}}$. The various formulas for b_n and $\hat{B}_n$ can be obtained from the corresponding formulas for $B_{n+\frac{1}{2}}$. Of particular interest is the fact that asymptotic expansions for $B_{n+\frac{1}{2}}$ become exact, giving

$$\hat{J}_n(x) = C_n(x) \sin\left(x - \frac{n\pi}{2}\right) + D_n(x) \cos\left(x - \frac{n\pi}{2}\right)$$
$$\hat{N}_n(x) = D_n(x) \sin\left(x - \frac{n\pi}{2}\right) - C_n(x) \cos\left(x - \frac{n\pi}{2}\right) \qquad \text{(D-22)}$$
$$\hat{H}_n{}^{(1)}(x) = j^{-n}[D_n(x) - jC_n(x)]e^{jx}$$
$$\hat{H}_n{}^{(2)}(x) = j^{n}[D_n(x) + jC_n(x)]e^{-jx}$$

where
$$C_n(x) = \sum_{m=0}^{2m \leq n} \frac{(-1)^m (n + 2m)!}{(2m)!(n - 2m)!(2x)^{2m}}$$
$$D_n(x) = \sum_{m=0}^{2m \leq n-1} \frac{(-1)^m (n + 2m + 1)!}{(2m + 1)!(n - 2m - 1)!(2x)^{2m+1}} \qquad \text{(D-23)}$$

Note that
$$\hat{H}_n{}^{(2)}(x) \xrightarrow[x \to \infty]{} j^{n+1} e^{-jx} \qquad \text{(D-24)}$$

which is of interest in radiation problems.

APPENDIX E

LEGENDRE FUNCTIONS

The associated Legendre equation is

$$\frac{1}{\sin\theta}\frac{d}{d\theta}\left(\sin\theta\frac{dy}{d\theta}\right) + \left[v(v+1) - \frac{m^2}{\sin^2\theta}\right]y = 0 \quad \text{(E-1)}$$

This can be put into another common form by using the substitution

$$u = \cos\theta \quad \text{(E-2)}$$

in Eq. (E-1). The result is

$$(1-u^2)\frac{d^2y}{du^2} - 2u\frac{dy}{du} + \left[v(v+1) - \frac{m^2}{1-u^2}\right]y = 0 \quad \text{(E-3)}$$

When $m = 0$, the associated Legendre equation reduces to the ordinary Legendre equation

$$(1-u^2)\frac{d^2y}{du^2} - 2u\frac{dy}{du} + v(v+1)y = 0 \quad \text{(E-4)}$$

We shall first consider solutions to this special case and later generalize to the associated Legendre equation.

In the spherical coordinate system, $0 \le \theta \le \pi$; so we shall be interested in solutions over the range $-1 \le u \le 1$. In particular, for $|1 - u| < 2$, the *Legendre function of the first kind* can be expressed as

$$P_v(u) = \sum_{m=0}^{N} \frac{(-1)^m (v+m)!}{(m!)^2 (v-m)!} \left(\frac{1-u}{2}\right)^m$$

$$- \frac{\sin v\pi}{\pi} \sum_{m=N+1}^{\infty} \frac{(m-1-v)!(m+v)!}{(m!)^2} \left(\frac{1-u}{2}\right)^m \quad \text{(E-5)}$$

where N is the nearest integer $N \le v$. As long as v is not an integer, $P_v(u)$ and $P_v(-u)$ are two independent solutions to Legendre's equation [Eq. (E-4)]. If $v = n$ is an integer, Eq. (E-5) becomes a finite series called the *Legendre polynomial* of degree n. In this case,

$$P_n(-u) = (-1)^n P_n(u) \quad \text{(E-6)}$$

and we no longer have two independent solutions. Another solution, called the *Legendre function of the second kind*, is defined as

$$Q_v(u) = \frac{\pi}{2} \frac{P_v(u) \cos v\pi - P_v(-u)}{\sin v\pi} \qquad \text{(E-7)}$$

When $v = n$ is an integer, the limit

$$Q_n(u) = \lim_{v \to n} Q_v(u) \qquad \text{(E-8)}$$

exists and defines a second solution to Legendre's equation.

The Legendre polynomials are of particular interest, because these are the only solutions finite over the entire range $0 \leq \theta \leq \pi$. In this case, only the first summation in Eq. (E-5) remains, which can be rearranged to

$$P_n(u) = \sum_{m=0}^{M} \frac{(-1)^m (2n - 2m)!}{2^n m!(n - m)!(n - 2m)!} u^{n-2m} \qquad \text{(E-9)}$$

where $M = n/2$ or $(n - 1)/2$, whichever is an integer. An alternative, and sometimes more convenient, expression for the Legendre polynomials is given by Rodrigues' formula

$$P_n(u) = \frac{1}{2^n n!} \frac{d^n}{du^n} (u^2 - 1)^n \qquad \text{(E-10)}$$

Some of the lower-degree polynomials are

$$\begin{array}{lll} P_0(u) = 1 & P_1(u) = u & P_2(u) = \tfrac{1}{2}(3u^2 - 1) \\ P_3(u) = \tfrac{1}{2}(5u^3 - 3u) & P_4(u) = \tfrac{1}{8}(35u^4 - 30u^2 + 3) \end{array} \qquad \text{(E-11)}$$

or, in terms of θ,

$$\begin{aligned} P_0(\cos \theta) &= 1 \quad P_1(\cos \theta) = \cos \theta \\ P_2(\cos \theta) &= \tfrac{1}{4}(3 \cos 2\theta + 1) \\ P_3(\cos \theta) &= \tfrac{1}{8}(5 \cos 3\theta + 3 \cos \theta) \\ P_4(\cos \theta) &= \tfrac{1}{64}(35 \cos 4\theta + 20 \cos 2\theta + 9) \end{aligned} \qquad \text{(E-12)}$$

Figure E-1 shows curves of the Legendre polynomials plotted against θ.

The Legendre functions of the second kind for integral $v = n$ are infinite at $\theta = 0$ and $\theta = \pi$, or at $u = \pm 1$. They can be expressed as

$$Q_n(u) = P_n(u) \left[\tfrac{1}{2} \log \frac{1+u}{1-u} - \phi(n) \right]$$

$$+ \sum_{m=1}^{n} \frac{(-1)^m (n+m)!}{(m!)^2 (n-m)!} \phi(m) \left(\frac{1-u}{2} \right)^m \qquad \text{(E-13)}$$

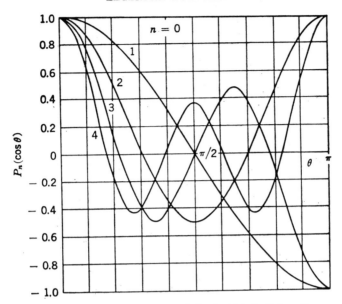

FIG. E-1. Legendre functions of the first kind, $P_n(\cos\theta)$.

where $\phi(m)$ is defined by Eq. (D-8). Some of the lower-order functions are

$$Q_0(u) = \tfrac{1}{2}\log\frac{1+u}{1-u} \qquad Q_1(u) = \frac{u}{2}\log\frac{1+u}{1-u} - 1$$
$$Q_2(u) = \frac{3u^2-1}{4}\log\frac{1+u}{1-u} - \frac{3u}{2}$$
(E-14)

or, in terms of θ,

$$Q_0(\cos\theta) = \log\cot\frac{\theta}{2} \qquad Q_1(\cos\theta) = \cos\theta\log\cot\frac{\theta}{2} - 1$$
$$Q_2(\cos\theta) = \tfrac{1}{4}(3\cos^2\theta - 1)\log\cot\frac{\theta}{2} - \tfrac{3}{2}\cos\theta$$
(E-15)

Figure E-2 shows curves of these functions plotted against θ.

Now consider the associated Legendre equation, Eq. (E-3). For simplicity, we first take m to be an integer. If Eq. (E-4) is differentiated m times, there results

$$\left[(1-u^2)\frac{d}{du^2} - 2u(m+1)\frac{d}{du} + (n-m)(n+m+1)\right]\frac{d^m y}{du^m} = 0$$

Letting $w = [(1-u)^{m/2}]\,d^m y/du^m$ in the above equation, we obtain Eq. (E-3) with y replaced by w. Hence, solutions to the associated Legendre equation are[1]

---

[1] Smythe and others omit the factor $(-1)^m$ on the right-hand side of these definitions.

$$P_n^m(u) = (-1)^m (1-u^2)^{m/2} \frac{d^m P_n(u)}{du^m}$$
$$Q_n^m(u) = (-1)^m (1-u^2)^{m/2} \frac{d^m Q_n(u)}{du^m} \qquad \text{(E-16)}$$

Note that all $P_n^m(u) = 0$ for $m > n$. Some of the lower-order associated Legendre polynomials are

$$\begin{array}{ll} P_1^1(u) = -(1-u^2)^{1/2} & P_3^1(u) = \tfrac{3}{2}(1-u^2)^{1/2}(1-5u^2) \\ P_2^1(u) = -3(1-u^2)^{1/2} u & P_3^2(u) = 15(1-u^2)u \\ P_2^2(u) = 3(1-u^2) & P_3^3(u) = -15(1-u^2)^{3/2} \end{array} \qquad \text{(E-17)}$$

while the $P_n^0(u) = P_n(u)$ are given by Eq. (E-11). Some of the lower-order associated Legendre functions of the second kind are

$$Q_1^1 = -(1-u^2)^{1/2}\left(\tfrac{1}{2}\log\frac{1+u}{1-u} + \frac{u}{1-u^2}\right)$$
$$Q_2^1 = -(1-u^2)^{1/2}\left(\tfrac{3}{2}u\log\frac{1+u}{1-u} + \frac{3u^2-2}{1-u^2}\right) \qquad \text{(E-18)}$$
$$Q_2^2 = (1-u^2)^{1/2}\left[\tfrac{3}{2}\log\frac{1+u}{1-u} + \frac{5u-3u^2}{(1-u^2)^2}\right]$$

while the $Q_n^0(u) = Q_n(u)$ are given by Eq. (E-14).

When m is not an integer, the situation becomes even more complicated. A standard formula for Legendre functions of the first kind,

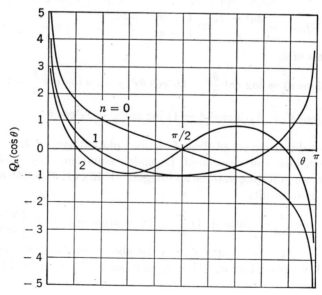

FIG. E-2. Legendre functions of the second kind, $Q_n(\cos\theta)$.

valid for $|1 - u| < 2$, is then

$$P_v^w(u) = \frac{\sin w\pi}{\pi}(w-1)!\left(\frac{u+1}{u-1}\right)^{w/2} F\left(-v, v+1, 1-w, \frac{1-u}{2}\right) \quad \text{(E-19)}$$

where F is the *hypergeometric function*

$$F(\alpha,\beta,\gamma,z) = 1 + \frac{(\gamma-1)!}{(\alpha-1)!(\beta-1)!} \sum_{m=0}^{\infty} \frac{(\alpha+m)!(\beta+m)!}{m!(\gamma+m)!} z^{m+1} \quad \text{(E-20)}$$

For real u, the associated Legendre function of the second kind is defined as

$$Q_v^w(u) = \frac{\pi}{2} \frac{P_v^w(u) \cos(v+w)\pi - P_v^w(-u)}{\sin(v+w)\pi} \quad \text{(E-21)}$$

The solutions $P_v^w(u)$ and $P_v^w(-u)$ are linearly independent, except when $u + w$ is an integer. In this latter case, the limit of Eq. (E-21) provides a second solution.

Perhaps the simplest way to calculate the Legendre functions is through the recurrence formulas. Letting $L_n^m(u)$ denote an arbitrary solution to the associated Legendre equation, we have

$$(m - n - 1)L_{n+1}^m + (2n+1)uL_n^m - (m+n)L_{n-1}^m = 0 \quad \text{(E-22)}$$

A recurrence formula in m also exists and is

$$L_n^{m+1} + \frac{2mu}{(1-u^2)^{1/2}} L_n^m + (m+n)(n-m+1)L_n^{m-1} = 0 \quad \text{(E-23)}$$

for the range $|u| < 1$. Many formulas for derivatives also exist, some of which are

$$\begin{aligned}
L_n^{m\prime}(u) &= \frac{1}{1-u^2}[-nuL_n^m + (n+m)L_{n-1}^m] \\
&= \frac{1}{1-u^2}[(n+1)uL_n^m - (n-m+1)L_{n+1}^m] \\
&= \frac{mu}{1-u^2} L_n^m + \frac{(n+m)(n-m+1)}{(1-u^2)^{1/2}} L_n^{m-1} \\
&= -\frac{mu}{1-u^2} L_n^m - \frac{1}{(1-u^2)^{1/2}} L_n^{m+1}
\end{aligned} \quad \text{(E-24)}$$

If $m = 0$ in the last formula, we have

$$L_n^1(u) = -(1-u^2)^{1/2} L_n'(u) \quad \text{(E-25)}$$

which is a useful special case.

Finally, some specializations of the argument will be of interest to us.

At $\theta = 0$, that is, at $u = 1$, the $Q_n{}^m$ functions are infinite and

$$P_n{}^m(1) = \begin{cases} 1 & m = 0 \\ 0 & m > 0 \end{cases} \qquad \text{(E-26)}$$

At $\theta = \pi/2$, that is, at $u = 0$,

$$P_n{}^m(0) = \begin{cases} (-1)^{(n+m)/2} \dfrac{1 \cdot 3 \cdot 5 \, \cdots \, (n+m-1)}{2 \cdot 4 \cdot 6 \, \cdots \, (n-m)} & n+m \text{ even} \\ 0 & n+m \text{ odd} \end{cases}$$

$$Q_n{}^m(0) = \begin{cases} 0 & n+m \text{ even} \\ (-1)^{(n+m+1)/2} \dfrac{2 \cdot 4 \cdot 6 \, \cdots \, (n+m-1)}{1 \cdot 3 \cdot 5 \, \cdots \, (n-m)} & n+m \text{ odd} \end{cases}$$

$$\text{(E-27)}$$

Some specializations involving derivatives are

$$\left[\frac{d^r}{du^r} P_n{}^m(u)\right]_{u=0} = (-1)^r P_n{}^{m+r}(0)$$
$$\left[\frac{d^r}{du^r} Q_n{}^m(u)\right]_{u=0} = (-1)^r Q_n{}^{m+r}(0) \qquad \text{(E-28)}$$

BIBLIOGRAPHY

A. Classical Books

1. Abraham, A., and R. Becker: "The Classical Theory of Electricity," Blackie & Son, Ltd., Glasgow, 1932.
2. Heaviside, O.: "Electromagnetic Theory," Dover Publications, New York, 1950 (reprint).
3. Jeans, J.: "Electric and Magnetic Fields," Cambridge University Press, London, 1933.
4. Maxwell, J. C.: "A Treatise on Electricity and Magnetism," Dover Publications, New York, 1954 (reprint).

B. Introductory Books

1. Attwood, S.: "Electric and Magnetic Fields," 3d ed., John Wiley & Sons, Inc., New York, 1949.
2. Booker, H. G.: "An Approach to Electrical Science," McGraw-Hill Book Company, Inc., New York, 1959.
3. Harrington, R. F.: "Introduction to Electromagnetic Engineering," McGraw-Hill Book Company, Inc., New York, 1958.
4. Hayt, W. H.: "Engineering Electromagnetics," McGraw-Hill Book Company, Inc., New York, 1958.
5. Kraus, J. D.: "Electromagnetics," McGraw-Hill Book Company, Inc., New York, 1953.
6. Neal, J. P.: "Electrical Engineering Fundamentals," McGraw-Hill Book Company, Inc., New York, 1960.
7. Page, L., and N. Adams: "Principles of Electricity," D. Van Nostrand Company, Inc., Princeton, N.J., 1931.
8. Peck, E. R.: "Electricity and Magnetism," McGraw-Hill Book Company, Inc., New York, 1953.
9. Rogers, W. E.: "Introduction to Electric Fields," McGraw-Hill Book Company, Inc., New York, 1954.
10. Sears, F. W.: "Electricity and Magnetism," Addison-Wesley Publishing Company, Reading, Mass., 1946.
11. Seely, S.: "Introduction to Electromagnetic Fields," McGraw-Hill Book Company, Inc., New York, 1958.
12. Shedd, P. C.: "Fundamentals of Electromagnetic Waves," Prentice-Hall, Inc., Englewood Cliffs, N.J., 1955.
13. Skilling, H. H.: "Fundamentals of Electric Waves," 2d ed., John Wiley & Sons, Inc., New York, 1948.
14. Spence, D., and R. Galbraith: "Fundamentals of Electrical Engineering," The Ronald Press Company, New York, 1955.
15. Ware, L. A.: "Elements of Electromagnetic Waves," Pitman Publishing Corporation, New York, 1949.
16. Weber, E.: "Electromagnetic Fields," John Wiley & Sons, Inc., New York, 1950.

C. *Intermediate and Advanced Books*

1. Jordan, E.: "Electromagnetic Waves and Radiating Systems," Prentice-Hall, Inc., Englewood Cliffs, N.J., 1950.
2. King, R. W. P.: "Electromagnetic Engineering," McGraw-Hill Book Company, Inc., New York, 1953.
3. Mason, M., and W. Weaver: "The Electromagnetic Field," University of Chicago Press, Chicago, 1929.
4. Ramo, S., and J. R. Whinnery: "Fields and Waves in Modern Radio," 2d ed., John Wiley & Sons, Inc., New York, 1953.
5. Schelkunoff, S. A.: "Electromagnetic Waves," D. Van Nostrand Company, Inc., Princeton, N.J., 1943.
6. Smythe, W. R.: "Static and Dynamic Electricity," 2d ed., McGraw-Hill Book Company, Inc., New York, 1950.
7. Stratton, J. A.: "Electromagnetic Theory," McGraw-Hill Book Company, Inc., New York, 1941.

D. *Books on Special Topics*

1. Aharoni, J.: "Antennae," Clarendon Press, Oxford, 1946.
2. Bronwell, A., and R. E. Beam: "Theory and Application of Microwaves," McGraw-Hill Book Company, Inc., New York, 1947.
3. Kraus, J. D.: "Antennas," McGraw-Hill Book Company, Inc., New York, 1950.
4. Lewin, L.: "Advanced Theory of Waveguides," Illiffe and Sons, London, 1951.
5. Marcuvitz, N.: "Waveguide Handbook," MIT Radiation Laboratory Series, vol. 10, McGraw-Hill Book Company, Inc., New York, 1951.
6. Mentzer, J. R.: "Scattering and Diffraction of Radio Waves," Pergamon Press, New York, 1955.
7. Montgomery, C. G., R. H. Dicke, and E. M. Purcell (eds.): "Principles of Microwave Circuits," MIT Radiation Laboratory Series, vol. 8, McGraw-Hill Book Company, Inc., 1948.
8. Moreno, T.: "Microwave Transmission Design Data," Dover Publications, New York, 1958 (reprint).
9. Reich, H. J. (ed.): "Very High Frequency Techniques," Radio Research Laboratory, McGraw-Hill Book Company, Inc., New York, 1947.
10. Schelkunoff and Friis: "Antennas, Theory and Practice," John Wiley & Sons, Inc., New York, 1952.
11. Silver, S.: "Microwave Antenna Theory and Design," Radiation Laboratory Series, vol. 12, McGraw-Hill Book Company, Inc., New York, 1950.
12. Slater, J. C.: "Microwave Electronics," D. Van Nostrand Company, Inc., Princeton, N.J., 1950.
13. Watkins, D.: "Topics in Electromagnetic Theory," John Wiley & Sons, Inc., New York, 1958.
14. Wait, J. R.: "Electromagnetic Radiation From Cylindrical Structures," Pergamon Press, New York, 1959.

INDEX

Boldface numbers in parentheses refer to problems

A-c phenomena, 1
Addition theorems, 232, 292
Admittance matrix, 119, 392
Admittivity, 19, 23–26
Ampère's law, 4
Antenna concepts, 81–85
Antenna gain, 83
  maximum, 307–311
Apertures, 11
  admittance of, 173, 428–431
  in cavities, 436–440
  in cones, 306
  in plane conductors, 11, 138(**17**), 139(**18, 19**), 180–186, 366–371, 428–431, 444(**21–23**)
  in spheres, 301–303
  transmission through, 366–371
  in waveguides, 174, 176
  in wedges, 250–254
Associated Legendre functions, 265, 468–470
Attenuation constant, 48, 66, 73, 86
  of biconical guides, 313(**13**)
  of circular guides, 255(**9**)
  of guides in general, 376(**19**), 441(**5**)
  intrinsic, 48
  of parallel-plate guides, 91(**30**)
  of rectangular guides, 86, 189(**4**)
  of transmission lines, 90(**24**)

Babinet's principle, 365–367
Bailin, L. L., 249, 303, 306
Berk, A. D., 346, 348
Bessel functions, 199–203, 460–464
  modified, 201, 463

Bessel functions, spherical, 265, 268, 464
  zeros of, 205, 270
Bibliography, 471–472
Biconical cavity, 284–286
Biconical waveguide, 281–283, 313(**13**)
Bierens de Haan, D., 194
Boundary conditions, 34, 55
Boundary-value problems, 103
Bounds, upper and lower, 335
Brewster angle, 59
Brown, G., 351

Capacitivity, 5
  a-c, 24
  relative, 6
Capacitor, 13, 30
Carter, P. S., 349, 351
Cavities (*see* Resonators)
Characteristic impedance, 62, 65
  of waveguides, 69, 152, 154, 385
Characteristic values, 67, 144
Chu, L. J., 278
Churchill, R. V., 231
Circuit elements, 13, 29
Circuit quantities, 3
Circular cavity, 213–216
  partially filled, 258(**23, 24**)
  with wedge, 259(**25**)
Circular polarization, 46, 88(**8**)
Circular waveguides, 204–208, 389
  with baffle, 208, 255(**6, 8**)
  partially filled, 220, 257(**20**), 321, 331
Circulating waves, 208, 256(**10**)
Circulator, 26

473

Closed contour, 2
Closed surface, 2
Coated conductor, 168, 219
Coaxial line, 65
  junction, with cavity, 434–436, 445(**24**–**27**)
    with waveguide, 179, 195(**33, 34**), 425–428, 444(**20**)
    opening onto plane conductor, 111
  spherical, 281
  waveguide modes of, 254(**5**)
Cohen, M. H., 362, 364, 430
Cohn, S. B., 328
Complementary antennas, 380(**43**)
Complementary solutions, 131
Complementary structures, 136(**7**), 365
Complex quantities, 13
  constitutive relationships, 18
  permittivities, 18, 451–455
  power, 19–23
  Poynting vector, 20
Concentric spheres, 321(**5, 6**)
Conducting cone, 279
  aperatures in, 306
  current element on, 316(**30**)
  as waveguide, 279–281
Conducting cylinder, 232–238
  and current elements, 262(**39**–**43**)
Conducting sphere, 292–297
  apertures in, 301–303
  and current element, 298–301, 315(**26**)
  and current loop, 316(**29**)
  with dielectric coating, 315(**25**)
Conduction current, 6, 27
Conductivity, 6
  complex, 18
Conductors, 6
  perfect, 34
Conical cavity, 284, 314(**18**)
Conical waveguide, 279–281
Conjugate problems, 64
Conservation, of charge, 2, 4
  of complex power, 21
  of energy, 10, 11
Constitutive relationships, 5
  complex, 18

Corrugated conductor, 170, 193(**25**)
  circular, 223
  radial, 219
Critical angle, 60
Crowley, T., 430
Current, 1, 7, 27, 34
  in cavities, 431–434
  conduction, 6, 27
  near cones, 303, 316(**30**)
  near cylinders, 262(**39**–**41**)
  displacement, 7, 27
  elements, 78–81, 287
  filament (*see* Filament of current)
  near half-plane, 263(**42, 43**)
  impressed, 7, 27
  loops, 93(**41, 42**), 100, 315(**28**)
  near planes, 103, 136(**12**–**14**)
  reactive, 27, 28
  ribbon of, 188, 260(**31**)
  sheets, 34
  source, 95, 118
  near spheres, 298, 315(**26, 27**), 316(**29**)
  surface, 33
  in waveguides, 97, 106, 134(**1**–**4**), 177–179, 194(**31, 32**), 425–428, 440(**2**)
  near wedges, 263(**44**)
Cutler, C. C., 171
Cutoff frequency, 68, 150, 166, 169, 206, 384
Cutoff wavelength, 68, 150, 206, 384
Cylinder of currents, 227, 260(**30**)
Cylindrical coordinates, 198, 447
Cylindrical waveguides, 381–391
Cylindrical waves, 85

Degenerate modes, 48, 150, 390
Delta function, 179, 459
Depth of penetration, 53
Diamagnetism, 6
Diaphragms, 414–420, 442(**13**–**16**)
Dicke, R. H., 392, 400
Dielectric, 6, 24, 451–455
Dielectric cylinders, 220, 261(**34, 35**)
Dielectric loss angle, 24

INDEX

Dielectric obstacles, 362–365
Dielectric rod guide, 221, 257(**21**)
Dielectric slab guide, 163–168, 192(**22**), 219
Dielectric spheres, 297
Differential scattering, 360
Dipole, 78
   antenna, 81–85
   in conducting wedge, 105
   near ground plane, 104
   magnetic, 259(**26, 27**)
   two-dimensional, 225
Directional coupler, 135(**3**)
Displacement current, 7, 27
Dissipated power, 11
Dissipative current, 27
Dominant mode, 69, 75
Dominant-mode source, 402
Duality, 98–100

Echo, 355, 363
Echo area, 116, 128, 357
Echo width, 358, 359, 364
Effective value, 15
Eigenfunctions, 144, 384
Eigenvalues, 67, 144, 383
Electric quantities, charge, 1, **3**
   current, 3, 7, 15, 27
   dipole, 78
   flux, 1, 3
   intensity, 1, 15
   scalar potential, 77
   vector potential, 99, 129
Elementary wave functions, 144, 200, 266
Elliptical polarization, 46
Emde, F., 272
Energy, 9, 10, 21, 23
   conservation of, 10, 11
   velocity of, 42
Equation of continuity, 2
Equiphase surfaces, 39, 85
Equivalence principle, 106–110
Equivalent circuit, 62
   of coax-to-waveguide feeds, 425
   of microwave networks, 401

Equivalent circuit, of obstacles in waveguides, 402
   of resonant cavities, 435, 437
   of spherical waves, 279
   of transmission lines, 62
   of waveguides, 386
Erdelyi, A., 245
Ether, 26
Euler's identity, 15
Evanescent field, 60, 147
Evanescent mode, 68

Faraday's law, 4
Ferromagnetism, 6, 25
Feshbach, H., 337, 432
Field coordinates, 80
Field quantities, 3
Filament of current, 34, 81, 223, 243
   near cylinder, 236–238
   near half-plane, 241–242
   near wedges, 238–242
Foster's reactance theorem, 396
Fourier series, 456–458
Fourier transforms, 145, 180, 458–459
Fourier-Legendre series, 275
Frank, N. H., 163
Free space, 5
Fundamental units, 1

Gain, 83
   antenna, maximum, 307–311
   normal, 309
   of dipole near ground plane, 104, 137(**12–14**)
   of dipoles, 84
   supergain, 309
Gauss' law, 4
Good dielectric, 24
Goubau, G., 223
Gray, M. C., 222
Green's functions, 120–123
   tensor, 123–125
Green's identities, 120
   modified vector analogue, 141(**28**)
   in two dimensions, 389

Green's identities, vector analogue, 121
Guide phase velocity, 68, 385
Guide wavelength, 68, 384

Hall effect, 35(2)
Hankel functions, 199–203, 462
　spherical, 266, 464
Harmonic functions, 144, 199
Harrington, R. F., 34, 128, 309
Helmholtz equation, 38
　in cylindrical coordinates, 198
　in rectangular coordinates, 77
　in spherical coordinates, 264
Helmholtz identity, 450
Hemispherical cavity, 284
Hildebrand, F. B., 332
Hogan, C. L., 26
Homogeneous matter, 37
Hu, Y. Y., 352, 354, 358
Hybrid modes, 154, 158

Im operator, 15
Impedance, of apertures, 428–431
　characteristic, 62, 65, 69, 152, 154, 385
　of circuit elements, 29
　of current loop, 93(**42**)
　input, 84
　intrinsic, 39, 48, 87
　of linear antenna, 82, 94(**44–46**), 352
　matrix, 119, 392, 398
　surface, 53, 371(**2**), 375(**18**)
　wave, 39, 55, 69, 86, 152
Impedivity, 19, 23–25
Impressed current, 7, 27
Impulse function, 179, 459
Incident field, 113
Index of refraction, 58
Induced emf method, 349
Induction field, 79
Induction theorem, 113–115
Inductivity, 5, 6, 25
Inductor, 13, 31
Input impedance, 84
Instantaneous phase, 85

Instantaneous quantity, 15
Insulators, 6
Integral equations, 125–128, 317
Intrinsic parameters, 39, 40, 48, 87
Isolator, 26
Isotropic matter, 37

Jahnke, E., 272
Junctions, coax-to-cavity, 434–436, 445(**24–27**)
　coax-to-waveguide, 179, 195(**33, 34**), 425–428, 444(**20**)
　waveguide-to-cavity, 436–440
　waveguide-to-waveguide, 172–177, 193(**27–29**), 420–425, 443(**17, 18**)

King, R., 351
Kirchhoff's laws, 4, 12

Legendre functions, 265, 465–470
　associated, 265, 468–470
LePage, W. R., 386
Levine, H., 113
Levis, C. A., 430
Linear antenna, 81–85, 94(**44–46**), 349–355
Linear matter, 6, 18
Linear polarization, 39, 45
Loop of current, 93(**41, 42**), 100
　in cavity, 445(**24**), 446(**27–29**)
　near conducting cone, 303–306
　near conducting sphere, 315(**28, 29**)
Loosely bound wave, 170
Lorentz reciprocity theorem, 117
Loss angle, of capacitor, 30
　of dielectric, 24
　of inductor, 32
　magnetic, 25
Lossy dielectric, 24

Macroscopic standpoint, 1
Magnetic quantities, conductor, 34
　current, 7, 27

INDEX

Magnetic quantities, dipole, 100
  flux, 1, 3
  intensity, 1, 15
  loss angle, 25
  vector potential, 77, 129
Magnetomotive force, 3
Magnitude, 85
Marcuvitz, N., 381, 389, 410, 411, 418, 420, 424
Matrix impedance, 119
Maxwell's equations, 2, 18
Mentzer, J. R., 122, 306
Microwave networks, 391–402
Mksc units, 1
Modal expansions, 171–177, 389–391
  in cavities, 431–434
Mode, 63, 68, 69, 75
Mode current, 72, 383
Mode function, 383
Mode patterns, 70
  for circular cavity, 215
  for circular guide, 207
  for coated conductor, 169
  for free space, 277
  in general, 387
  for rectangular cavity, 75, 157
  for rectangular guide, 70, 151, 155
  for slab waveguide, 168
  for spherical, cavity, 272
Mode voltage, 72, 383
Modified Bessel functions, 201, 463
Monopole antenna, 138(**15**)
Montgomery, C. G., 392, 400
Morse, P. M., 337, 371, 432
Multipoles, cylindrical, 226, 259(**29**)
  spherical, 286–289, 314(**19**)

Neumann's number, 172
Nonpropagating mode, 68
Normal gain, 309
Normalization, 383, 432
Notation, 16

Obstacles in waveguides, 402–418
Ohm's law, 13

Oliner, A. A., 413
Orthogonality, 273, 390, 432

Papas, C. H., 113
Parallel-plate waveguide, 90(**28**), 189(**6, 7**), 440(**1**)
  and coaxial feed, 378(**35–37**)
  opening onto ground plane, 181–186
  partially filled, 190(**12, 13**), 257(**18**)
  radially, 209, 256(**13**)
Paramagnetism, 6
Parseval's theorem, 182, 458
Partially filled cavities, circular, 258(**23, 24**)
  perturbational formulas for, 321–326, 371–373
  rectangular, 191(**16–18**), 325
  spherical, 313(**7, 8**), 326
  variational formulas for, 331–345, 376–377
Partially filled waveguides, circular, 220, 257(**20**), 321, 331
  perturbational formulas for, 326–331, 374–375
  rectangular, 158–163, 191(**14–16**), 345, 348
  variational formulas for, 345–348, 377–378
Particular solution, 131
Pattern, mode, 70
  radiation field, 83
  receiving, 119
  standing-wave, 44
Perfect conductor, 34
Perfect dielectric, 24
Permeability, 5, 6, 18
Permittivity, 5, 18
Perturbational methods, 73, 76, 317–331, 371–377
Phase, 85
Phase constant, 48, 85
Phase velocity, 39, 40, 68, 86, 385
Phasor, 15
Physical optics method, 127
Pincherle, L., 158
Plane waves, 39, 85, 143, 145–148

Polarization, of matter, 27
   of waves, 39, 45, 88(**8**)
Polarizing angle, 59
Ports, 391
Posts in waveguides, 406–411, 442(**12**)
Potentials, 77, 99, 129
Power, 9, 19, 22
Poynting vector, 10, 20
Probes, in cavities, 434–436, 446(**26**)
   in waveguides, 178, 425–428
Propagating mode, 68
Propagation constant, 62, 68, 86, 384
   stationary formulas for, 346–348, 378(**32**)
Purcell, E. M., 292, 400

Q (*see* Quality factor)
Quadrupole, cylindrical, 226, 259(**28**)
   spherical, 288, 314(**20**)
Quality factor, of biconical cavity, 285
   of cavities in general, 372(**3, 4, 6, 7**)
   of circular cavity, 216, 257(**16**)
   dielectric, 28
   of hemispherical cavity, 285
   of loss-free antenna, 309
   magnetic, 29
   minimum antenna Q, 310, 316(**31**)
   of rectangular cavity, 76, 190(**10**)
   of spherical cavity, 272, 312(**4**)
   of spherical waves, 279
Quasi-static, definitions, 79, 298, 419, 420

Radar cross section, 116
Radar echo, 115, 355
Radial waveguides, 208–213, 216–219, 279–283
Radiation, 77–81, 242–245
   two-dimensional, 228–230
Radiation conductance, 112
Radiation field, 79, 81, 132–134
Radiation resistance, 82, 93(**42, 44**), 94(**46**)
Ramo, S., 309
Rayleigh scattering, 295

Rayleigh-Ritz procedure, 339
Re operator, 15, 16
Reaction, 118, 340
Reactive current, 27, 28
Reactive power, 22
Realizability conditions, 400
Receiving pattern, 119
Reciprocity, 116–120
   for antennas, 120
   for circuits, 119
   for microwave networks, 392
Rectangular cavity, 74–76, 155–157
   partially filled, 191(**16–18**), 325, 373
Rectangular waveguide, 66–74, 148–155, 387
   partially filled, 158–163, 191(**14–16**), 192(**19**), 348, 374
Reference conventions, 3
Reflection of waves, 54–61
Reflection coefficient, 55, 421
Resonance, 74
Resonant antennas, 94(**45, 46**)
Resonant slots, 444(**21**), 445(**23**)
Resonators, circular cavity, 213–216
   concentric spheres, 284
   one-dimensional, 44
   rectangular cavity, 74, 155
   sources in, 431–434
   spherical cavity, 269–272
   spherical sector, 284
Ribbon of current, 188, 260(**31**)
Ridge waveguide, 327, 374(**12**)
Right-hand rule, 2
Ritz procedure, 338, 344
Rodrigues' formula, 466
Rubenstein, P. J., 371
Rumsey, V. H., 118, 340, 365

Saddle point, 335
Saunders, W. K., 245
Scattered field, 113
   reciprocity for, 141(**24**)
Scattering, by conducting plate, 115, 128, 140(**20, 21**)
   by conductors, 355–361
   by cylinders, 232–236, 261(**34, 35**), 364

INDEX

Scattering, by dielectrics, 362–365
  differential, 360
  by half-planes, 241–242, 261(**37, 38**)
  by magnetic obstacles, 380(**22**)
  by ribbons, 350, 378(**38**)
  by spheres, 292–298
  stationary formulas for, 355–365
  by wedges, 238–242
  by wires, 357, 379(**39–41**)
Scattering matrix, 399
Schelkunoff, S. A., 222, 268, 286
Schwartz, L., 459
Secondary units, 1
Sectoral horn, 213
Seely, S., 386
Segmental cavity, 284
Seidel, H., 222
Self-reaction, 118
Separation of variables, 143, 198, 264, 381
Silver, S., 245, 303, 306
Simple matter, 6, 18
Singular field, 32
Skin depth, 53
Slot in ground plane, 138(**17, 18**), 181–186, 261(**32**), 370, 430, 444(**21, 22**), 445(**23**)
Slotted cone, 306
Slotted cylinder, 238
Slotted sphere, 302
Smythe, W. R., 324, 419, 420, 467
Sneddon, I. N., 252
Snell's law, 58
Source coordinates, 80
Source-free regions, 37
Sources, 7, 12, 19, 95, 96
Spherical Bessel functions, 265, 268, 464
Spherical cavity, 269–273
  partially-filled, 313(**7, 8**), 326
Spherical coordinates, 265, 447
Spherical waves, 79, 85, 276, 286–289
Standing wave, 42–47, 69
Standing-wave pattern, 44
Standing-wave ratio, 45, 55
Static mode, 338

Stationary formulas, 317, 341
  for aperture admittance, 428–431
  for cavities, 331–345
  for cavity feeds, 434–440
  for impedance, 348–355
  for obstacles in waveguides, 402–406
  for scattering, 355–365
  for transmission, 365–371
  for waveguide feeds, 425–428
  for waveguide junctions, 420–425
  for waveguides, 345–348
Storer, J. E., 354
Stratton, J. A., 121, 324
Supergain antennas, 309
Surface of constant phase, 85
Surface currents, 33
Surface guided waves, 168–171, 219
Surface impedance, 53, 371(**2**), 375(**18**)

Tai, C. T., 358
TE, TM, TEM, 63, 67, 130, 202, 267, 382
Tector, R. J., 430
Teichmann, T., 434
Tensor Green's functions, 123–125, 356
Tesseral harmonics, 273
Tightly bound wave, 170
Total reflection, 59
Transmission, 360
Transmission area, 368
Transmission coefficient, 55, 368
Transmission lines, 61–66
  biconical, 284–286, 313(**13**)
  equivalent, 386
  modes, 63
  parallel-plate, 90(**28**), 91(**31**), 189(**6, 7**), 440(**1**)
  radial, 211
  twin-slot, 135(**7**)
  wedge, 212
Transmission matrix, 399
Transverse field vector, 382
Transverse fields, 63, 67, 130, 202
Traveling waves, 39
Trial field, 332
Twin-slot line, 135(**7**)

Uniform plane wave, 39, 147
Uniform waves, 85
Uniqueness, 100–103
Units, 1

Van Valkenburg, M. E., 397, 400, 435
Variation, 332
Variational methods, 317, 331–380
Vector analysis, 447–450
Vector Green's theorems, 121, 141(**28**)
Vector potential, 77, 99
Velocity, of energy, 42
  of light, 5
  of phase, 39, 40, 68, 86, 385
Voltage, 3, 15
Voltage source, 96, 118
Von Hipple, A., 23

Wait, J. R., 240, 242
Wall impedance, 371(**2, 3**), 375(**18**)
Wave equation, 37
  for inhomogeneous matter, 88(**2**)
Wave functions, 85
  cylindrical, 199–204
  plane, 143–145
  spherical, 264–269
Wave impedance, 39, 55, 86
  characteristic, 69, 152
Wave number, 37
Wave potentials, 77, 129
Wave transformations, 230–232, 289–292
Waveguide feeds, 179, 195(**33, 34**), 425–428, 444(**20**)

Waveguide junctions, 172–177, 193 (**27–29**), 420–425, 443(**17, 18**)
Waveguides, 66
  biconical, 284–286, 313(**13**)
  circular (*see* Circular waveguides)
  corrugated conductor, 170, 193(**25**)
  corrugated wire, 223
  dielectric slab, 163, 192(**22**)
  in general, 381–391
  parallel-plate (*see* Parallel-plate waveguide)
  posts in, 406–411, 442(**12**)
  probes in, 178, 425–428, 446(**26**)
  radial, 208, 279
    partially filled, 216
  rectangular (*see* Rectangular waveguide)
Wavelength, 40
  cutoff, 68, 150, 206, 384
  guide, 68, 384
  intrinsic, 40
Waves, in dielectrics, 41–48
  in general, 85–87
  in lossy matter, 51–54
Wedge cavity, 284
  waveguide, 208, 255(**7**), 256(**14**)
Whinnery, J. R., 309
Wigner, E. P., 434
Windows, 414

Zeros, of Bessel functions, 205
  of spherical Bessel functions, 270
Zonal harmonics, 273